Introduction to
ISAAC NEWTON'S *PRINCIPIA*

I. BERNARD COHEN

INTRODUCTION TO
NEWTON'S 'PRINCIPIA'

I. BERNARD COHEN

CAMBRIDGE UNIVERSITY PRESS
CAMBRIDGE • LONDON • MELBOURNE

Published by the Syndics of the Cambridge University Press
The Pitt Building, Trumpington Street, Cambridge CB2 1RP
Bentley House, 200 Euston Road, London NW1 2DB
296 Beaconsfield Parade, Middle Park, Melbourne 3206, Australia

Published in the United States by Harvard University Press

First published 1971
First paperback edition 1978

Library of Congress Catalogue Card Number 78–28770

ISBN 0–521–07648–X hard covers
ISBN 0–521–29338–3 paperback

First printed in Great Britain at the University Press, Cambridge
Reprinted in the United States of America

This *Introduction* is dedicated to

HERBERT BUTTERFIELD

with great respect and warmth

I.B.C.

PREFACE

Isaac Newton published his great treatise, *Philosophiae Naturalis Principia Mathematica*, in three editions (London, 1687; Cambridge, 1713 [reprinted Amsterdam, 1714, 1723]; and London, 1726), differing from one another in choice of language, in technical content, and in expressed philosophical position. But there has never been till now, some two and a half centuries later, an edition with variant readings (or 'variorum'[1] edition) enabling the reader to see at a glance the successive alterations made by Newton during the span of about four decades from the completion (in 1686) of the manuscript for the first edition up to the printing of the third and ultimate authorized edition in 1726.

The present edition—to which this volume constitutes the Introduction—primarily displays for comparison the variants uncovered by a *verbatim et litteratim* collation of the three authorized editions and the manuscript from which the first edition was printed—thus presenting Newton's concepts, biases, tastes, knowledge, and mastery of the subject, at three successive times: 1685–7, 1713, 1726. At the same time there are made readily accessible to the reader the corrections, improvements, and additions entered by Newton himself into one or both of a pair of copies of the first and of the second edition in his own library, thus giving some clue to the stages of development during the intervening period.

This edition of the *Principia* with variant readings was conceived by the late Professor Alexandre Koyré and me in 1956. The textual part was completed and delivered to Harvard University Press just one decade later, followed by the Introduction within the next year. The edition as a whole will consist of a number of parts, of which these first two comprise (i) the one-volume Introduction with its Supplements and (ii) the Latin text together with the Apparatus Criticus and Appendices, in two volumes. (Information concerning the further parts of this edition is given below.)

The heart of our edition is, of course, the text: here we display in facsimile—page by page—the third and ultimate approved version of the *Principia* (London, 1726), ultimate in the sense of representing the final revised authorized edition, completed and seen through the press just before the aged Newton's death (in March 1726/7).[2] To this facsimile there has been joined an Apparatus Criticus, giving—line by line—the prior versions from the first two authorized editions and the manuscript of the first edition, plus all textual modifications entered by Newton himself in the above-mentioned personal copies of the first and second

[1] On the pure or original sense of 'variorum', as contrasted with current usage today, see Chapter I, §1, note 1.

[2] A bibliography of all the editions of the *Principia* is given in Appendix VIII.

editions.[3] Because the Apparatus Criticus contains all textual differences found by collating the *editio princeps* of 1687 and the manuscript from which it was printed, the reader may see for himself the final alterations made just as the book was going to press and even while it was being printed. The printer's manuscript is written in the hand of an amanuensis, Humphrey Newton; Edmond Halley saw the book through the press (see Chapter III, §§1, 7; Chapter IV; Chapter V, §3; Supplements II and VII). To make the record complete, all textual changes that may be positively identified as in Newton's hand or in Halley's hand are so indicated in the Apparatus Criticus.

During the course of our labours in preparing this edition, Professor Koyré and I often asked ourselves how it could have happened that no one before us had undertaken this task. Certain obvious answers suggested themselves at once: the difficulty and magnitude of the undertaking, and the fact that editions with variant readings have not generally been made of modern scientific works. Furthermore, the history of science is a newcomer to the scholarly specialties, and hence the demand for detailed information on the textual growth of the *Principia* has not been appreciable until very recently (Chapter I, §9; Chapter II, §§1, 2). We found, once our own edition with variant readings was under way, that at least three suggestions had been made in the past that the *Principia* be edited so as to show some (if not all) of the changes made in successive editions: the first, seven decades ago by W. W. Rouse Ball (Chapter I, §§6–8). No doubt one of the reasons why scholars did not carefully study the variations among the three authorized editions was simply that they were unaware of how illuminating or significant the changes made by Newton actually are. Indeed, both Professor Koyré and I became interested in undertaking to produce an edition with variant readings only when, independently and wholly by chance, we had encountered a striking difference between the first and the later two authorized editions in the Regulae Philosophandi and Phaenomena which occur at the beginning of Book III (see the Introduction, Chapter I, §§1, 9; Chapter II, §§1, 2). Not only were the actual sense and content modified from edition to edition, but in the first presentation a single rubric of Hypotheses embraces the statements later presented separately as Regulae Philosophandi and Phaenomena.[4] We were thus made aware that in the case of the *Principia* any conceptual analysis would be inadequate if it did not take account of the changes introduced by Newton in successive editions.

The fourth Regula Philosophandi occurs only in the third edition, while the third Regula occurs in the second and the third editions. The concluding Scholium Generale does not appear in the first edition, but in the second and—with some

[3] These four copies from Newton's personal library are described below in Chapter I, §4, and in the Guide to the Apparatus Criticus, §1. The onetime existence of other copies of the first edition with annotations by Newton is discussed in Chapter VIII, §§4, 5.

[4] Actually, the transformation of the original set of Hypotheses is even more complex than this. One of the original Hypotheses remained a Hypothesis in all editions, and a Lemma of the first edition became Hypothesis II in the later editions. For details, see the Apparatus Criticus, {387–394}, {408.1–5}, {476.1–7}.

revisions[5]—in the third edition. Not only is it anachronous to lump together the Preface (1687), the first two Regulae (1687), the third Regula (1713), the Scholium Generale (1713, revised 1726), and the fourth Regula (1726), as if they all represented Newton's state of mind at one and the same time: to do so is also to deny the dynamic quality of a creative mind by assuming that there were no changes on fundamental questions during a span of some forty years (Chapter II).

Not all textual modifications from edition to edition represent a shift in Newton's avowed philosophic position; some are of a technical (i.e. purely scientific or mathematical) character. Thus, Newton was always on the look-out for new data of observation and experiment, to be introduced in both of the revised editions. In the second edition, the Corollaries to Prop. I, Book I, were (with alterations) shifted to Prop. II, and a wholly different set of Corollaries was added to Prop. I; this change enabled Newton to present a completely fresh and streamlined proof of Prop. IV, on the magnitude of the centripetal force in uniform circular motion. Almost all of Sec. VII of Book II, on the resistance of fluids, is new in the second edition. The proof of Prop. X, Book II, completely redone in the second edition, is of special interest because of the criticism by Johann Bernoulli, brought to Newton's attention by Johann's nephew, Nicolaus (Chapter IX, §4). The alterations in the so-called Leibniz Scholium (following Lemma II in Book II) are similarly attractive to study in the light of the great controversy between Newton and Leibniz concerning the first invention of the calculus.

The many textual alterations do not equally excite the reader's attention. Nevertheless, we have presented in the Apparatus Criticus every proper textual variant, including each alteration and addition found by collating the three printed editions, the printer's manuscript, and the two annotated copies of the first and of the second editions from Newton's library, mentioned above. The reasons for giving all proper textual variants,[6] and not trying to eliminate any supposedly 'insignificant' ones, are spelled out at length in Chapter II. Suffice it to say here that the exact borderline between the 'significant' and the 'insignificant' would not only differ from one scholar to another, but also from one age to the next. Because it was our intention that students of the *Principia*, now and in future, should not have to wonder what 'significant' variant readings had been omitted by us as supposedly 'insignificant', we did not pick and choose. The end we had

[5] For instance, in the second edition Newton says that to discourse about God on the basis of phenomena belongs to 'experimental philosophy', whereas in the third edition such discourse is said to belong to 'natural philosophy'.

[6] By proper textual variants I mean any difference in the actual words or mathematical expressions. We have not included printing or spelling variants, such as capitalization or non-capitalization of the same word, or the abbreviation or non-abbreviation of a given word (as 'Corol.' for 'Corollarium'), or the use of italic or roman type, or the change from roman to arabic numerals, or the change of spelling (as from 'caelo' to 'coelo'). Similarly, we have not included printer's misprints unless the misprint is a different word, or another and distinct form of accidence from the one intended; but we have given all printer's misprints considered to have been of sufficient importance to warrant inclusion in the printed Errata. (For further details see the Guide to the Apparatus Criticus.)

in mind was that scholars could readily find in the Apparatus Criticus every textual variant and every manuscript annotation of the original texts on which this edition with variant readings is based, and not be required to have constant recourse to the originals, save on very special occasions. Professor Koyré and I hoped from the very start that the textual part of our joint labours might stand the test of time, that this editing would not soon have to be undertaken again in its entirety—although the discovery of new manuscripts could of course require supplements.

Throughout this Introduction and its Supplements, superscript letters are printed in small letters above the line ('y^e' rather than 'ye' or 'the', 'y^n' rather than 'yn', 'then', or 'than', and so on); abbreviations have not been spelled out, and no attempt has been made to introduce a modern editorial consistency; punctuation, capitalization, and italicization generally follow the style of the manuscript as closely as possible. The major departures from the actual form of manuscripts and printed sources are the following: (1) accents on Latin words have, for the most part, been ignored; (2) the special abbreviation (sometimes represented by a semi-colon) for 'ue' in the enclitic 'que' has been consistently spelled out; (3) the 'j' frequently written and sometimes printed for the final 'i' in the combination 'ii' (for example, 'ij', 'medij', 'medijs', 'distantijs', 'spatij', 'spatijs') is printed throughout as 'i'; and (4) the ligatures 'æ' and 'œ' have always been rendered by the pairs of separate letters 'ae' and 'oe'. The omission of accents on Latin words has made it unnecessary to introduce variant readings when manuscripts or the several printed editions of the *Principia* are word-for-word identical in all respects save for accents (in such cases, how could one choose which accents to print, when their use varies among the several manuscripts and printed versions?). In quoting any single passage, however, from a manuscript or printed text, we have for the most part included the accents. The 'j' presents similar problems, since (although Newton himself wrote 'ij') the same printed text may use both 'ii' and 'ij' for the identical word: for instance, 'spatii' on pages 5, 6, 7 of E_1 but 'spatij' on pages 10, 11, or both 'ijsdem' and 'iisdem' on page 7. The ligatures 'æ' and 'œ' were avoided because of the obvious additional burden on compositors; we had encountered erroneous introductions of ligatures resulting from the difficulties of setting up Latin in type and of proof-reading. Lynx-eyed readers may observe that many printed documents are presented in the Introduction in new transcriptions from the manuscripts. All extracts are printed wholly in roman type, even though portions of the original may be in italics.[7]

The occurrence of even the most minor textual alterations may be not wholly devoid of scholarly interest. One very important function of any textual modifica-

[7] All titles—whether of books, of articles in journals, of names of chapters, or of statements of propositions—have generally been given in lower-case letters, without regard to whether individual words (other than initial words, proper names, and German nouns) are capitalized in the original. This statement applies equally to text references, footnotes, and the bibliography. But in actual quoted extracts from manuscripts and printed sources I have, as mentioned above, attempted to follow the style of the original as closely as possible with regard to capital letters, spelling, and punctuation, with the exceptions noted.

tion is to enable us to prove that Newton actually re-read (since he later worked over) a particular passage, proof, or section of the *Principia*. At first glance the many changes in 'pure' terminology may not seem to be very significant, but it must be kept in mind that the whole subject of scientific and methodologic or philosophic terminology is only just beginning to be studied seriously. In this new area of inquiry, the changes in terminology on Newton's part are likely to prove a most valuable component of our understanding of a significant aspect of the history of thought.

In Chapter II (§§2–4) the reasons are given for including in the Apparatus Criticus a certain number of Newton's manuscript alterations and additions, so as not merely to present the *Principia* at three discrete stages: 1685–7, 1713, 1726. Once we had decided to extend the Apparatus Criticus beyond the results of a *verbatim et litteratim* collation of the three printed editions and the printer's manuscript, and to include certain intermediate manuscript versions, the question of how to set the limits on the latter proved most vexatious. Professor Koyré and I devoted many long months of thought, and made a number of trials, before we found the solution adopted here: of including in the Apparatus Criticus only those manuscript annotations actually entered by Newton in his own copies of the first and second editions of the *Principia*. (This topic is discussed further in Chapter II, §§5–7.) In a sense, having recognized the practical impossibility of printing every known manuscript alteration, we decided that it would be better to have Newton himself make the choice of what to include.

Another major problem that Professor Koyré and I had to face from the very start was how to deal with the editorial and explanatory comments which we would wish to make. Our first thought was that they would take the form of footnotes, as a running commentary: explaining a difficult point; giving the significance of each major topic; presenting the antecedent history or the subsequent effect of a given concept or method or result; even expanding the mathematical proofs as needed; explaining the cause, occasion, or significance of a given alteration proposed or made by Newton; and even bringing in relevant material from Newton's manuscripts and published writings, and the writings of others.[8]

But we soon realized that it would be of most advantage to have our commentary appear separately from the text. It had been our experience as scholars that commentaries become dated much sooner than well-edited texts. Again and again, in the course of our separate and our joint researches, we had found

[8] In preparation for this task we began to assemble the various commentaries that had been written to aid the reader or student of the *Principia* during the eighteenth and nineteenth centuries. We contemplated giving the history of such commentaries, in some cases, problem by problem, as a means of providing an interesting cross-section of the development of the philosophy of science. This plan was adumbrated in the report on our progress which we wrote for the *Yearbook of the American Philosophical Society* (1960), pp. 516–20. Before long, however, two great difficulties arose. First we saw that the sheer bulk of the commentary, so conceived, would overwhelm the text and Apparatus Criticus: a rough estimate gave us some 1,500–2,400 pages of small type! Clearly any such opus could not conveniently be published along with the text and Apparatus Criticus.

ourselves still using for our purposes a text edited a century or more ago—often when the commentary to that same edition was no longer fully acceptable because of the discovery of new sources, the changes that had occurred in scholarly perspective or points of view, or even simply the normal march of scholarship. As I have said earlier, we hoped that our text—within the limits of human frailty—might serve all scholars for a long time and not be subject to doubt, disagreement, and replacement (save in details); but we cherished no such illusion about our own comments, happily recognizing the fact of progress in knowledge and understanding, even in historical studies. Furthermore, it seemed to us to be of decided advantage to separate physically the presentation of what Newton wrote and our own interpretations. For these reasons, both the Commentary and the Introduction have been planned so as to appear as separate volumes.[9]

While the Apparatus Criticus was taking shape, Professor Koyré and I discussed the Introduction to it at great length. The form adopted here evolved slowly. Part I of the Introduction (Chapters I and II), a sort of extended preface, was agreed on from the start: it was to be a presentation of the history of changes made in the *Principia* in its several editions, and of those contemplated by Newton between editions, so far as they would be of interest, together with an account of earlier attempts or proposals to produce such an edition as ours. We also decided that there should be a documentary 'biography' of the *Principia*, starting from Halley's famous visit in the summer of 1684, and recounting how Newton actually wrote his *magnum opus*; then some presentation seemed required of the *Principia*'s critical reception, its subsequent revisions and editions, and the way in which each of the two later authorized versions was received. Although Professor Koyré and I had more than once discussed what kind of things should go into this Introduction, we had not at the time of his death actually begun to write out any part of it, nor had we even made a complete or detailed outline. Hence this Introduction, written by me alone, could not profit from his elegant taste, his shrewd critical perception, or his vast learning.

I have, in so far as possible, tried to follow the general plan that we had worked out together, and—above all—to keep the Introduction a history of a book, and not too much a critical interpretation or analytic history of Newton's concepts, reserving that task for the commentary volume to come. Of course, it is impossible to write such a documentary 'biography' of the *Principia* as this without presenting interpretations of concepts and their history. Anyone who knew Professor Koyré, or who is at all familiar with his writings, will appreciate how much I missed his collaboration during the writing of this Introduction, and in the research that I had to undertake as the Introduction began to take form, as new and exciting documents came to light, and as new problems arose and new interpretations were

[9] Similar considerations led to the decision to separate any English translation or paraphrase from the Latin text and Apparatus Criticus and to combine it with the interpretative commentary and explanations. This is explained further below.

conceived. It is not out of place to add that the opportunity of having been able to work closely with Professor Koyré was a rare privilege and one of the greatest experiences in my scholarly life.

Readers will observe that I have attempted in every case to give as complete a documentation as is reasonably possible, so that all scholars may be aware of the wealth of manuscripts available and their often tantalizing, perplexing, and mutually contradictory contents. To help the reader find his way through this long Introduction (to a long book), I have divided each chapter into numbered sections, briefly described in the Table of Contents. This method of presentation enables the reader to skip any parts (such as those making up Chapter IV) that are necessarily technical and deal with precise and nice topics of manuscripts, and that might possibly interrupt the flow of the critical narrative. In order to lessen the textual burden of the Introduction as a whole, certain major documents have either been wholly relegated to the set of nine Supplements, or dealt with only partially in the main body of the Introduction, and then presented more completely in the Supplements.

Despite the amount of documentary material available in manuscripts and in printed works, there are many interesting and important questions that cannot be answered fully or definitively, and some others that cannot be answered at all. I have, throughout, tried to keep all points of doubt and conjecture distinct from those on which our certainty is reasonably great. Considering the vast extent of Newton's manuscript remains of all sorts, he would be a bold man indeed who would confidently assert that there exists no document related to a given topic.

Since this Introduction has been written specifically for the present edition of the *Principia* with variant readings, certain abbreviations and symbols used in the text appear here also. I have listed these prominently in the Table of Contents, so that the reader may be able to identify them readily. I certainly hope that the present volume may be useful for a wider class of readers than only those scholars interested in using the text-volumes, for there is presented here a documentary life-history of the *Principia*, the single most important book in the Scientific Revolution of the seventeenth century, and one of the greatest classics of science.

It is contemplated that the present Introduction (with its Supplements) and the text volumes (with the Apparatus Criticus and Appendices) will, in the coming years, be complemented by other related volumes currently in active preparation (and in various stages of completion) or being planned. One of these is a presentation of Newton's *De mundi systemate*, an early form of what eventually became Book III of the *Principia* (following Books I and II, 'De motu corporum') but which, when written, was entitled simply *De motu corporum, liber secundus*. Published posthumously in both English and Latin, this tract is most fascinating, not only as a first work by Newton on the 'System of the World', but also because the evolution of the text shows a number of exciting changes that merit our scholarly attention. This will be the first time that anyone will have undertaken to print this work exactly as Newton wrote it, and there will be an Apparatus Criticus, containing the cancelled prior versions and cancelled passages (still readable in Newton's original manuscript). The variations between the final manuscript and the printed editions will also be indicated. It is planned to include a critical text of the two contemporaneous English versions as well as of the Latin original, because there are major differences between the two English presentations and also divergences from the Latin original that may have a significance of their own.

Another related work is to be an 'index verborum' or analytical corcordance. This is currently being put on a computer tape, word by word and line by line, as a separate project under the general direction of Dr Owen Gingerich of Harvard University and the Smithsonian Astrophysical Observatory. We hope to make copies of this tape available at cost to any researchers for whatever projects they may wish to undertake. We plan, furthermore, to produce a separate publication for general use, which will be either an alphabetical table of the occurrence-in-context of all major words, or a tabulation of the occurrences of all words save for conjunctions, prepositions, pronouns, numerals, and other words of little interest.

With respect to the Commentary, it is my intention to follow closely the plans discussed at length with Professor Koyré over a number of years of our work together and brought to final form shortly before his death. In minimum form, it is my hope that the Commentary will, accordingly, contain primarily an analytic presentation of the Prefaces, the Definitions, the Laws of Motion, the first three Sections of Book I, most of Book III, and some other selected topics—in short, the parts that are today of interest to the greatest number of historians of science and to students of philosophy and the history of ideas. Those who know the *Principia* will recognize that this plan follows closely Newton's own instructions to all readers at the beginning of Book III,[10] repeated in detail in his instructions to Richard Bentley on how to read the *Principia*.[11] For each of these selections, it is intended to provide an English version, together with a modern paraphrase (using mathematical symbols for Newton's 'equations' or 'proportions' in words), accompanied by a similar treatment of the relevant portions of the Apparatus Criticus plus other related material from Newton's manuscripts and from published works (by Newton himself and by others). In addition to all of the text of the *Principia* up to (and including) Sec. III of Book I, together with the front matter, and a major part of Book III, certain other Sections of the *Principia*, or Lemmas, Propositions, and

[10] At the beginning of Book III, Newton wrote that he would not 'advise any one to the previous study of every proposition of those [two preceding] books. For they abound with such [things] as might cost too much time, even to readers of good mathematical learning. It is enough if one carefully reads the definitions, the laws of motion, and the first three sections of the first book. He may then pass on to this [third] book, and consult such of the remaining propositions of the first two books, as the references in this, and his occasions, shall require.'

[11] Given to Bentley presumably in about July 1691. See Newton (1959–), *Correspondence*, vol. 3, p. 156.

Scholia, will be presented in a similar (or possibly somewhat abbreviated) fashion; the choice of this supplementary matter will be based either on its intrinsic interest or on the student's need to be aware of yet further results or methods in order to understand the argument of Newton's System of the World.[12]

It is planned, furthermore, to have in this Commentary a complete Table of Contents (or Summary) of the rest of the *Principia*,[13] in English Translation, so that the reader will have a view of the whole work. Finally, there will be a glossary, and a number of short essays on such topics as the astronomical and physical data used by Newton; the mathematical background to the understanding of the *Principia*; the general methods of proof; the history and analysis of key concepts, methods, and propositions, and their influence. Eventually, the remainder of the *Principia* should be treated in a similar fashion, perhaps requiring two additional volumes of commentary and thus producing a new kind of English version; but at this stage of the job I am not so rash as to promise to produce these myself.

From the very first days of our preparation of an edition of Newton's *Principia*, Professor Koyré and I gave much thought to the requirements of an English version of Newton's treatise. The basic question was whether to retranslate the whole text or to make some use of Andrew Motte's contemporaneous translation (London, 1729). Again and again, when projected new versions were made and compared with Motte's, it proved to be the case that Motte had produced a sound, literate, and generally accurate work, one that conveyed admirably—in its style and language—the spirit of the Newtonian age.[14] Motte's text has, however, several disadvantages. It uses older and now-unfamiliar expressions such as 'motion in antecedentia' or 'in consequentia', 'sesquialterate', 'subduplicate', and the like. Furthermore, it is difficult for the reader of the present day to remember the special sense of 'toward the same parts' (or 'contrary parts'), a literal translation of Newton's Latin phrases for indicating the same (or the opposite) direction. Equally difficult for today's reader is the use of the vinculum. Furthermore, on occasion Motte introduces a word or phrase from the second (1713) rather than the third and ultimate (1726) edition, and there are even occasional traces of the first

12 These will include, among others, Props. XXXIX–XLII of Book I; and Props. XXIII, XXIV, XLI, LIII, and the last two scholia of Book II.

13 At one time we had planned to give a translation into English of the complete table of contents given in Appendix X to the text. It was almost immediately apparent, however, that each of the very first three paragraphs contains expressions every one of which would require a short essay to justify the particular interpretative translation to be adopted. Among these expressions are 'est mensura ejusdem' (Def. I and Def. II), 'quantum in se est' (Def. III), and 'vis insita' (Def. III). On the last two, see my article (1964), '"Quantum in se est"', and the discussion in the Introduction, Chapter III, §5. Since all of the volumes associated with the present edition with variant readings are conceived as a single working unit, however complex, it seemed better to put the translation of the table of contents in the Commentary volume, along with other translations and interpretations. The suggested English versions are: 'quantum in se est' = 'of and by itself, as far as it can'; 'vis insita materiae' = 'the force inherent in matter' or possibly 'the inherent force of matter'.

14 This topic (in relation to the possibility of producing a wholly fresh translation of the *Principia*) has been studied carefully by Professor George Huxley of Queen's University, Belfast, and by the late Professor I. E. Drabkin of City College, New York.

edition (1687).[15] Finally, there are instances of a radical departure from a strict rendering of the Latin original, and there also occur occasional interpolations which may mislead the reader.[16]

The unsatisfactory result achieved by the late Florian Cajori serves as a warning to anyone who may be tempted merely to modernize the text of Motte's contemporaneous translation of 1729;[17] and yet the existence of Motte's generally first-rate English version is apt to discourage even the heartiest translator from setting out to do the job anew, especially when his net achievement may be only to help the uninitiated modern reader to avoid the pitfalls of Motte's sometimes archaic English, of Newton's obsolete mathematical language, and of outdated technical expressions in general.

A trial is now under way of a method of both preserving Motte's original version and presenting an easily readable text for the student of our day. In this scheme, no attempt is being made to improve or correct or rewrite Motte's version. Rather, there would be a reprint of Motte's version of 1729 with an apparatus criticus consisting of two parts. One of these would contain—line by line, as needed—the mathematical and technical modern equivalents of any expressions no longer in current usage and would record every indication of a radical departure of Motte's text from the printed version of E_3 (1726); the other would present in a new English translation those entries of greatest interest in the Apparatus Criticus of the present edition. In this way, the student of Newton or of the history of the exact sciences could have available to him a readable and tolerably correct English text and he would also be able to study Newton's *Principia* in the same English version that has served readers continuously during the approximately two and a half centuries since 1729. It is too early to tell whether or not this proposal has sufficient merit to be put into execution.

When Professor Koyré and I made plans to undertake an edition of Newton's *Principia* with variant readings, I agreed to take the major responsibility for the supervision and preparation of the variant readings. Needless to say, Professor Koyré worked at and closely scrutinized every stage of the preparation of the Latin text, and we spent many long hours over the problem of how best to present the results of our collation in the final printed version. There were, of course, a large number of vexing questions, such as whether or not to display the manuscript (or possibly the first edition of 1687) as the main text and show the later stages in the Apparatus Criticus; or possibly to display the ultimate authorized version, the

15 These are discussed by me in Cohen (1963), 'Pemberton's translation...,with notes on Motte's translation'. See also my Introduction to the facsimile reprint of Motte's translation (London: Dawsons of Pall Mall, 1968).

16 A most significant example is discussed below in Chapter II, §3.

17 Cajori's modernization of Motte's translation introduces so many infelicities that it may serve as a cautionary object-lesson in not tampering with older translations. Again and again scholars have been misled by Cajori's version, which must always be used with the greatest caution and always checked against the original. See Cohen (1963), 'Pemberton's translation', Appendix 2.

third edition of 1726. Many of our decisions are described in the Introduction, Chapter II; perhaps the two most significant ones—arrived at after we had made a considerable number of tentative investigations—were to extend our original concept beyond merely the three printed editions, and to limit the manuscript material to the annotations in Newton's own copies of the *Principia* and to the printer's manuscript. The ultimate form of this edition of the *Principia* profited enormously from Professor Koyré's rich experience in dealing with texts.[18]

The critical apparatus of the edition of Newton's *Principia* was all but completed—requiring only typing and the final checking and the last revision—by the summer of 1963. In June 1963 I spent a week or so in Paris, so as to have the opportunity to talk with Professor Koyré about Newtonian problems and our edition of the *Principia*. Chiefly, however, we worked out the form (which I have described earlier in this Preface) for the Commentary volume.

Professor Koyré died in the spring of 1964. During the following summer I organized the materials for the prospective Introduction, which it was now my assignment to write alone, as I shall now have to produce by myself the Commentary we had always hoped to have written together. But, of course, both the Introduction and the Commentary, and whatever else I may ever have to say about Newton, or about Galileo and Descartes, or about any other figures of that time, will necessarily bear the impress of his learning and insight and will be conditioned by points of view deriving from the studies we carried out together. To be able to work so closely with Professor Koyré was a very real privilege, an honour for which I shall ever be grateful.

In the spring term of 1965, I spent a sabbatical leave of absence from Harvard University in Cambridge, England, where for part of that time the Governing Body of Peterhouse made available a study in which I composed the first draft of this Introduction. During the next two years, back home in Cambridge, Massachusetts, I completed and revised the Introduction and its Supplements, and supervised the final typing of the Apparatus Criticus; and I undertook the preparation of the Appendices. The ultimate stages of revision, following a critical reading by two friendly colleagues, were completed in Rome.

Anne (Mrs Cedric H.) Whitman, the assistant editor of the Apparatus Criticus, has worked with me at every stage of the preparation of this Introduction and its

18 In preparing as complex an Apparatus Criticus as this one, many stages of trial were required before Professor Koyré and I were satisfied that we had achieved a form of presentation that would be reasonably accurate and easy for anyone to use without much prior experience. During several stages of preparation the Apparatus Criticus contained the emendations occurring in John Locke's copy of the first edition, and those in another copy of the first edition which had at one time belonged to W. W. Rouse Ball, and in which some early owner has entered almost every change occurring in the second edition, so as to make this particular copy the equivalent of the second edition. Both of these copies are at present in the library of Trinity College, Cambridge. At another time we had included a significant number of spelling and typographical variants, and even punctuation variants. It was only after we had examined carefully what possible use the Apparatus Criticus might be put to that Professor Koyré and I finally adopted the present form.

Supplements, as well as of the Appendices. I was certainly fortunate to obtain the collaboration, in this enterprise, of so gifted a classicist and so conscientious and meticulous an editor. Week in and week out, she and I have read Newton's manuscripts together, and our discussions have clarified the meaning of difficult passages. Together and separately we have run through reels of microfilm and stacks of photocopies, searching for a missing document or fragment, or seeking information in relation to some topic, or attempting to identify a manuscript. I have profited again and again by my good fortune in thus having a second pair of eyes. In particular I owe Mrs Whitman a special debt of gratitude for her painstaking checking of the Introduction and its Supplements, especially in the matter of extracts from Newton's writings and other manuscripts, and English versions. Above all, her mastery of Latin has been indispensable in the preparation of the Apparatus Criticus; and her skill in conjecturing the true sense of almost illegible passages, and in restoring missing words and phrases, has been attested to again and again when her readings have later found confirmation in yet other manuscripts.

I have also a deep sense of gratitude to Dr Derek Thomas Whiteside of Cambridge, England. Dr Whiteside has rapidly become our acknowledged master in all questions relating to Newton's mathematical writings and mathematical thought. Through years of intensive work among the Newton papers in Cambridge he has acquired an intimate knowledge of the contents and style of Newton's manuscripts. Again and again I have turned to him in questions of dating a manuscript of Newton's, or in the matter of identifying an unknown hand, or of seeking verification of my own conjectures or conclusions in matters of handwriting. For over a decade I have happily and profitably discussed with him a variety of Newtonian questions. Over these years, I have enjoyed the pleasure of a fruitful correspondence with him on all aspects of Newtonian scholarship, and he has very generously answered many questions, both from his personal store of knowledge and from his ability to find solutions to my problems by a renewed contact with the manuscripts. Finally, Dr Whiteside read carefully the penultimate typescript of this Introduction and gave me the considerable benefit of his great erudition and critical acumen. I have acknowledged individually those contributions depending on his intimate knowledge of the sources and his special competence in dealing with mathematical questions; but the reader must not therefore conclude that these represent the totality of the improvements that he suggested and I have adopted in the final revision. He certainly exemplifies the principle that great generosity to other scholars is a concomitant of the highest learning.

Much of the research embodied in this Introduction, and in the text (with its Apparatus Criticus and its Appendices) could not have been undertaken without the possibility of extended periods of study among the manuscript sources (chiefly in Cambridge, England, and in London). I am, in consequence, more than ordinarily grateful to the National Science Foundation (U.S.A.) for its liberal support of my work, following an initial subvention by the American Philosophical Society and the Rockefeller Foundation, and a fellowship awarded by the John Simon Guggenheim Memorial Foundation. Research in preparation for the Commentary volume was begun in Spring 1968, while I was a Visiting Overseas Fellow of Churchill College, Cambridge, supported by a grant from the Ford Foundation.

The dramatic growth, during the last two decades, of the history of science as an academic subject in America, attracting an ever-growing company of students and professional scholars (see Chapter I, §9; Chapter II, §1), owes much to the enlightened support given by the National Science Foundation.

ACKNOWLEDGEMENTS

The preparation of this edition with variant readings has depended heavily on many institutions, whose treasures of manuscripts and rare books have constantly been made available to me. For over a decade, I have enjoyed the privilege of being able to study (and to make photocopies of) manuscripts and books by or associated with Newton in the University Library, Cambridge, and the libraries of Trinity College and King's College, and the Library of the Royal Society of London and the British Museum. In the University Library, Cambridge, I am especially grateful for the constant kindness shown me by the staff of the Anderson Room, and in particular Mr P. J. Gautrey; I am also grateful to Mr W. G. Rawlings, for his unfailing skill in photographing books and manuscripts needed for my studies. Mr Arthur Halcrow, Assistant Librarian of Trinity, has again and again far exceeded the bounds of duty in order to help me to solve a problem or to locate a book or a document. At King's, I have had the good fortune to know Dr A. N. L. Munby, the scholarly Librarian and Bursar, who not only guided me through the Keynes Collection but helped me to solve many a complex bibliographical problem. One of the pleasant aspects of visits to London was the time spent in the old Library of the Royal Society, in Burlington House, where I learned so much from the late Mr Henry W. Robinson, and where I used to meet such colleagues and friends as Douglas McKie, E. N. da C. Andrade, J. F. Scott, Rupert and Marie Hall, and W. P. D. Wightman. I am especially indebted to Mr I. Kaye, the present Librarian, and Mr H. W. Robinson, Jr, the Assistant Librarian, for many favours. Other libraries, or colleges and universities, which have kindly made their resources available to me are: Christ Church and the Bodleian, Oxford; the Universities of Edinburgh and St Andrews; the Royal Greenwich Observatory at Herstmonceaux Castle, where I was guided through the Flamsteed archives by Mr P. M. Laurie; the library of the Académie des Sciences, Paris, where I have profited by the wisdom and have enjoyed the courteous assistance of Mme Gauja, and also the Bibliothèque Nationale; the libraries of the Universities of Geneva, Zurich, and Basel; and, in my own country, the University of Texas (Austin), the Public Library of Cincinnati (Ohio), the Institute for Advanced Study (Princeton,

New Jersey), and the Newton Collection in the Babson Institute, in Babson Park (Wellesley, Massachusetts). I am, furthermore, greatly indebted to the following individuals and institutions either for information or for photocopies of books: Mr Joseph Halle Schaffner of New York City; Professor Wladimir Seidel of Wayne University; Mr Warren Howell of Los Angeles, California; Professor René Taton of Paris; Miss Margaret Norman of Cremorne, New South Wales, Australia; the Library of Congress, Washington, D.C.; the National Library of Scotland; and the libraries of the University of Sydney, Australia, Columbia University, Yale University, and the Massachusetts Institute of Technology.

I happily and gratefully record, in a separate category, my special debt to the Harvard College Library (the Widener and Houghton Libraries), and to the History of Science Library (Harvard). So many of the staff have answered questions and have solved problems for me, have helped to clarify obscure bibliographical references, and have located books or serials needed for my research, that it would be impossible to list them all here. But I cannot mention the Harvard College Library without saying a word about a dear friend, the late Professor William A. Jackson, Director of the Houghton Library, who was—ever since my student days—my constant mentor on every question relating to bibliography and who opened my eyes to many of the intriguing questions of old books, and of manuscripts, which are discussed in this Introduction.

Two scholars to whom I am especially grateful for assistance are Professor Marshall Clagett of the Institute for Advanced Study (Princeton) and Professor Richard Samuel Westfall of Indiana University. At an early stage Professor Koyré and I began turning to Professor Clagett for advice about the technical questions of producing an Apparatus Criticus. He styled our Apparatus, even suggesting the form of sigla we have used. We showed him the various stages of our work for his critical comments, since we both admired his skill as a textual editor and were aware of his vast experience in dealing with scientific texts with variant readings. Finally, he very kindly looked over the completed Apparatus Criticus for Book I, to make sure that all was clear and comprehensible and relatively clean and free from error. I am grateful to Professor Westfall for a reading of an all-but-final draft of the Introduction, the Supplements, and the Appendices, and a portion of the Apparatus Criticus. The final version has benefited from his comments.

Among Newton scholars not mentioned above with whom I have had the opportunity for many fruitful discussions of questions relating to Newton and the *Principia* are: Professor E. N. da C. Andrade (London), Professor A. Rupert Hall and Professor Marie Boas Hall (Imperial College, London), Mr John Herivel (Queen's College, Belfast), Mr J. E. McGuire (Leeds University), Dr J. F. Scott (current editor of Newton's *Correspondence*), and the late Professor Herbert W. Turnbull (initial editor of the *Correspondence*). To this list I add the name of Johannes A. Lohne of Flekkefjord, Norway, because—although I have never met him in person (our contact being, until now, limited to the exchange of letters and reprints)—I have so profited by reading his articles and letters that it has often seemed as if I had been having fruitful personal discussions with him.

I am particularly glad that the bibliography of the three substantive editions of the *Principia* (Appendix VIII, Part 1) could be given over to the professional hands of Professor William B. Todd, of the University of Texas. Mrs Virginia Harrison, of the Newton Collection, Babson Institute, made freely available to me all the information she had collected about the editions of the *Principia*, and has kindly checked the bibliography (Appendix VIII, Part 2) for completeness and for possible error. Mr John Neu, Science Bibliographer of the Library of the University of Wisconsin, gave me a valuable key to the sequence of the mid-nineteenth-century American editions of the *Principia*. I should like also to record my gratitude to Dr Margaret Glover Foley of New York City, who prepared, as a pilot project, a preliminary collation of the three printed editions of the *Principia*.

What a fortunate author I am in having been able to profit from the concern and wisdom of the heads of the two great university presses that have combined forces to produce this edition: the late Thomas J. Wilson, Director of the Harvard University Press, and Richard W. David, Secretary of the Board of Syndics of the Cambridge University Press. I am especially grateful to Mr Joseph D. Elder, Science Editor of the Harvard University Press. Not only did his help to me by far exceed the normal bounds of editing, but in addition he was forced to assume extra responsibilities in the final stages, owing to my being abroad. This is the fourth book of mine to which he has contributed, and my indebtedness to him increases exponentially, book by book. I am notably indebted to Mr A. Prag and Dr D. T. Whiteside for having helped in the arduous work of proof-reading.

During my frequent visits to Cambridge, England, I have had the splendid good fortune to have as my sponsor and guide Herbert Butterfield, Regius Professor of Modern History (emeritus), former Master of Peterhouse, and sometime Vice-Chancellor of Cambridge University. To him, friend and promoter of the history of science, I have dedicated that part of the edition of the *Principia* which is mine alone, this Introduction.

As ever, my most personal debt is to my cruellest and most passionate critic, Frances Davis Cohen.

I.B.C.

Cambridge, England—Cambridge, Massachusetts—Rome
1965–1967

CONTENTS

PART ONE

AN EDITION OF THE 'PRINCIPIA' WITH VARIANT READINGS

PART TWO

THE WRITING AND FIRST PUBLICATION OF THE 'PRINCIPIA'

PART THREE

REVISING THE 'PRINCIPIA'

PLATES

(following p. 48)

ABBREVIATIONS AND SPECIAL SYMBOLS

E_1, E_2, E_3	The first (1687), second (1713), and third (1726) editions of the *Principia*
E_1a, E_2a	The annotated copies of E_1, E_2 from Newton's personal library
E_1i, E_2i	The interleaved (and annotated) copies of E_1, E_2 from Newton's personal library
LL (LL_α, LL_β)	The draft deposited by Newton of the text of his professorial lectures ('Lucasian Lectures') for 1685, 1686, containing the Definitions, Laws of Motion, and a portion of Book I of the *Principia*
M	The manuscript from which E_1 was printed (where two overlapping versions of the same passage occur, M_{i} is the earlier and M_{ii} the later)
⌊ . . . ⌋	An indication, in a quotation from a manuscript source, that a word or group of words is a later addition or insertion

AN EDITION OF NEWTON'S 'PRINCIPIA'

(with variant readings)*

* The text of the edition with variant readings and the Appendices are printed in volumes separate from the Introduction.

PART ONE

AN EDITION OF THE 'PRINCIPIA' WITH VARIANT READINGS

CHAPTER I

INTEREST IN THE DEVELOPMENT OF THE 'PRINCIPIA'

1. INTRODUCTION

NEWTON's *Principia* needs no ordinary introduction to tell who its illustrious author was. Nor does an editor have to explain to the reader that this book is one of the glories of the human intellect, a founding document of our modern exact science. And yet a special kind of introduction is required for the presentation in the mid-twentieth century of an edition of a scientific treatise in Latin. This is all the more necessary in that the present edition, an edition with variant readings (known often as a 'variorum' edition[1]), appears in a form usually reserved for the learned presentation of literary or philosophical works, or of ancient and medieval scientific treatises, rather than books associated with the modern exact sciences.

[1] Commonly, today, 'variorum' is used to designate an edition of a text with variant readings, even though the Latin word 'variorum' is the genitive plural of the adjective 'varius', and hence in association with 'edition' means literally 'of various persons'; that is, the Latin phrase 'editio cum notis variorum' means an edition with notes of various commentators or editors.

The expression 'variorum edition' appears to have gained its modern currency through the editions of Shakespeare, notably the one edited by Horace Howard Furness in the last century, *A new variorum edition of the works of Shakespeare*, of which vol. 1, devoted to *Romeo and Juliet*, was published in Philadelphia in 1871. Many scholars have forgotten or are unaware that this 'variorum edition' was both an 'editio cum notis variorum' and an 'editio cum variis lectionibus'. The modern loose usage of 'variorum' apparently derives from the mistaken idea that the main feature of Furness's 'variorum edition' was intended to be the 'variae lectiones' rather than the 'notae variorum'. But, in general, in the volumes of this edition, the variant textual readings occupy far less space on any page than the selection of comments or annotations by various previous editors. Hence the merest casual examination would at once convey the impression that the main purpose of this edition had been to collect critical comments of previous editors rather than to produce a set of variant readings. Should there be any doubt on this score, the editor's preface would cast it out of consideration; it begins with a reference to 'the last so-called Variorum Edition of Shakespeare', and the notes it contains, derived from many 'Shakespearian commentators'. Furness then refers to 'the Cambridge Edition', in which, 'at the foot of every page, is given a thorough and minute collation of the Quartos and Folios and a majority of the *variae lectiones* of many modern editors, together with many conjectural emendations...'. Indeed, Furness's 'variorum edition' of Shakespeare had been undertaken as 'a New Variorum, which, taking the Third Variorum, that of 1821, as a point of departure, should contain the notes of the editors since that date only; in other words, to form a supplement to the Third Variorum'. I do not know who first referred to such an edition of Shakespeare as a 'variorum', but according to Sir E. K. Chambers, 'The book-sellers have chosen to call the 1803 and 1813 editions of Johnson and Steevens the *First* and *Second Variorum Shakespeares*, and the 1821 edition of Malone...the *Third Variorum*.' In these editions, the word 'variorum' does not appear on the title-page.

The chief feature of the present edition is an integral reproduction of the third and ultimate edition of the *Principia*, together with a display of all the alterations that occurred after the completion of the manuscript in 1686–7. These alterations, determined by a word-for-word and letter-for-letter collation, are assembled in the Apparatus Criticus accompanying the text proper. They are not 'variant readings' in the sense of being alternative readings which derive from more or less independent manuscripts, as in classical or medieval texts, where the author's original holograph manuscript is usually not available. Ours are primarily successive rather than alternative versions, and they exhibit a sequence of ideas and methods, and forms of expression from 1685–7 to the final revision by Newton for the third edition in 1726, one year before his death. These variant readings include not only every difference among the three printed editions (E_1, 1687; E_2, 1713; E_3, 1726),[2] but also the textual differences between the manuscript prepared for the printer[3] and the first edition; furthermore, the Apparatus Criticus includes all the long-hand corrections, alterations, and additions entered by Newton into one or both of two copies of the first edition and of the second edition, which he kept in his own library especially for this purpose.[4] In all, therefore, there are presented the fruits of collating eight separate texts or versions, resulting in an Apparatus Criticus containing several thousand variant readings.

The merest glance at the Apparatus Criticus of the present edition shows at once how much care was lavished on this book by its author. There is hardly a single page that was reprinted from the first edition (E_1, 1687) in the third edition (E_3, 1726) exactly as it was.[5] Throughout forty years of his mature creative life Isaac Newton worked to make his *Principia* a better book. The purpose of the present edition is to exhibit the successive states of Newton's expressions, concepts, methods, and data, as the *Principia* advanced from manuscript to print and from edition to edition. In the present edition with variant readings, it is possible for the first time to read the *Principia* as a living document, in the sense of one that grows and matures along with its author, rather than merely as a static expression as of 1687, 1713, or 1726.

[2] No account is taken in the present edition of the variants in the two unauthorized reprints of the second edition (Amsterdam, 1714, 1723), largely the introduction of errors in both the text and the diagrams.

[3] The manuscript is written out in the hand of an amanuensis, Humphrey Newton, with alterations and corrections made by Isaac Newton and by Edmond Halley. In the Apparatus Criticus, all such emendations are given, together with the earlier versions and an indication as to whether each such change occurs in Newton's hand or Halley's or was made by Humphrey.

[4] There exist from Newton's library an annotated copy of the first and of the second edition, and an interleaved (and annotated) copy of each of these two editions. The sigla we have used to designate these four texts are, respectively, E_1a, E_2a and E_1i, E_2i, while the manuscript is designated by M. Thus the eight texts on which the Apparatus Criticus is based are: M = manuscript used in printing E_1; E_1 = first edition (1687); E_1a = annotated copy of E_1; E_1i = interleaved (and annotated) copy of E_1; E_2 = second edition (1713); E_2a = annotated copy of E_2; E_2i = interleaved (and annotated) copy of E_2; E_3 = third edition (1726). These texts are described more fully in the 'Guide to the Apparatus Criticus' of the present edition.

[5] I refer here to substantive changes, not mere alterations in spelling, abbreviation, capitalization, punctuation, italicization; of course these changes are sometimes very minor (for example, a change in word order, in tense, or in choice of word).

The improvement of the printed text began as soon as parts of the *Principia* had been printed, when errors or misprints were found.[6] Before long Newton was altering the form of his expression, changing the order of words, choosing better words to convey the niceties of his thought. He made some more important changes too, introducing new proofs of existing propositions and substituting fresh propositions (and almost a whole new section) for existing ones. Further, some new scholia were added, while many of the existing scholia were enlarged. In some cases the number of corollaries was increased, the corollaries to a given proposition were altered and then transferred to a different proposition. Constantly, Newton strove to introduce the latest data of experiment and observation, and the results of his most recent deliberations. The 'Hypotheses' disappeared from the beginning of Book III, becoming transformed into 'Regulae Philosophandi' and 'Phaenomena'; one that remained an 'Hypothesis' was transferred elsewhere. By the time of the second edition (E_2, 1713), a concluding 'Scholium Generale' had been written to give the book the resounding finish it deserved.[7] In the third edition (E_3, 1726) the 'Regulae' were increased from three to four, and the 'Scholium Generale' was revised. I do not know of any other scientific book of that age that after publication was so continuously revised by the author, over so long a period of time. Certainly, there are no such examples among the works of Galileo, Descartes, Kepler, Huygens, or Hooke. In fact, there are few scientific treatises of the same rank in any age that were subject to this kind of extensive successive revisions by the author.[8]

2. CONTEMPORANEOUS INTEREST IN NEWTON'S SUCCESSIVE REVISIONS OF HIS 'PRINCIPIA'

The fact that this is the first edition of the *Principia* with variant readings—that is, the first attempt to make a systematic presentation of the successive stages of Newton's text—should not be taken as an indication that in earlier times there had been no interest in such alterations. Quite the contrary! Newton's contemporaries became attentive to the improvements he was making in his *Principia* as far back as the early 1690s, just after the publication of the first edition. A concern to discover just what changes Newton was making was not limited to his private circle of friends and acquaintances; even Continental savants wanted to find out what was the current state of revision.

[6] The earliest such corrections to any printed-off pages of the first edition seem to have been suggested to Newton by Edward Paget in October 1686; it was not until the end of March of the next year that Newton sent to Halley the final manuscript of Books II and III. See Chapter III, §6, and page 137 below.

[7] Although Newton had composed a 'Conclusio' for the first edition, he chose not to have it printed; it lay unnoticed among his manuscripts in the Portsmouth Collection (University Library, Cambridge) until it was published by A. R. and Marie B. Hall, *Unpublished papers* (1962), pp. 320–47.

[8] One notable exception is Charles Darwin's *Origin of species*; see Darwin (1959). Of course, handbooks of data and works of classification (or catalogues) of animals, plants, or minerals are constantly undergoing revision; but these are not scientific treatises like the *Principia* or the *Origin of species*.

Although normally secretive, Newton seems to have been quite willing to allow others to see the alterations he had been making in the *Principia*. We shall see below that he had certain of these emendations entered into a copy of the *Principia* which he presented in the early 1690s to the philosopher John Locke; and he permitted intimates (notably David Gregory and N. Fatio de Duillier) to transcribe selected improvements which he had written in one or another copy of the *Principia* in his library. A portion of Fatio's transcriptions, together with some notes of his own, were sent to Huygens and were eventually printed by J. C. Groening in 1701.[1] Even one who was a stranger to Newton saw the annotations in an interleaved copy of the *Principia*, on a visit in 1702. Many alterations were made by Newton directly in his personal annotated and interleaved (and annotated) copies of the first edition,[2] on text pages, interleaves, and end pages. Others were written out on sheets or on scraps of paper (many of which are now preserved in the Portsmouth Collection, U.L.C.), and then selectively entered in one or both of the above-mentioned two examples of the *Principia* in his library. The most widely disseminated improvements of the early 1690s appear to have been taken largely from the notes written out by Newton at the end of his annotated copy of the *Principia* (E_1a), on the page containing the printed Errata and succeeding blank pages (which we have referred to throughout this edition as MS Errata to E_1a).

Naturally enough, when the second edition of the *Principia* (E_2) appeared in 1713 (reprinted in Amsterdam in 1714), readers wanted to know what features distinguished the new edition from the old one. By far the most significant and extensive review of this new edition aimed at giving just such information. The author of this review, which was published anonymously in the *Acta Eruditorum* of Leipzig, remains unknown to this day; most likely he was a close associate of Leibniz's (if indeed he was not Leibniz himself).[3] For the most part the reviewer took the position that the general aspects of 'the incomparable work of [this] most excellent man' had been discussed abundantly in the previously published account of the first edition: 'Now, therefore, it will have sufficed for us to have set forth in what respects the new edition differs from the previous one.'[4] There follow some dozen pages of critical exposition of some of the novelties to be found in the new edition. The reviewer warns the reader 'that we are not going to descend

[1] See Chapter VII, §10, and Appendix IV to the text.

[2] As mentioned in §1, note 4, these are denoted in the present edition by E_1a and E_1i. We do not know how many other such annotated or interleaved (and annotated) copies of E_1 Newton may have had in his library. But he evidently had prepared yet another interleaved copy of E_1 for the press. Whether it still survives, and, if so, where it may presently be located, I have been unable to discover.

[3] It has been suggested (Koyré and Cohen (1962), 'Newton and the Leibniz–Clarke correspondence', p. 68, n. 16) that Leibniz's participation in, if not authorship of, this review may be detected in the reference to the 'hylarchical principle of Henry More'. This particular review is not included in the (admittedly incomplete) list drawn up by Ravier (1937, chap. 3) of 'comptes rendus anonymes' written by Leibniz. Further information concerning this review is given below in Chapter X, §2.

[4] *Acta Eruditorum* (March 1714), p. 131, here translated from the Latin.

to any minutiae, especially if they shall have seemed of rather slight importance, such as we judge them to be if here and there certain things are changed in diction'. Nevertheless, he does give a few examples which show that in fact he had made a most careful comparison of at least part of the two editions, word for word. Thus he points out that in E_2 Newton says 'that air, its density having been doubled, in a space also having been doubled, is quadruple', whereas the former edition merely states 'that air in a double space is quadruple'; he even observes 'that now a ratio is called *subduplicate* which before was said to be *dimidiate*' and that now one reads '*it is resisted to bodies* where before there stood *bodies are resisted*'.[5]

3. EIGHTEENTH-CENTURY EDITIONS OF THE 'PRINCIPIA'

By 1726, when Newton published the third edition, the *Principia* had become a classic: it was forty years old! No longer could an edition command attention for its novelty, as in 1687, or even in 1713–14. Readers then, and for some time thereafter, were apparently content to accept Newton's final version without overly concerning themselves with the rejected earlier stages. Thus, the two great eighteenth-century editions of the *Principia*—the Marquise du Chastellet's French translation and commentary, revised by the mathematician Clairaut, and the Latin edition with a continuous or running commentary by Fathers Le Seur and Jacquier—exhibit concern only for the explication of the most mature expression of the Newtonian principles; one may search through these two editions in vain for any information concerning the development of Newton's thought or the difference between the several editions of the *Principia*. (All the editions of the *Principia* are listed in Appendix VIII.)

Save for the single eighteenth-century printing of Andrew Motte's translation in 1729, the *Principia* was not republished in its entirety in England after 1726[1] until, in the 1770s, Samuel Horsley undertook the preparation of an edition of Newton's *Opera*. By then scholars generally had no interest in the actual stages of development of the *Principia*. When, on 1 January 1776, Horsley announced his *Proposals for publishing by Subscription, dedicated by permission to the King, ISAACI NEWTONI*

[5] These variant readings occur at {1.8–9}, {44.17–19}, {230.5}, and frequently in Book II. (Number pairs within curly braces stand for page and line of E_3. Thus {1.14} means page 1, line 14, {1.8–9} means page 1, lines 8–9, {1.3/4} means page 1, between line 3 and line 4, and {1.8, 9} would mean page 1, line 8 and also line 9. The page and line numbers of the present edition are those of the third edition (E_3: London, 1726).) The first change is notable since it occurs on p. 1. The second is one of a number of alterations made throughout the book; it and similar alterations are discussed in the 'Guide to the Apparatus Criticus'. The third is a grammatical improvement, it being more strictly correct to use the impersonal construction with the dative, 'resistitur corporibus' (as in E_2) than the passive 'corpora resistuntur' (as in E_1). Another change of this grammatical type is the alteration of expressions using the transitive verb 'moveo'; thus in E_1 we find such constructions as 'corpus movens', which are altered in E_2 to 'corpus motum' {104.7, 105.21, 108.9, etc.}.

[1] In many ways Robert Thorp's revision of Motte's translation, with lavish notes, is still one of the best texts to use for anyone who wishes to make a careful study of Newton's *Principia*. Although Thorp announced a two-volume set, only the first one (containing Book I) was ever published (London, 1777). A second edition appeared in 1802. See Appendix VIII, part 2.

OPERA quae extant omnia, commentariis illustrata, opera et studio Samuelis Horsley, he declared his intention:

IV. Every piece will be published in the language in which it was originally written, or in which it appeared in the latest edition made in the Author's life, and under his own correction.[2]

Only the latest and most authentic version was wanted.[3] And, in fact, neither in the prospectus nor in the edition itself is there any mention made of the simple fact that the *Principia* had been subject to a kind of continual revision from the appearance of the first edition (E_1) in 1687 to the completion of the third edition (E_3) in 1726.[4]

During the century or so following Newton's death, however, there was an occasional sign of interest in the development of the *Principia*, rather than in the final version of the doctrines to be found therein: a conspicuous example is the great French mathematician Joseph Louis Lagrange. In the *Théorie des fonctions analytiques*, Lagrange exhibits an awareness of the fundamental alteration of the proof of Prop. X (Book II) between the first and second editions of the *Principia*, after Johann Bernoulli had pointed out in 1710 that the original version leads to an incorrect result in the case of a circular trajectory (see below, Chapter IX, §4). Lagrange says: 'Newton, without replying [to Bernoulli], abandoned his first method entirely, and in the second edition of the *Principia* gave a different solution of the same problem.'[5] Lagrange then demonstrates correctly, for the first time in print, just how Newton had erred in the proof of Prop. X, published in 1687. But Lagrange does not show himself to have been concerned with the growth of the *Principia* in general.

4. NINETEENTH-CENTURY NEWTONIAN SCHOLARS: RIGAUD, EDLESTON, BREWSTER, DE MORGAN

It was only in the second third of the nineteenth century, however, that scholars became seriously interested in the actual sources, birth, and later development of Newton's ideas. The new scholarship may be seen in the publication in 1838 by Stephen Peter Rigaud, Savilian Professor of Astronomy at Oxford, of an *Historical*

[2] Quoted from the prospectus for Horsley's edition, of which a copy exists in the library of Christ Church, Oxford.

[3] Horsley, however, in printing the *Opticks*, did indicate rather carefully which Queries in their entirety and which parts of other Queries were introduced successively in the Latin edition of 1706 and the later English and Latin editions; see [Newton] (1779–85), *Opera* [Horsley], vol. 4, pp. 216–64. But Horsley gave no evidence that he even knew that any major changes had occurred from edition to edition of the *Principia*.

[4] Horsley's notes on the *Principia* are largely explanatory in the sense of working out mathematical details, although some (for example those to the Scholium to Prop. IV, Book I) are historical. They refer the reader to other writings of Newton's, or such books as Keill's *Introductio ad veram physicam*, Kepler's *Harmonice mundi* and *Epitome*, and Galileo's *Discorsi*, and there are many references to 'Patres doctissimi Le Soeur & Jacquier' (from whose commentary a considerable portion of the notes was derived). Horsley (vol. 2, p. 51) quoted from the manuscript *Notae* of David Gregory ('ex notis MSS Davidis Gregorii') which he must have seen in the library of Christ Church, Oxford.

[5] Lagrange (1797), 1[e] partie, §8, pp. 5–6.

essay on the first publication of Sir Isaac Newton's "Principia" (Oxford, 1838), which set a new high standard to the degree that it was based on an examination of manuscript sources. Indeed, about half of the book is made up of primary documents, of which a number were published by Rigaud for the first time, including an early draft of Newton's system of dynamics, *Propositiones de motu*,[1] a kind of earnest of the *Principia* to come, which Newton apparently sent the Royal Society in the 1680s. Rigaud's volume contains a very brief account of the production of the second edition, and does not give any information concerning the differences between the first two editions. Rigaud also prepared other important works in the history of science, among them the especially precious *Correspondence of scientific men of the seventeenth century...printed from the originals in the collection of the Right Honourable the Earl of Macclesfield* (Oxford, 1841), based on the famous Macclesfield Collection, which is not available to general public inspection.

A dozen years after Rigaud's *Essay*, J. Edleston published his exemplary edition of the *Correspondence of Sir Isaac Newton and Professor Cotes* (London, 1850), with various supplementary letters and notes. Edleston, Fellow of Trinity College, Cambridge, edited this volume from the manuscripts which were in the Trinity College Library,[2] and revealed for the first time the intimacies of the final stages of revision of the *Principia* in 1709–13. Edleston not only thus documented Cotes's role in the improvements made in the second half of the *Principia*, but included much new information on many other aspects of Newton's life and thought. As Newtonian scholars are aware, Edleston's volume is a never-failing source of accurate information on all aspects of Newton's life and career; especially valuable is the 'Synoptical view of Newton's life', which (together with the documentation therefor) occupies sixty-one pages of tiny print.[3]

Within five years, in 1855, Sir David Brewster issued his two-volume set of *Memoirs of the life, writings, and discoveries of Sir Isaac Newton* based on new or unpublished documents: the first full-scale life of Newton.[4] Not only did Brewster describe some of the alterations made in the second and third editions, but he included in an appendix a list (drawn up by the astronomer John Couch Adams) of the principal alterations made by Newton in the third edition of the *Principia*.

[1] For this tract *De Motu* see Chapter III, §3. Rigaud also printed an 'Early notice of fluxions', 13 November 1665, from 'an original paper, in Newton's handwriting, belonging to the Earl of Macclesfield', and six letters from Newton to Halley during the writing of the *Principia*.

[2] Edleston used the letters sent by Newton to Cotes and Cotes's draft letters to Newton, then available to him in the Trinity College Library. At that time, the final versions of Cotes's letters to Newton were still in Hurstbourne Castle, in the possession of the Earl of Portsmouth. They are now in the Portsmouth Collection, U.L.C., MS Add. 3983. But Edleston did not seek out other manuscript letters of Cotes's in the Cambridge libraries; these remain unpublished to this day.

[3] Edleston (1850), pp. xxi–lxxxi.

[4] Brewster (1855); reprinted, 1965, with a new introduction by R. S. Westfall. Although many scholars have joined with Augustus De Morgan (1914, pp. 117–82, notably p. 127) in dismissing this work because of its tone of continuous adulation and its stuffy mannerisms (or 'Brewsterisms'), it is a mine of useful information and leads, although it must be used today with a greater degree of caution than the books of Rigaud, Edleston, or Rouse Ball. De Morgan himself wrote a number of useful studies on Newton, but these largely do not bear directly on the *Principia*; see S. E. De Morgan (1882), pp. 401 ff., for a partial bibliography.

From his own acquaintance with the Portsmouth Collection,[5] Brewster was able to add to Adams's list and also to give some indication as to which of the alterations were due to Henry Pemberton, who had prepared the third edition (E_3) for the press under Newton's direction.[6] A year later, in 1856, owing primarily to the efforts of J.-B. Biot, there was published in France an edition of that famous document in the Newton–Leibniz controversy, the *Commercium epistolicum...réimprimée sur l'édition originale de 1712 avec l'indication des variantes de l'édition de 1722...*,[7] the second time, to my knowledge, that any work of Newton's was printed in which an attempt was made to include the variants from one edition to another, the first being the *Opticks* in Horsley's edition of Newton's *Opera.*

5. THE PORTSMOUTH COLLECTION

Materials for the full-scale investigation of Newton's scientific thought became readily available to scholars in 1872, when the Earl of Portsmouth presented the 'scientific portion' of Newton's manuscripts and some personal copies of Newton's books to the University Library, Cambridge.[1] The decision as to which letters, books, or manuscripts in the collection should be assigned to the 'scientific portion' was evidently made in 1888 by a syndicate appointed by the University 'to examine, classify, and divide' the material, and consisting of H. R. Luard, G. G. Stokes, J. C. Adams, and G. D. Liveing, who together drew up *A catalogue of the Portsmouth Collection of books and papers written by or belonging to Sir Isaac Newton.*[2] The *Catalogue* describes, albeit all too briefly, both the 'scientific portion' which is today in the University Library and 'the papers relating to Theology, Chronology, History, and Alchemy', which, in accordance with the wishes of the Earl of Portsmouth, were to 'be returned to him at Hurstbourne, where they would be carefully preserved' together with 'all the papers relating to private, personal, and family matters'.[3] The *Catalogue* has a short but important Preface discussing the new light cast by the manuscripts on 'three subjects, viz. 1st, the Lunar Theory, 2nd, the Theory of Atmospheric Refraction, and 3rd, the Determination of the

[5] D. T. Whiteside has recently analysed carefully the actual limited access to the Portsmouth manuscripts afforded to Brewster; see [Newton], *Mathematical papers* (1967–), vol. 1, pp. xxix–xxx.

[6] A complete list of Pemberton's contributions may be found in Appendix VI to the text.

[7] [Collins *et al.*] (1856), *Commercium.*

[1] The Portsmouth Collection consists of Newton's papers and some of his books, which after his death came into the hands of John Conduitt, an associate of Newton's and his successor as Master of the Mint, who had married Newton's niece Catherine Barton ten years before Newton's death. Their daughter, Catherine Barton Conduitt, married the Hon. John Wallop, son of the John Wallop (1690–1762) who was created Viscount Lymington in 1720 and first Earl of Portsmouth in 1743. This collection remained more or less intact in the possession of the Earls of Portsmouth until the gift to Cambridge University in the 1880s; see [Sotheby and Co.], *Newton papers* (1936). The most complete account available of the history of the Portsmouth Collection, and its presentation to the University, is that just published by D. T. Whiteside: [Newton], *Mathematical papers* (1967–), vol. 1, pp. xxiv–xxxiii.

[2] [Portsmouth Collection] (1888), *Catalogue*, p. ix.

[3] The latter were sold at public auction in 1936; see Munby (1952*a*); [Sotheby and Co.] (1936), *Newton papers*. See § 9 below.

Form of the Solid of Least Resistance';[4] an Appendix gives some new textual extracts.[5] In the Preface some indication is also given of the riches of the collection for the dispute with Leibniz and his followers concerning the invention of the calculus.

For the editing of the *Principia* the Portsmouth Collection in the University Library, Cambridge, contains treasures of vast extent. There is correspondence of the three editors of the successive editions of the *Principia*—Halley, Cotes, Pemberton—and also correspondence with Flamsteed on topics discussed in the *Principia*. There are the interleaved (and annotated) copies of both the first and second editions (E_1i, E_2i), and more than 1,000 pages of notes and calculations relating to the growth of the *Principia*, together with some preliminary tracts on dynamics, early drafts, versions of the Prefaces and the concluding Scholium Generale, and the unused Conclusio written for the first edition.

The Preface to the *Catalogue of the Portsmouth Collection*, and its appendix, would seem to show that no study of the *Principia* can be considered complete if it does not take account of these papers; yet one of the surprising aspects of Newtonian scholarship prior to World War II has been the relatively little use made of this vast resource of books and papers. A reason may be that the *Catalogue* itself all too often tends to mask or to conceal rather than to reveal the nature of the treasures the Portsmouth Collection contains, as two examples will make clear. 'Section VII. Books' contains as its final entry:

> 18. A common-place book, written originally by B. Smith, D.D., with calculations by Newton written in the blank spaces. This contains Newton's first idea of Fluxions.[6]

Who would guess that this is Newton's now-famous 'Waste Book' (U.L.C. MS Add. 4004), in which he wrote out some of his earliest complete essays on dynamics and mathematics, including the solution of geometric problems 'by motion'? Far from being a book into which Newton has crammed some calculation into the odd blank space here or there, the book is still mostly blank. The former owner has numbered the folios from 1 to 1194, but omitting several hundreds in the sequence, and has written out headings on the tops of pages, all relating to theology: as 1 *Deus*, 11 *Abnegatio Christi*, 13 *Accusator*, 14 *Accusatio*, 15 *Academia* (*Academici*), 47 *Affinitas Spiritualis*, 53 *Allegoria*, 72 *Annulus*, 73 *Antichristus*, 77 *Anthropomorphitae*, 81 *Relapsus* [*sic*], 97 *Audacia*, 98 *Auditores*, 99 *Aulicus*, 101 *Aurum*, 102 *Authoritas*, 103 *Baptismus*,..., and ending with 1129 *Unctio*, 1130 *Uxor*, 1131 *Zelotypia*, 1132 *Zelus*. When Newton obtained the notebook the verso pages were blank, as were some recto pages, but most recto pages had one of the above words, for the most part roughly in alphabetical order; occasionally there was a very short paragraph

[4] [Portsmouth Collection] (1888), *Catalogue*, pp. xi, xxi–xxx.

[5] These extracts related to: 'I. The form of the Solid of Least Resistance. *Principia*, Lib. II, Prop. 35, Schol.; II. A List of Propositions in the Lunar Theory intended to be inserted in a second edition of the *Principia*; III. The motion of the Apogee in an elliptic orbit of very small eccentricity, caused by given disturbing forces.'

[6] [Portsmouth Collection] (1888), *Catalogue*, p. 48.

or other entry under the heading. The earliest pages contain Newton's youthful notes on dynamics and an extended treatise on motion dating from the mid-1660s.[7] Page '1191' contains an attempt by Newton to deal with 'Kepler's problem', by considering the motion of a planet according to the simplifying assumption (often associated with the name of Seth Ward) that the planet moves along an elliptical orbit in such a fashion that a radius vector drawn from the empty focus to the planet sweeps out equal angles in equal times.[8] This 'Waste Book' (the title written on the cover) also contains some of Newton's early tracts on mathematics, including the drawing of tangents and other aspects of geometry, and the 'method de Maximis et minimis'. There are discussions of optical Theorems, a new telescope, 'Problems of Gravity & Levity, &c.', computations of areas, observations of comets, and (p. 81^{v}) a memorandum dated October 1676, giving the full transcription of the cipher message sent 'in my second epistle to M. Leibnitz'.[9] Although this Waste Book is reckoned today to be one of the primary manuscript repositories of Newton's early scientific and mathematical ideas, there is almost no mention of it in print (other than entry in the *Catalogue*) from 1727 to 1959, when Professor H. W. Turnbull drew attention to it in vol. 1 of his edition of Newton's *Correspondence*. (It is, however, mentioned *en passant* in de Villamil's essay on *Newton the man*.)

As a second example, consider the Conclusio written by Newton for the first edition and then rejected. Recently published for the first time by A. R. and M. B. Hall,[10] it may be found in the Portsmouth Collection (together with an unused portion of the Preface to the *Principia*) in 'Section VIII. Miscellaneous Papers'. It is described in the *Catalogue* merely as '6. Systema Mundi'. This title not only is inaccurate, but even hides the fact that this group of manuscripts contains 'Papers connected with the *Principia*' and so should have been put in that classification in Sec. I, viii.

It is useless to speculate whether more attention might have been paid to the Portsmouth Collection if there had been a better *Catalogue*. Yet the fact remains that the Collection was not much used by scholars prior to World War II, a major exception being Louis Trenchard More, in the preparation of his biography (1934). He was evidently unable to study the contents of the Portsmouth Collection as fully as they would seem to have required.

We may now easily trace the evolution of Newtonian scholarship year by year from 1908 to 1959, thanks to the chronological bibliography prepared by Dr Clelia Pighetti.[11] The most cursory look at her work proves how little the

[7] See Herivel (1965), *Background*, pp. 128 ff.

[8] This approximation, in which the empty focus is treated as an 'equant' (in the sense of Ptolemaic astronomy), was first discussed by Kepler himself, but he later rejected it.

[9] The mathematical parts are to be included in Whiteside's edition of Newton's mathematical writings (in progress).

[10] *Unpublished papers* (1962), pp. 320 ff.

[11] Pighetti (1960), 'Studi newtoniani'.

Portsmouth Collection had been used, and how few major new documents were published in their entirety in the first half of the twentieth century. How things have changed since then![12]

6. ROUSE BALL'S 'ESSAY' AND CALL FOR A COMPLETE EDITION OF THE 'PRINCIPIA'

In 1893 W. W. Rouse Ball published his now-classic *Essay on Newton's "Principia"*, in which he gave a full-scale exposition of the history and development of Newton's ideas on mechanics and delineated some of the stages by which the *Principia* had grown. Rouse Ball, a member of the bar, Fellow and Tutor of Trinity College, Cambridge, is still well known today for his ever-popular *Mathematical recreations and essays* (1892 and many later editions). He also wrote a series of excellent essays on Newton and a most valuable history of mathematics at Cambridge.[1]

Rouse Ball's *Essay* remains to this day an indispensable handbook for the student of the *Principia*. It contains chapters on Newton's 'Investigations in 1666', '...in 1679', and '...in 1684', discusses the 'Preparation of the Principia, 1685–1687', and has a concluding chapter on 'Investigations from 1687 to 1726'. Of the greatest value is 'Chapter VI. Analysis of the Principia', in which each major proposition is quoted or paraphrased, with some indication of any alterations made in successive editions; this is at present the only analytical table of contents of the *Principia* available in English. A set of Appendices gives the full texts of the correspondence then available between Hooke and Newton, 1678–80, and between Halley and Newton, 1686–7, and includes 'Memoranda on the correspondence concerning the production of the second [and also of the third] edition of the *Principia*'.

So far as I know, Rouse Ball was the first person to express positive opinions on the need for a critical edition *cum variis lectionibus*, which he had planned to produce himself. On the first page of his *Essay* he tells the reader:

> I had at one time hoped to publish a critical edition of the work, with a prefatory account of its origin and history, notes showing the form in which it was printed originally and the changes introduced in 1713 and 1726, accompanied where desirable by an analytical commentary, together with various other propositions on the subject which are extant among Newton's papers though they were not incorporated in any of the editions he issued. I am unlikely in the immediate future to find time to carry out this plan.

Later on in the book he returned to this subject, in introducing his own analytical table of contents of the *Principia*:

> A complete edition of the *Principia* should, I think, show the changes made in the second edition, and the further changes introduced in the third edition. I possess in manuscript a list of the additions

[12] The scholarship of the last two decades is far too varied to permit a simple summary. Two recent publications—Cohen (1960), 'Newton scholarship', and Whiteside (1962), 'Newtonian research'—are devoted to a survey of recent trends and current literature.

[1] Ball (1918), *Cambridge papers*; (1889), *Mathematics at Cambridge.*

and variations made in the second edition.... I have not formed a list of the changes introduced into the third edition, but I believe that the bulk of them are given in the list by Adams which is printed by Brewster.[2]

It is not clear how extensively Rouse Ball had used the Portsmouth Collection. For instance, he does not ever say explicitly which manuscripts he deemed worthy of being printed *in extenso* in a critical edition of the *Principia*, as he might have done had he known the Collection well. He did print Halley's letters to Newton 'in their original form as shown in the Portsmouth copies',[3] and he studied the drafts of Newton's tract *De Motu* (for which see Chapter III, §3, below) in that Collection and printed excerpts from them. But we do not know how thorough an examination he made of the large mass of manuscripts relating to the *Principia*,[4] although I have no doubt that he would eventually have done so had he ever undertaken the edition.

Among Rouse Ball's papers preserved in Trinity College there are no notes in preparation for such an edition, nor even a plan. Moreover, I have not been able to find any 'list of the additions and variations made in the second edition', which Rouse Ball said he 'possess[ed] in manuscript'.

Can it be that Rouse Ball had in mind only a most extraordinary copy of the *Principia*, then in his possession, but now in the Trinity College Library? This is a copy of the first edition, in which the owner—who has not proved to be identifiable—has with rather scrupulous care entered almost every one of the 'additions and variations made in the second edition'.[5] The owner of this book even copied out the revised dedication (mentioning Queen Anne), although he did not include the revisions in Halley's poem. Such a copy might be useful at a preliminary stage

[2] Ball (1893), *Essay*, p. 74.

[3] The full texts of the Newton–Halley correspondence on the *Principia* were first published by Brewster in his *Memoirs*.... Rouse Ball did not have access to the originals of these letters and revised Brewster's texts from the copies made by the cataloguers of the Portsmouth Collection (now MS Add. 4007, U.L.C.).

[4] In his *Essay*, Rouse Ball does not give much exact information concerning the contents of those portions of the Portsmouth Collection (especially MSS Add. 3965, 3966, 3967) relating to the *Principia*. With the notable exception of the tract *De Motu* and the Halley letters, almost all the material quoted from the manuscripts in the Portsmouth Collection is taken from the *Catalogue*, whence also derive such errors as the statement that the 'Portsmouth Collection contains holograph manuscript copies of the *Principia*' (Ball, *Essay*, p. 3). But Rouse Ball did publish manuscript material of Newton's from the Portsmouth Collection in other studies, e.g. those collected in his *Cambridge essays*. His 'Newtonian fragment' (Ball (1892*a*)) is a careful restoration of Newton's argument about centripetal forces in MS Add. 3965, §2.

[5] To do so there were needed many interleaved pages for such new material as Cotes's preface, the new Section in Book II and the additional material on comets at the end of Book III, the 'Regulae Philosophandi' at the beginning of Book III, the concluding 'Scholium Generale', and the Index (lacking in E_1 but added by Cotes to E_2). In this copy the owner has even introduced the running heads 'Liber Primus', 'Liber 2dus'—but not 'Liber Tertius', save on one page (403). He has, furthermore, even written out the new title-page: 'Philosophiae Naturalis/Principia Mathematica./Auctore/Isaaco Newtono Equite Aurato./Editio secunda auctior et emendatior./Cantabrigiae 1714.' While not copied out line for line, this text of the title-page corresponds exactly to the second edition, save for the omission of two commas, after 'Newtono' and after 'Cantabrigiae', and an erroneous translation of the date from roman to arabic numerals, MDCCXIII to 1714. The first edition had used the form *Autore Is. NEWTON* rather than *Auctore Isaaco Newtono*, and had described Newton as Fellow of Trinity College, Lucasian Professor of Mathematics, and Fellow of the Royal Society. This copy is press-marked C.15.159.

of preparing a critical edition, since it gives a good idea of the kind and extent of the alterations made in 1713. But it would never serve for a critical text because of the omissions made by the copyist, who also committed a number of minor errors.[6]

7. AN EDITION OF NEWTON'S SCIENTIFIC WORKS AND CORRESPONDENCE, PROPOSED (1904) UNDER THE AUSPICES OF THE CAMBRIDGE PHILOSOPHICAL SOCIETY

A decade after the publication of Rouse Ball's *Essay*, a plan was put forward 'to produce a complete edition of Newton's Scientific Works and Correspondence, to be prepared under the auspices of the Cambridge Philosophical Society, which was represented by a strong editorial committee of residents'. Professor R. A. Sampson 'was charged with the duties of Editor-in-Chief. The Syndics of the [Cambridge University] Press undertook the whole financial responsibility, including the provision of expert clerical work in transcribing MSS., etc., as well as necessary personal expenses of the editor.' These quotations come from an article written by Sampson in 1924, some twenty years later, bringing to light the abortive efforts of 1904 to produce an edition of Newton's scientific writings.

Reading over the scheme for the edition after a lapse of two decades, Sampson experienced a feeling 'of some surprise that with so prosperous an opening I let it go'. He then offered the following explanation 'to those who were more than willing to collaborate in the production':

> Perhaps I ought not to have undertaken it. In 1904, when the scheme was approved, I had upon my hands a very heavy piece of work, the Theory and Tables of Jupiter's Satellites, which I might never have finished had I allowed any other task to take priority of it. As often happens, it took more time than was anticipated. The conclusion was assured in 1910; and, had nothing intervened, I should then have taken up actively the edition of Newton, though in the meantime I had learned to think rather temperately of my own qualifications for the work. But in the same year I was transferred to Edinburgh where different demands upon my time have never allowed me to think of it again.

True, no doubt! But it is of more than passing interest that neither the Syndics of the Press nor the Cambridge Philosophical Society were able to find a substitute editor-in-chief, if indeed they ever sought one.

Sampson's edition was planned 'to embrace the whole of the scientific works edited from the originals as far as these are accessible, including all letters and any fragments of interest that are among the Portsmouth papers or elsewhere'. There was to have been included 'an English edition of the *Principia* in addition to the Latin original'. It was envisaged that the whole collection might occupy six quarto volumes (containing some 320 words per page) of about 600 pages each, an estimate that would need to be revised if the figures were to be printed 'in their

[6] Further information about this particular copy may be found in Chapter VIII, §5.

proper places in the midst of the text with the repetitions that this entails'. The first volume of the edition was to contain:

Principia reprinted from Latin of Edition III., with bibliography, recension of textual variations, notes of brief extent, and full index, as mentioned above.

Sampson did not plan merely to reprint Newton's major works, correcting them by reference to the manuscripts. He believed that his text should 'be accompanied by marginal analysis, recension of textual changes where these occur, notes mainly bibliographical and historical, and full index'. Even so, he estimated that 'this additional matter' would occupy only about one-tenth of the space of the text.[1]

[1] A practising astronomer, neither a trained historian nor a classical scholar, Sampson planned to have assistance in such jobs as collating and transcribing manuscripts; he early on called attention to the expenses involved in the 'large amount of collating, which would probably best be done by the occasional services of a clerk expert at such work'.

Evidently, the only new manuscript material (other than letters) that Sampson planned to include was the Lucasian Lectures, which he described as '*De Motu Corporum*, delivered 1684–1685. MS. of these in University Library.' This work is a predraft of a good bit of Book I of the *Principia*, most of which is so nearly identical to the *Principia* as to make an edition *per se* of little value. (This manuscript had been described by Edleston (1880), xcv–xcvii; see below, Chapter IV, §2 and Supplement IV.) Thus, the proposed edition would have consisted almost entirely of correspondence plus reprints of treatises and tracts that had already been printed, but in more correct versions than had been previously available. One gathers that the riches of the Portsmouth Collection would in considerable measure have remained untapped, save for their possible use in the preparation of editorial comments and introductions.

I assume that Sampson had not done much (if any) work in the Portsmouth Collection. If he had, he would surely have known that Newton's posthumously published 'System of the World' exists in that collection in the very manuscript, written out by Humphrey Newton and then corrected and partly rewritten by Newton, from which the Latin edition had been printed (on this topic see Chapter IV, §6, below). As a matter of fact, this manuscript (Add. 3990) is described plainly in Sec. VII of the *Catalogue of the Portsmouth Collection* (p. 47) as follows: '(4) De motu Corporum Liber Secundus. This is the treatise De Mundi Systemate. Horsley, iii. pp. 180–242.' The reason for the two names is that this work was originally intended to be the second book of the *Principia* at a time when the *Principia* was still being called *De Motu Corporum* and was to have consisted of two books; hence it was entitled by Newton 'De Motu Corporum Liber Secundus'. But when Newton later rewrote this material, he had already decided to entitle the whole work *Philosophiae naturalis principia mathematica*, and to add the present Book II. Hence the new version became Book III of the *Principia* and was entitled 'De Mundi Systemate Liber Tertius'. When the early version was found among Newton's papers and published posthumously, it was then given the confusing title *System of the world*, identical to the title of Book III of the *Principia*. See, further, Supplement VI.

Sampson, unaware of the existence of the authentic manuscript in the Portsmouth Collection, planned to edit this work by reprinting one of the eighteenth-century printings, correcting it by comparison with a manuscript version in Trinity College. His note reads: '(4) *De Motu Corporum Liber*, commonly called *De Mundi Systemate*, delivered 1687– . MS. in Trinity College Library, from which printed copies may be corrected. It contains some 20,000 words.' Now this Trinity College manuscript (press-marked R.16.39) is a poor choice, since it is only a copy made (in part) by Roger Cotes—who did not even make a copy of the original (which was unavailable to him), but made a copy of a copy. The first part was copied by Cotes from a version (in the hand of Humphrey Newton) deposited in the University Library as Newton's professorial lectures for 1687; this manuscript (U.L.C. MS Dd.4.18) corresponds to about one-half of the whole work. The remainder was apparently taken from a copy made by (or for) Charles Morgan, and now in the library of Clare College; Morgan obtained the text from Martin Folkes, a secretary of the Royal Society.

The Trinity College manuscript is fully described by Edleston, in his account of Newton's professorial lectures, in the beginning of his edition of the Newton–Cotes correspondence. Edleston did not know of the copy in the Portsmouth Collection, which at that time had not yet been presented to Cambridge University. Either Sampson himself or one of his associates or assistants picked up the description by Edleston without exploring the possibilities of a better manuscript being in either the Portsmouth Collection or the general manuscript collection in the University Library.

This estimate is so small that one can only wonder how it was arrived at: for instance, our variant readings occupy considerably more than a tenth of the text, and they include neither the marginal analyses nor the historical and bibliographical notes that Sampson proposed.

8. DREYER'S PROPOSAL (1924) FOR A CRITICAL EDITION OF NEWTON'S WORKS

In 1924 the subject of an edition of the *Principia* was raised publicly in an address by the President of the Royal Astronomical Society, J. L. E. Dreyer, 'On the desirability of publishing a new edition of Isaac Newton's collected works'. Dreyer, despite his prominence in the world of science and his eminence as an historian of astronomy, had evidently never heard of the earlier project of 1904, which had been so well kept a secret that news of it had never leaked out. It was only on the publication of Dreyer's address in 1924 that Sampson decided to print the 1904 *Outline of proposed critical edition of Newton's scientific works and correspondence*. Probably Sampson, made aware by Dreyer's address that his countrymen were ignorant of the abortive earlier attempt to produce an edition of Newton's works, took this method of informing Dreyer and others that the need for the edition being advocated in 1924 had actually been recognized twenty years earlier.

In this presidential address, published in February 1924 in the *Monthly Notices of the Royal Astronomical Society*, Dreyer called attention to 'fine editions...published within the last sixty years of the works of Copernicus, Tycho Brahe, Kepler, Galileo, Torricelli, Descartes, Fermat, Huygens, Leibniz, Euler, Lagrange, Laplace, William Herschel', and observed correctly that 'the greatest name of all is missing from the list, that of Isaac Newton'.[1] Criticizing Samuel Horsley, whose edition of Newton's *Opera* was published in 1779–85, Dreyer was nevertheless fair enough to concede that Horsley's work 'is not worse than' other similar ventures of that period, such as Albèri's edition of Galileo. Dreyer's chief complaints were that Horsley made little use 'of unpublished material, while there is an entire absence of introductions and notes'.[2] Above all, Dreyer regretted the want of 'any lists of variants of the different editions published during the lifetime of Newton' as well as information concerning 'when and under what conditions the various books and papers were originally issued'.

Dreyer's plan for a new edition was drawn up with some knowledge of the mathematical and scientific papers in the Portsmouth Collection in the University Library, Cambridge. It would not appear, however, that Dreyer had ever made a personal study of the manuscripts in this collection. Most likely, his information had been drawn from the *Catalogue of the Portsmouth Collection*, published in 1888, since he mentions only matters described in the introduction to the *Catalogue*. As to

[1] It was Dreyer himself who had edited the works of Tycho Brahe and of William Herschel!

[2] True there is no introduction, but there are many notes; on this subject see footnote 4 to §3 above.

the changes made in the printed editions, Dreyer was content to quote a paragraph from Rouse Ball's *Essay* (given in § 6 above) with the comment:

All these variants are of great importance, and should be clearly indicated in any critical edition of the *Principia*. Perhaps the proper thing to do would be to reprint the first edition on the right-hand pages of the open book, giving the alterations and additions on the left-hand pages.

Astronomer, historian of astronomy, biographer, and editor, Dreyer may be presumed to have spoken with considerable authority, augmented by his presidency of the Royal Astronomical Society. He concluded his address with a note of regret concerning 'the want of respect to the memory of Isaac Newton implied in the neglect to publish a proper edition of his works'. He appealed to the 'distinguished members of Newton's old University' to undertake this great project, and he ended on an optimistic note: 'Surely the pecuniary means will be forthcoming, if the proper amount of talent and enthusiasm can be secured.'

9. ASPECTS OF NEWTONIAN SCHOLARSHIP IN THE FIRST HALF OF THE TWENTIETH CENTURY

Dreyer's noble address had no immediate practical consequences. At least, no steps were at once undertaken to prepare an edition of Newton's scientific writings, nor did anyone set out to prepare a scholarly edition of any single major scientific work of Isaac Newton's. Undoubtedly, Dreyer—like Sampson—had been overly sanguine, perhaps not fully aware of the magnitude and complexity of the task, which, as we shall see presently, does entail tremendous though not insuperable difficulties. In a letter to *Nature*, published in 1924, in reply to Dreyer, the physicist Sir Joseph Larmor (who had been a secretary of the Royal Society from 1901 to 1912) called attention to the special types of problem that would confront the would-be editor of Newton. These, he said, were so different in kind from those faced by the editors of Galileo, Huygens, and Laplace that they could readily serve to explain why an edition of Newton's scientific works would appear almost too difficult to undertake. Larmor ended by expressing 'the view that a systematic collection of the letters [of Newton] remains most desirable' and he was firm in the 'hope that some day' at least that part of the original scheme of 1904 relating to the correspondence might be realized. Eventually, in 1938, a committee was constituted by the Council of the Royal Society to 'discuss the question of the publication of the Newton letters';[1] under the very able and distinguished chairmanship of Sir Charles Sherrington, the noted physiologist whose writings on subjects of the history of science were outstanding, the committee moved ahead with the business and proposed Professor H. C. Plummer as editor. Professor Plummer accepted this assignment but died in 1946 'with the main work still to do'.

During the first half of the twentieth century the only major work[2] of Newton's

[1] [Newton] (1959–), *Correspondence*, vol. 1, pp. xiii–xiv, foreword by C. N. Hinshelwood.

[2] In 1919 Duncan Fraser published Newton's *Methodus differentialis* (and again in 1927), but this tract—however interesting—is not a major work in the sense of the *Principia* or the *Lectiones opticae* or even *De quadratura*.

to have been reprinted (although not, properly speaking, edited) in the original language was Newton's *Opticks*, reissued in 1931 in London as a mere reprint of the fourth edition (1730), but with a new introductory essay by E. T. Whittaker and a foreword by Albert Einstein.[3] Two major collections of essays on Newton had appeared in the late twenties, however, both of them commemorative volumes issued on the two-hundredth anniversary (1927) of Newton's death;[4] and in the following decade Louis Trenchard More (1934) published his biography of Newton, to some extent based on new (or theretofore unpublished) manuscript material. But the general lack of concern for the history of science as a subject and for Isaac Newton as a dominant figure is illustrated by the 1936 sale at public auction and dispersal of Newton's papers and personal documents. This collection (sold in order to enable the owners to pay death duties) had remained in the family of the Earl of Portsmouth after the magnificent gift to Cambridge University of the scientific papers. In 1936 not only was the collection permitted to be disassembled and scattered to the four winds at the whim of purchasers, but in point of fact the total sum fetched by the entire collection was trivial, a mere £9,030 10*s*.[5] At a recent sale of a famous history of science collection, belonging to the eminent physicist and Newtonian scholar Professor E. N. da C. Andrade, two examples of the first edition of the *Principia* alone fetched half that sum[6]—a fact which cannot be explained solely on the grounds of monetary inflation and the decline in purchasing power of the pound. For the truth of the matter is that the first edition of the *Principia*, which today commands a price of some £3,000 at auction, was regularly advertised in booksellers' catalogues in the 1930s for £60–70,[7] and no one would attribute such a fiftyfold increase in value entirely to a supposed fiftyfold decrease in the purchasing power of the pound during the last thirty years. Many factors, of course, including the normally changing habits and interests of collectors, are responsible for the dramatic shift in the monetary value of Newtoniana. But two primary causes are: (1) a constantly growing interest in 'antiquarian' science—books, manuscripts, scientific instruments—owing in considerable measure to an increasing inescapable awareness of the importance of science as a primary force moulding our society and our lives, and (2) a concomitant rise of the history of science as a full-time subject to occupy university teachers and researchers.

No doubt Professor Sampson had been quite correct in saying (in 1924):

> The purpose [of such an edition as had been planned], I presume, is to place in the hands of scholars a critical text of what Newton actually wrote, complete at least in its scientific bearings. It would be but little read by ordinary students and investigators. It is an historical task, and, in a sense, one of national importance, but for the advancement of science it matters only in moderate degree.

[3] Reprinted, 1952, with a preface by I. B. Cohen and an analytical table of contents by D. H. D. Roller.
[4] Greenstreet (1927), Brasch (1928).
[5] See Munby (1952*a*), p. 41.
[6] [Sotheby and Co.] (1965); see Andrade, 'Newton collection' (1953).
[7] See Munby (1952), pp. 38–9.

Sampson was not an historian, but a scientist, and no doubt was fully justified in concluding that 'one might well hesitate before advising anyone except a professional historian to tie himself to such a piece of work'. And surely, in 1924, Sampson appraised the situation accurately when he observed that 'a professional historian would be only too likely to feel a merely tepid interest in such a statement as "a ray of light has four sides" '. What Sampson did not foresee was the rise of a special class of scientifically trained historians or historically minded scientists who would consider the scientific work of a man such as Isaac Newton to be a primary subject of research worthy of the highest consideration. Nor did he envisage a day when the general historian as well as the historian of ideas (or the 'intellectual' historian) would consider the history of science a worthy and even essential topic.

CHAPTER II

THE PLAN OF THE PRESENT EDITION WITH VARIANT READINGS AND THE PROBLEMS OF NEWTONIAN SCHOLARSHIP

1. THE NEW HISTORY OF SCIENCE: THE NEED TO KNOW THE CHANGES IN THE 'PRINCIPIA' AND 'OPTICKS'

THE present edition of Newton's *Principia*, conceived in the late 1950s, illustrates the new concern of the history of science for texts that show the growth of ideas rather than their mere final expression. It was a better time to undertake this assignment than 1924, 1904, or 1897, when previous editions had been suggested or planned—to no avail. By the 1950s the history of science had grown tremendously, to take its place as a young but well-established academic subject supported by philanthropic foundations and by national councils, committees, or government foundations for scientific research.

The decades following World War II witnessed a dramatic change in the goals or the fundamental concepts of the history of science. No longer is it considered sufficient to develop and recount chronologies of the great discoveries. A fundamental methodological precept of the present historians of science is that research be based on primary documents, whether books, manuscripts, artifacts, or instruments. The ideal of today's scholarship is to seek constantly for any possible sources or antecedents of any given set of scientific ideas and to be always on the look-out for clues to the growth of concepts and methods in the lifework of a single individual.

The new programme of the history of science has set into motion explorations in depth into periods that hitherto have been studied only in a general way, and it has focused attention on the actual texts and scientific instruments that mark the progress of scientific thought. Today's scholars demand that studies be conceived from a dynamic rather than a static point of view. Whereas previous generations might have been content to admire the architectural majesty of the final version of a scientific classic, historians of science at present want to know also about the pre-drafts, the early versions, the stages of successive alteration—and they are eager to see the creative mind of the scientist actually at work, to study the attempted revisions which the author discarded as well as those which proved acceptable and thus entered the ultimate text.

It is not too long ago that in the standard biography of Isaac Newton (More, 1934) the whole set of Queries at the end of Newton's *Opticks* was presented as a single unit following the discussion of his paper on light and colours of 1672. The otherwise uninformed reader would never guess that these Queries represent Newton's thoughts (or, at least, the expression of his thoughts) over an interval of more than a dozen years, beginning in 1704, and hence should not be regarded as a mere appendix to Newton's views on the subject of light as of the 1670s. Even Dreyer (1924), who was aware of the need of comparing the different editions of the *Principia*, seems to have given no heed to the possibility that the *Opticks*, too, might have been revised or altered in successive editions, although this information was already available in the notes to Horsley's edition in the eighteenth century. These Queries did not appear simultaneously, but were published in three separate stages. There were 16 Queries in the original edition of 1704 in English; the new ones in the Latin edition of 1706 increased the number to 23; and finally in the second edition of 1717 (1718) 8 additional Queries were added to bring the total up to 31.[1] Today no scholar worthy of the name would write about the *Opticks* without taking account of the gradual unfolding of the Queries.

The *Principia* has had a curious history with regard to the changes made in successive editions. Ever since 1850 (when Edleston published Newton's correspondence with Cotes), there has been a general appreciation that the second edition (E_2, 1713) differs markedly from the first (E_1, 1687), but not too much attention has been paid to the actual changes themselves and their significance.[2] Because of the acrimony of the Newton–Leibniz controversy, the 'Leibniz Scholium' and its alterations were given some attention.[3] Although many scholars have become more or less aware that the concluding Scholium Generale was written for the second edition and was absent from the first, the literature concerning Newton of the fairly recent past shows an ignorance of this simple fact and other changes. In all too many works, there has been a general tendency to view the *Principia* as a single entity, as if the final version had been conceived in every aspect at the time of the first edition in 1687.[4]

[1] But the eight 'final' Queries added in 1717 are not the terminal Queries 24–31. What Newton did was to revise and renumber the set of seven Queries 17–23 introduced in the Latin edition of 1706 so as to make *them* the final seven Queries 25–31 of the 1717 (1718) English edition, and to place the new set of eight Queries, appearing for the first time, in the intermediate position of Queries 17–24.

[2] For instance, Brewster listed the major alterations made in the third edition, but not those made in the second edition.

[3] Cf. De Morgan (1852*a*), 'Early history of infinitesimals'; Cajori (1919), *Conceptions of limits and fluxions*; [Newton] (1934), *Principia* [Motte–Cajori], appendix, p. 654.

[4] Alternatively, some scholars have written that the Scholium Generale was composed for the edition of 1713 (E_2) and have then discussed the final version published in 1726 (E_3) as if it had appeared in 1713! One

But although it was known to some degree that the concluding Scholium Generale, famous for its slogan of 'Hypotheses non fingo', was composed only in 1712 for the second edition of the *Principia*, scholars were not generally aware until very recently that the set of 'Regulae Philosophandi' and of 'Phaenomena' at the beginning of Book III were largely part of a single set of 'Hypotheses' in the first edition (E_1, 1687). It was in fact only on learning purely by chance about the latter change that Professor Koyré and I became interested in producing an edition of the *Principia* with variant readings, one in which the reader might learn at a glance whether any part of the *Principia* had been altered during the four decades between the first and the third editions.[5] Accordingly, with the aid of a grant from the American Philosophical Society and a subsidiary grant from the Rockefeller Foundation, a collation *verbatim ac litteratim* was undertaken of the three authorized editions of the *Principia* to appear during Newton's life.[6] Such were the beginnings of the present work.

2. THE INADEQUACIES OF AN APPARATUS CRITICUS BASED SOLELY ON THE PRINTED EDITIONS, SHOWN BY AN EXAMPLE: 'REGULA PHILOSOPHANDI III'

The new demands of the history of science directly affected the present edition, tending to make our task more complex, particularly because of the ideal of having as complete a documentation as possible of the growth of thought and expression. As editors, Professor Koyré and I could not for long avoid the question of whether a critical apparatus would be adequate to the needs of other scholars if it were limited only to variant readings deriving from a collation of the three authorized printed editions: E_1 (1687), E_2 (1713), E_3 (1726). We therefore undertook a number of ancillary inquiries in an attempt to make precise just how our edition might be used in specific investigations of Newtonian problems. Almost at once it became clear that an edition with variant readings embodying only the alterations made in the printed editions would not fulfil even our own requirements and expectations. The interim product of our labours, while extremely interesting in many ways, had an obvious fault in that it displayed the state of the *Principia* only at three discrete epochs—1687, 1713, 1726—without so much as a hint of the

notable alteration was the substitution of the word 'natural' (E_3) for 'experimental' (E_2) in the statement that to discuss God 'from phenomena belongs to experimental philosophy'. See the variant readings on pages {526–530} of this edition of the *Principia*.

[5] Professor Koyré (1955) came upon this alteration when he happened to be leafing through a copy of E_1 in the Bibliothèque Nationale for another purpose. I came upon the same alteration while examining a facsimile reprint of E_1 preparatory to writing a book review of it; see *Isis*, vol. 44 (1953), pp. 287–8. Rouse Ball (1893, p. 106) is one of the few modern commentators to have been aware of this alteration but mentioned the fact only *en passant*: following a summary of Regulae I–IV, he wrote, 'The above rules are taken from the third edition. The hypotheses enumerated in the first edition are less clear and less full.'

[6] This first collation was in the nature of an exploratory exercise. The final collation (on which the present edition is based), on a much greater scale, embraced manuscript alterations as well as printed variants, and was supported by a grant from the National Science Foundation (U.S.A.).

intermediate stages. The resulting loss, to the potential user of our edition, may best be seen in one or two examples.

I have referred earlier to the 'Regulae Philosophandi' at the beginning of Book III {387.1 ff.}, which grew out of the 'Hypotheses' in E_1. The reader who had available to him information based solely on the printed editions would know that there are only three 'Regulae' in E_2, that the fourth appeared first in E_3; and that 'Regula I' and 'Regula II' are respectively the 'Hypothesis I' and 'Hypothesis II' of E_1 practically unaltered,[1] while 'Regula III' in E_2 and E_3 deals with a topic entirely different from the 'Hypothesis III' printed in E_1.[2] Indeed, when both Professor Koyré and I had written about this transformation,[3] before we undertook to collaborate in preparing the present edition of the *Principia* with variant readings, this was the boundary of our knowledge.

From a collation of the printed editions alone it would not be possible to learn that when Newton first wrote out a version of what appears in E_2 as 'Regula III' he had not yet divided the 'Hypotheses' into 'Regulae' and 'Phaenomena'. The fact is that 'Regula III' began life as an 'Hypothesis III' intended as a substitute for that 'Hypothesis III' in the original set of 'Hypotheses' in E_1. This particular transformation is especially fascinating because of the great importance of 'Regula III' as a primary statement of Newton's philosophy of science.[4]

Our attention was all the more sharply focused on 'Regula III' as a consequence of examining the copy of E_1 given by Newton to Locke in the early 1690s, at present in the Trinity College Library,[5] in which there is entered in the margin in a scribal hand:

Hypoth. III. Qualitas [!] corporum quae intendi et remitti nequeunt, quaeque corporibus omnibus competunt in quibus experimenta instituere licet, sunt qualitates corporum universorum.

Hypoth. III. The qualities of bodies that cannot be intended and remitted,[6] and that belong to all bodies in which one can set up experiments, are the qualities of bodies universally.

[1] A glance at the Apparatus Criticus {387.6–7} will show that in E_2 a new sentence was added to introduce the previous explanation of 'Regula I'.

[2] Such a reader would know, furthermore, that, of the other 'Hypotheses', 'Hyp. IV' ('Centrum Systematis Mundani quiescere') remains an 'Hypothesis' in both E_2 and E_3, while Hypotheses V–IX become 'Phaenomena I, III–VI', 'Phaenomenon II' appearing for the first time in E_2.

[3] Koyré (1955), 'Pour une édition'; (1960*a*), 'Les "Regulae philosophandi"', reprinted in *Newtonian studies* (1965), chap. 6. Cohen (1953), Review of facsimile edition of Newton's *Principia*; (1956), *Franklin and Newton*, pp. 131 ff.

[4] Cf. Mandelbaum (1964), *Philosophy*, chap. 2; also Cohen (1971), *Scientific ideas*, chap. 2, 'From hypotheses to rules'; Cohen (1966), 'Hypotheses in Newton's philosophy'; McGuire (1967), 'Transmutation and immutability'.

[5] Press-marked Adv. b.1.6.

[6] The expressions 'intendi' and 'remitti' are associated with the medieval theory known as the latitude of forms, a theory concerning those 'qualities' that may be quantified, that—to use the language of the theory—may undergo either an intension ('intensio') or a remission ('remissio') of degrees. Motte, in his eighteenth-century English version, rendered 'quae intendi et remitti nequeunt' by 'which admit neither intension nor remission of degrees' (Newton (1729), vol. 2, p. 203), modernized by Cajori so as to read 'which admit neither intensification nor remission of degrees' (Newton (1934), p. 398)—which is not very helpful and perhaps serves chiefly to illustrate the bewilderment that may arise when Newton himself uses the obsolescent technical

Where could this have come from? It is not the text of the 'Regula III' as published in E_2 and E_3; the final 'pro qualitatibus corporum universorum habendae sunt' ('are to be taken as the qualities of bodies universally') found in E_2 and E_3 is given here as 'sunt qualitates corporum universorum'.[7] Far more significant is the fact that this form of 'Regula III' appears in Locke's copy as 'Hypoth. III'.

A search among the various Newtoniana in Cambridge, England, disclosed in Newton's own annotated copy of the first edition an apparent source of the emendation to be found in Locke's copy. Newton's annotated copy of E_1 (designated by us as E_1a), formerly in his personal library, is now among Newton's books in the Trinity College Library.[8] The version in E_1a, however, begins correctly with 'Qualitates' rather than the erroneously transcribed 'Qualitas' of the Locke copy.[9] In his own copy, Newton had at first begun to write out the new 'Hypoth. III' with the opening word 'Proprietates'; he then scratched out 'Proprietates' and began again with 'Qualitates'. (See Plate 1.) In the MS Errata which Newton wrote out in the back of the above-mentioned annotated copy of E_1, this 'Hypoth. III' appears as

Hypoth. III. Leges et proprietates corporum omnium in quibus experimenta instituere licet sunt leges et proprietates corporum universorum.	Hypoth. III. The laws and properties of all bodies on which experiments can be made are the laws and properties of bodies universally.

The words 'et proprietates' occurring twice in this form of the 'Hypothesis' were inserted by Newton afterwards.[10] In this version, Newton goes on to explain what he means (beginning 'Nam proprietates corporum...'), much as in E_2 (save that in E_2 the explanation begins 'Nam qualitates corporum...'). Evidently, of the four versions, the one in the MS Errata is earliest, dealing with 'leges corporum' (there changed to 'leges et proprietates corporum'), followed by the version written out on the printed page, in which Newton dropped the 'leges' and began with 'proprietates', which he then altered to 'qualitates' while at the same time expanding the middle portion. This is the version copied into Locke's copy. But as printed in E_2, two further changes were made: (1) the designation of 'Hypothesis' was altered to 'Regula', and (2) the final clause was rewritten.

This final stage of transformation may be traced through another copy of E_1 that was in Newton's personal library, an interleaved copy that contains his

language of an earlier age. In the French version, the Marquise du Chastellet modernized Newton's expressions so as to make them more fully in accord with the spirit of the Newtonian (rather than the medieval) philosophy, and wrote of 'les qualités des corps qui ne sont susceptibles ni d'augmentation ni de diminution' (Paris, 1759, vol. 2, p. 3).

[7] Although 'universorum' modifies 'corporum' rather than 'qualitates', I have above translated Newton's final 'qualitates corporum universorum' by 'properties of bodies universally' as being close to Newton's actual words. It would be misleading to have written 'properties of universal bodies', although 'properties of all bodies' would be permissible. In the version as printed in E_2 and E_3, 'pro qualitatibus corporum universorum' is translated by Motte as 'the universal qualities of all bodies whatsoever'.

[8] Press-marked NQ. 16. 200.

[9] For other examples of scribal error and ignorance in Locke's copy of the *Principia*, see Chapter VIII § 2, below.

[10] These variants are all to be found in the text of this edition {387. 15 ff.}; see Appendix IV.

alterations on both printed pages and interleaves, now in the Portsmouth Collection of the University Library, Cambridge[11] (and which we designate as $E_1 i$). (See Plates 2 and 3.) On the printed page Newton has crossed out the heading 'Hypotheses' and written above it 'Regulae Philosophandi', while the designations 'Hypoth. I' and 'Hypoth. II' have been changed in hand to become 'Reg. I' and 'Reg. II'. The whole of 'Hypoth. III' has been cancelled without any prior alteration. On the interleaf, Newton has written out the new 'Regula' as 'Hypoth. III'—in a version agreeing with the one entered on the printed page of $E_1 a$ and on the page of Locke's copy. Then, later, he has changed the designation 'Hypoth. III' to 'Reg. III' and has also altered the final clause to the form found in E_2 and E_3. At the same time he has changed the opening of the discussion from 'Nam proprietates corporum...' to 'Nam qualitates corporum...'. It is apparent from the ink and handwriting that these changes were made at a later time, when Newton also decided to add a second paragraph to his explanation, as is found in E_2.[12]

'Regula III' achieved its final form by successive stages. Obviously, it seemed desirable to have the Apparatus Criticus of our edition present these and other manuscript emendations from Newton's own copies of the *Principia*. We even contemplated introducing those from Locke's copy. The scholarly value of these early variant forms is great because a knowledge of the development of 'Regula III' leads to a better comprehension of this very important statement of Newton's philosophy of science.

Furthermore, by knowing such emendations, it becomes possible to trace the dissemination of Newton's manuscript alterations as they appear and reappear in one or another marked copy of E_1.[13] At once the source is apparent of such references in print, as in J. C. Groening's book of 1701, where we find: 'Hypoth. III. Leges & proprietates corporum omnium...sunt leges & proprietates corporum universorum', followed by a single long paragraph beginning, 'Nam proprietates corporum...'. This text is identically that of the MS Errata in Newton's annotated copy of E_1.[14]

3. A SECOND EXAMPLE: NEWTON'S 'ELECTRIC AND ELASTIC SPIRIT'

The desirability of making available the intermediate manuscript versions was presented to us in an interesting discovery by Professors A. R. and M. B. Hall: the puzzling presence of the words 'electric and elastic' in the last sentence of Andrew Motte's English version of the concluding 'Scholium Generale'.[1] Newton

[11] Press-marked Adv. b.39.1.

[12] The evolution of this Regula III is traced in full in Cohen (1971), *Scientific ideas*, chap. 2, 'From hypotheses to rules', and in Cohen (1966), 'Hypotheses in Newton's philosophy'. Cf. also two important studies: McGuire (1966), 'Body and void', and McGuire and Rattansi (1966).

[13] See Chapter VIII, § 3, below, and Appendix IV to the text.

[14] Groening's publication is discussed below in Chapter VII, § 10, and a guide to the list of emendations which he published is given in Appendix IV.

[1] Hall and Hall (1959), 'Newton's electric spirit'.

has been discussing 'a certain most subtle Spirit which pervades and lies hid in all gross bodies'. By 'the force and action' of this spirit, he says, a variety of physical phenomena are produced: the mutual attraction of the particles of bodies and their possible coherence, the attraction and repulsion of 'neighbouring corpuscles' by 'electric bodies', the emission, refraction, and inflection of light and the heating of bodies by radiant light, the excitation of sensation and the exercise of the commands of the will upon the muscles 'by the vibrations of this Spirit, mutually propagated along the solid filaments of the nerves'. Alas, says Newton—in Motte's version—we do not have sufficient experiments to determine and to demonstrate 'the laws by which this electric and elastic spirit operates'. The Halls showed that this widely quoted statement differs from Newton's concluding clause, 'neque adest sufficiens copia experimentorum, quibus leges actionum hujus spiritus accurate determinari & monstrari debent'. In neither the second nor the third edition is the 'spiritus' said by Newton to be 'electric and elastic'. The Halls, however, did find (and publish) some early drafts of the 'Scholium Generale' that show that Newton had been thinking about electricity in relation to attractive forces.[2]

We know next to nothing about Andrew Motte, nor the circumstances under which he made his translation of the *Principia*—published in 1729, two years after Newton's death.[3] But it is difficult to conceive that Motte would have taken such extreme liberty with the concluding lines of Newton's book without some source or authority. Professor Koyré and I found that in one of Newton's own copies of E_2 he himself had added these words in hand—'electrici & elastici'—although he did not then actually go on to introduce them into E_3,[4] as may be seen in the Apparatus Criticus {530.34}.

4. A THIRD EXAMPLE: NEWTON AND KEPLER'S CONCEPT OF INERTIA

A final example has to do with one of the central concepts of the *Principia*: inertia. It is now becoming generally known that the word 'inertia', introduced into science in its modern technical sense by Newton, was first used in a physical context by Johannes Kepler. Newton does not mention Kepler's name in discussing inertia in any of the printed editions of the *Principia*. Could it be that he was unacquainted with Kepler's use of this term?

Kepler used 'inertia' in its original and literal sense of 'laziness', implying, in the older pre-Galilean–Cartesian–Newtonian physics, that a force is always required to maintain motion, that—owing to the inherent laziness of matter—a body will come to rest whenever the *vis motrix* ceases to act. Although this quality of 'inertia' is in many ways like the medieval *inclinatio ad quietem*, Kepler's concept is revolutionary as a half-way stage toward Newtonian inertia: for instance, it

[2] Hall and Hall (1962), *Unpublished papers*, pp. 207–11.

[3] See, however, Cohen (1963), 'Pemberton's translation'.

[4] Koyré and Cohen (1960), 'Newton's "electric...spirit"'.

implies that a body's 'inertia' will cause it to come to rest wherever it happens to be when the force stops acting, and hence that there can be no 'natural places' as held by the Peripatetic and Scholastic philosophers. Newton's possible awareness of Kepler's physical concepts was very much on our minds since Professor Koyré had completed a lengthy study of the science of Kepler,[1] while I had undertaken a complementary examination of the spread of Kepler's ideas. In particular, we were both anxious to find the exact sources from which Newton might have learned of Kepler's laws of planetary motion[2] and also his physical concepts. One of the major questions that Professor Koyré had raised was whether or not Newton might have read Kepler's writings. How else could he have learned of Kepler's concept of inertia? How had he come upon this Keplerian technical expression?

In the *Principia* Newton was ungenerous in his references to Kepler; for example, he mentioned Kepler's name in connection with the third law (or harmonic law) of planetary motion, but not the first two,[3] and omitted any mention of Kepler whatsoever prior to Book III. In Book I Newton introduces inertia in 'Definitio III' and 'Lex Motus I' without any reference to Kepler, or for that matter Descartes (to whom he certainly was indebted for the Law of Inertia).[4] Perhaps Newton does not mention Kepler because Kepler's concept of inertia differed so fundamentally from his own, which he defines as a 'force inherent in matter' ('materiae vis insita'[5]) that keeps bodies either at rest or in motion in a straight line at constant speed; rest and uniform rectilinear motion are inertially equivalent states in Cartesian–Newtonian physics.[6]

The reader may easily imagine how exciting it was to find on page 2 of Newton's interleaved copy of E_2 a slip of paper containing a long-hand statement contrasting the Keplerian and Newtonian concepts of inertia. (See Plate 4.) This sentence also appears, slightly revised, as one of Newton's long-hand emendations on the margin of the printed page, as follows:

Non intelligo vim inertiae Kepleri qua corpora ad quietem tendunt sed vim manendi in eodem seu quiescendi seu movendi statu.

I do not mean Kepler's force of inertia by which bodies tend toward rest, but the force of remaining in the same state whether of resting or of moving.

[1] Koyré (1961*a*), *La révolution astronomique*.

[2] Cohen (1971*a*), *Newton and Kepler*; cf. Whiteside (1964*a*), 'Newton on planetary motion'.

[3] In the tract *De Motu*, written just before the *Principia* (see Chapter III, § 3, below), Newton did mention Kepler in a scholium that states what we generally know as Kepler's first two laws of planetary motion (elliptical orbits and equal areas in equal times): 'Gyrant ergo Planetae majores in ellipsibus habentibus umbilicum in centro solis, et radiis ad solem ductis describunt areas temporibus proportionales, omnino ut supposuit Keplerus' ('Therefore the major Planets revolve in ellipses having a focus in the centre of the Sun; and by radii drawn to the sun describe areas proportional to the times, altogether as Kepler supposed'). Hall and Hall (1962), *Unpublished papers*, pp. 253, 277; cf. Herivel (1965), *Background*.

[4] See Cohen (1964), 'Quantum in se est'; Koyré (1965), *Newtonian studies*, chap. 3, 'Newton and Descartes'; Herivel (1965), *Background*, chap. 2.2.

[5] I have translated Newton's 'vis insita' as 'inherent force' rather than the customary 'innate force', which would be the translation of 'vis innata', an expression also used by Newton. The choice of 'inherent' for 'insita' comes indirectly from Newton; see Chapter III, §5.

[6] No doubt Newton was more impressed by the basic differences between his concept and Kepler's than by the similarities.

If the Apparatus Criticus of our edition were limited to the variants in the printed versions, there would be no place for this fascinating allusion to Kepler's concept of inertia.

I have shown elsewhere that Newton's declaration of the difference between the two concepts of inertia was almost certainly a response to a remark by Leibniz.[7] In 1726, when Newton published the third edition of the *Principia*, Leibniz had been dead for a dozen years or so and Leibnizian physics was no longer an issue; understandably, Newton never printed the remark just quoted.

The foregoing three examples well illustrate the importance of knowing of, and being able to consult, the manuscript emendations made by Newton in his own copies of the *Principia*.

5. THE POSSIBLE INCLUSION OF EVERY MANUSCRIPT OF NEWTON'S RELATING TO THE 'PRINCIPIA'

Once the decision had been made to include in the Apparatus Criticus all manuscript notes from E_1a, E_1i, E_2a, E_2i,[1] yet another problem presented itself forthwith. For there are other interesting manuscript variant versions or revisions of the *Principia* apart from those entered by Newton in his personal copies of the first and second editions. These latter are to be found in all major collections of Newton's manuscripts, such as those in the libraries of Trinity College and King's College, Cambridge, and of the Royal Society. Notable in the first of these collections are the drafts sent on to Cotes for E_2, which were either edited and altered by Cotes before being transmitted to the printer, or changed by Cotes at Newton's direction, usually following an epistolary exchange between them. Of enormous fascination to any Newton scholar are the projected revisions to Book III in the library of the Royal Society, among the papers of David Gregory. These contain, in Newton's hand and in Gregory's copies, a series of projected scholia to the propositions at the beginning of Book III, 'De Mundi Systemate', in which Newton showed his intentions of quoting long passages from Lucretius, or of referring back to a partly mythological tradition of supposed ancient sages who knew the basic principles of Newtonian celestial dynamics, including the concept (if not the very quantitative expression) of universal gravitation.[2]

But the greatest store of manuscript versions of, or emendations to, the *Principia* occurs in the Portsmouth Collection in the University Library, Cambridge. Here, in MS Add. 3965, comprising about 800 folios, many of them with writing on both sides of the page, there are many varieties of *Principia* manuscripts. These are but one group of the 'Papers connected with the Principia. (Mostly Holograph.)', as

[7] Cohen (1971 *a*), *Newton and Kepler*.

[1] These are the sigla of, respectively, the annotated copy of the first edition (E_1), the interleaved (and annotated) copy of E_1, the annotated copy of the second edition (E_2), and the interleaved (and annotated) copy of E_2.

[2] Cf. McGuire and Rattansi (1966).

they are described in the *Catalogue of the Portsmouth Collection*, the other two being Add. 3966 ('Lunar Theory') and Add. 3967 ('Mathematical Problems'). The papers in MS Add. 3965 contain a variety of types of manuscript changes proposed for the *Principia* by Newton, including bare lists of errata and corrigenda, early tracts, pages of computations, tables of observations and data of experiments, drafts and pre-drafts of alterations, corrections, and additions intended for the second and third editions of the *Principia*,[3] together with material written for the first edition and then rejected, including part of the original 'Praefatio'.[4] There are other fragments relating to the *Principia* scattered here and there throughout the great mass of manuscripts in the Portsmouth Collection. Professors A. R. and M. B. Hall found the suppressed 'Conclusio' to E_1 in MS Add. 4005.[5] I searched in vain again and again through the 800 pages or so of MS Add. 3965 for the missing final part of a new conclusion to the *Principia* written by Newton for a revision of E_1 before he actually wrote the 'Scholium Generale'; the missing page finally turned up in a wholly different part of the Portsmouth Collection.[6]

Among the treasures of the Portsmouth Collection there are a number of preliminary (and discarded) drafts of the concluding 'Scholium Generale', and even an unused 'Regula V'. The latter was found by Professor Koyré, and printed by him from the manuscript, in a study of the evolution of the 'Regulae Philosophandi':

Reg. V. Pro hypothesibus habenda sunt quaecunque ex rebus ipsis vel per sensus externos, vel per sensationem cogitationum internarum non derivantur. Sentio utique quod Ego cogitem, id quod fieri nequiret nisi simul sentirem quod ego sim. Sed non sentio quod Idea aliqua sit innata.

Et pro Phaenomenis habeo, non solum quae per sensus quinque externos nobis innotescunt, sed etiam quae in mentibus nostris intuemur cogitando: Ut quod, Ego sum, Ego credo, Ego intelligo, Ego recordor, Ego cogito, volo, nolo, sitio, esurio, gaudeo, doleo etc. Et quae ex Phaenomenis nec demonstrando nec per argumentum Inductionis consequuntur, pro Hypothesibus habeo.

Rule V. Whatever things are not derived from objects themselves, whether by the external senses or by the sensation of internal thoughts, are to be taken for hypotheses. Thus I sense that I am thinking, which could not happen unless at the same time I were to sense that I am. But I do not sense that some idea is innate.

And I do not take for phenomena only things which are made known to us by the five external senses, but also those which we contemplate in our minds when thinking: such as, I am, I believe, I understand, I remember, I think, I wish, I am unwilling, I am thirsty, I am hungry, I am happy, I am sad, etc. And those things which follow from the phenomena neither by demonstration nor by the argument of induction, I hold as hypotheses.[7]

This 'Reg. V' must have been written after 1713, since E_2 contains only three 'Regulae'. 'Regula IV' appears on an interleaf in E_2i {389.6ff.}, copied from the last version in a sequence on the recto and verso of the very sheet containing the

[3] Of these, some appear in E_2 and E_3 much as they appear on such sheets, while others were used in a considerably altered form, and yet others were never used at all.

[4] A fair copy, differing somewhat, is to be found in MS Add. 3963.

[5] Hall and Hall (1962), *Unpublished papers*, pp. 320–47.

[6] Cohen (1971*b*), *Newton's principles of philosophy*, chap. 7, 'A preliminary new conclusion to the *Principia*'.

[7] Koyré (1960*a*), 'Les "Regulae philosophandi" '.

text of 'Regula V'. Apart from its significance as a major key to understanding Newton's philosophy of science, 'Regula V' is of enormous interest in displaying his continuing battle with Cartesian ideas and—in particular—discussing specifically the slogan of the Cartesian philosophy, 'Cogito ergo sum'.

MS Add. 3965 also contains an important group of statements on the nature of matter, the vacuum, and atoms.[8] There are theorems and corollaries on a great variety of topics, whose existence would not even be suspected but for their appearance in these pages; there are also notable contributions to astronomy, and plans sketched for revisions never made. Interestingly enough, there is even a revision by Newton of that remark quoted in §4, contrasting the Keplerian concept of inertia and his own, in which he began, as before, by saying that he did 'not mean the force of inertia of Kepler' but then crossed out 'of Kepler' and wrote instead 'of some men'.[9] This is an isolated sentence, occurring on part of a letter (or envelope), immediately following a fragment of a sentence on the relative densities of air and water (MS Add. 3965, f. 423).

However desirable it might seem to include all these many varieties of emendation in our Apparatus Criticus, the sheer bulk indicated at once that the task of preparing an edition with absolute completeness was beyond our capacities as editors.[10] An edition on this scale would, furthermore, be too huge to be printed at a reasonable cost, and would prove to be far too detailed and complex to be used profitably by any but the most dedicated specialist.[11] If our goal was to produce a critical edition to be used generally, the 'ideal' could not be to include every

[8] See below, §7. These statements have been studied in Cohen (1966), 'Hypotheses in Newton's philosophy', in Cohen (1971), *Scientific ideas*, chap. 2, 'From hypotheses to rules', and in McGuire (1966), 'Body and void'.

[9] The actual sentence reads: 'Non intelligo vim inertiae [Kepleri *del.*] ⌊aliquorum⌋ qua corpora ad quietem tendunt sed vim manendi in eodem quiescendi [statu *del.*] vel movendi statu.' That is, 'I do not mean the force of inertia of some men [!], by which bodies tend toward rest, but the force of remaining in the same state of resting or of moving.' (The symbols ⌊ ⌋ indicate an insertion.)

[10] The whole corpus of relevant manuscript material, for example, would have to be transcribed or photocopied and then the transcriptions cut up so that each page, or paragraph, or line, or jotting could be classified according to the page and line of the *Principia* to which it pertains. Such identification is not always easy and is apt to require much research and study. A job like this might well occupy a team of a dozen scholars for more than a dozen years—an unwise expenditure of our presently available scholarly resources of both manpower and funds.

It would, however, be a very useful project for some team of scholars to make a detailed and analytical catalogue of the Portsmouth Collection, page by page. All scholars who use the Portsmouth Collection would profit if several copies were made of every page, of which a first set could then be rearranged in new and better classifications, in which material on the same topic would be found together. A second set could be reclassified so as to order all the manuscripts chronologically, as far as possible. It would be wise not to alter the present old system of numbering the originals, however, because that would make references to particular manuscripts in the future inconsistent with those of the past and present. In such a system, copies of manuscripts in other collections could be added for completeness.

[11] For those who may wish to see just how difficult and costly an end-product would result, let me make a guess at its possible size. I believe it is reasonable to suppose that such an edition would require four or five quarto pages for each page of the text of E_3, since so many of the versions differ to the degree that they would have to be printed separately and could not always be easily combined. If this is a reasonably sensible guess, it would imply an edition of some 2000–4000 pages, most of which would be uninteresting and of questionable value even to Newton specialists.

possible line or scrap textually related to the *Principia*.[12] In any event, even if we concluded that we should strive for the maximum inclusiveness possible, we were fully aware that we would never have been able to tell with any certainty the degree of completeness of the result. There is just no way of knowing accurately how much additional material related to the *Principia* Newton may have thrown away or lost or destroyed;[13] or how much may simply have disappeared from view, or have been scattered and buried in private collections, or have perished since Newton's day.[14]

6. LIMITING THE APPARATUS CRITICUS TO THE PRINTED EDITIONS AND NEWTON'S MANUSCRIPT ANNOTATIONS IN HIS OWN COPIES OF THE 'PRINCIPIA'

Once we had recognized the sheer physical impossibility of a complete Apparatus Criticus embracing every line Newton wrote relating to the *Principia*, Professor Koyré and I explored the desirability of printing a generous selection from Newton's manuscript notes, largely MSS Add. 3965, 3966, 3967 (Portsmouth Collection, U.L.C.). This policy, if adopted, would have enabled us to eliminate systematically much that is repetitive in these manuscripts, a reflection of Newton's well-known penchant for repeated (and often seemingly endless) copying and recopying of certain of his own writings and even those of others. These manuscripts contain overlapping neat lists of errata and corrigenda (by page and line), perhaps prepared for some more or less public distribution, from which excerpts might be taken. Minor changes of a purely grammatical or stylistic kind could be eliminated. The selection from the manuscripts could thus be confined to *variae lectiones* of definite substantive interest.[1]

Thanks to the kindness of the University Library, Cambridge, and with the aid of a grant from the National Science Foundation (U.S.A.), we were able to obtain

[12] Furthermore, only a few of the sheets in MS Add. 3965 are dated and so this material can be ordered chronologically only in a very rough fashion (as by variations in handwriting, by the kinds of paper, and by possible correlation with other documents).

[13] Of these the most significant would appear to be a draft of Books II and III of the *Principia* which I have not found as such among the papers in the Portsmouth Collection, or elsewhere. The evidence on the basis of which I presume that there must have been such a draft is presented below in Chapter III, §7, Chapter IV, §8, and Supplement VII. A draft of Book I (in part) has survived in the so-called Lucasian Lectures, for which see Chapter IV, §2, and Supplement IV.

[14] I have not mentioned another factor that would even further increase the scale of an ideally 'complete' edition and add to the complexity and magnitude of the editorial problems. I have in mind that the fund of possible *variae lectiones* would not be wholly exhausted by adding the contents of Newton's manuscripts to an Apparatus Criticus based on the collation of the printer's manuscript, the three printed editions, and Newton's annotations in his own copies of E_1 and E_2. There would be other sources of *variae lectiones*, such as the contents of Newton's correspondence, primarily with the editors of E_1, E_2, E_3, respectively Halley, Cotes, Pemberton. Then there would be the various *corrigenda et addenda* recorded by such intimates as N. Fatio de Duillier and David Gregory. Finally, there would be remarks of those who saw annotations that no longer survive, as in a now lost copy of E_1.

[1] Of course, the method of eliminating this much from the manuscripts would not need to be applied to the printed versions, where the reader would most likely expect a high degree of completeness.

two Xerox copies of the whole of MS Add. 3965, so that we could study it carefully in detail and at leisure, page by page, in order to determine the feasibility of choosing a limited number of annotations for inclusion in the Apparatus Criticus of our critical edition of the *Principia*. This examination of the manuscripts enabled us to gain a more fully documented view of the development of the text of the *Principia* from E_1 through E_2 to E_3.

It was at once clear that these manuscript papers contain an extraordinary wealth of both potential textual annotations and possible material for a commentary or many sets of articles. The difficulty in attempting to incorporate into our Apparatus Criticus a portion of this body of manuscript was not so much to find some entries worthy of consideration as to establish a clear and easily applicable rule to guide us in rejecting some manuscript emendations while keeping others. We greatly feared that, once we began to pick and choose, we would never go beyond merely reflecting what we as editors had found 'interesting' (that is, interesting to us).

In a volume of commentary, or the presentation of 'selected works', it is generally understood that the editor's taste, predilections, intellectual concerns, and special competences are the chief factors determining the subjects selected for presentation or discussion, and indeed this is the reason why one commentary is so apt to differ from others that deal with the same work. But we had undertaken the labour of preparing this critical edition in the hope that it would serve all scholars, that it might in fact transcend the personalities and predilections of the editors. Our constant hope had been that scholars wishing to ascertain what differences, if any, might exist between one edition and the next would be able to turn to our edition for any given word, sentence, paragraph, proposition, ..., and would not have to consult the originals. But a body of selections does not obviate the need of consulting the originals; it rather invites the reader to investigate the manuscripts, since it tells him what sorts of documents the manuscript collection contains. Were we to include a fascinating part of a manuscript on electrical attraction or on the nature of atoms, the reader would at once ask himself what other manuscript material there might be on these and on similar topics which we had not selected for inclusion in the Apparatus Criticus of our edition. Such an effect is not necessarily bad, but it differs greatly from the aim of completeness we expected a critical edition to have. We were also mindful that foci of scholarly interest change from time to time, that our concerns would be different not only from those of some contemporaries but certainly from those that would be current perhaps as soon as a decade or so hence. We might very well omit just those topics that would be of primary scholarly concern to others, now and in the future, wholly apart from our disregard for grammatical constructions, or certain computations or collections of data.

Our resolution of the problem has been to reserve the great mass of manuscripts for use in the commentary volume to accompany this edition and in separate studies or monographs. But in order to give the reader a glimpse of the state of the *Principia* between E_1 and E_2 and between E_2 and E_3, we have included in the Apparatus Criticus every note, correction, change, and proposed alteration that was entered by Newton in the annotated and the interleaved (and annotated) copies of E_1 and E_2 which were in his personal library (and which we have referred to throughout the edition under the sigla: E_1a, E_1i, E_2a, E_2i); and—let me repeat—we have also introduced every variant found by collating *M* (the printer's manuscript) and the first edition, E_1.[2] Not only does this reveal the presence of certain alterations that have never been printed till now, but it enables the reader to see the form and substance of other alterations as they were entered by Newton on the printed pages or interleaves and prior to their final redaction for the press by R. Cotes and H. Pemberton respectively or by their author.

It is true that the entries in long-hand in E_1a, E_1i, E_2a, and E_2i are often only a selection made from a greater mass of material, but in each case the selection was made by Newton himself—and it is indeed fascinating to see what he actually did include in these copies. These four texts, furthermore, are complete in themselves, with their own integrity. All of them are Newtonian documents of considerable interest and importance in their own right; because their contents may be found in the Apparatus Criticus in their entirety, they may be fully and easily reconstructed by any reader from the present edition. No longer do scholars interested in determining whether a given topic occurs in one of the four 'codices' have to pore over the originals; they may turn to the part of the Apparatus Criticus in question, just as they may do for information concerning the printed editions or *M*.[3] This edition with the Apparatus Criticus thus makes available to every reader the contents of eight integral texts: *M*, E_1, E_1a, E_1i, E_2, E_2a, E_2i, and E_3.

7. NEWTON'S DEGREE OF COMMITMENT TO THE ANNOTATIONS IN HIS COPIES OF THE 'PRINCIPIA'

The decision to include in our Apparatus Criticus only the manuscript alterations in E_1a, E_1i, E_2a, and E_2i,[1] to the exclusion of all other manuscript sources, was not made on the basis of convenience alone, that is, by having Newton make the selection for us. Our researches showed that Newton treated the emendations entered into the copies he annotated as having a degree of finality not necessarily associated with those he wrote out on separate sheets or scraps of paper. Of course, some of the latter might have been no more tentative than those entered in his copies of the *Principia*, but we cannot ever be sure. We do know for certain, however, that Newton was apt to try out certain modifications—large or small—

[2] The variants in the Apparatus Criticus do not include changes in spelling, capitalization, punctuation, or italicization. See the 'Guide to the Apparatus Criticus'.

[3] The reasons for including *M* are given in Chapter II, §10.

[1] The only annotated copies of the *Principia* from Newton's library that we were able to locate.

without fully committing himself to making that particular change. How different this is from actually altering the printed page!

Nowhere is this difference shown more clearly than in the 'Regulae Philosophandi' at the beginning of Book III. When Newton crossed out the original 'Hypoth. III' and wrote in the new 'Hypoth. III' he was in fact as definite as could be that this was the way he wanted the next edition of the *Principia* to appear. It does not matter that at some other time he changed the wording, that he later altered the designation 'Hypoth. III' to 'Reg. III'. He had been firm, not tentative; he had treated the old 'Hypoth. III' as he would a mistake in calculation or a printer's typographical error: with his own pen he had cancelled the unwanted sentence and replaced it by one he conceived to be more suitable.

Compare this action with his treatment of 'Reg. V' (see § 5 above), which he tried out in various drafts on a sheet of paper but never entered into E_2a or E_2i. He had firmly decided to enlarge the number of 'Regulae Philosophandi' in a third edition; and in E_2i he wrote out the new 'Regula IV' on an interleaf. Surely, there can be no doubt of a fundamental difference between his treatment of the tentatively proposed fifth rule and of the fully accepted third and fourth rules.[2]

We have seen that on at least two bits of paper Newton essayed a comment about Kepler's concept of inertia and his own, explaining that by 'inertia' he did not mean inertia in the sense used by Kepler of a tendency of bodies to come to rest, but rather a force tending to keep bodies in whatever state they happened to be —and whether a state of rest or of uniform rectilinear motion. Whatever degree of tentativeness these two versions may have had for Newton, he did at one time envisage that this addition should be part of his explanation of 'vis insita' in Definition III, in the next revised edition of the *Principia*. Accordingly, he took his pen and produced a page that reads just as if this change had actually been made.

Every author of a published work is aware that there is a difference between a correction or alteration actually made in a typescript or a manuscript or a printed book and a tentative alteration or substitute version essayed on a scrap of paper but not on the finished draft. Can it be doubted that the handwritten annotations made by Newton in E_1i and E_1a and E_2i and E_2a stand in general in a different relation to the degree of finality of his intentions than many of the pages composing MS Add. 3965 in the University Library, Cambridge, or in other manuscripts of Newton's relating to the revision of the *Principia*? The question here is not only the psychological one of the extent of an author's commitment when he enters an alteration in hand on the pages of his book. The fact is that Newton had one or possibly two special interleaved copies of E_1, bound up for the express purpose of receiving emendations for any future edition. And he did so in preparation for E_3

[2] I certainly do not wish to imply that Newton was at all times so orderly in his habits that he entered every definite alteration into the pages of his copies of the *Principia*. But I would argue that there is a stage of definiteness or of commitment of which an entry into E_1i, E_1a, E_2i, E_2a is a sufficient if not always a necessary condition.

just as he had previously done in preparation for E_2. There is considerable evidence that Newton himself actually did consider that these copies of E_1 and E_2 in which he entered revisions contained the revised state of the *Principia* for a possible new edition. For instance, even before he had actually made a firm decision to produce a third edition, he wrote out a brief memorandum (U.L.C. MS Add. 3968, f. 596), stating that he had been correcting a copy of the *Principia* for just such a new edition:

Hujus libri editio tertia forte lucem videbit et in exemplari quod in hunc finem correxi, in fine Prop. XVII addidi haec verba Nam si corpus [in his casibus *del.*] revolvatur in Sectione Conica sic inventa, demonstratum est in Prop. XI, XII et XIII quod vis centripeta erit reciproce ut quadratum distantiae corporis a centro. Caetera quae correxi, ad D. Bernoullium nil spectant.[3]

The third edition of this book will perhaps be published and in a copy which I have corrected for this purpose, at the end of Prop. XVII I have added these words: 'For if a body be revolved in a conic section thus found, it has been demonstrated in Props. XI, XII, and XIII that the centripetal force will be reciprocally as the square of the body's distance from the centre.' The other things that I have corrected have nothing to do with Mr Bernoulli.

David Gregory, who visited Newton in Cambridge in the 1690s, described how Newton was correcting the *Principia*, getting the text of the new edition ready for the printer, by revising an interleaved copy (for details, see Chapter VII, §15).

One or two examples of a different kind may illustrate the degree of tentativeness of many of the alterations that Newton proposed to himself but never put into a copy of the *Principia*. The Apparatus Criticus shows that Newton once conceived that in Book I the order of Props. IV and V be reversed {43.26 ff.}. But on manuscript sheets (U.L.C. MS Add. 3965, f. 36) Newton essayed a truly radical recasting of the beginning of this Sec. II of Book I, in which a wholly new 'Prop. I. Theor. I.' would deal with the forces in uniform circular motion. There were to have been four new corollaries, followed by 'Corol. 5–7' which were to have been the 'Corol. 5–7' of the old 'Prop. IV' of Book I. Then would come a scholium consisting of the first and third paragraphs of the Scholium to 'Prop. IV' of Book I in E_1.[4]

Another striking example of a radical change that was tried out on separate sheets of paper, but not copied into one of Newton's copies of the *Principia*, occurs

[3] There is no such addition in E_2a, but a variant form appears in E_2ii (see Apparatus Criticus {64.28–33}). There is also in E_2ii an earlier version in the form of a fifth corollary, later cancelled (see Apparatus Criticus {65.18/19}), of which the foregoing sentence, 'Nam si corpus...a centro', is a close paraphrase, but not an exact quotation.

[4] This plan appears on one side (f. 36) of a double folded sheet in U.L.C. MS Add. 3965. On another side Newton has written out (f. 36^v) a number of corollaries, whose numbers he has then changed. Corollaries 6, 7, 8, 9, and 10 correspond to Corols. 3, 4, 5, 6, and 7 of E_2 and E_3; only one of them, Corol. 3, corresponds to a Corollary of E_1 (Corol. 3). Since on f. 36^v these corollaries are further worked over before attaining the form in which they finally appear in E_2, they seem to be an early version of the revised text of E_2. Another part of this double sheet (f. 36^r) contains a Prop. V, Theor. IV, together with Corols. 1, 2, 3, 4, 5 (evidently to go with Corols. 6, 7, 8, 9, 10 of f. 36^v, and a Corol. 11 ['Eadem omnia...applicata'] which is equivalent in text to Corol. 7 to Prop. IV of E_1). A discussion of these proposed emendations may be found below, Chapter VII, §3.

in Book III. This is the series of attempts by Newton to introduce a discussion of what he means by 'body' and by 'vacuum', together with an explanation of what he intends the reader to understand by 'Hypothesis', 'Phaenomenon', and 'Rule'.[5] In one version (MS Add. 3965, f. 437^{v}), Newton indicates that a set of 'Definitiones' is to be placed just before the 'Regulae Philosophandi' and 'Phaenomena'. There were to be two:

DEFINITIO I.

Corpus voco rem omnem tangibilem qua tangentibus resistitur, et cujus [actio *del.*] ⌞resistentia⌟, si satis magna sit, sentiri potest.

Hoc enim sensu vulgus vocem corporis semper accipit. Et hujus generis sunt [Tellus *del.;* Tellus, Planetae, Cometae *del.*] metalla, lapides, arena, argilla, lutum, ⌞terra,⌟ salia, ligna, ossa, carnes, aqua, oleum, lac, sanguis, aer, ventus, fumus, exhalatio, flamma, & quicquid sub elementis quatuor comprehendi potest, vel ab his exha⌞la⌟ndo manare & in haec per condensationem redire: ⌞Addo [Tellurem & *del.*] corpora coelestia...⌟[6]

DEFINITIO II.

Vacuum voco locum omnem in quo corpus sine resistentia movetur.

Sic enim vulgus loqui solet...

DEFINITION I.

I call 'body' every tangible thing that resists things touching it, and whose [action] resistance, if it be great enough, can be felt.

For it is in this sense that ordinary people always take the word 'body'. And of this kind are metals, stones, sand, clay, mud, earth, salts, woods, bones, meats, water, oil, milk, blood, air, wind, smoke, exhalation, flame, and whatever can be comprehended under the four elements, or can flow [arise or emanate] from these by evaporating and return into them by condensation. I add the celestial bodies...

DEFINITION II.

I call 'vacuum' every place in which a body moves without resistance.

For thus ordinary people are accustomed to speak...

Another sequence defines 'Corpus', 'Vacuum', 'Phaenomenon', 'Hypothesis', 'Regula' (MS Add. 3965, f. 420). Among these revisions are some new propositions for the beginning of Book III on the forces acting on the moon,[7] on planets, and on planetary satellites, and on the weights of planets and the moon, as well as a definition (in several versions) of the force of gravitation (MS Add. 3965, f. 417).

Later Newton embodied a discussion of these topics in a scholium, evidently intended to follow the 'Regulae'[8] in a revision of E_2 (MS Add. 3965, f. 430). In one of the versions of his definition of 'Phaenomenon' the discussion is very much like the discarded 'Regula V'.

The foregoing group of alterations may be said to have been tentative in that they never were used in a later edition of the *Principia*, and also were never written into a copy of the *Principia* by Newton. We may see from the several versions that he was working to improve his statement; perhaps he never was fully satisfied. Finally he gave it up, as witness the last state of 'Definitions I and II' in which

[5] These are printed in full and discussed in their relation to the whole Newtonian philosophy in two separate publications; see §5, note 8.

[6] In this extract, as throughout the Apparatus Criticus, the symbols ⌞ ⌟ indicate an insertion.

[7] A 'List of Propositions' apparently intended to be inserted in a second edition of the *Principia* is printed in an Appendix to the Preface of the *Catalogue of the Portsmouth Collection*, pp. xxiii–xxvi.

[8] The scholium is said by Newton to go in 'pag. 360'. This would put it among the 'Phaenomena' of E_2. Perhaps he meant 358.

both are cancelled. But there was some degree of commitment to these definitions, for some of them advanced from odd sheets to take their place in tables of regularly ordered emendations by page and line, no doubt prepared by Newton for private circulation. But not one of them ever was entered on the text pages or interleaves of E_1a, E_1i, E_2a, or E_2i. Possibly in this instance, in Newton's mind there never was the same degree of certainty that these definitions really should go into the next edition as there was about the alterations he actually made on the printed pages of his own copies or wrote out on the interleaves he had had bound into his working copies for this very purpose.[9]

8. AN INTERLEAVED (AND ANNOTATED) COPY OF THE FIRST EDITION SHOWN TO GREVES: A SECOND INTERLEAVED COPY

In the previous sections of this chapter I have explained that Newton's marked-up copies of the *Principia* constitute well-defined texts, representing the current state of revision for the next edition. Newton himself seems to have conceived that the alterations he had entered into his personal annotated or interleaved copies had a more or less public character. By this I mean that Newton was perfectly willing to show one or more such copies to others, and even permitted the emendations to be copied out—although how extensive this practice was I do not know. At least this was the case for the alterations in E_1. For instance, in November 1702, Bd.[1] Greves wrote to Lord Aston that Sir Edward Southcote had taken him 'in his chariot to wait upon Sr Isaack Newton'. Greves discussed with Newton 'Mr Cassini's dispute with Mr Flamsteed' and asked Newton 'if the absolute distance of the moon might not be demonstrated'.[2] Greves evidently discussed the *Principia* with Newton and even saw a corrected interleaved copy, for he reports on it as follows:

> He [Newton] owns there are a great many faults in his booke, and has cross'd it, and interlevd it, and writ in the margin of it, in a great many places. It is talkd he designs to reprint it, tho he

[9] According to the editors of the *Catalogue of the Portsmouth Collection* (p. 5; IX.B: 1, 4), certain loose sheets were originally found in one of Newton's copies of the *Principia*. This topic is discussed in Appendix III to the text. The possibility of there having been at least one other personal (and annotated) copy of both E_1 and E_2 is discussed below in §8 and in Chapter VIII, §§4, 5. Had we found any such copies, we would have included their annotations in our Apparatus Criticus (thus altering the status of certain alterations that are not in E_1a or E_1i or E_2a or E_2i from 'possibly only tentative' to 'definite'). But this would not have caused us to eliminate any entry now in the Apparatus Criticus.

It should also be stated explicitly that some of the alterations on loose sheets of manuscript may have been just as definite as those to be found in E_1a or E_1i or E_2a or E_2i; we simply don't know! Certainly there are instances when Newton would have made a change or a correction right on the *Principia* because this particular alteration could be made more simply by changing a word or two in a line of type than by writing out a new sentence in whole or in part. Also, some of the changes may have been made as a matter of record during the last stages of preparing E_2 or E_3 for the press. Thus, when Pemberton suggested certain revisions that proved acceptable to Newton, Newton might have proceeded to write them into his own copy of E_2. To this degree it is possible that some (though not very many) of the variant readings in E_1i, E_1a, E_2i, E_2a may derive directly from Cotes or Pemberton.

[1] Perhaps 'B^d^' and so possibly Bernard.

[2] [Tixall letters] (1815), vol. 2, pp. 149, 152.

would not owne it. I asked him about his proofe of a vacuum...I finde he designs to alter that part, for he has writ in the margin, Materia Sensibilis; perceiving his reasons do not conclude in all matter whatsoever.[3]

Alas! the letter then closes, 'I fear I have tired your ldp's patience, and therefore will only beg leave to present my humble respects...and subscribe myself... Your...humble servant, Bd. Greves.' We are thus denied any further details of the visit, but what is said is enough to demonstrate Newton's readiness on occasion to show his working copy of the *Principia* to a stranger.

Evidently, there must have been yet another interleaved and annotated copy of E_1, in addition to E_1i (which now survives in the Portsmouth Collection in the University Library, Cambridge), since Greves says that the words 'Materia Sensibilis' were written out in the margin. But in the surviving interleaved copy (E_1i), these words do not appear in the margin. On the other hand, a similar alteration does appear in both the interleaved copy (E_1i) and the annotated copy (E_1a): 'sensibilium' is inserted after 'corporum' at {402.8}, but is cancelled in E_1i. Presumably, then, if this account is accurate, Greves saw yet another interleaved copy, which may be identified by the words 'Materia Sensibilis' appearing in the margin. Of course we cannot tell whether Greves meant a copy of E_1 with an interleaf bound between each two successive printed leaves (as in E_1i) or a copy with many loose or unbound interleaves.

Various references to this other interleaved copy appear hauntingly through the literature, to tantalize the scholar, and we shall study some of them below. Clearly, this copy must have embodied some or all of the annotations entered into E_1a and E_1i, and it may have ended up by serving as the printer's copy for E_2, with some supplements entered on separate manuscript pages and with some entries crossed out (which Newton had decided not to use).

We shall see (Chapter VII, §13; Chapter VIII, §5) that David Gregory saw an interleaved copy, which—as he described it—does not exactly conform to E_1i. Furthermore, someone, very likely Gregory, had entered in pencil some corrections in the margin of an interleaved copy, which were in fact in error (see Chapter VIII, §5); these do not occur in E_1i, nor in E_1a. It is also clear that Cotes received an interleaved copy of E_1 to be used as printer's copy; this may or may not have been the same interleaved copy seen by Greves, but it was undoubtedly the one seen by Gregory, in the margin of which Gregory presumably had made the above-mentioned (incorrect) annotation. On the principle of not multiplying hypotheses unnecessarily, we may suppose the copy seen by Greves to have been the same one seen by Gregory and eventually used by the printer for the composition of E_2.

The existence of such a copy (or copies) is particularly frustrating, because—other than E_1i or E_1a—no copy of the first edition of the *Principia* extensively annotated in Newton's hand has been found since the nineteenth century (see Chapter VIII, §5). Mr Henry Macomber, the librarian of the Newton Collection

[3] See below, Chapter VIII, §3.

at the Babson Institute of Business Administration (Massachusetts), made a census of all the copies of the first edition of the *Principia* that he could discover, but none of these was an interleaved copy nor did any correspond to the descriptions of the missing copy. Perhaps this copy (or copies) may still exist somewhere, either in private hands or in some public collection. The missing copy, at least if it were the one used for printing E_2, would no doubt conform more closely to E_2 than either E_1i or E_1a does.

One of the major values of this copy would be to enable us to determine exactly how much of the early part of the *Principia*—the introductory matter, Book I, and the early part of Book II (see Chapter IX)—Cotes had actually altered. It would also serve, in the same sense that M does for E_1, to let us see what the last-minute changes were—those made after the copy had been delivered to the printer. Furthermore, we must not discount the possibility that such a copy might show some stages, presently unknown to us, whereby Newton had arrived at some of the final new versions. It is doubtful, however, whether such a copy would, in any real sense, appreciably alter the view of the development of the *Principia* shown in the present edition.

9. THE DISSEMINATION BY FATIO AND GREGORY OF THE MANUSCRIPT ANNOTATIONS TO E_1

Of all who saw Newton's manuscript annotations, the most important was probably the Genevese mathematician and mystic N. Fatio de Duillier, since he was responsible for transmitting a portion of them directly to Huygens, and thence indirectly to Leibniz: they were in fact even published in 1701.[1] Fatio was an especially intimate protégé of Newton's, known today for having fired the first salvo in the battle between Leibniz and Newton and their respective followers. He is the author of the epigram on Newton's tomb: 'Sibi gratulentur Mortales, Tale tantumque extitisse Humani Generis Decus' ('Let mortals rejoice that there has existed such and so great an ornament of the human race').[2] At one time Fatio planned to bring out a new edition of the *Principia*, and E_2 actually contains several minor improvements which he suggested.[3] Newton allowed Fatio to copy out some of the annotations in E_1a or some other similar copy, and in his own hand he entered some alterations in Fatio's personal copy of E_1, which survives in the Bodleian Library, Oxford. Fatio's transcripts are especially precious because some of them are dated or datable and thus enable us to find out exactly when Newton had made certain very important changes in the *Principia*. But, as mentioned above, the chief reason why Fatio's copies of Newton's emendations have such an importance is that through Fatio they were circulated and eventually published by Groening.

[1] On the dating and dissemination of Fatio's transcripts, see Chapter VII, §10, and Appendix IV.
[2] Cf. Fatio de Duillier (1949).
[3] These are discussed below in Chapter VII, §9; cf. Rigaud (1838), *Essay*, pp. 91–3.

But it was not only through Fatio that Newton permitted these early emendations of E_1 to be circulated. I have already mentioned the fact that Newton himself presented a copy of E_1 to the philosopher John Locke and had entered into that copy a number of corrections such as are to be found in E_1a. Furthermore, Newton also allowed David Gregory, the Scottish astronomer and mathematician, to copy out the improvements he had been making in E_1. Gregory was apparently a true favourite of Newton's; he was highly recommended by Newton for his post of Savilian Professor of Astronomy at Oxford.[4] Among Gregory's manuscripts there are extensive accounts of visits to Newton, containing detailed notes on Newton's progress in revising the *Principia* and his plans for further changes and for publication. Gregory quoted extracts from these revisions by Newton in his 'Notae in Isaaci Newtoni Principia philosophiae', a manuscript commentary that he never published, though it must have been fairly widely read. These 'Notae' exist today in at least four copies, three being more or less identical fair copies in amanuenses' hands, and the fourth Gregory's own holograph copy.[5] In the last of these, now in the library of the Royal Society, the emendations made by Newton in his own copy or copies of the *Principia* largely appear on little slips pasted in after the 'Notae' had been written[6] and thus must be of a later date than the text itself.[7]

Clearly, then, there are very compelling reasons for including in any edition of Newton's *Principia* all of Newton's manuscript improvements of E_1 that were made public through the agency of Fatio, Locke, and Gregory, and thence, in a derivative way, through Huygens, Leibniz, and Groening. Many come from Newton's marked-up copies of E_1, which he treated as if they were at any time the latest version of the *Principia*, and as such fit to be shown to others. Hence it may be seen from another point of view why we have included in the present edition all the emendations entered by Newton in E_1i and E_1a and in E_2i and E_2a. The corrections entered into the copy presented to Locke, those in Gregory's 'Notae' and in Fatio's list (published by Groening), and those in certain other annotated copies of E_1 have been partially tabulated in Appendix IV to the text of this edition, so that they may be available to scholars. They have not as such been put into the Apparatus Criticus, however, since they were not made by Newton in his own hand, nor are they all to be found in any of Newton's own copies of the *Principia*. But, of course, they almost entirely derive from such annotations by Newton in his own copies, of which the originals are given from E_1i and E_1a in the Apparatus Criticus.

Now it becomes possible to make certain studies that would not otherwise have been readily feasible. For instance, throughout the world there exist a certain number of copies of E_1 with contemporaneous annotations. These may now be

4 Newton's relations with Gregory are discussed in Chapter VII, §§ 12–14.

5 These are to be found in the libraries of the Royal Society of London, Aberdeen University, the University of Edinburgh, and Christ Church (Oxford). A variant title is: 'Notae in Newtoni Principia mathematica philosophiae naturalis'.

6 Cf. Wightman (1953), 'Gregory's "Notae"'; (1957) 'Gregory's commentary'.

7 See, further, p. 42 below and Appendix IV to the text.

studied by collation with the final alterations of E_2 or with the entries in E_1i and in E_1a, or with the list published by Groening in 1701, or with the emendations to be found in Gregory's 'Notae'. Hence one may tell at once whether a given manuscript alteration of E_1 (or set of alterations) is apt to be original with the owner or derived from E_2, from Groening's printed list, or from E_1a or E_1i—and in that latter case via Gregory or Fatio. For example, a page of an annotated copy of E_1 has been reproduced in the *Catalogue of... the Babson Institute*,[8] in which the demonstration of Lemma VI has been cancelled and a new demonstration written in the margin. A glance at our Appendix IV shows at once that this appears in Groening's published list, but is not in Gregory's 'Notae'.

Turning next to the Apparatus Criticus, we see that the new demonstration is all but identical to the one in E_2 and E_3 {31.12–14}; apart from the use of capital letters, the only differences are that the word 'continebit' follows 'arcus AB cum tangente AD' whereas in E_3 'continebit' precedes these words and 'arcus AB' is 'arcus ACB'. Since this change appears in E_1i and E_1a, and is noted in the MS Errata to E_1a (where Newton introduces the new demonstration with 'deleatur demonstratio Lemmatis sexti, vel legatur'), this particular copy of E_1 was emended by a copyist who had direct access to E_1i or E_1a, or to a copy that contains emendations from one of those, but a copyist who either was careless or who copied from another version not done carefully, or possibly who had access to a variant text of Newton's.[9]

The evidence of the importance of the annotated and interleaved copies of the *Principia* should not be construed as implying that these were the sole sources from which Newton's contemporaries learned of emendations made or planned. In Appendix VIII, it may be seen that E_1a (especially the MS Errata) had special significance in the dissemination of such emendations, and this was the exclusive source of all such emendations that appear in (and thus were circulated by) Gregory's 'Notae'. But Gregory's memoranda of the 1690s show that he saw other manuscript emendations than those in Newton's copies of the *Principia*. And there can be no doubt that certain carefully written-out lists of emendations must have been intended for private distribution or circulation among members of Newton's circle. Hence, any special role that may properly be assigned to the annotations in Newton's copies of the *Principia* should not be interpreted as negating the possibility of an equal (or at least a similar) role for certain other manuscript documents.

10. THE PRINTER'S MANUSCRIPT AND THE FIRST EDITION

We have set the terminal date of our editorial assignment at 1726, when E_3 issued from the press; for after that date the aged author (*aet.* 83) did no further work

[8] [Babson Collection] (1950), *Catalogue*, plate facing p. 97.

[9] In Appendix IV it is shown that there are at least two other copies of E_1 in which this particular emendation has the form of the copy in the Babson Institute.

on the *Principia* and in fact he died within the year. But the choice of starting date (or text) presented something of a problem. Clearly, the first edition marks the beginning in the same sense that the third edition marks the end. But there are at least two states of the text of E_1: the printed version and the manuscript (M) from which the printers composed E_1. Thanks to the courtesy of the Council of the Royal Society, we were enabled to make a photocopy of the manuscript, written in the hand of Newton's amanuensis, Humphrey Newton (not a close relation of our author). This manuscript has corrections in Newton's hand and in Halley's, and contains proof-readers' and printers' marks as well. (See Plates 5 and 6.) The variant readings of M prove to be of special interest, in part owing to the rapidity with which the whole *Principia* was written.[1] These variants are of three kinds. Some are afterthoughts of the author's which are entered into the manuscript text, following either a re-reading of Humphrey's copy or later reflection—in which case they would have been forwarded to Halley in London for inclusion. Then there are changes made in the manuscript by Halley, acting as all good editors do to eliminate slips of the pen, infelicitous expressions, or downright mistakes. Lastly, there are the differences between the final version of the manuscript (M)[2] and the printed text of the first edition (E_1), giving the reader a measure of the alteration and rewriting that was done in proof. A description of this manuscript is given in the preliminary chapter to the text part of this edition.

Since the final version of the *Principia* was begun in November or December 1684, the variant readings in the Apparatus Criticus to the present edition span a period of approximately four decades, from late 1684 (or early 1685) to the completion of the printing of the third edition in 1726.

[1] Apparently all three Books, except for the end of Book III on comets, were completed within about a year and a half. See Chapter III, §6.

[2] Although M is referred to throughout this work as 'the printer's manuscript', this expression is used in a general sense and does not imply that there was only one printer. The manuscript M was actually sent out to two different printers; see p. 137, below.

PART TWO

THE WRITING AND FIRST PUBLICATION OF THE 'PRINCIPIA'

CHAPTER III

STEPS TOWARD THE 'PRINCIPIA'

1. THE BEGINNINGS: HALLEY'S VISIT TO NEWTON IN 1684

THE history of the *Principia* begins with a definite event: a trip from London to Cambridge to see Newton, made by Edmond Halley, one of the secretaries of the Royal Society, famous today primarily for his contributions to astronomy and for the comet named after him, but then also considered an able geometer. The date is 1684, presumably August. Newton, aged 41, is Lucasian Professor of Mathematics at Cambridge and a Fellow of Trinity College, known at large for his published discoveries concerning light and colour, but admired by the cognoscenti for his work in mathematics (not as yet published, but to some degree circulated in manuscript) and his grasp of the fundamentals of dynamics. During the visit, Halley asks Newton what path the planets would describe if they were continually attracted by the sun with a force varying inversely as the square of the distance. Newton replies that the path would be an ellipse. This is not a mere guess on Newton's part, but a result he has obtained by mathematics, a proved statement.

Newton—to continue with the story—cannot lay his hands on the 'calculation', but promises to send it on to Halley in London. He reworks the material and composes a short tract which Edward Paget takes to London to give Halley. On reading it, Halley gets so excited that he returns to Cambridge to see Newton once again. One effect of his visit is that Newton promises to make public his work. He accordingly sends a tract (presumably the one previously shown to Halley) to the Royal Society. Encouraged by Halley, and by the warm commendations of the Royal Society, Newton eventually completes a manuscript for publication. Such are the beginnings of the *Principia.*

Looking back, it seems to us that an inverse-square law of force must have appeared to Newton's contemporaries as a very likely cause of the planetary motions. After all, anything that spreads out uniformly in all directions from a centre will diminish in concentration or intensity with the square of the distance as light does.[1] Furthermore, by the 1680s no phenomenal skill in mathematics was required to deduce that if all the planets move with uniform circular motion,

[1] It is interesting to observe, however, that Kepler, who discovered that the intensity of illumination diminishes as the inverse square of the distance, did not hold that the Sun's force follows the same law. He supposed that the Sun's force diminishes inversely with distance and accordingly is confined more or less to the plane of the ecliptic. Cf. Koyré (1961*a*), *Révolution astronomique*, pp. 199 ff.

in conformity with Kepler's third law, then if there is a single central force that keeps them in such orbits, it must be inversely proportional to the square of the radius. In 1673 Huygens had published the rule of 'centrifugal force':[2] that in uniform circular motion, this 'force' is directly proportional to the square of the speed and inversely proportional to the radius of the circle. Thereafter, the job of computing the magnitude of the planetary force for circular motion was a mere exercise in applied algebra.[3] What was required was a conceptual shift from centrifugal to centripetal force, and the recognition that Huygens's rule could give the magnitude of the centripetal force, so conceived. Understandably, Newton did not esteem it a discovery of any extraordinary significance to conclude that the planetary force (whether centripetal or centrifugal) 'must be' or 'probably is' an inverse-square force.[4] But to show that an inverse-square type of force is at the basis of elliptical rather than circular orbits, and then to show that one and the same force maintains the planets in their orbits and the planetary satellites in theirs, and causes the falling of terrestrial bodies in the manner we observe, and makes the tides ebb and flow—that is something else again!

In particular, we may stress the elliptical orbits—even at the expense of the universality of the gravitating force, radical as the latter was. In the *Principia* there are two separate treatments of the motions that may be found in the planets. Propositions I, II, and III deal with motion according to Kepler's law of areas as a necessary and as a sufficient condition for a central force (that is, a force directed toward a point as centre). Then, in Prop. IV, Newton proves Huygens's rule for 'centripetal [!] forces of bodies, which by equable motions describe different circles'; then Corol. VI states that Kepler's third law is both a necessary and a sufficient condition that the central force vary inversely as the square of the radius. But Prop. IV and its corollaries deal only with the strictly limited case of circular motion; even so, in the scholium that follows, Newton says that the 'case of the 6th corollary obtains in the celestial bodies (as Sir *Christopher Wren*, Dr. *Hooke*, and Dr. *Halley* have severally observed)'.

But it is not until the next section (Sec. III) in the opening Prop. XI that, by a further and quite different sequence altogether, Newton shows that in an elliptical orbit the force tending to a focus also varies inversely as the square of the distance.[5] This level of mathematical understanding and control of celestial mechanics went

[2] Huygens (1888–), *Oeuvres*, vol. 16, pp. 315–18; vol. 18, pp. 366–8; this is the appendix to Huygens's *Horologium oscillatorium* (1673).

[3] For, in modern terms, let A be the acceleration. Suppose the planets to move along circular orbits. Let v be the speed, r the radius, and T the periodic time. Then $A = v^2/r = (2\pi r/T)^2/r = (4\pi^2/r^2)\,(r^3/T^2)$, whence it follows from Kepler's third law (r^3/T^2 is a constant) that $A \propto 1/r^2$. If it is supposed that forces are as accelerations, then $F \propto A \propto 1/r^2$.

[4] Of course, there was the necessary condition that one had to believe that there are such forces, spreading out through space so as to 'act at a distance'. In Newton's day many scientists, among them Huygens, did not believe in the existence of such forces.

[5] A somewhat similar sequence in the tract *De Motu*, preceding the *Principia*, is discussed a few pages below; see also §3.

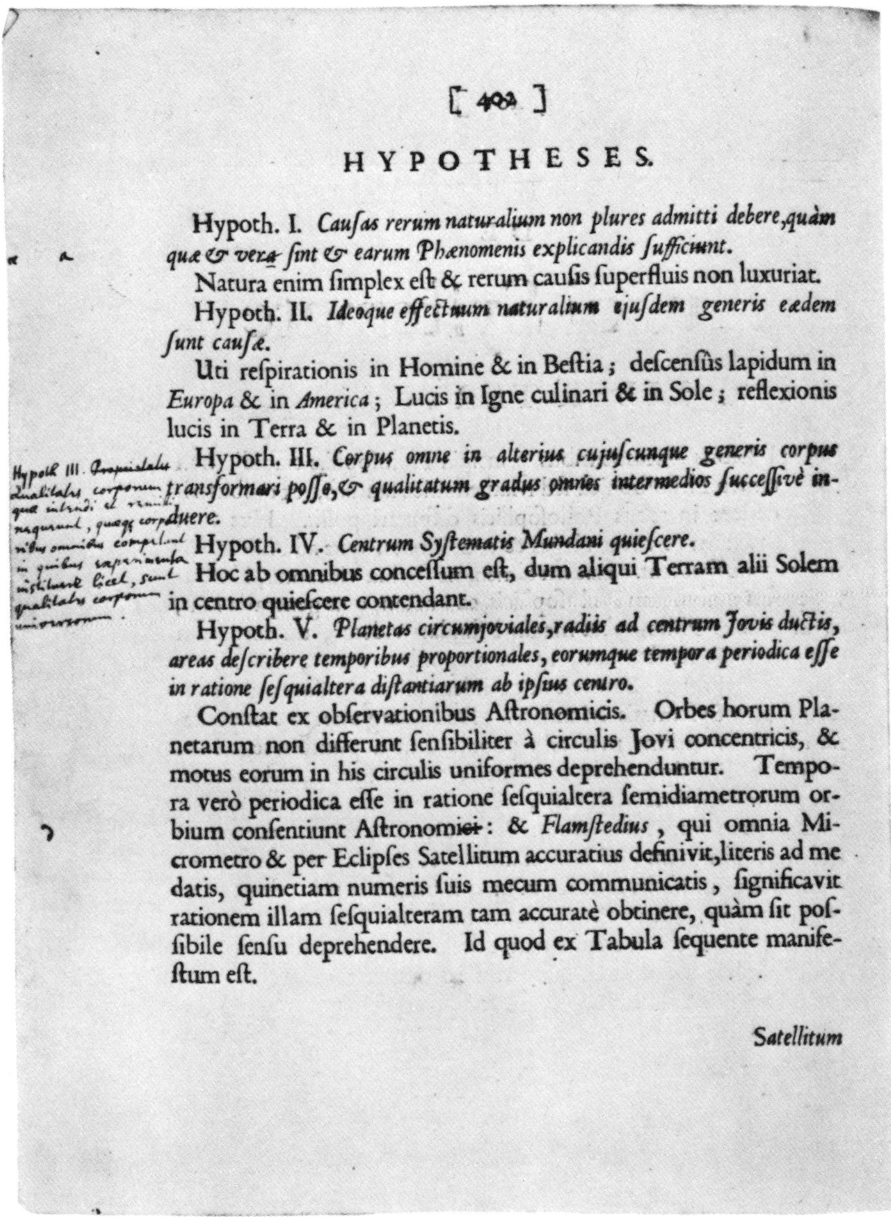

[402]

HYPOTHESES.

Hypoth. I. *Cauſas rerum naturalium non plures admitti debere, quàm quæ & vera ſint & earum Phænomenis explicandis ſufficiunt.*

Natura enim ſimplex eſt & rerum cauſis ſuperfluis non luxuriat.

Hypoth. II. *Ideoque effectuum naturalium ejuſdem generis eædem ſunt cauſæ.*

Uti reſpirationis in Homine & in Beſtia; deſcenſûs lapidum in *Europa* & in *America*; Lucis in Igne culinari & in Sole; reflexionis lucis in Terra & in Planetis.

Hypoth. III. *Corpus omne in alterius cujuſcunque generis corpus transformari poſſe, & qualitatum gradus omnes intermedios ſucceſſivè induere.*

Hypoth III. ~~Proprietates~~ Qualitates corporum quæ intendi et remitti nequeunt, quæq; corporibus omnibus competunt in quibus experimenta instituere licet, sunt qualitates corporum universorum.

Hypoth. IV. *Centrum Syſtematis Mundani quieſcere.*

Hoc ab omnibus conceſſum eſt, dum aliqui Terram alii Solem in centro quieſcere contendant.

Hypoth. V. *Planetas circumjoviales, radiis ad centrum Jovis ductis, areas deſcribere temporibus proportionales, eorumque tempora periodica eſſe in ratione ſeſquialtera diſtantiarum ab ipſius centro.*

Conſtat ex obſervationibus Aſtronomicis. Orbes horum Planetarum non differunt ſenſibiliter à circulis Jovi concentricis, & motus eorum in his circulis uniformes deprehenduntur. Tempora verò periodica eſſe in ratione ſeſquialtera ſemidiametrorum orbium conſentiunt Aſtronomi: & *Flamſtedius*, qui omnia Micrometro & per Eclipſes Satellitum accuratius definivit, literis ad me datis, quinetiam numeris ſuis mecum communicatis, ſignificavit rationem illam ſeſquialteram tam accuratè obtinere, quàm ſit poſſibile ſenſu deprehendere. Id quod ex Tabula ſequente manifeſtum eſt.

Satellitum

Plate 1. Newton's alteration of Hypoth. III in E_1a. (See Chapter II, §2.)

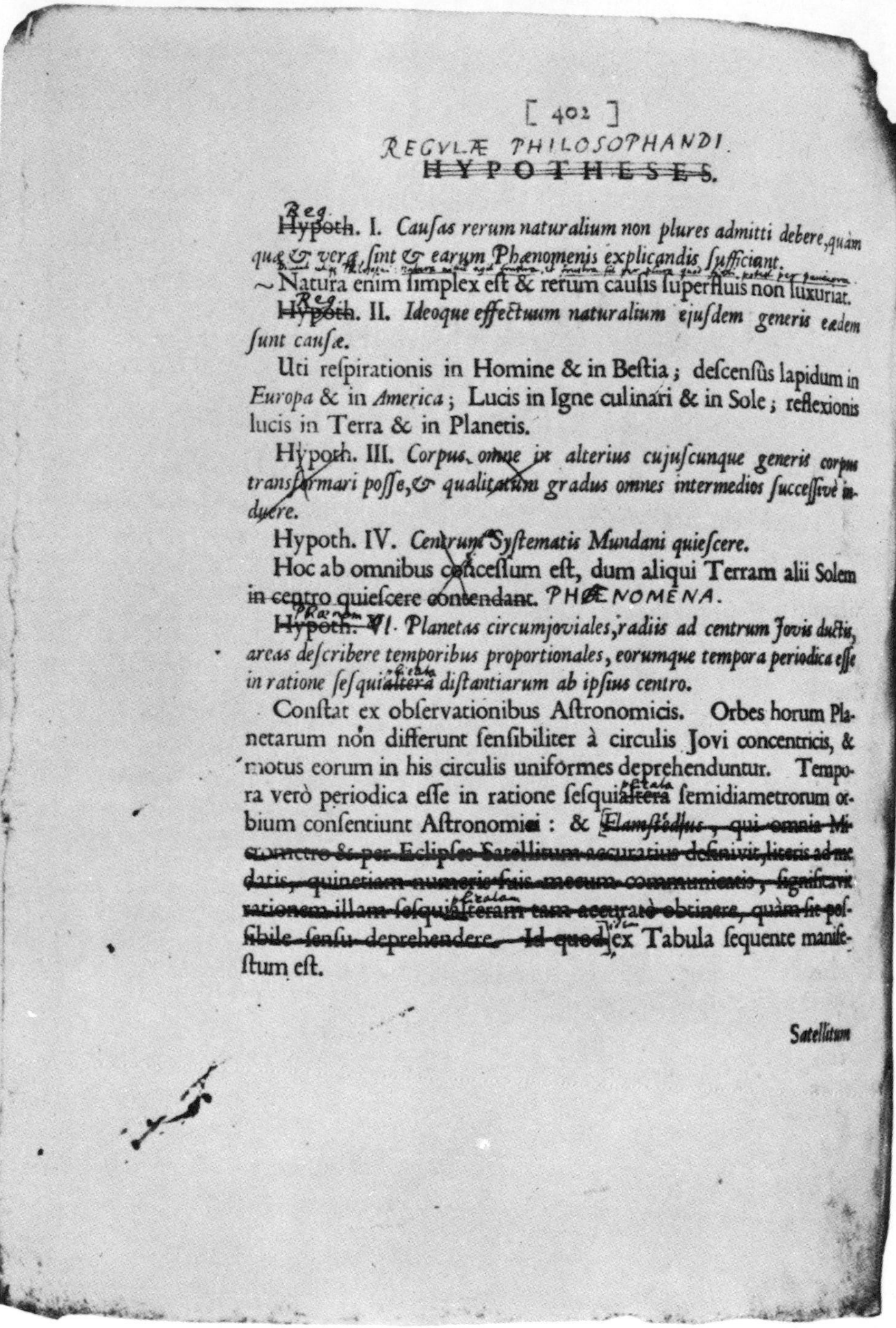
[402]

REGULÆ PHILOSOPHANDI.

~~HYPOTHESES.~~

Reg. ~~Hypoth~~. I. *Causas rerum naturalium non plures admitti debere, quàm quæ & vera sint & earum Phænomenis explicandis sufficiant.*

Natura enim simplex est & rerum causis superfluis non luxuriat.

Reg. ~~Hypoth~~. II. *Ideoque effectuum naturalium ejusdem generis eædem sunt causæ.*

Uti respirationis in Homine & in Bestia; descensûs lapidum in *Europa* & in *America*; Lucis in Igne culinari & in Sole; reflexionis lucis in Terra & in Planetis.

Hypoth. III. *Corpus omne in alterius cujuscunque generis corpus transformari posse, & qualitatum gradus omnes intermedios successivè induere.*

Hypoth. IV. *Centrum Systematis Mundani quiescere.*

Hoc ab omnibus concessum est, dum aliqui Terram alii Solem ~~in centro quiescere contendant.~~ PHÆNOMENA.

Phæn. ~~Hypoth. V~~I. *Planetas circumjoviales, radiis ad centrum Jovis ductis, areas describere temporibus proportionales, eorumque tempora periodica esse in ratione sesquiplicata ~~sesquialtera~~ distantiarum ab ipsius centro.*

Constat ex observationibus Astronomicis. Orbes horum Planetarum non differunt sensibiliter à circulis Jovi concentricis, & motus eorum in his circulis uniformes deprehenduntur. Tempora verò periodica esse in ratione sesquiplicata ~~sesquialtera~~ semidiametrorum orbium consentiunt Astronomi : & ~~Flamstedius, qui omnia Micrometro & per Eclipses Satellitum accuratius definivit, literis ad me datis, quinetiam numeris suis mecum communicatis, significavit rationem illam sesquialteram tam accuratè obtinere, quàm sit possibile sensu deprehendere. Id quod~~ ex Tabula sequente manifestum est.

Satellitum

Plate 2. Newton's alteration of the 'Hypotheses' of Book III in E_1i, p. 402. (See Chapter II, §2.)

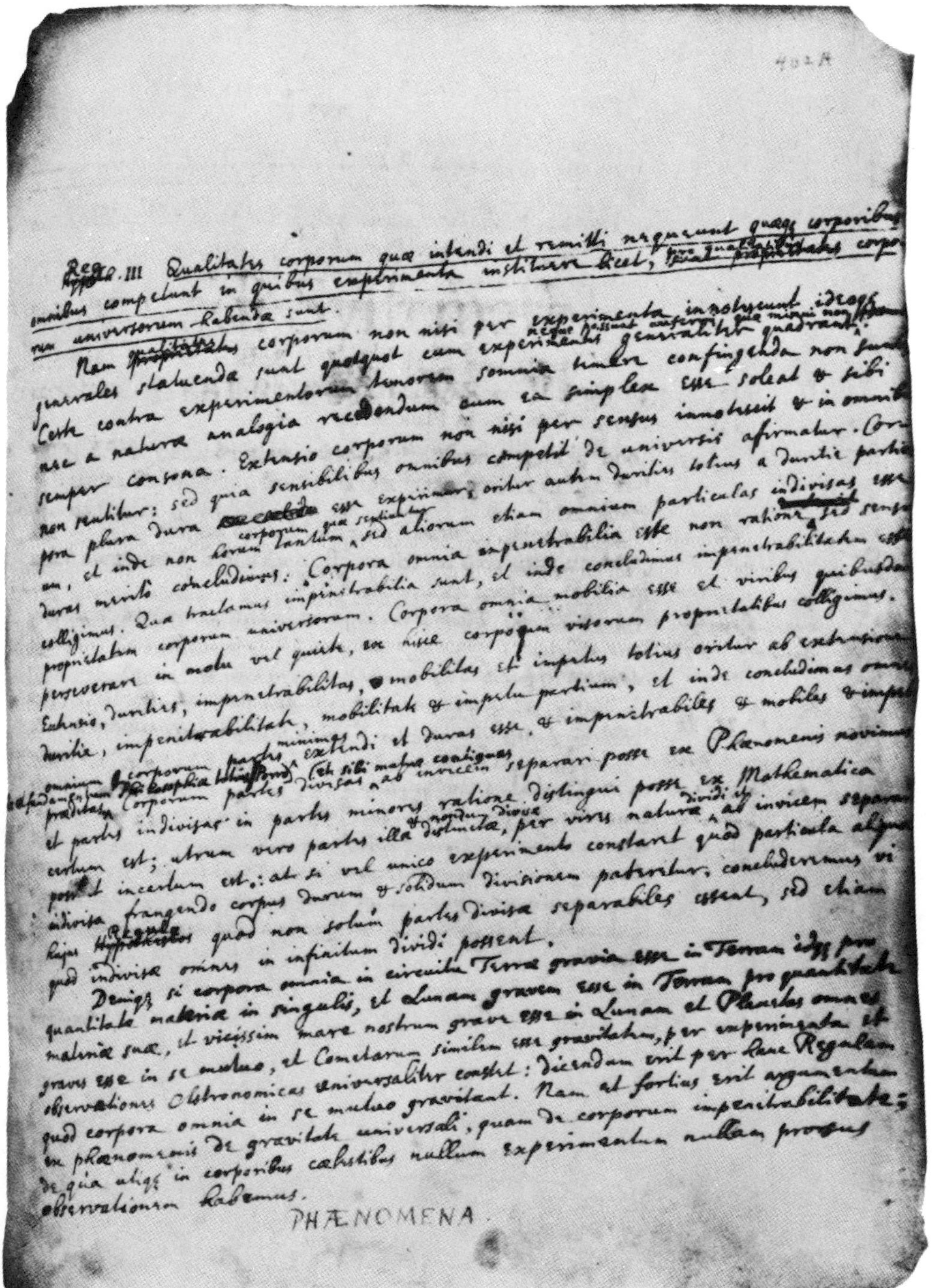

402A

Reg. III Qualitates corporum quæ intendi et remitti nequeunt quæque corporibus omnibus competunt in quibus experimenta instituere licet, pro qualitatibus corporum universorum habendæ sunt.

Nam qualitates corporum non nisi per experimenta innotescunt ideoque generales statuendæ sunt quotquot cum experimentis generaliter quadrant. Certe contra experimentorum tenorem somnia temere confingenda non sunt nec a naturæ analogia recedendum cum ea simplex esse soleat & sibi semper consona. Extensio corporum non nisi per sensus innotescit & in omnibus non sentitur; sed quia sensibilibus omnibus competit de universis affirmatur. Corpora plura dura esse experimur, oritur autem durities totius a duritie partium, et inde non horum tantum corporum quæ sentiuntur sed aliorum etiam omnium particulas indivisas esse duras merito concludimus. Corpora omnia impenetrabilia esse non ratione sed sensu colligimus. Quæ tractamus impenetrabilia sunt, et inde concludimus impenetrabilitatem esse proprietatem corporum universorum. Corpora omnia mobilia esse et viribus quibusdam perseverare in motu vel quiete, ex hisce corporum visorum proprietatibus colligimus. Extensio, durities, impenetrabilitas, mobilitas et impetus totius oritur ab extensione, duritie, impenetrabilitate, mobilitate & impetu partium, et inde concludimus omnes omnium corporum partes minimas extendi et duras esse, & impenetrabiles & mobiles & impetu præditas. Corporum partes divisas et sibi mutuo contiguas ab invicem separari posse ex Phænomenis novimus et partes indivisas in partes minores ratione distingui posse ex Mathematica certum est; utrum vero partes illæ distinctæ et nondum divisæ per vires naturæ dividi et ab invicem separari possint incertum est. At si vel unico experimento constaret quod particula aliqua indivisa frangendo corpus durum & solidum divisionem pateretur, concluderemus vi hujus Regulæ quod non solum partes divisæ separabiles essent, sed etiam quod indivisæ omnes in infinitum dividi possent.

Denique si corpora omnia in circuitu Terræ gravia esse in Terram idque pro quantitate materiæ in singulis, et Lunam gravem esse in Terram pro quantitate materiæ suæ, et vicissim mare nostrum grave esse in Lunam et Planetas omnes graves esse in se mutuo, et Cometarum similem esse gravitatem, per experimenta et observationes Astronomicas universaliter constet: dicendum erit per hanc Regulam quod corpora omnia in se mutuo gravitant. Nam et fortius erit argumentum ex phænomenis de gravitate universali, quam de corporum impenetrabilitate; de qua utique in corporibus cælestibus nullum experimentum nullam prorsus observationem habemus.

PHÆNOMENA.

Plate 3. Newton's alteration in the 'Hypotheses' of Book III in $E_1 i$, interleaf facing page 402. (See Chapter II, §2.)

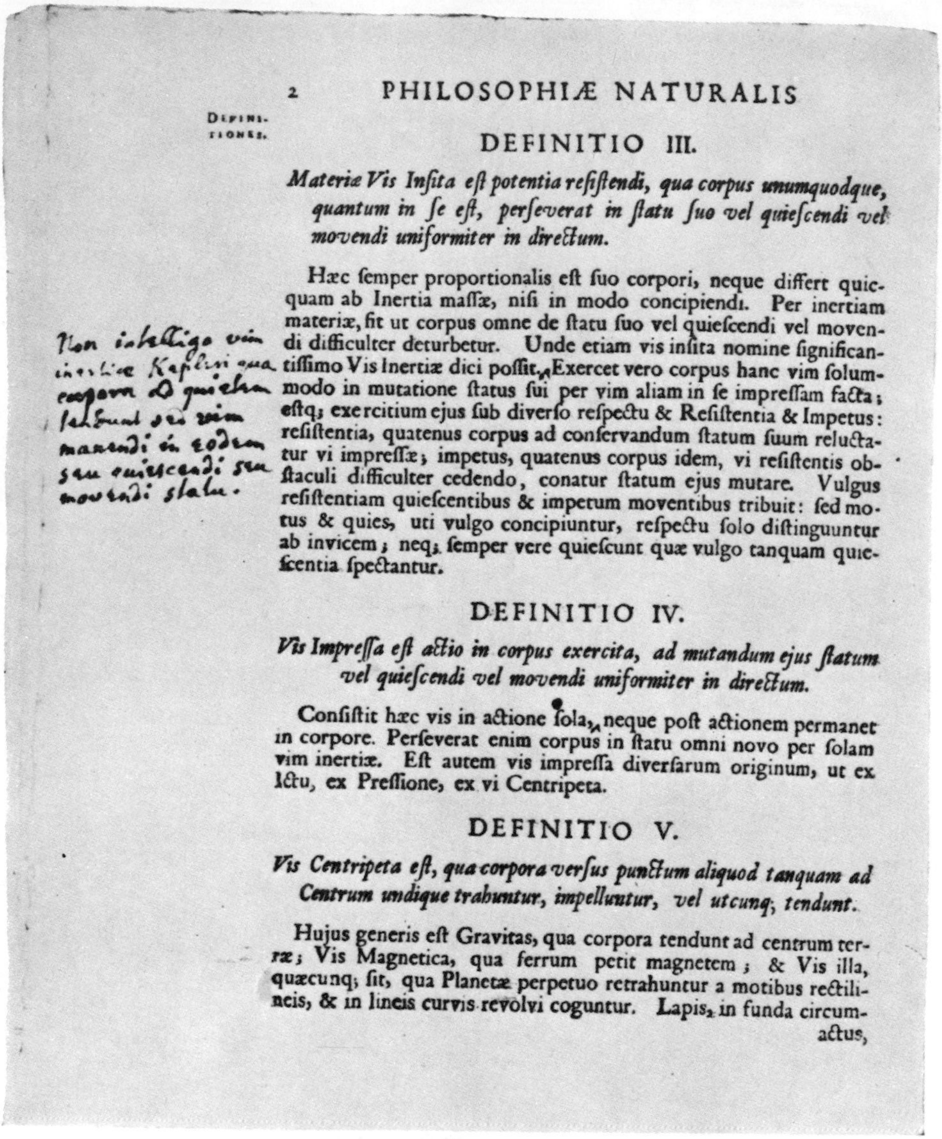

2 PHILOSOPHIÆ NATURALIS

DEFINI-TIONES.

DEFINITIO III.

Materiæ Vis Inſita eſt potentia reſiſtendi, qua corpus unumquodque, quantum in ſe eſt, perſeverat in ſtatu ſuo vel quieſcendi vel movendi uniformiter in directum.

Hæc ſemper proportionalis eſt ſuo corpori, neque differt quicquam ab Inertia maſſæ, niſi in modo concipiendi. Per inertiam materiæ, fit ut corpus omne de ſtatu ſuo vel quieſcendi vel movendi difficulter deturbetur. Unde etiam vis inſita nomine ſignificantiſſimo Vis Inertiæ dici poſſit. Exercet vero corpus hanc vim ſolummodo in mutatione ſtatus ſui per vim aliam in ſe impreſſam facta; eſtq; exercitium ejus ſub diverſo reſpectu & Reſiſtentia & Impetus: reſiſtentia, quatenus corpus ad conſervandum ſtatum ſuum reluctatur vi impreſſæ; impetus, quatenus corpus idem, vi reſiſtentis obſtaculi difficulter cedendo, conatur ſtatum ejus mutare. Vulgus reſiſtentiam quieſcentibus & impetum moventibus tribuit: ſed motus & quies, uti vulgo concipiuntur, reſpectu ſolo diſtinguuntur ab invicem; neq; ſemper vere quieſcunt quæ vulgo tanquam quieſcentia ſpectantur.

Non intelligo vim inertiae Kepleri qua corpora ad quietem tendunt sed vim manendi in eodem seu quiescendi seu movendi statu.

DEFINITIO IV.

Vis Impreſſa eſt actio in corpus exercita, ad mutandum ejus ſtatum vel quieſcendi vel movendi uniformiter in directum.

Conſiſtit hæc vis in actione ſola, neque poſt actionem permanet in corpore. Perſeverat enim corpus in ſtatu omni novo per ſolam vim inertiæ. Eſt autem vis impreſſa diverſarum originum, ut ex Ictu, ex Preſſione, ex vi Centripeta.

DEFINITIO V.

Vis Centripeta eſt, qua corpora verſus punctum aliquod tanquam ad Centrum undique trahuntur, impelluntur, vel utcunq; tendunt.

Hujus generis eſt Gravitas, qua corpora tendunt ad centrum terræ; Vis Magnetica, qua ferrum petit magnetem; & Vis illa, quæcunq; ſit, qua Planetæ perpetuo retrahuntur a motibus rectilineis, & in lineis curvis revolvi coguntur. Lapis, in funda circumactus,

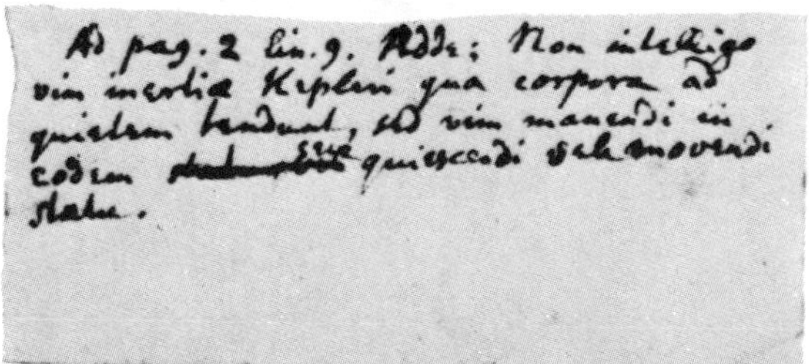

Ad pag. 2 lin. 9. Adde; Non intelligo vim inertiæ Kepleri qua corpora ad quietem tendunt, sed vim manendi in eodem ~~statu~~ seu quiescendi seu movendi statu.

Plate 4. Newton's contrast of his own concept of inertia and Kepler's, in E_2i. (See Chapter II, §4.) The insert is a small slip of paper (not the regular interleaf), bound into E_2i.

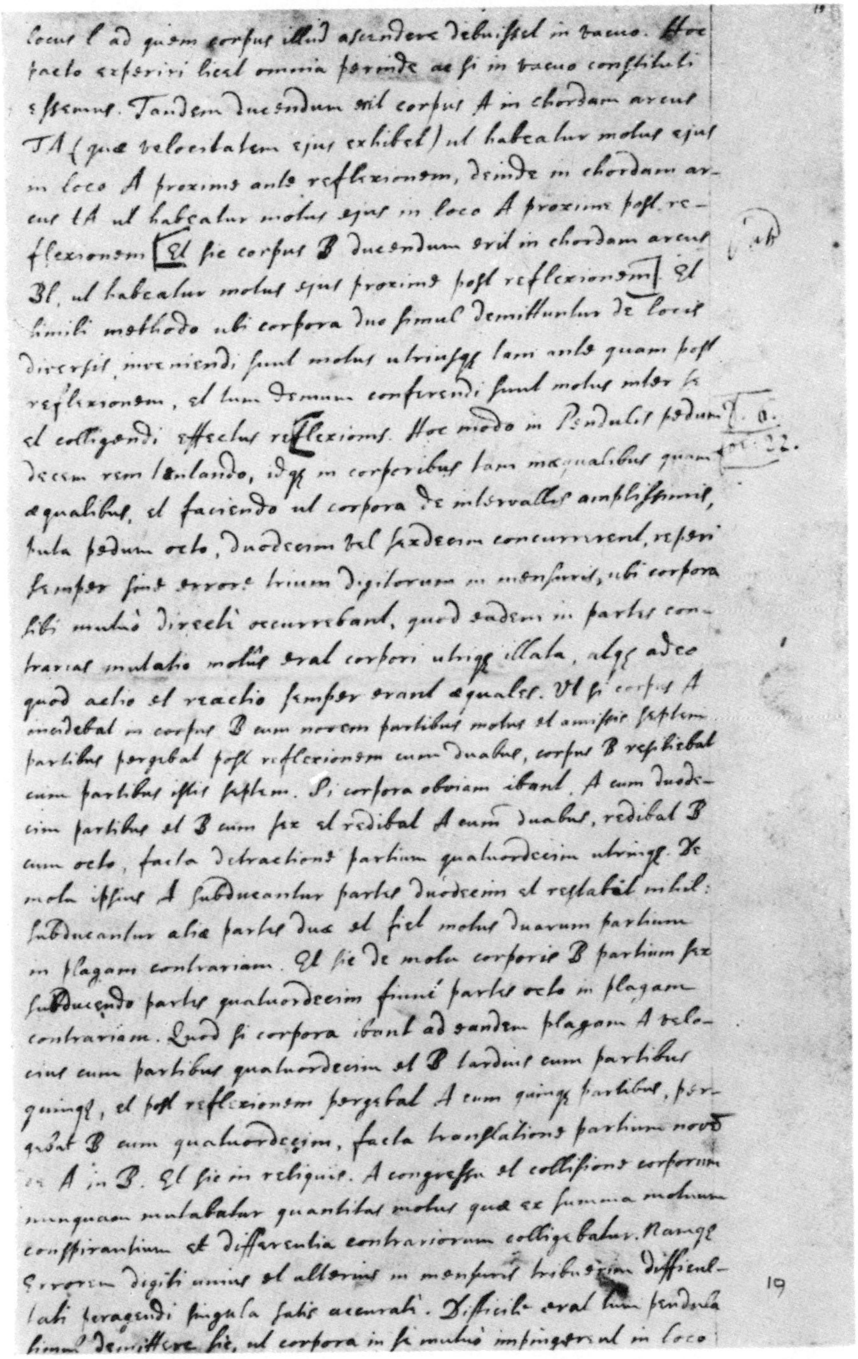
10

locus l ad quem corpus illud ascendere debuisset in vacuo. Hoc
pacto experiri licet omnia perinde ac si in vacuo constituti
essemus. Tandem ducendum erit corpus A in chordam arcus
TA (quae velocitatem ejus exhibet) ut habeatur motus ejus
in loco A proxime ante reflexionem, deinde in chordam arcus tA ut habeatur motus ejus in loco A proxime post reflexionem. [Et sic corpus B ducendum erit in chordam arcus out
Bl, ut habeatur motus ejus proxime post reflexionem.] Et
simili methodo ubi corpora duo simul demittuntur de locis
diversis, inveniendi sunt motus utriusque tam ante quam post
reflexionem, et tum demum conferendi sunt motus inter se
et colligendi effectus re|flexionis. Hoc modo in Pendulis pedum D.6. fol. 22.
decem rem tentando, idque in corporibus tam inaequalibus quam
aequalibus, et faciendo ut corpora de intervallis amplissimis,
puta pedum octo, duodecim vel sexdecim concurrerent, reperi
semper sine errore trium digitorum in mensuris, ubi corpora
sibi mutuo directe occurrebant, quod eadem in partes contrarias mutatio motus erat corpori utrique illata, atque adeo
quod actio et reactio semper erant aequales. Ut si corpus A
incidebat in corpus B cum novem partibus motus et amissis septem
partibus pergebat post reflexionem cum duabus, corpus B resiliebat
cum partibus istis septem. Si corpora obviam ibant, A cum duodecim partibus et B cum sex et redibat A cum duabus, redibat B
cum octo, facta detractione partium quatuordecim utrinque. De
motu ipsius A subducantur partes duodecim et restabit nihil:
subducantur aliae partes duae et fiet motus duarum partium
in plagam contrariam. Et sic de motu corporis B partium sex
subducendo partes quatuordecim fiunt partes octo in plagam
contrariam. Quod si corpora ibant ad eandem plagam A velo-
cius cum partibus quatuordecim et B tardius cum partibus
quinque, et post reflexionem pergebat A cum quinque partibus, pergebat B cum quatuordecim, facta translatione partium novem
de A in B. Et sic in reliquis. A congressu et collisione corporum
nunquam mutabatur quantitas motus quae ex summa motuum
conspirantium et differentia contrariorum colligebatur. Namque
errorem digiti unius et alterius in mensuris tribuerim difficul-
tati peragendi singula satis accurate. Difficile erat tum pendula
simul demittere sic, ut corpora in se mutuo impingerent in loco

19

Plate 5. A page of *M*, in Humphrey Newton's hand, corresponding to {23.15–24.18}, Scholium to Laws of Motion. (See Chapter II, §10, and Supp. II.) The folio number 10 is in Newton's hand. The square brackets and the word 'out' indicate an omission in proof, restored by the time of printing. The single bracket between *re-* and *flexionis* marks the top of a page 'fol. 22.', 'D.6.'—the 6th page of gathering D. In E_1, however, the line beginning with *flexionis* is the second line on page 22; an insertion into the previous page caused a line to be carried over from page 21.

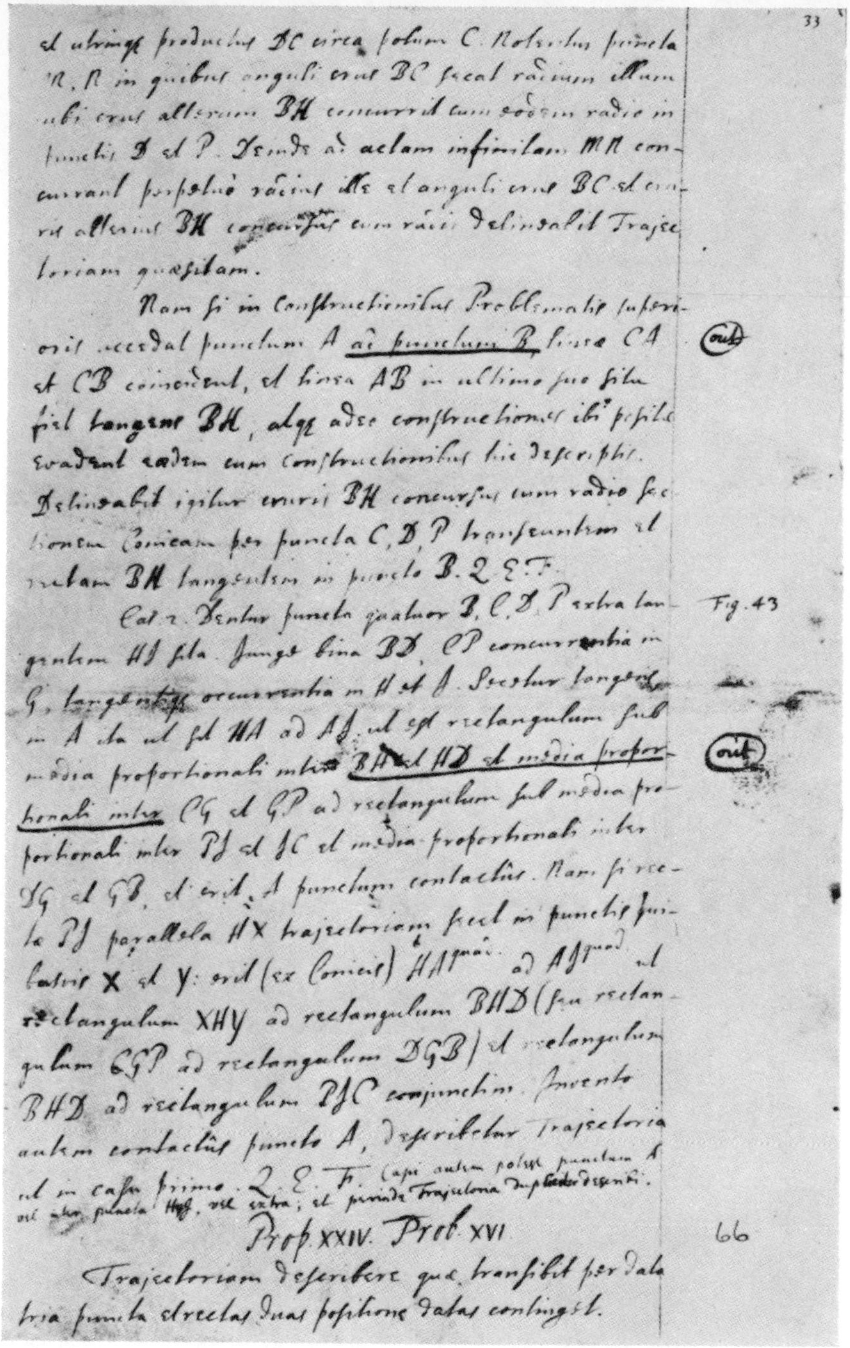

33
et utrinque productus BC circa polum C. Notentur puncta N, n in quibus anguli crus BC secat radium illum ubi crus alterum BH concurrit cum eodem radio in punctis D et P. Deinde ad actam infinitam Mn concurrant perpetuo radius ille et anguli crus BC et cruris alterius BH concursus cum radio delineabit Trajectoriam quaesitam.
Nam si in Constructionibus Problematis superioris accedat punctum A ad punctum B, lineae CA et CB coincident, et linea AB in ultimo suo situ fiet tangens BH, atque adeo constructiones ibi positae evadent eaedem cum constructionibus hic descriptis. Delineabit igitur cruris BH concursus cum radio sectionem Conicam per puncta C, D, P transeuntem et rectam BH tangentem in puncto B. Q.E.F.
out
Cas. 2. Dentur puncta quatuor B, C, D, P extra tangentem HI sita. Junge bina BD, CP concurrentia in G, tangentique occurrentia in H et I. Secetur tangens in A ita ut sit HA ad AI, ut est rectangulum sub media proportionali inter BH et HD et media proportionali inter CG et GP ad rectangulum sub media proportionali inter PI et IC et media proportionali inter DG et GB, et erit A punctum contactus. Nam si rectae PI parallela HX trajectoriam secet in punctis quibusvis X et Y: erit (ex Conicis) HA quad. ad AI quad. ut rectangulum XHY ad rectangulum BHD (seu rectangulum CGP ad rectangulum DGB) et rectangulum BHD ad rectangulum PIC conjunctim. Invento autem contactus puncto A, describetur Trajectoria ut in casu primo. Q.E.F. Capi autem potest punctum A vel inter puncta H et I, vel extra; et perinde Trajectoria duplex describi.
Fig. 43
out
Prop. XXIV. Prob. XVI
66
Trajectoriam describere quae transibit per data tria puncta et rectas duas positione datas continget.

Plate 6. A page of *M*, with additions by Newton. (See Chapter II, §10, and Supp. II.) The folio number 33, the marginal indication 'Fig. 43', and the addition made at the end of Case 2, Prop. XXIV, Book I, are in Newton's hand. The underscored words, accompanied by the marginal word 'out', were evidently omitted in proof but restored by the time of printing.

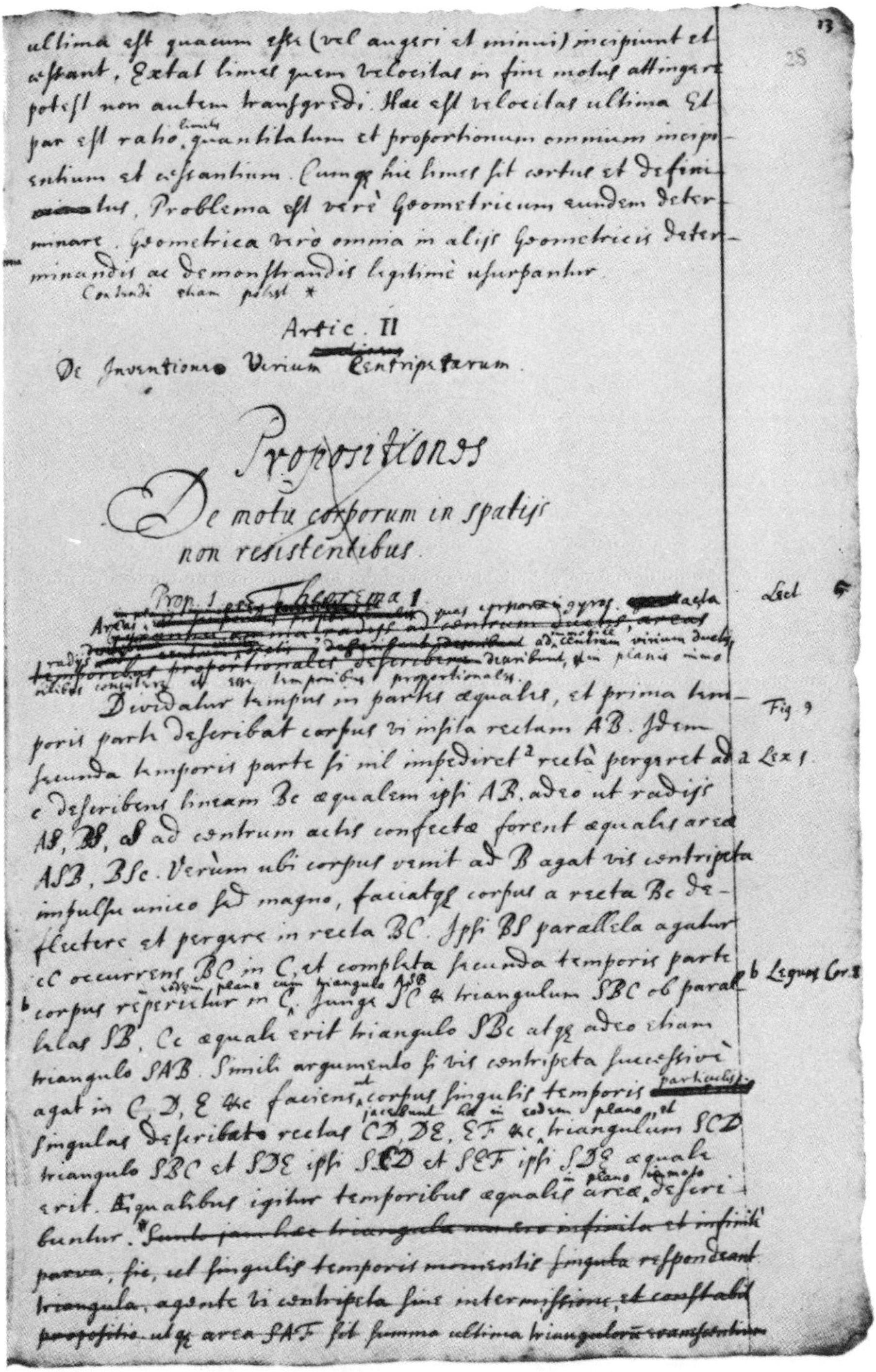
13
28

ultima est quacum esse (vel augeri et minui) incipiunt et cessant, Extat limes quem velocitas in fine motus attingere potest non autem transgredi. Haec est velocitas ultima. Et par est ratio limitis quantitatum et proportionum omnium incipientium et cessantium. Cumque hic limes sit certus et definitus, Problema est vere Geometricum eundem determinare. Geometrica vero omnia in aliis Geometricis determinandis ac demonstrandis legitime usurpantur.
Contendi etiam potest *

Artic. II
De Inventione Virium Centripetarum.

Propositiones
De motu corporum in spatiis non resistentibus.

Prop. 1 Theorema 1 — Lect 6
Areas quas corpora in gyros acta radiis ad centrum virium ductis describunt, et in planis immobilibus consistere et esse temporibus proportionales.

Dividatur tempus in partes aequales, et prima temporis parte describat corpus vi insita rectam AB. Idem secunda temporis parte si nil impediret recta pergeret ad [a] c describens lineam Bc aequalem ipsi AB, adeo ut radiis AS, BS, cS ad centrum actis confectae forent aequales areae ASB, BSc. Verum ubi corpus venit ad B agat vis centripeta impulsu unico sed magno, faciatque corpus a recta Bc deflectere et pergere in recta BC. Ipsi BS parallela agatur cC occurrens BC in C, et completa secunda temporis parte corpus [b] reperietur in C, in eodem plano cum triangulo ASB. Junge SC & triangulum SBC ob parallelas SB, Cc aequale erit triangulo SBc atque adeo etiam triangulo SAB. Simili argumento si vis centripeta successive agat in C, D, E &c faciens ut corpus singulis temporis particulis singulas describat rectas CD, DE, EF &c, jacebunt hae in eodem plano et triangulum SCD triangulo SBC et SDE ipsi SCD et SEF ipsi SDE aequale erit. Aequalibus igitur temporibus aequales areae in plano immoto describuntur. *Sunto jam haec triangula numero infinita et infinite parva, sic ut singulis temporis momentis singula respondeant triangula, agente vi centripeta sine intermissione, et constabit propositio, utque area SAF sit summa ultima triangulorum consentium

Fig. 9
a Lex 1
b Legum Cor. 1

Plate 7. A page from *LL*, showing corrections made by Newton. The text is written by Humphrey Newton, but the textual and marginal additions (including the folio number 13, above the librarian's folio number 28) are by Newton. (See Chapter IV, §2, and Supp. IV.)

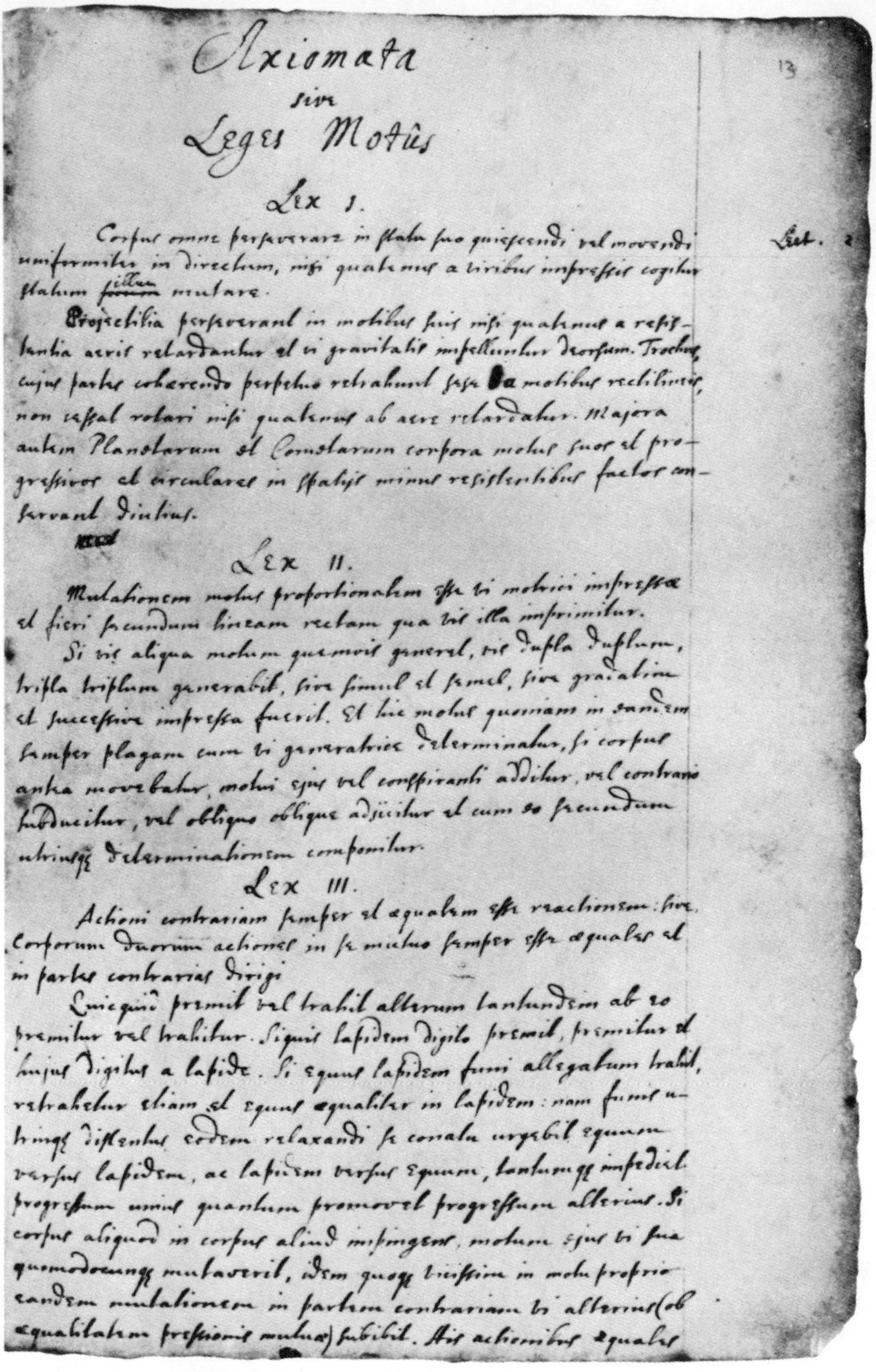

13

Axiomata
sive
Leges Motûs

Lex 1.

Corpus omne perseverare in statu suo quiescendi vel movendi uniformiter in directum, nisi quatenus a viribus impressis cogitur statum ~~suum~~ illum mutare. Gal. 2

Projectilia perseverant in motibus suis nisi quatenus a resistentia aeris retardantur et vi gravitatis impelluntur deorsum. Trochus, cujus partes cohaerendo perpetuo retrahunt sese a motibus rectilineis, non cessat rotari nisi quatenus ab aere retardatur. Majora autem Planetarum et Cometarum corpora motus suos et progressivos et circulares in spatiis minus resistentibus factos conservant diutius.

Lex II.

Mutationem motus proportionalem esse vi motrici impressae et fieri secundum lineam rectam qua vis illa imprimitur.

Si vis aliqua motum quemvis generet, vis dupla duplum, tripla triplum generabit, sive simul et semel, sive gradatim et successive impressa fuerit. Et hic motus quoniam in eandem semper plagam cum vi generatrice determinatur, si corpus antea movebatur, motui ejus vel conspiranti additur, vel contrario subducitur, vel obliquo oblique adjicitur et cum eo secundum utriusque determinationem componitur.

Lex III.

Actioni contrariam semper et aequalem esse reactionem: sive Corporum duorum actiones in se mutuo semper esse aequales et in partes contrarias dirigi.

Quicquid premit vel trahit alterum tantundem ab eo premitur vel trahitur. Siquis lapidem digito premit, premitur et hujus digitus a lapide. Si equus lapidem funi allegatum trahit, retrahetur etiam et equus aequaliter in lapidem: nam funis utrinque distentus eodem relaxandi se conatu urgebit equum versus lapidem, ac lapidem versus equum, tantumque impediet progressum unius quantum promovet progressum alterius. Si corpus aliquod in corpus aliud impingens, motum ejus vi sua quomodocunque mutaverit, idem quoque vicissim in motu proprio eandem mutationem in partem contrariam vi alterius (ob aequalitatem pressionis mutuae) subibit. His actionibus aequales

Plate 8. A page from *LL*, showing corrections made by Newton. The text is written by Humphrey Newton, but the textual alteration and marginal addition are by Newton. (See Chapter IV, §2, and Supp. IV.)

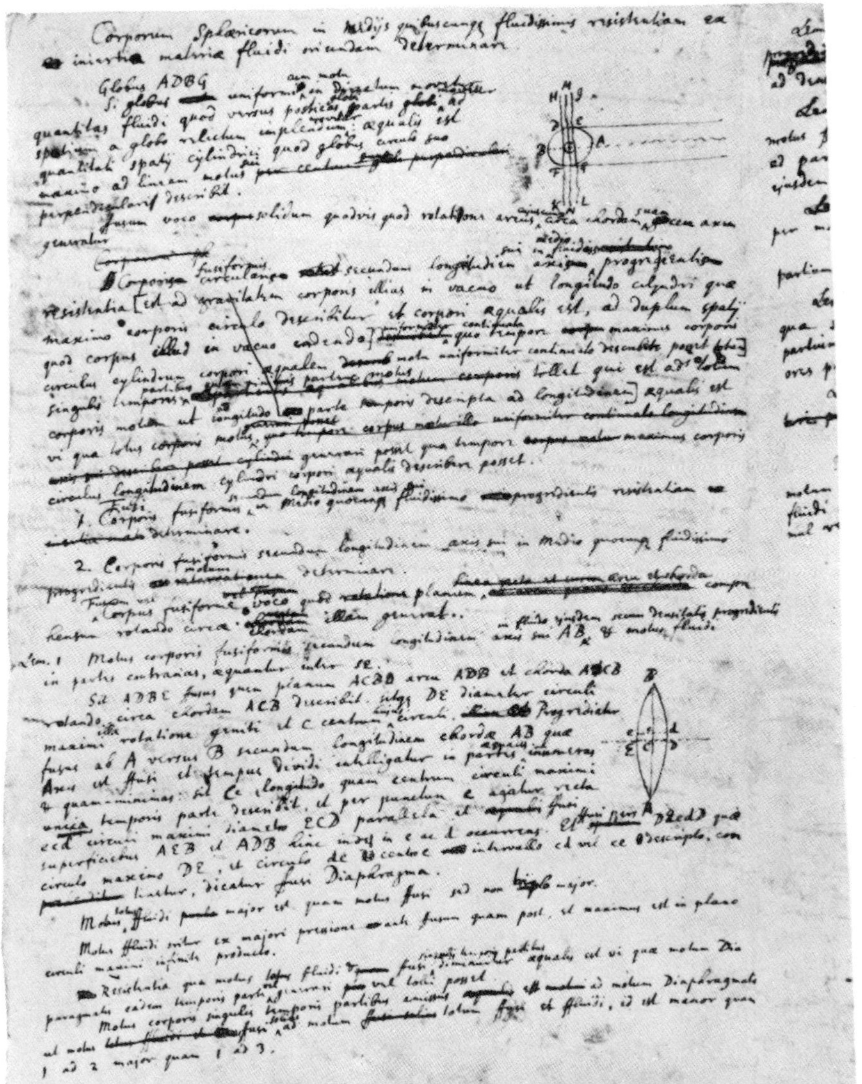

Plate 9. Newton's draft essay on spindle-shaped (or fusiform) bodies. (See Chapter IV, §5, and Supp. V.)

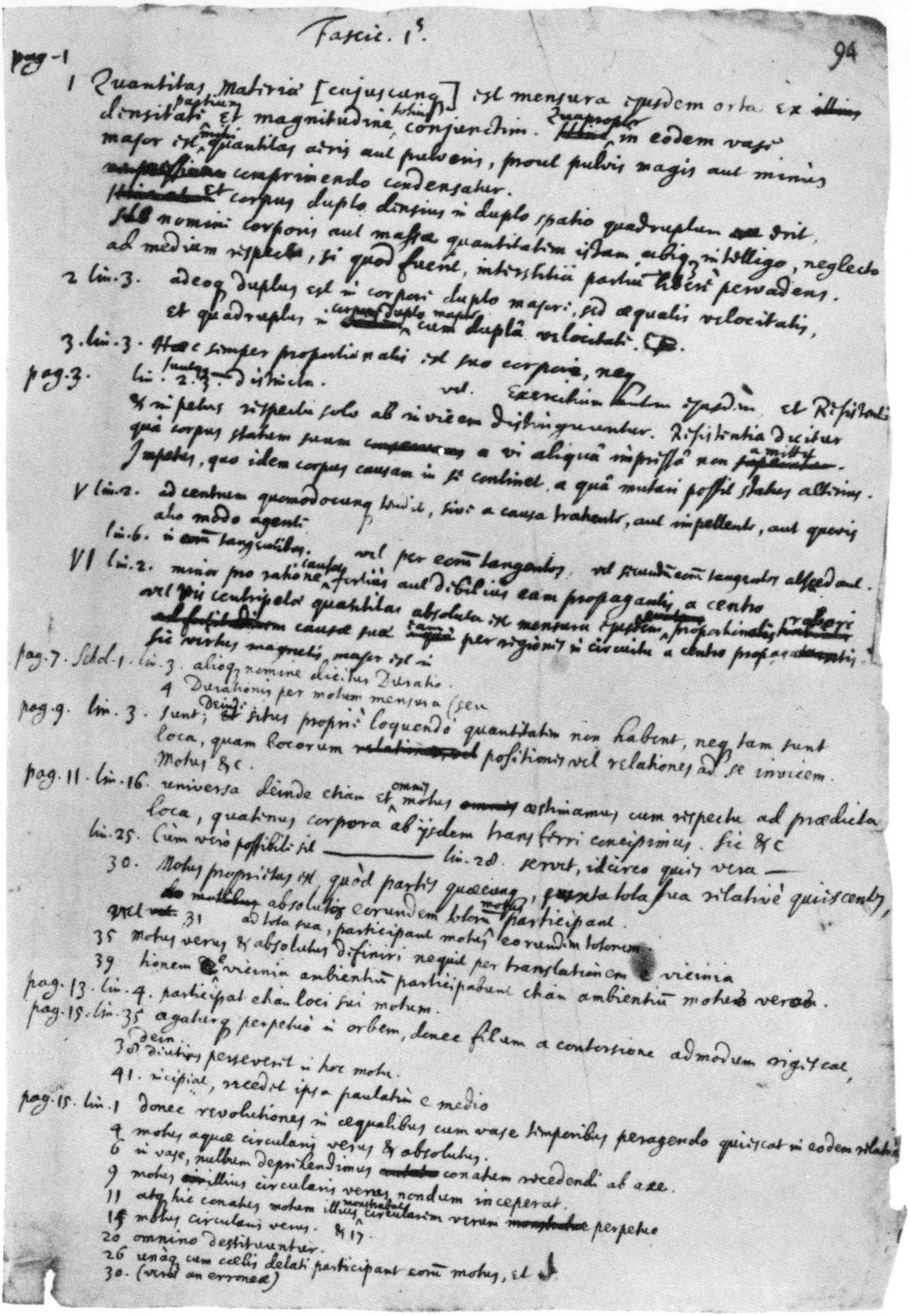

Fascic. 1s.

pag. 1 94

1 Quantitas Materiæ [cujuscunq] est mensura ejusdem orta ex partium densitate et magnitudine totius conjunctim. Quapropter in eodem vase major est multo quantitas aeris aut pulveris, prout pulvis magis aut minus comprimendo condensatur. Et corpus duplo densius in duplo spatio quadruplum erit. Nomine corporis aut massæ quantitatem istam ubiq intelligo, neglecto ad medium respectu, si quod fuerit, interstitia partium libere pervadens.

2 lin. 3. adeoq duplus est in corpore duplo majori sed æquali velocitati, et quadruplus in corpore duplo majori cum duplâ velocitate. &c.

3. lin. 3. Hæc semper proportionalis est suo corpori, neq

pag. 3. lin. 2. 3. districta. vel. Exercitium [illegible] ejusdem et Resistentia et impetus respectu solo ab invicem distinguuntur. Resistentia dicitur quâ corpus statum suum [illegible] a vi aliquâ impressâ non amittit. Impetus, quo idem corpus causam in se continet, a quâ mutari possit status alterius.

V lin. 2. ad centrum quomodocunq tendit, sive a causa trahente, aut impellente, aut quovis alio modo agente.

lin. 6. in eorum tangentibus. vel per eorum tangentes. vel secundum eorum tangentes abscedant.

VI lin. 2. minor pro ratione causæ fortius aut debilius eam propagantis a centro. vel Vis centripetæ quantitas absoluta est mensura ejusdem, proportionalis [illegible] causæ seu [illegible] per regiones in circuitu a centro propagati. [illegible] sic virtus magnetis major est in

pag. 7. Schol. 1. lin. 3. alioq nomine dicitur Duratio.

4 Durationis per motum mensura (seu

pag. 9. lin. 3. sunt; Denique situs proprie loquendo quantitatem non habent, neq tam sunt loca, quam locorum positiones vel relationes ad se invicem. Motus &c.

pag. 11. lin. 16. universa deinde etiam et omnes motus æstimamus cum respectu ad prædicta loca, quatenus corpora ab ijsdem transferri concipimus. Sic &c

lin. 25. Cum vero possibile sit ——— lin. 28. servet, idcirco quies vera —

30. Motus proprietas est, quod partes quæcunq, juxta tota sua relative quiescentes, absoluti eorundem totorum motus participant.

vel 31 ad tota sua, participant motus eorundem totorum.

35 Motus verus & absolutus definiri nequit per translationem e vicinia

39 tionem e vicinia ambientium participabunt etiam ambientium motus veros.

pag. 13. lin. 4. participat etiam loci sui motum.

pag. 15. lin. 35 agaturq perpetuo in orbem, donec filum a contorsione admodum rigescat,

38 deinde diutius perseverit in hoc motu.

41. incipiat, recedet ipsa paulatim e medio

pag. 15. lin. 1 donec revolutiones in æqualibus cum vase temporibus peragendo quiescat in eodem relative

4 motus aquæ circularis verus & absolutus.

6 in vase, nullum deprehendimus conatum recedendi ab axe.

9 motus illius circularis verus nondum inceperat.

11 atq hic conatus motum illius circularem verum manifestat perpetuo

15 motus circularis verus. & 17.

20 omnino destituentur.

26 unàq cum cœlis delati participant eorum motus, et

30. (verbum erroneum)

Plate 10. A critique of a preliminary version of the *Principia*, apparently written by Edmond Halley, dealing with the Definitions (MS Add. 3965, f. 94). (See Chapter IV, §8, and Supp. VII.)

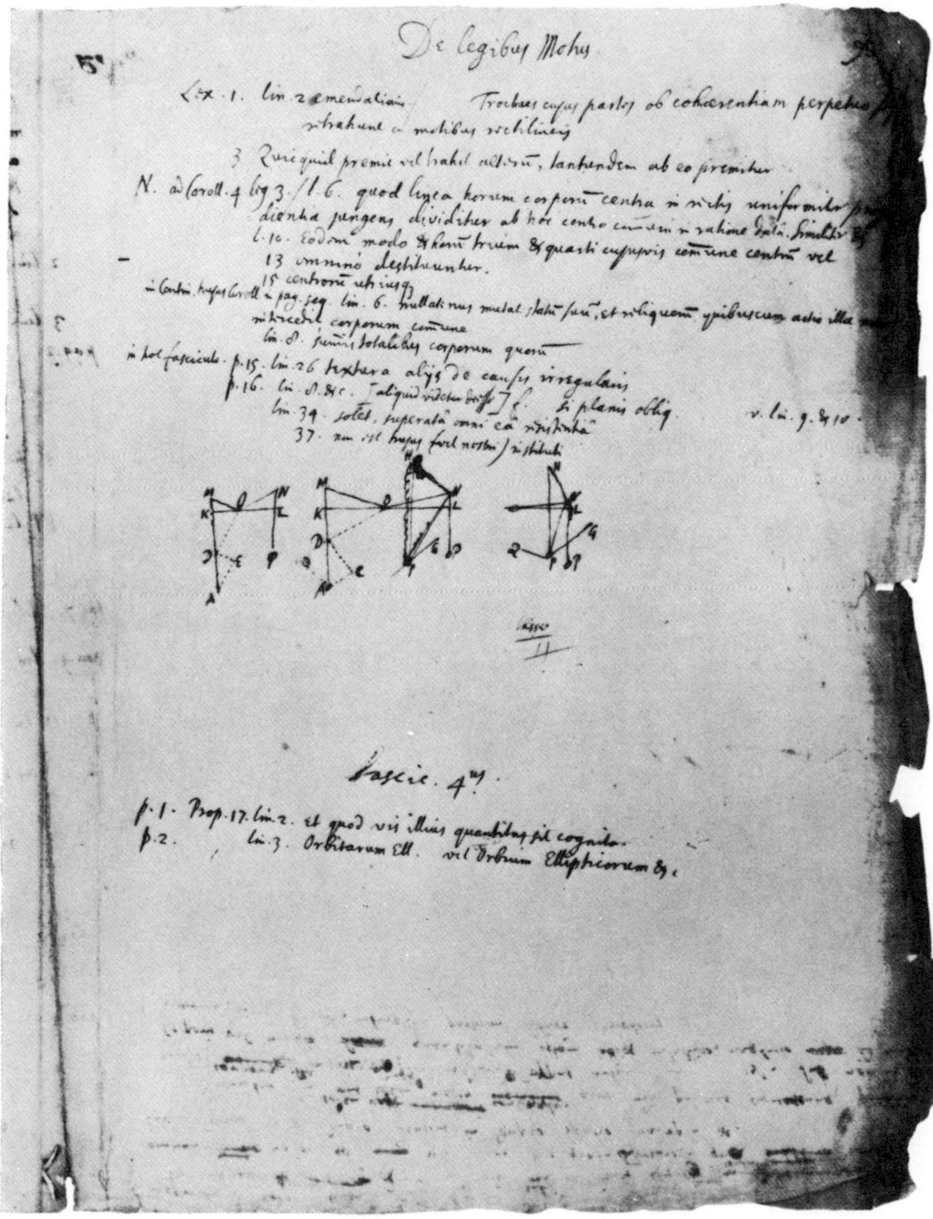

De legibus Motus

Lex. 1. lin. 2 emendatione / Trochus cujus partes ob cohaerentiam perpetuo retrahunt a motibus rectilineis

3 Quicquid premit vel trahit alterum, tantundem ab eo premitur

N. ad Coroll. 4 leg 3. l. 6. quod linea horum corporum centra in rectis uniformiter pro-dientia jungens, dividitur ab hoc centro communi in ratione data. Similiter

l. 10. Eodem modo & horum trium & quarti cujuspiis commune centrum vel

13 omnino destituuntur.

15 centrorum utriusq

in Centri hujus Coroll. in pag. seq. lin. 6. nullatenus mutat statum suum, et reliquorum quibuscunque actio illa introcedit corporum commune

lin. 8. summis totalibus corporum quorum

in hoc fasciculo. p. 15. lin. 26 textura aliis de causis irregularis

p. 16. lin. 8. &c. [aliquid videtur deesse] &. in planis oblig. v. lin. 9. & 10.

lin. 34. solet, superata omni ea resistentia

37. non est hujus fori nostri) restituti

Fascic. 4.

p. 1. Prop. 17. lin. 2. et quod vis illius quantitas sit cognita.

p. 2. lin. 3. Orbitarum Ell. vel Orbium Ellipticorum &c.

Plate 11. A critique of a preliminary version of the *Principia*, apparently written by Edmond Halley, dealing with the Laws of Motion (f. 95). (See Chapter IV, §8, and Supp. VII.)

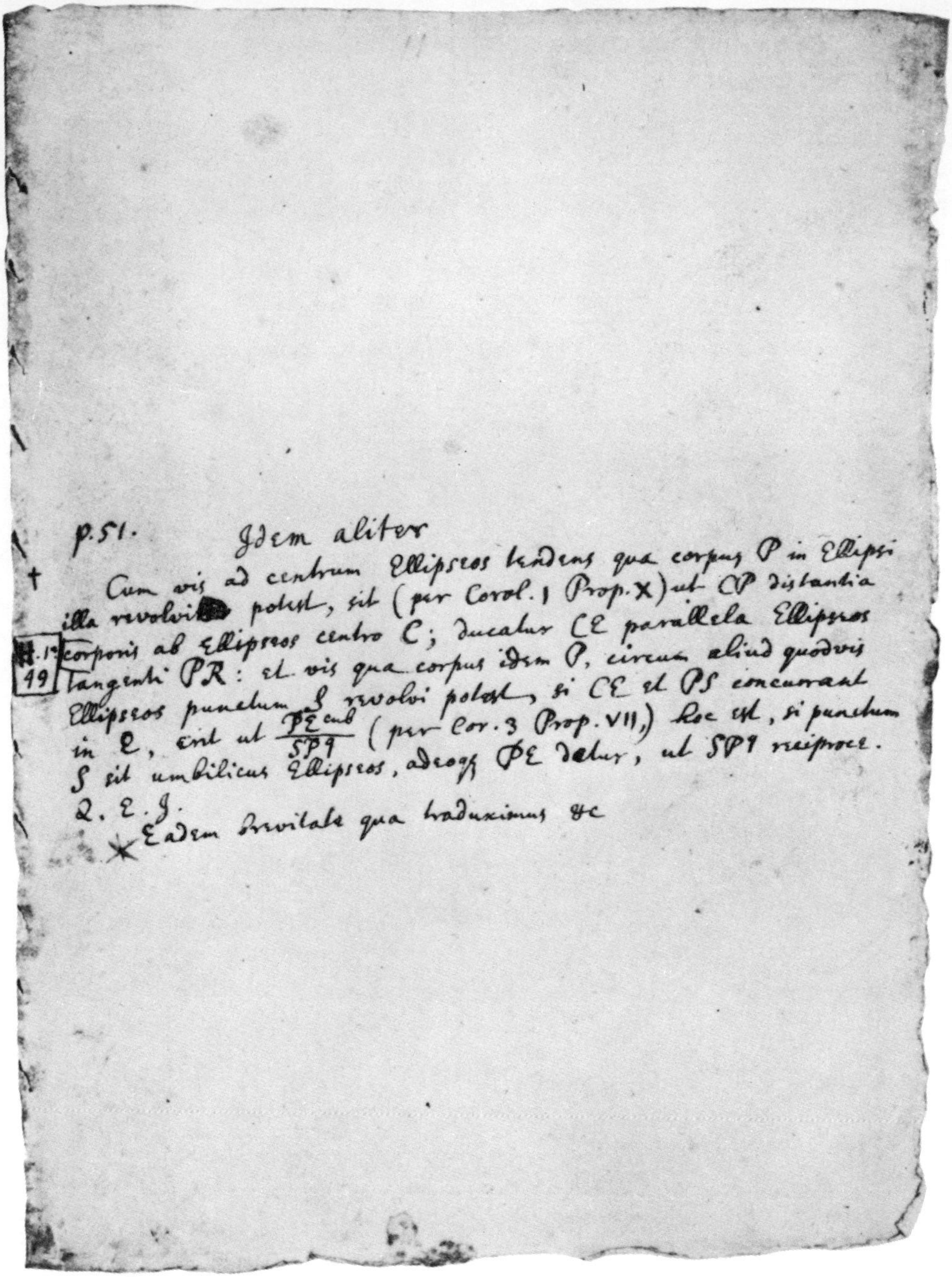
p. 51. Idem aliter

Cum vis ad centrum Ellipseos tendens qua corpus P in Ellipsi illa revolvi potest, sit (per Corol. 1 Prop. X) ut CP distantia corporis ab Ellipseos centro C; ducatur CE parallela Ellipseos tangenti PR: et vis qua corpus idem P, circum aliud quodvis Ellipseos punctum S revolvi potest, si CE et PS concurrant in E, erit ut $\frac{PE^{cub}}{SP^{q}}$ (per Cor. 3 Prop. VII,) hoc est, si punctum S sit umbilicus Ellipseos, adeoque PE detur, ut SP^{q} reciproce. Q.E.I.

* Eadem brevitate qua reduximus &c

H. 1a
49

Plate 12. Alternative proof of Prop. XI, Book I, sent by Newton to Cotes for E_2. (See Chapter VIII, § 5, and Chapter IX.) The addendum is written out by Newton on an interleaf so as to be just opposite that part of the printed page (p. 51, E_1) where it is to be inserted. The printer's mark refers to E_2, p. 49, the first leaf (1^a) of gathering H. Although Newton has indicated powers by superscript 'cub' and 'q', these are printed in E_2 on the line, $\frac{PE\ cub.}{SP\ q}$.

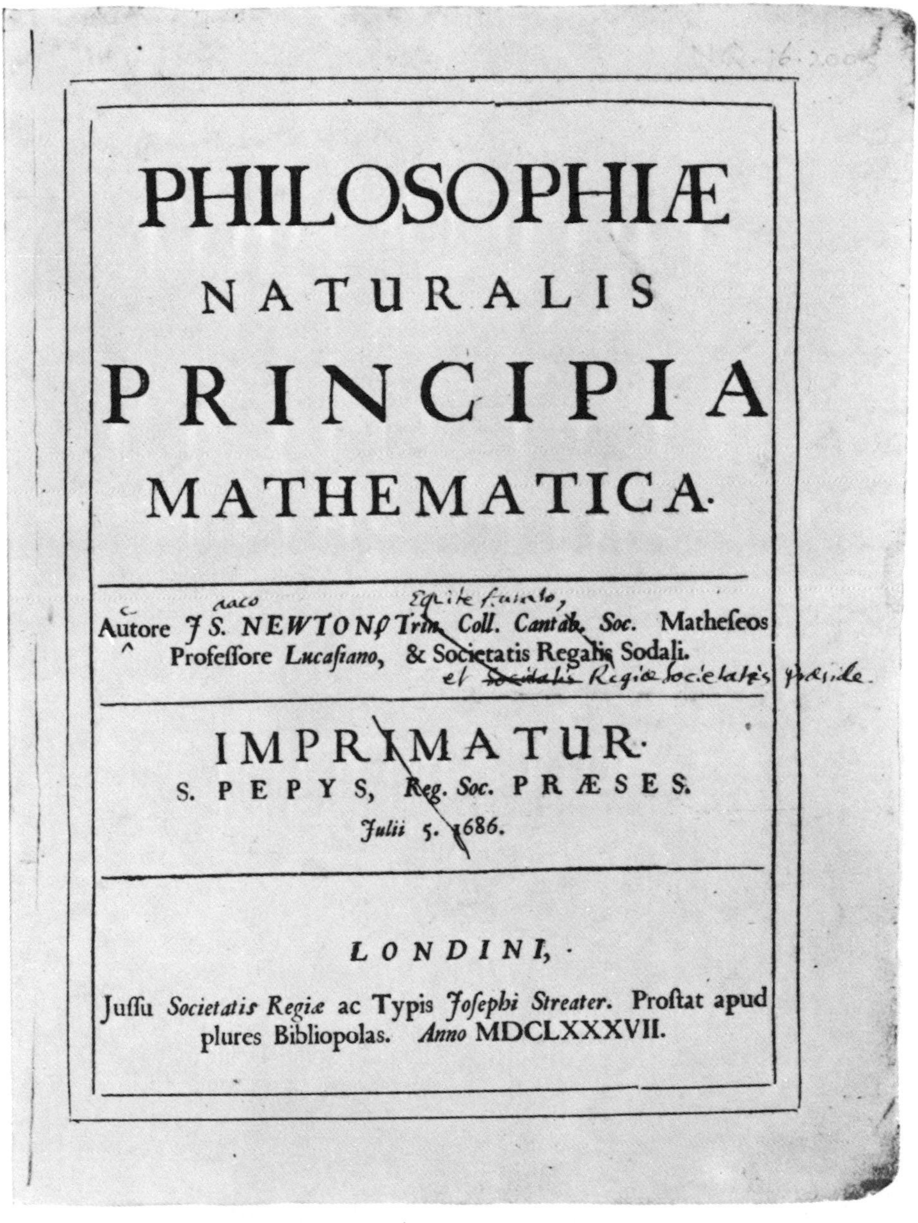

PHILOSOPHIÆ
NATURALIS
PRINCIPIA
MATHEMATICA.

Autore J S. NEWTON, *Trin. Coll. Cantab. Soc.* Matheseos Professore *Lucasiano*, & Societatis Regalis Sodali.

IMPRIMATUR.
S. PEPYS, *Reg. Soc.* PRÆSES.
Julii 5. 1686.

LONDINI,

Jussu *Societatis Regiæ* ac Typis *Josephi Streater*. Prostat apud plures Bibliopolas. *Anno* MDCLXXXVII.

Plate 13. Title-page of E_1a, with alterations made by Richard Bentley.
(See Chapter VIII, §6, and Guide to the Apparatus Criticus, §1.)

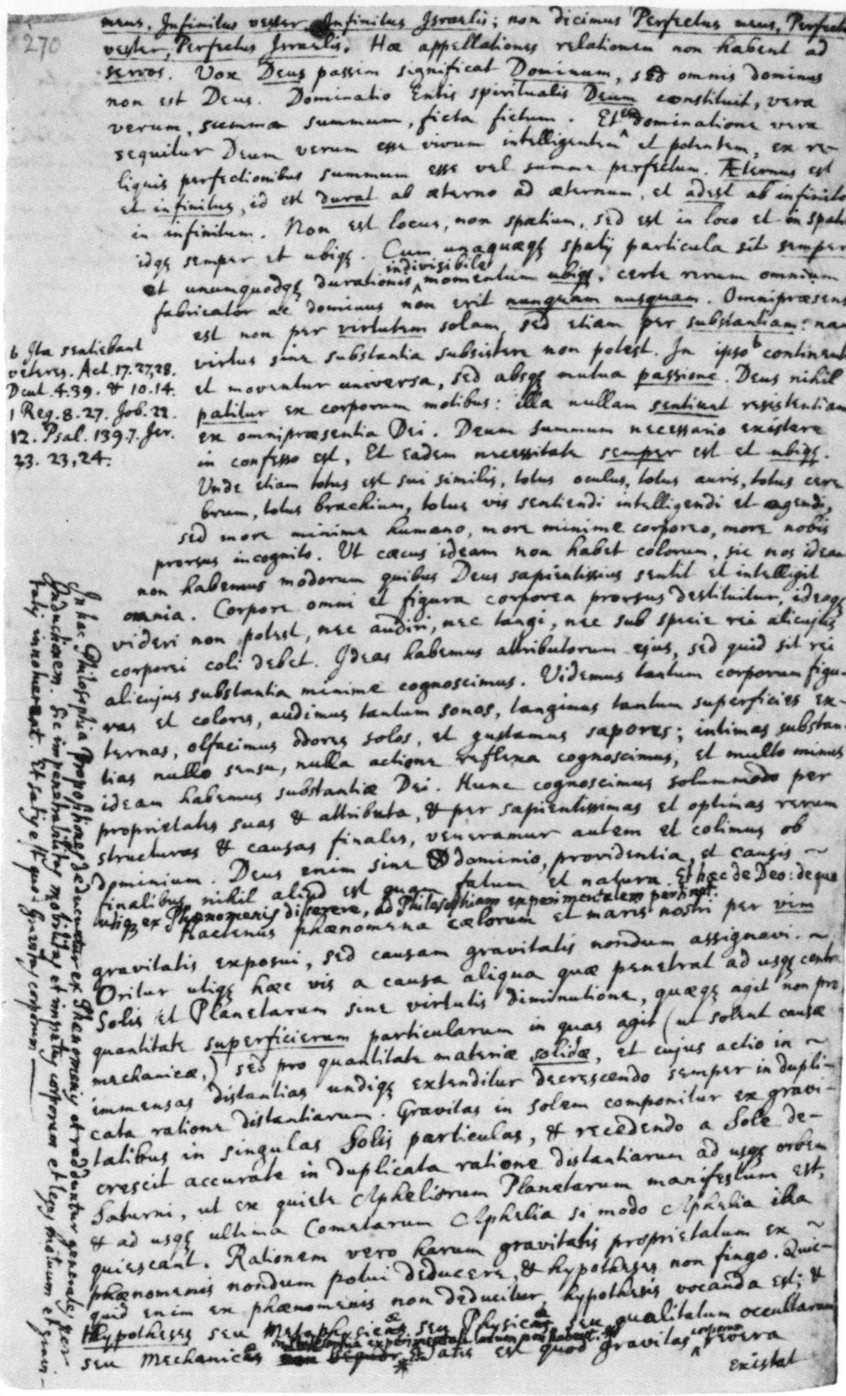

270

meus, Infinitus vester, Infinitus Israelis; non dicimus Perfectus meus, Perfectus vester, Perfectus Israelis. Hæ appellationes relationem non habent ad servos. Vox Deus passim significat Dominum, sed omnis dominus non est Deus. Dominatio Entis spiritualis Deum constituit, vera verum, summa summum, ficta fictum. Et ex dominatione vera sequitur Deum verum esse vivum intelligentem et potentem, ex reliquis perfectionibus summum esse vel summe perfectum. Æternus est et infinitus, id est durat ab æterno ad æternum, et adest ab infinito in infinitum. Non est locus, non spatium, sed est in loco et in spatio idque semper et ubique. Cum unaquæque spatii particula sit semper, et unumquodque durationis indivisibile momentum ubique, certe rerum omnium fabricator ac dominus non erit ~~nunquam nusquam~~. Omnipræsens est non per virtutem solam, sed etiam per substantiam: nam virtus sine substantia subsistere non potest. In ipso[b] continentur et moventur universa, sed absque mutua passione. Deus nihil patitur ex corporum motibus: illa nullam sentiunt resistentiam ex omnipræsentia Dei. Deum summum necessario existere in confesso est, Et eadem necessitate semper est et ubique. Unde etiam totus est sui similis, totus oculus, totus auris, totus cerebrum, totus brachium, totus vis sentiendi intelligendi et agendi, sed more minime humano, more minime corporeo, more nobis prorsus incognito. Ut cæcus ideam non habet colorum, sic nos ideam non habemus modorum quibus Deus sapientissimus sentit et intelligit omnia. Corpore omni et figura corporea prorsus destituitur, ideoque videri non potest, nec audiri, nec tangi, nec sub specie rei alicujus corporei coli debet. Ideas habemus attributorum ejus, sed quid sit rei alicujus substantia minime cognoscimus. Videmus tantum corporum figuras et colores, audimus tantum sonos, tangimus tantum superficies externas, olfacimus odores solos, et gustamus sapores; intimas substantias nullo sensu, nulla actione reflexa cognoscimus, et multo minus ideam habemus substantiæ Dei. Hunc cognoscimus solummodo per proprietates suas & attributa, & per sapientissimas et optimas rerum structuras & causas finales, veneramur autem et colimus ob dominium. Deus enim sine ⊗ dominio, providentia, et causis finalibus, nihil aliud est quam fatum et natura. Et hæc de Deo: de quo utique ex Phænomenis disserere, ad Philosophiam experimentalem pertinet.

b Ita sentiebant veteres. Act. 17. 27, 28. Deut. 4. 39. & 10. 14. 1 Reg. 8. 27. Job. 22. 12. Psal. 139. 7. Jer. 23. 23, 24.

Hactenus phænomena cœlorum et maris nostri per vim gravitatis exposui, sed causam gravitatis nondum assignavi. Oritur utique hæc vis a causa aliqua quæ penetrat ad usque centra Solis et Planetarum sine virtutis diminutione, quæque agit non pro quantitate superficierum particularum in quas agit (ut solent causæ mechanicæ,) sed pro quantitate materiæ solidæ, et cujus actio in immensas distantias undique extenditur decrescendo semper in duplicata ratione distantiarum. Gravitas in Solem componitur ex gravitatibus in singulas Solis particulas, & recedendo a Sole decrescit accurate in duplicata ratione distantiarum ad usque orbem Saturni, ut ex quiete Apheliorum Planetarum manifestum est, & ad usque ultima Cometarum Aphelia si modo Aphelia illa quiescant. Rationem vero harum gravitatis proprietatum ex phænomenis nondum potui deducere, & hypotheses non fingo. Quicquid enim ex phænomenis non deducitur, hypothesis vocanda est; & hypotheses seu Metaphysicæ seu Physicæ, seu qualitatum occultarum seu Mechanicæ ~~non sequor~~ in Philosophia experimentali locum non habent. * Satis est quod gravitas corpore revera existat

In hac Philosophia Propositiones deducuntur ex Phænomenis et redduntur generales per Inductionem. Sic impenetrabilitas, mobilitas et impetus corporum et leges motuum et gravitatis innotuerunt. Et satis est quod gravitas corporum —

Plate 14. A page of the Scholium Generale, sent by Newton to Cotes, in Newton's hand. (See Chapter IX, §§7 and 11.) Cotes has introduced corrections at the end of the top paragraph, at the bottom of the page, and in the left-hand margin, in accordance with Newton's instructions.

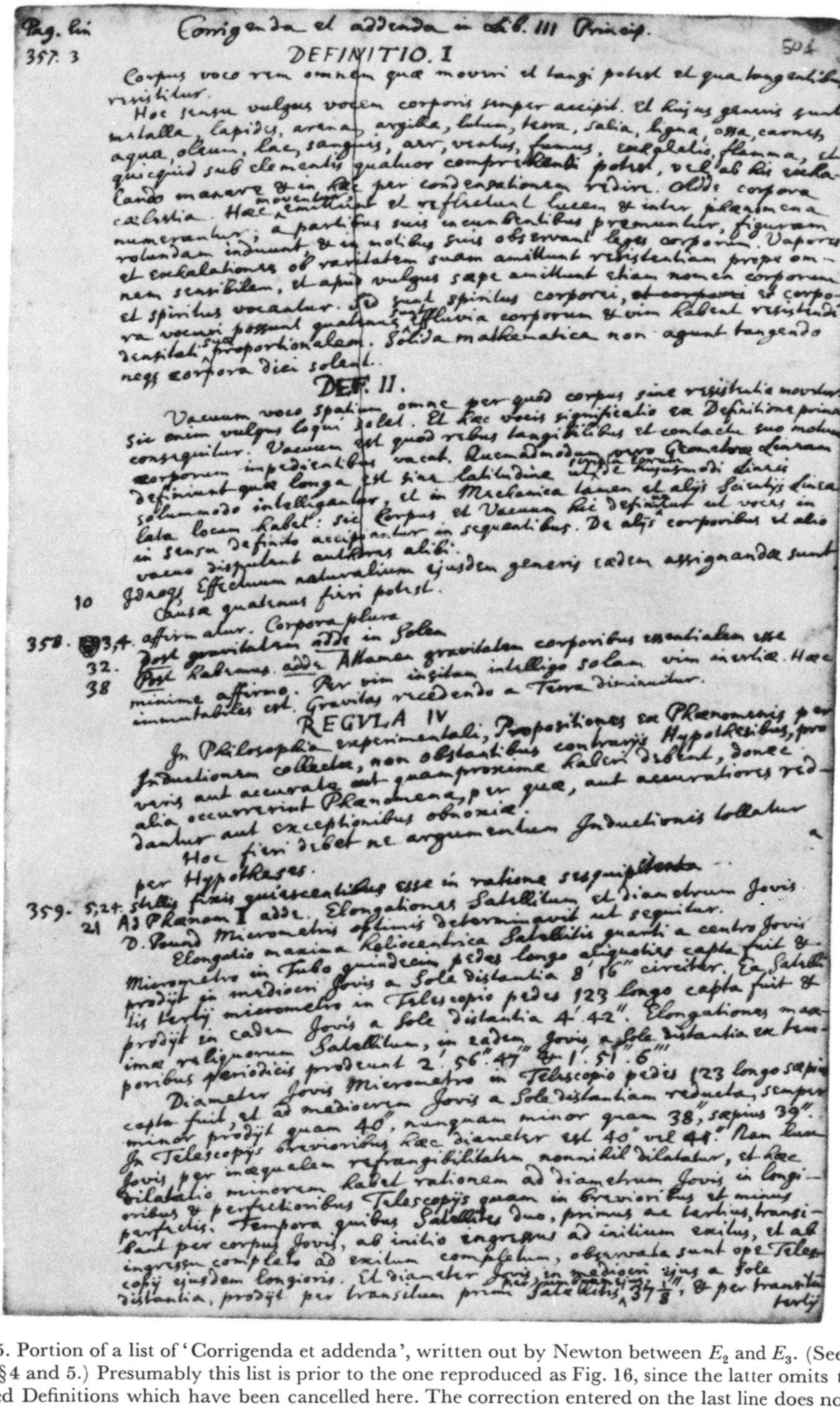

Pag. lin. Corrigenda et addenda in Lib. III Princip. 504

357. 3 DEFINITIO. I

Corpus voco rem omnem quæ moveri et tangi potest et qua tangentibus resistitur. Hoc sensu vulgus vocem corporis semper accipit. Et hujus generis sunt metalla, lapides, arena, argilla, lutum, terra, salia, ligna, ossa, carnes, aqua, oleum, lac, sanguis, aer, ventus, fumus, exhalatio, flamma, et quicquid sub elementis quatuor comprehendi potest, vel ab his exhalando manare & in hæc per condensationem redire. Olim corpora cælestia. Hæc moventur emittunt et refrahunt lucem & inter planetas numerantur, a partibus suis incumbentibus premuntur, figuram rotundam induunt & in motibus suis observant leges corporum. Vapores et exhalationes ob raritatem suam amittunt resistentiam prope omnem sensibilem, et apud vulgus sæpe amittunt etiam nomen corporum et spiritus vocantur. Sed sunt spiritus corporei, ~~et corporei~~ et corpora vocari possunt quatenus effluvia corporum & vim habeat resistendi densitati suæ proportionalem. Solida mathematica non agunt tangendo neque corpora dici solent.

DEF. II.

Vacuum voco spatium omne per quod corpus sine resistentia movetur. Sic enim vulgus loqui solet. Et hæc vocis significatio ex Definitione prima consequitur. Vacuum est quod rebus tangibilibus et contactu suo motum corporum impedientibus vacat. Quemadmodum vero Geometræ Lineam definiunt quæ longa est sine latitudine [illegible] ut de hujusmodi lineis solummodo intelligantur, et in Mechanica tamen et alijs Scientijs linea lata locum habet: sic Corpus et Vacuum hic definitur ut voces in sensu definito accipiantur in sequentibus. De alijs corporibus et alio vacuo disputent authores alibi.

Effectuum naturalium ejusdem generis eædem assignandæ sunt

10 Idemque Causæ quatenus fieri potest.

358. 3,4. affirmatur. Corpora plura

32. Post gravitatem adde in Solem. Attamen gravitatem corporibus essentialem esse minime affirmo.

38 Post Radiorum adde Per vim insitam intelligo solam vim inertiæ. Hæc immutabilis est. Gravitas recedendo a Terra diminuitur.

REGULA IV

In Philosophia experimentali, Propositiones ex Phænomenis per Inductionem collectæ, non obstantibus contrarijs Hypothesibus, pro veris aut accurate aut quamproxime haberi debent, donec alia occurrerint Phænomena, per quæ aut accuratiores reddantur aut exceptionibus obnoxiæ.

Hoc fieri debet ne argumentum Inductionis tollatur per Hypotheses.

359. 5,24. stellis fixis quiescentibus esse in ratione sesquiplicata.

21 Ad Phænom I adde. Elongationes Satellitum et diametrum Jovis D. Pound Micrometris optimis determinavit ut sequitur.

Elongatio maxima heliocentrica Satellitis quarti a centro Jovis Micrometro in Tubo quindecim pedes longo aliquoties capta fuit & prodijt in mediocri Jovis a Sole distantia 8′.16″ circiter. Ea Satellitis tertij micrometro in Telescopio pedes 123 longo capta fuit & prodijt in eadem Jovis a Sole distantia 4′.42″. Elongationes maximæ reliquorum Satellitum, in eadem Jovis a Sole distantia ex temporibus periodicis prodeunt 2′.56″.47‴ & 1′.51″.6‴.

Diameter Jovis Micrometro in Telescopio pedes 123 longo sæpius capta fuit, et ad mediocrem Jovis a Sole distantiam reducta, semper minor prodijt quam 40″, nunquam minor quam 38″, sæpius 39″. In Telescopijs brevioribus hæc diameter est 40″ vel 41″. Nam lux Jovis per inæqualem refrangibilitatem nonnihil dilatatur, et hæc dilatatio minorem habet rationem ad diametrum Jovis in longioribus & perfectioribus Telescopijs quam in brevioribus et minus perfectis. Tempora quibus Satellites duo, primus ac tertius, transibant per corpus Jovis, ab initio ingressus ad initium exitus, et ab ingressu completo ad exitum completum, observata sunt ope Telescopij ejusdem longioris. Et diameter Jovis in mediocri ejus a Sole distantia, prodijt per transitum primi Satellitis $37\frac{1}{8}$″, & per transitum tertij

Plate 15. Portion of a list of 'Corrigenda et addenda', written out by Newton between E_2 and E_3. (See Chapter X, §§4 and 5.) Presumably this list is prior to the one reproduced as Fig. 16, since the latter omits the two proposed Definitions which have been cancelled here. The correction entered on the last line does not occur on the second list or in E_3.

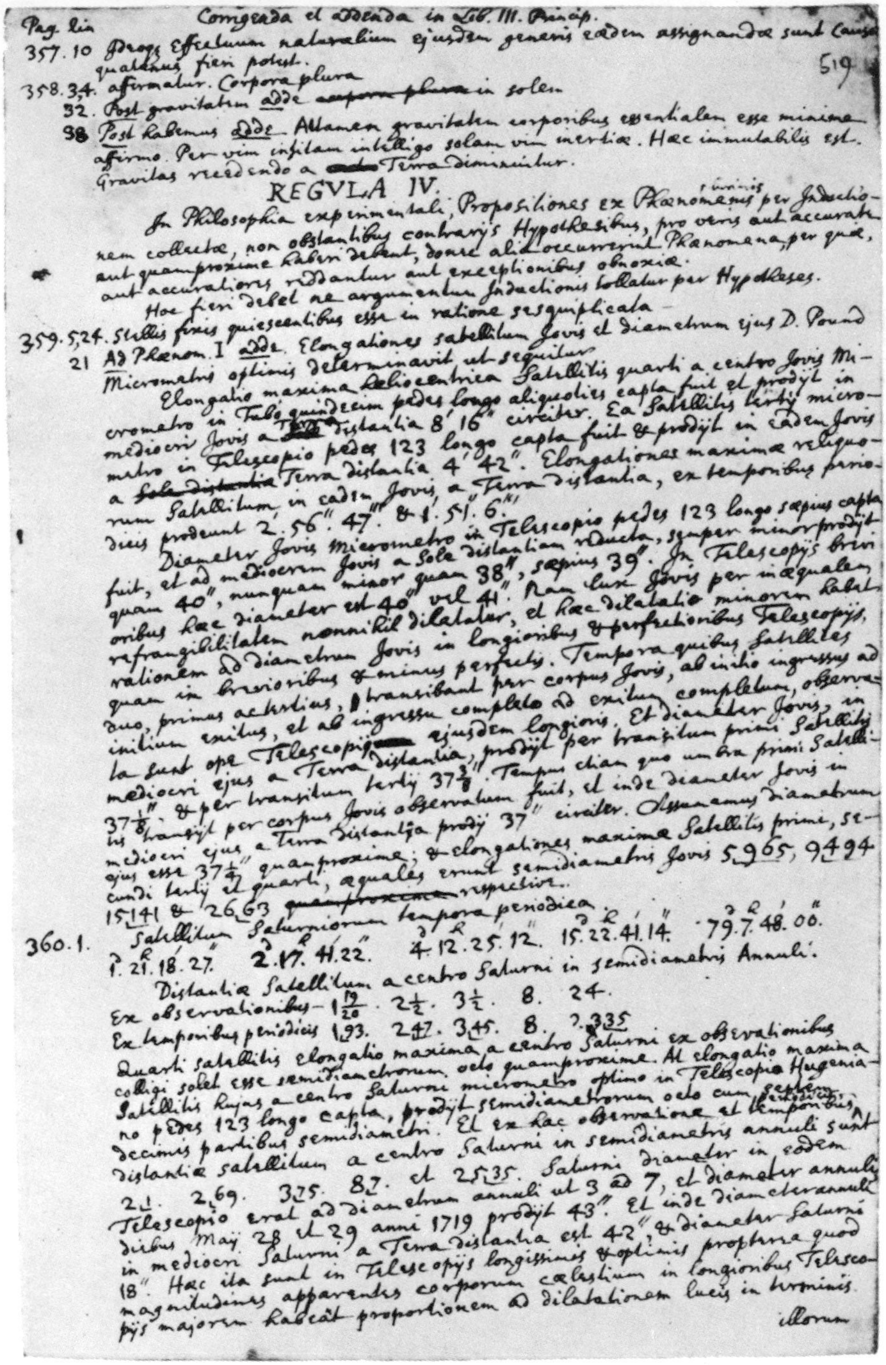
Pag. lin Corrigenda et addenda in Lib. III. Princip.
357. 10 Propterea Effectuum naturalium ejusdem generis eædem assignandæ sunt Causæ
quatenus fieri potest. 519
358. 3,4. affirmatur. Corpora plura
32. Post gravitatem adde ~~corpora plura~~ in solem
38 Post habemus adde Attamen gravitatem corporibus essentialem esse minime
affirmo. Per vim insitam intelligo solam vim inertiæ. Hæc immutabilis est.
Gravitas recedendo a ~~[illegible]~~ Terra diminuitur.
REGULA IV.
In Philosophia experimentali, Propositiones ex Phænomenis per Inductionem collectæ, non obstantibus contrarijs Hypothesibus, pro veris aut accurate
aut quamproxime haberi debent, donec alia occurrerint Phænomena, per quæ
aut accuratiores reddantur aut exceptionibus obnoxiæ.
Hoc fieri debet ne argumentum Inductionis tollatur per Hypotheses.
359. 5,24. Stellis fixis quiescentibus esse in ratione sesquiplicata
21 Ad Phænom. I adde. Elongationes satellitum Jovis et diametrum ejus D. Pound
Micrometris optimis determinavit ut sequitur.
Elongatio maxima ~~heliocentrica~~ Satellitis quarti a centro Jovis Micrometro in Tubo quindecim pedes longo aliquoties capta fuit et prodijt in
mediocri Jovis a ~~[illegible]~~ Terra distantia 8′ 16″ circiter. Ea Satellitis tertij micrometro in Telescopio pedes 123 longo capta fuit & prodijt in eadem Jovis
a ~~Sole distantia~~ Terra distantia 4′ 42″. Elongationes maximæ reliquorum Satellitum, in eadem Jovis a Terra distantia, ex temporibus periodicis prodeunt 2′. 56″. 47‴ & 1′. 51″. 6‴.
Diameter Jovis Micrometro in Telescopio pedes 123 longo sæpius capta
fuit, et ad mediocrem Jovis a Sole distantiam reducta, semper minor prodijt
quam 40″, nunquam minor quam 38″, sæpius 39″. In Telescopijs brevioribus hæc diameter est 40″ vel 41″. Nam lux Jovis per inæqualem
refrangibilitatem nonnihil dilatatur, et hæc dilatatio minorem habet
rationem ad diametrum Jovis in longioribus & perfectioribus Telescopijs,
quam in brevioribus & minus perfectis. Tempora quibus Satellites
duo, primus ac tertius, transibant per corpus Jovis, ab initio ingressus ad
initium exitus, et ab ingressu completo ad exitum completum, observata sunt ope Telescopij ~~[illegible]~~ ejusdem longioris. Et diameter Jovis, in
mediocri ejus a Terra distantia, prodijt per transitum primi Satellitis 37 1/8″. & per transitum tertij 37 3/8″. Tempus etiam quo umbra primi Satellitis transijt per corpus Jovis observatum fuit, et inde diameter Jovis in
mediocri ejus a Terra distantia prodijt 37″ circiter. Assumamus diametrum
ejus esse 37 1/4″ quamproxime, & elongationes maximæ Satellitis primi, secundi tertij et quarti, æquales erunt semidiametris Jovis 5,965, 9,494
15,141 & 26,63 ~~quamproxime~~ respective.
360. 1. Satellitum Saturniorum tempora periodica.
1ᵈ. 21ʰ. 18′. 27″. 2ᵈ. 17ʰ. 41′. 22″. 4ᵈ. 12ʰ. 25′. 12″. 15ᵈ. 22ʰ. 41′. 14″. 79ᵈ. 7ʰ. 48′. 00″.
Distantiæ Satellitum a centro Saturni in semidiametris Annuli.
Ex observationibus — 1 19/20. 2 1/2. 3 1/2. 8. 24.
Ex temporibus periodicis 1,93. 2,47. 3,45. 8. 23,35
Quarti satellitis elongatio maxima a centro Saturni ex observationibus
colligi solet esse semidiametrorum octo quamproxime. At elongatio maxima
Satellitis hujus a centro Saturni micrometro optimo in Telescopio Hugeniano
pedes 123 longo capta, prodijt semidiametrorum octo cum septem
decimis partibus semidiametri. Et ex hac observatione et temporibus periodicis
distantiæ satellitum a centro Saturni in semidiametris annuli sunt
2,1. 2,69. 3,75. 8,7. et 25,35. Saturni diameter in eodem
Telescopio erat ad diametrum annuli ut 3 ad 7, et diameter annuli
diebus Maij 28 et 29 anni 1719 prodijt 43″. Et inde diameter annuli
in mediocri Saturni a Terra distantia est 42″, & diameter Saturni
18″. Hæc ita sunt in Telescopijs longissimis & optimis propterea quod
magnitudines apparentes corporum cælestium in longioribus Telescopijs majorem habeant proportionem ad dilatationem lucis in terminis
illorum

Plate 16. Portion of another list of 'Corrigenda et addenda', written out by Newton between E_2 and E_3. (See Chapter X, §§4 and 5.) This list is presumably later than the one reproduced as Fig. 15. In the paragraph beginning 'Elongatio maxima...' (to be added to Phaenom. I, Book III) the corrected version here corresponds to the final text as printed.

far beyond Newton's contemporaries; they might possibly have arrived at Corol. VI to Prop. IV, but surely not at Prop. XI. We may thus understand that Newton was willing (in the Scholium to Prop. IV) to give Hooke credit (along with Halley and Wren) for the relatively minor work of analysing circular orbits,[6] but refused to allow anyone else on that account a major share in his own discovery. No predecessors are mentioned in the presentation of Prop. XI; and indeed there was no other contemporary who really had been able to deal with the dynamics of the true heavenly motions (in ellipses) and not mere simplified or idealized motions (as in circles). To see the gulf between Newton and Hooke, and even Leibniz, it is only necessary to study Props. XXXIX–LII of Book I of the *Principia.*

In other words, when Halley came to see Newton there was a general awareness that planetary orbits are elliptical, and there was good ground for suspecting that the force (if any) exerted by the Sun on the Earth must vary inversely as the square of the distance,[7] but no one had yet shown—so far as Halley knew—a necessary (that is, a logico-mathematical or causal) relation between the two. There was only a surmise or a reasonable guess, of the kind that has no place in science save to serve as an inspiration for research. So we may see why Newton's immediate reply to Halley's question was so impressive: the problem had been solved! Halley would not have been impressed merely by the fact that Newton said the orbits would be ellipses; he himself was already convinced of this, as were his colleagues at the Royal Society, Hooke and Wren. But it was a feat of extraordinary consequence to have 'calculated' this result, not merely because to have done so was a demonstration of more than ordinary skill in mathematics, but because this feat implied that Newton was so perfect a master of mathematical celestial physics that he had been able to solve an outstanding scientific problem of the age: what makes the planets go?

As to Halley's visit, regrettably, we have no contemporaneous details. Newton kept no diary in which the event is recorded, nor did he then write out a memorandum concerning his conversations with Halley. Our information derives from later documents, the earliest being Halley's correspondence with Newton two years later, in 1686. Other references to this event occur in manuscripts written by Newton *c.* 1713 (during the controversy with Leibniz over priority in the discovery of the calculus). Finally, there is a memorandum, dated November 1727, by the Anglo-Huguenot mathematician Abraham De Moivre, often quoted from Brewster's version of the transcript made by John Conduitt, who married Newton's niece in 1717 and who succeeded Newton at the Mint.[8]

The De Moivre–Conduitt account of this episode is the most complete one and may be presumed to have been based on conversations with Newton. But it must

[6] But, on the reference to Hooke, and to Wren and Halley, in this scholium, see Chapter V, §1.

[7] Perhaps only Hooke had a clear idea of a solar force. Professor R. S. Westfall points out (in a private communication to me) that the concept of such a force is radically different from the belief 'that some sort of mechanical arrangement produced an effect corresponding to an inverse-square force'.

[8] See Brewster (1855), vol. 1, pp. 296–7; also De Morgan (1885).

be read with some caution since Conduitt wrote out these notes after Newton's death in 1727, more than forty years after the original event. Furthermore, we do not know whether De Moivre was recasting notes he had made immediately or soon after hearing Newton tell the story, or was setting down his recollections entirely from memory. Furthermore, although I have presumed that De Moivre may have heard about this event directly from Newton, he does not actually say so, and in fact he may have got the details from Halley rather than Newton, or even from a third party. Indeed, he may only have been recounting the kind of anecdote that is always circulated among the intimates of any great man. Even apart from our uncertainty as to the actual source from which De Moivre learnt about the beginnings of the *Principia*, we must keep in mind that any such story is subject to all the unreliability of oral history: happy recollections perhaps, but not of themselves to be taken necessarily as gospel truth. But all such accounts are very precious to the historian—not for details but for the clues they provide to the significance of events. Here is what we are told:

> In 1684 Dr Halley came to visit him at Cambridge,   after they had been some time together, the D^r asked him what he thought the Curve would be that would be described by the Planets supposing the force of attraction towards the Sun to be reciprocal to the square of their distance from it. Sr Isaac replied immediately that it would be an Ellipsis,   the Doctor struck with joy & amazement asked him how he knew it,   Why saith he I have calculated it,   Whereupon Dr Halley asked him for his calculation without any further delay,   Sr Isaac looked among his papers but could not find it, but he promised him to renew it, & then to send it him...[9]

This presentation of Halley's visit has one curious aspect. Newton would have been able to prove that, if an object moves along an ellipse under the action of a central force directed to a focus, then that force must vary as the inverse square of the distance. But, as told, Halley's question and Newton's answer imply that Newton could prove the converse. Surely this is wrong, since the converse is not true. That is, under the action of an inverse-square force, an object will not necessarily move along an ellipse, but its path will be either a straight line directed toward the centre of force, or a curve that may be a circle, a parabola, an ellipse, or even an hyperbola. It is possible, of course, that the question attributed to Halley is given incorrectly, and that Halley actually asked Newton the converse: what the

[9] Conduitt's 'Memorandum relating to Sr Isaac Newton given me by Mr Demoivre in Novr 1727', part of the collection sold at auction in 1936 (see Chapter I, §9), is currently in private hands. A nineteenth-century transcript, made by H. R. Luard, may be found in the University Library, Cambridge (MS Add. 4007, ff. 706–7), and a portion was copied out by L. T. More before the auction and printed in his biography of Newton (1934), p. 299. Conduitt incorporated De Moivre's information into a manuscript account of Newton's life which has been the source of most presentations ever since it was used by Brewster in his two-volume *Memoirs* (1855), vol. 1, pp. 21–2, 24; vol. 2, pp. 296–8. Part of Conduitt's account ('Newton's life and work at Cambridge') has been printed by Whiteside from the original in the King's College Library (Keynes MS 130.4) as appendix 1 to his Introduction to [Newton] (1967–), *Mathematical papers*, vol. 1, pp. 15 ff.; he has given an account of the history of this document on p. 5, n. 9. In the version printed here, made available through the kindness of the owner, Mr Joseph Halle Schaffner, the text is based on the original manuscript. See, further, Supplement I to this Introduction, for a full text of that portion of De Moivre's Memorandum relating to Halley's visit. Wide spacing between some words has been introduced editorially to separate sentences.

force would be *if a planet moves along an ellipse*.[10] Alternatively, the De Moivre–Conduitt account may presuppose that only the ellipse would have been considered an admissible answer. The other conic sections might have been removed from consideration by Halley's specific reference to planets. In any event we may, I think, be sure that neither Halley nor Newton would have discussed such a major question in a loose and inexact fashion.

We may gain an insight into the state of Newton's thinking at this time from the tract *De Motu*,[11] written soon after Halley's visit. In it there is no theorem or problem devoted to Halley's question in its fullest generality, namely, given a centripetal inverse-square force, what will the trajectory be? But the converse, for a single conic section, appears as Prob. 3: 'A body revolves in an ellipse; the law of the centripetal force tending to a focus of the ellipse is sought.' Newton shows that 'the centripetal force is reciprocally as...the square of the distance'.[12] This proof is similar to the one that Newton worked out two years later in the *Principia*, where this same 'problem' is presented as Prop. XI (Book I). But in *De Motu*, 'Prob. 3' is immediately followed by a scholium, containing neither proof nor discussion, beginning: 'Therefore the major planets revolve in ellipses having a focus in the centre of the Sun...' ('Gyrant ergo Planetae majores in ellipsibus habentibus umbilicum in centro solis...'). The logic of proof and sequence is hardly to be taken as unexceptionable. But Newton moves on at once, via a Theor. 4, to Prob. 4:

> Suppose that the centripetal force is reciprocally proportional to the square of the distance from the centre, and the quantity of that force is known; there is required the ellipse which a body describes in moving from a given place, with a given velocity, along a given straight line.[13]

Newton evidently has merely excluded any other possibility save the ellipse (or circle). But he is aware, as he states clearly in a paragraph following the conclusion of the proof, that what he has been describing 'happens thus when the figure is an ellipse', which is not always the case: 'For it may be that the body will move in a parabola or hyperbola.' In his own mind, then, he had imposed the limitation to *planetary* orbits, say to recurrent orbits, or curves more or less closed,[14] thus

[10] I am not at all certain that Conduitt would have fully grasped the difference between the two, but surely an able mathematician like De Moivre would have been quite aware of the difference. In at least one of Newton's autobiographical statements written during the second decade of the eighteenth century, he said that Halley asked him if he 'knew what figure the Planets described in their Orbit about the Sun' (see Supplement I, §2), which sounds as if Halley had asked Newton specifically what the orbits would be *around the Sun*; that is, he assumed more or less closed orbits, thus excluding parabolas, hyperbolas, or even straight lines. But see note 14 below.

[11] See, below, §§2, 3.

[12] Hall and Hall (1962), *Unpublished papers*, p. 277.

[13] *Ibid*. p. 281. Theorem 4 of *De Motu* states that if the centripetal force is inversely proportional 'to the square of the distance from the centre; the squares of the periodic times about the ellipses are as the cubes of the transverse axes'.

[14] In *De Motu*, Newton writes: 'By the displacement of the Sun from the centre of gravity it may happen that the centripetal force does not always tend to that immobile centre, and thence that the planets neither revolve exactly in ellipses nor revolve twice in the same orbit. Each time a planet revolves it traces a fresh orbit, as happens also with the motion of the Moon, and each orbit is dependent upon the combined motions of all the planets, not to mention their actions upon each other. Unless I am much mistaken, it would exceed the

excluding as solutions to the original problem contained in Halley's question all open curves such as straight lines, parabolas, or hyperbolas. But in the tract *De Motu* any such assumptions of exclusion are at most implicit.

Our inquiry into what Halley actually said to Newton, and what Newton replied, is not much advanced by Halley's letters to Newton in 1686, on the eve of printing the *Principia*, although these letters do confirm the fact of the visit and also allow us to fill in the background of events responsible for Halley's decision to make the trip to Cambridge in order to see Newton. Halley, in these letters, was writing to specify the chronology of the recognition that the observed motions of the planets may be due to a solar force varying as the inverse square of the distance. The question at issue was how much credit, if indeed any at all, should be given to Hooke, who expected some 'mention...in a preface', if there should be one. Hooke had made great claims, but—according to Halley—he had not produced his alleged demonstration of the law of planetary force, although challenged to do so by Sir Christopher Wren: architect, mathematician, physicist, physiologist, and patron of the arts and sciences. After Hooke's manifest failure, Halley set out for Cambridge in the hope that Newton might be able to solve the problem, little knowing that he had already done so but had kept silent about his accomplishment.

One of Halley's letters (29 June 1686, first printed by Rigaud from the guard-book of the Royal Society[15]) relates how he himself had come to conclude that the centripetal force must vary as the inverse square of the distance:

> And this I know to be true, that in January 83/4, I, having from the consideration of the sesquialter [3/2] proportion of Kepler, concluded that the centripetall force decreased in the proportion of the squares of the distances reciprocally, came one Wednesday to town, where I met with Sr Christ. Wrenn and Mr Hook, and falling in discourse about it, Mr Hook affirmed that upon that principle all the laws of the celestiall motions were to be demonstrated, and that he himself had done it.[16]

I take it that Halley meant that he had applied Kepler's 'proportion' only to circular orbits, and that, using the result published by Huygens in 1673 concerning centrifugal force, he had concluded that the force in uniform circular motion must vary inversely as the square of the distance. Supposedly, 'all the laws of the celestiall motions', in Hooke's affirmation, must imply at least Kepler's law of elliptical orbits, and perhaps also the harmonic law (or 'sesquialter proportion'), and possibly even the law of areas.[17]

force of human wit to consider so many causes of motion at the same time, and to define the motions by exact laws which would allow of an easy calculation.' Quoted from Hall and Hall (1962), *Unpublished papers*, p. 281.

[15] Rigaud (1838), *Essay*, appendix, p. 36.

[16] [Newton] (1959–), *Correspondence*, vol. 2, p. 442.

[17] In the seventeenth century many writers on astronomy tended to omit the law of areas from their accounts of planetary motion (see Russell, 1964) and to substitute therefor an approximation making use of some kind of mean motion with a correction factor. The chief devices of this kind are associated with the names of Ismael Bullialdus (or Boulliau), Seth Ward, and Nicolaus Mercator (or Kaufmann); see Whiteside (1964*a*). There is no documentary evidence that Newton was consciously aware of Kepler's law of areas as an astronomical principle prior to 1676, when he might well have encountered it in Mercator's *Institutionum astronomicarum libri duo*. By the end of his correspondence on motion with Hooke in 1679–80, or possibly soon thereafter, but at least by 1684, he used the law of areas in finding the forces producing elliptical planetary motions. He showed (Prop. XI, Book I, Sec. III) that elliptical motion results from the action of a centrally directed inverse-square force acting on a body with an initial component of inertial motion. The significance of area conservation, as implying a centre of

In this same letter (29 June 1686), in a passage immediately preceding the above quotation, Halley reported to Newton that he had waited upon Wren, according to Newton's desire,

to inquire of him, if he had the first notion of the reciprocall duplicate proportion from Mr Hook, his answer was, that he himself very many years since, had had his thoughts upon the making out the planets motions by a composition of a Descent towards the sun, & an imprest motion; but that at length he gave over, not finding the means of doing it. Since which time Mr Hook had frequently told him that he had done it, and attempted to make it out to him, but that he never satisfied him, that his demonstrations were cogent.[18]

This statement is extraordinarily interesting, because it shows how profound a change had occurred in physical thinking in only a very few decades. Kepler had believed that the inertia of bodies tended to bring them to rest, and thus that a force was required to produce motion and to keep up motion.[19] Galileo, however, held that motion of certain kinds could continue without a force being constantly exerted, although such motion was not necessarily for Galileo the uniform rectilinear motion that it was for Newton.[20] But Descartes, and after him Gassendi, had written that such continued motion without an external force (which, following Newton, we call inertial motion) can only be linear. By the 1680s physicists had begun to accept the concept that uniform rectilinear motion is a 'state' (to usc the word adopted from Descartes by Newton[21]) and no more needs a force in order to be maintained than does that other 'state', rest. Hence it was conceived that in circular (or any curvilinear) motion there must be a combination of a linear (tangential) or inertial component and a central or accelerated component: an 'impressed force' and an accelerating force. This is basically Huygens's analysis of circular motion. But Huygens wrote of centrifugal force ('vis centrifuga'). The expression 'centripetal force', used by Halley in his letter to Newton, was brand new; it was, in fact, invented by Newton and named by him 'vis centripeta' in honour of Huygens,[22] and Halley had undoubtedly picked it up either from Newton's tract *De Motu* or from the *Principia*, which he was then editing for the press.

Wren's resolution of the forces producing planetary motion, as described by

force, was not known before Newton's analysis, and was one of the novelties of the *Principia*, being the subject of Props I and II, Book I (Sec. II).

18 [Newton] (1959–), *Correspondence*, vol. 2, pp. 441–2.

19 See [Kepler], *Dream* [Rosen] (1967), pp. 222 ff.; Aiton (1965); this and other aspects of the Laws of Motion are dealt with in the Wiles Lectures, Cohen (1971), *Scientific ideas*, chap. 1, 'Laws of motion and principles of philosophy'.

20 Galileo seems to have put forward the view, rather explicitly expressed in his book on sunspots, his *Dialogue on the two great world systems*, and implied in his *Discourses and demonstrations concerning two new sciences*, that such 'inertial' motion could be and was uniform circular motion. Only near or at the surface of the Earth did he conceive of a true linear inertial motion, and even that was apt to be a small arc of a very large circle (save for the terminal speed of a body falling in a resisting medium). See Drake (1964); Koyré (1939), *Etudes*, part 3; also [Galileo], *Discoveries and opinions* [Drake] (1957), p. 113.

21 See Koyré (1965), *Newtonian studies*, pp. 66 ff., amplified in Cohen (1964), '"Quantum in se est"'.

22 In a manuscript written at about the time of E_2, printed in Supplement I, Newton said, in part, 'Mr Hygens gave the name of vis centrifuga to the force by wch revoling [*sic*] bodies recede from the centre of their motion⟨.⟩ Mr Newton in honour of that author retained the name & called the contrary force vis centripeta'; U.L.C. MS Add. 3968.28, f. 415v; see Koyré and Cohen (1962), 'Newton and the Leibniz–Clark correspondence', p. 122.

Halley, shows him to have grasped the principles of the new inertial physics. It was upon making just such an analysis that Wren had, apparently independently, 'had his thoughts' about 'the reciprocall duplicate proportion'.[23]

As to Hooke (to complete the story from Halley's letter to Newton), we have seen that he affirmed that he had demonstrated 'upon that principle [of an inverse-square force] all the laws of the celestiall motions'. Then, according to Halley,

I declared the ill success of my attempts; and Sr Christopher to encourage the Inquiry, s^d^ that he would give Mr Hook or me 2 months time to bring him a convincing demonstration therof, and besides the honour, he of us that did it, should have from him a present of a book of 40s. Mr Hook then s^d^ that he had it, but he would conceale it for some time that others triing and failing, might know how to value it, when he should make it publick; however I remember Sr Christ⟨opher⟩ was little satisfied that he could do it, and tho Mr Hook then promised to show it him, I do not yet find that in that particular he has been as good as his word. The August following when I did my self the honour to viset you, I then lear̄t [=learnt] the good news that you had brought this demonstration to perfection, and you were pleased, to promise me a copy therof, which the November following I received with a great deal of satisfaction from Mr Paget; and therupon took another Journy [*possibly* Jorney] down to Cambridg, on purpose to conferr with you about it, since which time it has been enterd upon the Register [*originally* Journall] books of the Society.[24]

This 'demonstration', which may still be found in the Register Books, is the first public—though not published—version of Newton's system of dynamics and celestial mechanics. Its contents are discussed below (§§2, 3, 4). Once Newton had begun to write and to communicate what he had been discovering, and was encouraged by Halley and the Royal Society to go on with his research into this subject, there was no stopping him until he had finished the *Principia*.

2. THE DATE OF HALLEY'S VISIT AND THE TRACT 'DE MOTU'

The endeavour to trace the actual sequence of events following Halley's visit is rendered difficult by the lack of contemporaneous evidence. Halley's letter of 29 June 1686 and the De Moivre–Conduitt memorandum are amply supported by many autobiographical documents (such as those printed in Supplement I to this Introduction), at least so far as the nature and importance of Halley's visit are concerned. But in some of those manuscript documents Newton puts the time of Halley's visit in the spring or in May 1684 (and he has sometimes even dated it in 1683!). I believe we may take Halley's date as correct, setting the month at August, since he was writing within two years of the event; Newton's statements in the autobiographical manuscripts were written some thirty years later, apparently without his being able to refer to any contemporaneous dated document.[1]

[23] Halley does not mention Wren's having referred specifically to either Kepler or Huygens; Halley, in relating how he himself found the inverse-square law, mentioned Kepler but not Huygens.

[24] Newton (1959–), *Correspondence*, vol. 2, p. 442.

[1] The argument for the earlier date (Herivel (1965), *Background*, p. 96) rests chiefly on the conclusion therefrom that Newton would have had more time after Halley's visit to complete the text of his lectures, which began in October 1684. But such an argument entails two assumptions: (1) that the manuscript deposited by

What occurred after the visit is told by Newton in a letter to Halley (14 July 1686), following the latter's narration of the sequence of events before his trip to Cambridge to ask Newton about the planetary force.[2] Newton recalled, first of all, how he had studied this problem five years earlier and how he 'threw the [first] calculation by being upon other studies & so it rested for about five yeares'. Then Halley came to see him, and 'upon your request I sought for y^t paper, & not finding it did it again and reduced it into ye Propositions shewed you by Mr Paget'.[3] These same events are described somewhat more fully in the De Moivre–Conduitt memorandum as follows:

> Sr Isaac in order to make good his promise [to send the calculation to Halley] fell to work again but he could not come to that conclusion which he thought he had before examined with care, however he attempted a new way which though longer than the first brought him again to his former conclusion. Then he examined carefully what might be the reason why the calculation he had undertaken before did not prove right and he found that, having drawn an Ellipsis cursorily with his own hand he had drawn the axes of the curve instead of drawing two diameters somewhat inclined to one another whereby he might have fixed his examination to any two conjugate diameters, which was requisite he should do. That being perceived he made both his calculations agree together.[4]

As the Italians say, *Se non è vero, è ben trovato*!

Whatever document it was that Newton finally sent to Halley by the agency of Paget—in November 1684, according to Halley's letter to Newton, quoted above at the end of §1—the reading of it had the effect that Halley thereupon (or soon after) set out once again for Cambridge, with the express purpose of conferring with Newton about it.

When Halley returned to London from this second visit to Newton, he made a public announcement of a work Newton had then shown him. The report of a meeting of the Royal Society of 10 December 1684 reads as follows:

> Mr. Halley...had lately seen Mr. Newton at Cambridge, who had shewed him a curious treatise, *De Motu*; which, upon Mr. Halley's desire, was, he said, promised to be sent to the Society to be entered upon their Register. Mr. Halley was desired to put Mr. Newton in mind of his promise for the securing his invention to himself till such time as he could be at leisure to publish it.[5]

Leaving aside for the moment the question of what the 'curious treatise, *De Motu*' might have been, let us move on with the documentary chronology. On 23 February 1684/5, Newton wrote to Francis Aston, Secretary of the Royal Society,

---

Newton (*LL*) contains the actual text of lectures delivered at that time, and (2) that Newton had not done any work in dynamics just before Halley's visit. The first of these assumptions is discussed later on in this section and also in Chapter IV, §2, and in Supplement III; I believe that we do not have to take it very seriously. The second is discussed below in §6. I see no reason not to accept Halley's date of August, as has been accepted traditionally. See, further, note 15 to §6 below.

Newton had access to Halley's letter dated 29 June 1686, but this was written some two years after Halley's visit.

² See §1. ³ Newton (1959–), *Correspondence*, vol. 2, pp. 444–5.

⁴ See §1, note 9.

⁵ Quoted from Birch (1756–7), *History*, vol. 4, p. 347; also in Rigaud (1838), *Essay*, p. 15; Ball (1893), *Essay*, p. 30.

thanking him 'for entring in your Register my notions about motion'. He further said,

> I designed them for you before now but the examining severall things has taken a greater part of my time than I expected, and a great deale of it to no purpose. And now I am to goe into Lincolnshire for a Month or six weeks. Afterwards I intend to finish it as soon as I can conveniently.[6]

Two days later, on 25 February 1684/5, the letter was communicated to the Royal Society at a weekly meeting.[7]

The next documents are from the spring of 1686. At a meeting of the Royal Society on 21 April 1686, Halley referred to an 'incomparable *Treatise on Motion* almost ready for the press' by our 'worthy country-man Mr. Isaac Newton',[8] and at the meeting of 28 April following, Dr Nathaniel Vincent, Senior Fellow of Clare Hall, 'presented to the Society a manuscript treatise intitled *Philosophiae Naturalis principia mathematica*, and dedicated to the Society by Mr. Isaac Newton'.[9] This was the manuscript of Book I of the *Principia*, bearing the full title of the whole work at the top of the first page.

The tract sent by Newton to the Royal Society some time between 10 December 1684 (after Halley made his second visit) and 23 February 1684/5 (when Newton thanked Aston for having entered his 'notions about motion' in the Register) was first identified and published by S. P. Rigaud,[10] who gave it the title

ISAACI NEWTONI

PROPOSITIONES DE MOTU

although this version has no heading whatsoever. Thus Rigaud made it appear that this was unquestionably that 'curious treatise, *De Motu*' shown to Halley by Newton.[11]

I have seen four copies or versions of this tract: the above-mentioned one in the Register of the Royal Society and three others in the Portsmouth Collection.[12] A fifth exists in the private collection of the Earl of Macclesfield. The text has been printed at least four times: first by Rigaud and then by Rouse Ball in his *Essay*;[13] it has since been re-edited and translated by A. R. and M. B. Hall in their *Unpublished scientific papers of Isaac Newton*, and edited and translated anew by J. Herivel in his volume called *The background to Newton's Principia*.[14] Unfortunately,

[6] Newton (1959–), *Correspondence*, vol. 2, p. 415.

[7] Birch (1756–7), *History*, vol. 4, p. 370.

[8] *Ibid.* p. 479. Cf. Halley (1686), 'Discourse concerning gravity', p. 6.

[9] Birch (1756–7), *History*, vol. 4, p. 479.

[10] Rigaud (1838), *Essay*, Appendix No. I. He found it in the Register of the Royal Society, vol. 6, pp. 218 ff., where it is still to be seen.

[11] In a different hand from the body of the text, there is written a date, 'Dec. 10, 1684'. Rigaud held this to be 'most probably a mere reference (as commonly is made in these books) to the place in the Journals where particulars respecting their several contents may be found'. And it was on 10 December that Halley reported to the Society (*see* Journal Book) on his visit.

[12] U.L.C. MS Add. 3965, ff. 55–62 bis, 63–70, 40–54.

[13] Ball (1893), *Essay*, pp. 35 ff.

[14] Hall and Hall (1962), *Unpublished papers*, pp. 239–92; Herivel (1965), *Background*, pp. 257–303.

we do not know the exact date of composition of any of these versions, although a sequence may be established.[15] But the texts, despite interesting variations, are all 'substantially the same'.[16] There is no question of the fact that this tract is to be held an earnest of the great work to come.[17]

We are at once faced with a series of puzzles which the available documents do not permit us to solve save by conjecture. The first of these arises from three references to a work or works by Newton following Halley's initial visit: are any two, or perhaps all three, of these the same? The first one is the 'demonstration' which Newton 'had brought...to perfection'—to quote from Halley's letter to Newton of 29 June 1686—and promised 'a copy thereof, which the November following I received...from Mr Paget'. The second is the 'curious treatise, *De Motu*' which, according to Halley's report to the Royal Society on 10 December 1684, he had seen when he went to Cambridge to visit Newton for the second time. The third is the untitled tract actually entered in the Register Book of the Royal Society. I believe that all three references are to be taken as to this one work, although not necessarily to one and the same state or version.[18] For in his letter Halley says explicitly concerning the copy 'received...from Mr Paget': 'since which time it has been enterd upon the Register books of the Society'. And as to the 'treatise, *De Motu*' which Halley saw, the Royal Society report states explicitly that Newton 'upon Mr. Halley's desire...promised' to send it 'to the Society to be entered upon their Register'.

As if to confirm the foregoing identification conclusively, at least two of the copies of this tract among Newton's papers[19] bear titles beginning with the words *De Motu*; but the third—a copy in Halley's hand—has no title whatsoever, in which characteristic it resembles the version in the Royal Society.[20] Hence at first glance there would seem to be no problem, save to ascertain which state or version of the five copies of this tract may correspond to the one sent up to London in November with Paget and to decide which one of them was shown to Halley on the occasion of his subsequent visit to Newton.

It must be kept in mind, however, that there are three other pre-*Principia* manuscripts called *De Motu*.... First of all there is a fragment, almost certainly

[15] This has been attempted in Hall and Hall (1962), *Unpublished papers*, pp. 242 ff., and in Herivel (1965), *Background*, pp. 96 ff., 292–4.

[16] Herivel (1965), *Background*, p. 257.

[17] What is perhaps most fascinating and bewildering about those versions of the tract *De Motu* is that Newton apparently would continue to make revisions on an early version even after he had had it copied in a later version. In a moment, we shall see him doing the very same thing in the so-called Lucasian Lectures, *De motu corporum*.

[18] These differences are discussed by Rouse Ball, the Halls, Herivel, and, in part, below in §§3 and 4. With regard to the text proper, apart from the introductory Definitions, Hypotheses or Laws, and Lemmas, these differences are not major.

[19] Portsmouth Collection, U.L.C. MS Add. 3965, ff. 40–54, 55–62 bis.

[20] The versions of this tract in the Portsmouth Collection (U.L.C. MS Add. 3965) bear titles as follows: [MS *B*] ff. 55–62 bis, *De motu corporum in gyrum* [in Newton's hand]; [MS *C*] ff. 63–70 [no title; in Halley's hand]; [MS *D*] ff. 40–54, *De motu sphaericorum corporum in fluidis* [in Humphrey Newton's hand]. The letters *B, C, D* are those assigned by the Halls. The version in the Royal Society's Register Book bears no title.

written later than the previously discussed tract, which is entitled *De motu corporum in mediis regulariter cedentibus* (discussed below, §4). Then there is a collection of definitions, apparently later than either of these writings, but definitely prior to the *Principia*, called *De motu corporum: Definitiones* (discussed in Chapter IV, § 3). Finally, there is a manuscript corresponding very closely to the beginning of the *Principia* (that is, the Definitions, Laws of Motion, and a great part of Book I), called *De motu corporum, liber primus* (discussed in Chapter IV, §2, and there denoted by *LL*), no doubt of later origin than any of the other manuscripts *De Motu* mentioned thus far. It is this manuscript that creates the greatest problem, since it was deposited by Newton in the University Library, allegedly as the text of the lectures he had read as Lucasian Professor beginning in 'Octob. 1684'—thus even before Halley's second visit, and even before Paget brought the demonstrations in November! Why would Paget have brought to London an early and imperfect work *De Motu* when Newton had already completed what is to all intents and purposes the text of the *Principia* containing the propositions on elliptic motion (for example, Prop. XI, Book I) and much else besides? And would not Halley have been shown the *magnum opus* on his second visit, rather than the prior tract, of which he had already seen a version that had been brought to him by Paget? Finally, as the last facet of the puzzle, why would Newton have sent to the Royal Society (some time between December 1684 and February 1684/5), to be registered, a version of a tract that had been superseded by a more mature presentation that had already been completed (at least through Sec. III, Book I) some time earlier?[21]

All of these questions are based on the assumption that the manuscript *De motu corporum, liber primus* is in fact the text of the lectures given by Newton as Lucasian Professor at the times indicated; that is, beginning in October 1684. But if this document, alleged by Newton to have been the text of his professorial lectures, was written some time (at least several months) later than October 1684, then not all of the foregoing questions need be raised. I shall indicate below (Chapter IV, §2, and Supplements III and IV) the reasons why I believe we need not take very seriously the dates entered by Newton on the pages of his alleged 'lectures'. He may very well, of course, have given lectures on these dates, but he did not necessarily read the text he later deposited as of those dates.

[21] It is possible, of course, that Newton sent in to the Royal Society an early state of his work, rather than the then-current version, simply to be able to assert his priority if the occasion were ever to arise: 'to secure his invention'. The text in the Register of the Royal Society does not, however, contain a date of composition, nor do the three versions now in the Portsmouth Collection. But, of course, if need be, Newton could have called upon at least two witnesses, Halley and Paget, who had seen a version of this document months before Newton sent it to the Royal Society.

3. THE STRUCTURE OF 'DE MOTU' IN RELATION TO THE 'PRINCIPIA'

The history of the *Principia*, considered apart from the development of Newton's dynamics and celestial mechanics,[1] may be traced from the several versions of the tract *De Motu*, through the two fragments *De motu corporum in mediis regulariter cedentibus* and *De motu corporum: Definitiones*,[2] to the manuscript deposited by Newton as the text of his Lucasian Lectures (*LL*).[3] The latter, closely resembling the manuscript (*M*) used for printing the *Principia*, will be discussed in Chapter IV, together with *De motu corporum: definitiones*, and certain other manuscript materials, all of which are closely related in textual content, form, and structure to the *Principia* as we know it from the printed editions (E_1, E_2, E_3).

But the tract *De Motu* and the fragment *De motu corporum in mediis regulariter cedentibus* are of a different sort, representing earlier conceptions of the definitions and axioms on which the Newtonian system of dynamics was to be erected. The tract *De Motu*, furthermore, is so extremely short that one could never guess from it what subjects the *Principia* was eventually to contain. Since both *De Motu* and *De motu corporum in mediis regulariter cedentibus* are readily available, I shall not devote much attention to them, except in so far as they reveal certain discrete stages of importance in the eventual unfolding of the concept of the *Principia*. A contrast in the systems of axiomatization will reveal at once the differences between these two works and show how far both are removed from the *Principia*. The latter is founded on three 'Axiomata sive Leges Motus', whereas the most complete version of *De Motu* has five 'Hypotheses'[4] later designated 'Leges' and *De motu corporum in mediis regulariter cedentibus* has six 'Leges Motus'.[5]

In *De Motu*, following certain definitions, laws, and lemmas, there are eleven propositions, presented in the form of four theorems and seven problems, together

[1] That is, I make a distinction between (1) the actual documents that were later (in whole or in part) either transferred directly to the text of the *Principia* or modified or enlarged and then so transferred, and (2) various texts that show the progress of Newton's thought—including letters, student notebooks, and other miscellaneous manuscripts—and which may be conceptually, though not textually, related to the *Principia*. Only the former are treated in this Introduction; the latter are reserved for the Commentary volume.

[2] Since these works are now readily available (see §2), I shall not here deal with them in much detail but shall only discuss them briefly in relation to the text of *M* (the manuscript of the *Principia* sent to the printer) with which the present assignment properly begins, and in relation to *LL* (the Lucasian Lectures) which immediately preceded *M*.

[3] The major documents for studying Newton's early work in dynamics and celestial mechanics have been made available in the aforementioned books by the Halls and Herivel; the classic account by Rouse Ball (1893), *Essay*, may be supplemented by the recent studies of Koyré (1965), *Newtonian studies*, the Halls, Herivel, Whiteside (1964*a*), 'Newton on planetary motion', Lohne (1960), 'Hooke vs Newton', myself, and others; see the Bibliography. I shall not attempt to list the many other authors whose recent studies are either directly or peripherally related to Newton's dynamics. Any further attempt to unravel the background and development of Newton's ideas prior to *De Motu* would take us too far afield and must be reserved therefore for the Commentary volume (in progress), but some further information may be found in autobiographical documents in Supplement I to this Introduction.

[4] These are discussed in §4.

[5] This sequence—from 'Hypotheses' to 'Leges' in *De Motu*, to 'Leges Motus' in *De motu corporum in mediis regulariter cedentibus*, and finally to 'Axiomata sive Leges Motus' in *LL*, *M*, E_1, E_2, and E_3—is as much of interest as the change in the contents of these 'Laws'.

with corollaries and scholia. Theorem 1, an early form of Prop. I (Theor. I) of Book I of the *Principia*, states that, when bodies revolve about a centre, they will, by radii drawn to the centre, describe areas proportional to the times.[6] Theorem 2, similar to Prop. IV (Theor. IV) of Book I of the *Principia*, states that for bodies revolving uniformly in circles the centripetal forces are 'as the squares of the arcs described in a given time divided by the radii of the circles'.[7] Five corollaries follow, of which the fifth states that if the 'squares of the periodic times are as the cubes of the radii, the centripetal forces are reciprocally as the squares of the radii'.[8] It is observed in a scholium that:

> The case of the fifth corollary obtains in celestial bodies. The squares of the periodic times are as the cubes of the distances from a common centre about which they revolve. That this obtains in the major planets revolving about the Sun and in the minor planets about Jupiter is now accepted by Astronomers.[9]

Theorem 3 is Corol. 1 to Prop. VI (Theor. V) of Book I of the *Principia*. A corollary to this Theorem 3 states that this Theorem enables one to compute the centripetal force if any figure is given and 'a point on it to which the centripetal force is directed'.[10] Then, in Probs. 1, 2, and 3, examples are given. In Prob. 1, a body revolves along the circumference of a circle, and the centre of force S is located on that circumference; the force S proves to be reciprocally proportional to SP^5, where P is the position of the body in the circle. In Prob. 2, the body revolves in an 'ellipse', and the centre of force is at the centre of the ellipse; the force is directly as the distance PC from the moving body to the centre. In Prob. 3 the body moves in an ellipse, and the centre of force is at a focus S of an ellipse; the centripetal force is proved to be reciprocally as the square of the distance. There follows a scholium:

> Therefore the major planets revolve in ellipses having a focus in the centre of the Sun; and the radii vectores to the Sun describe areas proportional to the times, exactly as Kepler supposed.[11]

This leads to Theor. 4. Newton supposes 'that the centripetal force is reciprocally proportional to the square of the distance from the centre',[12] and takes it for granted that the orbit is an ellipse; he proves that 'the squares of the periodic times about the ellipses are as the cubes of the transverse axes'.[13] A scholium then shows how 'in the celestial system, from the periodic times of the planets may be known the proportions of the transverse axes of the orbits'. Prob. 4 (quoted above in §1)

[6] That this is a centre of force is not specified but only implied. The proposition is general in that the path need not be circular.

[7] From the translation in Hall and Hall (1962), *Unpublished papers*, p. 272.

[8] I have no regard at this point for any differences among the several versions, being concerned only to present the general structure of *De Motu*.

[9] Hall and Hall (1962), *Unpublished papers*, p. 273.

[10] *Ibid.* p. 274. [11] *Ibid.* p. 277. [12] *Ibid.*

[13] What he does in fact is quite neat. He shows by the law of areas how the 'revolutions are performed in the same time in ellipses and circles whose diameters are equal to the transverse axes [axibus transversis] of the ellipses' [the centre of the circle coincides with the focal centre of force of the ellipse]. Then from Corol. 5 to Theor. 2, that 'the squares of the periodic times in circles are as the cubes of the diameters', it follows that the same relation holds for ellipses.

shows how to determine the elliptical path of a body 'moving from a given place, with a given velocity, along a given straight line', on the supposition of an inverse-square force whose 'quantity...is known'. In a scholium, Newton shows how the solution to Prob. 4 is of use in enabling 'the planetary orbits to be defined, and thence their times of revolution'; also in letting us 'know whether the same comet returns time and again'. Then Prob. 5 completes the subject; it is equivalent to Prop. XXXII, Book I, of the *Principia.* A scholium then takes up the problem of possible resistance to motion in the planetary spaces, leading to the general topic of 'the Motion of Bodies in Resisting Mediums'. This topic is discussed briefly in Probs. 6 and 7 and a concluding scholium.

Clearly, *De Motu* is only one small step toward Newton's *magnum opus.* Most of the work remained to be done, and even the final formal structure of the *Principia* is not here revealed. While *De Motu* may represent the state of Newton's researches in dynamics at about the time of Halley's second visit in November 1684, it by no means exhibits the outer boundaries of his knowledge at that time (for which see §6). Even more important, *De Motu* does not yet bear the mark of genius. There is no hint of the grandeur of the *Principia* to come; for example, Prop. XLI and Secs. IX and XI of Book I, or even the universality of the gravitating force and the theory of the tides, much less the mechanics of celestial perturbations and the magnificent demonstration of the shape of the earth as a consequence of the observed precession of the equinoxes. Hence, when we talk of the real beginning of the *Principia,* we cannot go back much before Halley's second visit: the true first steps of the *Principia* as a treatise must be dated in November or December of 1684.[14]

In the light of Halley's importance in getting Newton started on the *Principia,* it is interesting to find one of the three copies of *De Motu* in the Portsmouth Collection unmistakably in his handwriting.[15] This copy lacks the final Probs. 6 and 7 on the motion of bodies in resisting media. The absence of the concluding part of the manuscript is not surprising and helps to confirm the identification of Halley as transcriber, since in 1686 he sent these two propositions to John Wallis.[16] Another version[17] is written in the hand of Humphrey Newton, Newton's amanu-

[14] See §6 below. I am assuming, of course, that Newton's lectures in October 1684 were not given from the text of *LL*; even if they had been, the date of starting the *Principia* would be pushed back by only a couple of months.

[15] MS Add. 3965, ff. 63–70, described in Hall and Hall (1962), *Unpublished papers,* p. 238, as MS *C.*

[16] On 11 December 1686, Halley wrote to Wallis, in part: 'Mr Isaac Newton about 2 years since gave me the inclosed propositions, touching the opposition of the Medium to a direct impressed Motion, and to falling bodies, upon supposition that the opposition is as the Velocity; which tis possible is not true; however I thought any thing of his might not be unacceptable to you, and I begg your opinion thereupon, if it might not be (especially the 7th problem) somewhat better illustrated.' See Halley (1939), *Correspondence,* p. 74; Newton (1959–), *Correspondence,* vol. 2, p. 456. It may be permitted to 'frame' an hypothesis as to how Halley's copy of *De Motu* ended up in the Portsmouth Collection. Newton evidently disliked having certain of his papers circulate in manuscript and so took any opportunity to get back any copies he could; but he was not apparently always able to do so. For instance, in 1720, James Wilson sent Newton a pirated manuscript copy of the 1671 tract, asking to have it returned with comments; Newton held on to the transcript (now in MS Add. 3960, §2, U.L.C.).

[17] MS Add. 3965, ff. 40–54.

ensis,[18] who wrote out the manuscript of the *Principia*. This fact by itself does not help much in the dating, since we do not know exactly when Humphrey came to work for Newton.[19] This manuscript, with corrections and additions in Newton's hand, is more or less the final copy, comprising the longest and most polished version.[20] It opens with a set of four definitions: (1) 'Vis centripeta', (2) 'Vis corporis seu corpori insita', (3) 'Resistentia', (4) 'Exponentes quantitatum'.[21] One has only to compare these with the set of definitions in the *Principia* to see once again how far removed from *De Motu* the *Principia* actually is.

4. THE LAWS OF MOTION: FROM 'DE MOTU' TO 'DE MOTU CORPORUM IN MEDIIS REGULARITER CEDENTIBUS'

Some time after *De Motu*, but before composing the text he deposited as his Lucasian Lectures, Newton began to write the work he entitled *De motu corporum in mediis regulariter cedentibus* ('On the motion of bodies in regularly resisting media'). Only a fragment now remains, in the Portsmouth Collection.[1] Whether it was ever completed, or how much more there ever was, we cannot at present tell. The presently available portion is in an amanuensis' hand, with corrections and emendations made by both the amanuensis and Newton; it consists of a set of 'Definitiones', 'Leges', and 'Lemmata', but no 'Theoremata' or 'Problemata'. I shall not discuss here either these 'Definitiones' or the 'Lemmata',[2] but shall rather concentrate on the 'Leges' in relation to *De Motu*, *LL*, *M*, and E_1, to show the progress of Newton's thought and expression. In this manuscript (which I shall refer to as *De motu corporum* to distinguish it from the tract *De Motu*), there are six 'Leges', called 'Lex 1'...'Lex 6' under the general head of 'Leges Motus'. The manuscripts of *De Motu* show a progression from two 'Hypotheses' to three and then to four, and then finally to five in the manuscript in the hand of Humphrey Newton, where there is also a transition from 'Hypothesis' to 'Lex'. In the manuscript *De motu corporum*, they begin and end as 'Leges Motus'; they have not yet become the 'Axiomata sive Leges Motus' of *LL*, *M*, E_1, E_2, and E_3.

Thus through this manuscript *De motu corporum*, we may see Newton at work during the interval between *De Motu* and *LL*, revising and stating the Laws of Motion, but not yet putting them into the form in which we know them today, which is already present in *LL* and in *M*. Furthermore, although in *De motu corporum* the first three 'Leges Motus' are substantially the three 'Leges Motus' of

18 Although the handwriting seems unmistakably Humphrey Newton's, it has recently been suggested (quite wrongly, I believe) that this manuscript 'may possibly be in the hand of David Gregory, who is known to have visited Newton at Cambridge in 1694'. See Herivel (1965), *Background*, pp. 94, 257.

19 This question is discussed in Supplement II.

20 According to Hall and Hall (1962), *Unpublished papers*, p. 239.

21 In the other two manuscripts in the Portsmouth Collection, and in the Royal Society manuscript, there are only the first three Definitions.

1 U.L.C. MS Add. 3965, ff. 23–6.

2 For a critical summary, and a new analysis of Newton's procedure, see Cohen (1967*a*), 'Second law'.

the *Principia*, there are three more, making six in all.[3] Evidently the 'Axiomata sive Leges Motus' required a number of restatements before achieving a form that was satisfactory to their critical author. 'Lex 1' occurs in *De Motu*, in two successive stages, then is transformed in *De motu corporum*, and undergoes still further revision before becoming an acceptable primary law. 'Lex 2' has an even more curious history; it is absent from the early version of *De Motu* and appears first in the copy of *De Motu* in Humphrey's hand. But significant changes are made when the law is presented in *De motu corporum* and *LL*. The initial absence of this 'Lex 2' in the early *De Motu* no doubt is related to Newton's lack of awareness that it needed express formulation and was not obvious; in the *Principia*, he apparently believed he needed 'Lex 2' as an explicitly stated axiom for impulsive forces, but assumed that a form of this law holding for continuous forces followed from Definitions VI–VIII.[4] 'Lex 3', positing the equality of action and reaction, is not in any of the manuscripts of *De Motu* in the Portsmouth Collection, nor in the version copied into the Register of the Royal Society. It appears first in *De motu corporum*, and is radically altered when presented again in *LL*. Like the other 'Leges Motus', it too went through at least two stages of development.

To see the actual states of evolution, let us turn to 'Lex 1', which in Humphrey's manuscript of *De Motu* reads:

Lex 1 [*altered from* Hypoth. 1]. Sola vi insita corpus uniformiter [*originally* motu uniformi] in linea recta semper pergere si nil impediat.

Law 1. That by its inherent force (*vis insita*) alone, a body always proceeds uniformly along a right line if nothing impedes it.[5]

In *De motu corporum*, the amanuensis has written 'Lex 1. Vi insita [*originally* innata] corpus semper...', which is much as in *De Motu*, with the omission of the opening word 'Sola' and the transposition of 'semper' so as to have it follow 'corpus' directly. This sentence is then altered in Newton's hand so that 'corpus' becomes 'corpus omne' and the word 'semper', thereby rendered superfluous, is deleted.

The remainder of this new 'Lex 1', however, differs significantly from the previous forms; the statement of 'Lex 1' now reads in full:

[3] In MS *B* (see §2, note 20), ff. 55–62 bis, there were at first only two 'Hypotheses'; later Newton altered the first, added a third, and began to write a fourth (but got no further than the title 'Hyp 4'). In MS *C*, ff. 63–70, there are four 'Hypotheses'. In MS *D*, ff. 40–54, there are five 'Hypotheses', but later—in each instance—the word 'Hypoth.' is cancelled and replaced by 'Lex'. In the most complete version (MS *D*), the third, fourth, and fifth 'Hypotheses' ('Leges') deal, respectively, with the relative motions of bodies at rest and in uniform rectilinear motion, the state of the common centre of gravity of a system of bodies being unaffected by the mutual actions of the bodies, and the resistance of a medium being proportional to its density and to the area of the moving (spherical) body and its velocity. These are the fourth, fifth, and sixth 'Leges Motus' in *De motu corporum*.

[4] This point has been discussed by a number of writers, notably Dijksterhuis (1961), *Mechanization*, pp. 464 ff.

[5] In another version of *De Motu*, presumably earlier, this reads: 'Hypoth 2. Corpus omne sola vi insita uniformiter secundum rectam lineam in infinitum progredi nisi aliquid extrinsecus impediat' ('Hypoth. 2. Every body by its inherent force alone progresses uniformly along a right line infinitely unless something external impedes it').

Lex 1. Vi insita corpus omne [omne *replaces* semper *del.*] perseverare in statu suo quiescendi vel movendi uniformiter in linea recta nisi quatenus viribus impressis [et impedimentis [*changed from* et impedientibus] *del.*] cogitur statum illum mutare.

In the previous version in *De Motu*, Newton had said that 'by its *vis insita* alone' a body 'uniformiter in linea recta semper pergere', but now such a body ('every body') will 'persevere in its state' ('perseverare in statu suo') whether that 'state' is one 'of resting [being at rest] or of moving uniformly in a right line'. Hence Newton has added a new concept, the state of a body being at rest, and has thus extended the law to embrace the two states in which a body may continue without an external force: rest and rectilinear motion. In *De Motu*, furthermore, the uniform rectilinear motion would continue 'si nil impediat' ('if nothing were to impede [it]'), but now the body will 'persevere in its state' 'nisi quatenus viribus impressis cogitur statum illum mutare' ('except in so far as by impressed forces it is compelled to change that state'). Newton is here using the Cartesian expression of 'state' ('status'), as he was to do in *LL* and in *M* and in all the editions of the *Principia*.[6]

In a comment, Newton has later written in the following:

Motus autem uniformis hic est duplex, progressivus secundum lineam rectam quam corpus centro suo aequabiliter lato describit & circularis circa axem suum quemvis qui vel quiescit vel motu uniformi latus semper manet positionibus suis prioribus parallelus.

Uniform motion, however, here is twofold, progressive along a right line which a body describes by the uniform translation of its centre and circular about any axis of it whatsoever, which either is at rest or by a uniform motion of translation always remains parallel to its previous positions.

There is no indication as to whether both kinds of 'uniform motion' are to be considered as 'states'. There is no doubt, however, that Newton already had learned that only uniform rectilinear motion is a 'state'.

Thus in *De motu corporum*, 'Lex 1' has all but attained the form encountered in *LL*, *M*, and E_1, and continued unchanged in E_2. In that final version, the further alterations made by Newton have been to drop the opening phrase 'Vi insita', to substitute 'in directum' for 'in linea recta', and to add 'a' before 'viribus impressis'. In E_3, Newton altered 'nisi quatenus a viribus impressis' to 'nisi quatenus illud a viribus impressis' and also replaced the concluding 'statum illum mutare' by 'statum suum mutare'.[7] [*LL* originally had 'suum', altered to 'illum'.]

In 'Lex 2', as in 'Lex 1', we may see various stages of development: from *De Motu* and *De motu corporum* on to *LL*, *M*, E_1, E_2, and E_3. This 'Lex 2' appears in the manuscript version of *De Motu* copied out by Humphrey, and is absent from the other two manuscripts of *De Motu* in the Portsmouth Collection and from the

[6] In the discussion of this law in *De motu corporum*, Newton alludes to the idea, developed explicitly in *LL*, in *M*, and in E_1, E_2, and E_3, that on proper analysis uniform circular motion may prove to exemplify in its tangential component the principle of uniform rectilinear motion.

[7] See the Apparatus Criticus, {13.6, 7}.

Royal Society copy. The text of the first part of the law appears in *De Motu* (in Humphrey's hand) as:

Hypoth. 2. Mutationem motus proportionalem esse vi impressae...

and has subsequently been changed by Newton in two stages. First:

Motum genitum vel Mutationem motus proportionalem esse vi impressae...

and then:

Mutationem status movendi vel quiescendi proportionalem esse vi impressae...

meanwhile becoming 'Lex 2'. This is a most significant difference because 'motus', or more properly 'quantitas motus', is a quantitative measure determined—in the *Principia*—by the product of the mass ('quantitas materiae') and velocity; the change ('mutatio') of this 'motus' (or 'quantitas motus') is found to be in a direct proportion to the impressed force. In the final version in *De Motu* Newton holds that it is the change in 'status' that is proportional to the impressed force, but this intuitively reasonable statement clearly cannot have its full meaning until 'mutatio status' is quantitatively defined. In this version of *De Motu* Newton defines only 'vis centripeta', 'vis corporis', 'resistentia', and 'exponentes quantitatum'; he does not define 'quantitas materiae' or 'quantitas motus'. Not only is the reader given no clue to the quantification of either 'status' or 'mutatio status', but the lack of definition means that nothing whatever is said concerning the possibility that a change in direction (as in uniform circular motion) might be a change in 'status', so that uniform circular motion is therefore very different from uniform rectilinear motion. But in *De motu corporum*, Newton transferred the concept of 'status' from 'Lex 2' to 'Lex 1' where it more properly belongs.

In *De motu corporum*, furthermore, the beginning of 'Lex 1' reads exactly like the first state of this law in *De Motu*:

Mutationem motus proportionalem esse vi impressae...

The second part of this 'Lex 2' is identical in both *De Motu* and *De motu corporum*:

et fieri secundum lineam rectam qua vis illa imprimitur.

The final version, in *LL* and *M* and in E_1, E_2, and E_3, is the same as in *De motu corporum* and the first state in *De Motu*, save only that 'vi impressae' is made more precise by the addition of the word 'motrici' to become 'vi motrici impressae'.

'Lex 3' in *De motu corporum* is a forerunner of Law 3 as found in *LL*, *M*, E_1, E_2, and E_3:

Lex 3. Corpus omne tantum pati reactione quantum agit in alterum.

Every body is acted upon by reaction to the same extent that it acts upon another (body).

Wholly absent from all the versions of *De Motu* that I have seen, this particular law may in fact be one of Newton's most original formulations. As found in *LL*, *M*, E_1, E_2, and E_3 it reads quite differently:

Actioni contrariam semper et aequalem esse reactionem: sive, corporum duorum actiones in se mutuo semper esse aequales et in partes contrarias dirigi.

To every action there is always opposed an equal reaction: i.e., the mutual actions of two bodies upon each other are always equal, and directed to contrary parts [that is, in opposite directions].

One of the most fascinating features of *De motu corporum* is the actual method of composition. It begins as a draft written out in the hand of Newton's amanuensis. Newton has then reworked the text in his own hand, revising, cancelling, adding much new material. Some part at least, or even the whole text may possibly have been dictated; or perhaps an earlier dictated version was copied. This conclusion is suggested to us by the occurrence of a scribal error of the kind that would seem to have come from dictation: when Newton spoke of 'hujus generis est... vis coelestis cohibens Planetas ne abeant in tangentibus orbitarum' ('Of this kind is...the celestial force preventing the Planets from going off along the tangents of their orbits'), the scribe (presumably Humphrey Newton) wrote 'in tangentibus orbi terram' [or possibly 'orbiterram'], and then crossed out 'orbi terram' [or 'orbiterram'] and made it 'orbitarum'. We shall see, below, other examples of errors in the *Principia* that may also suggest dictation.[8]

5. NEWTON'S USE OF 'VIS INSITA'

In both of the versions of *De Motu* containing introductory 'Hypotheses' or 'Leges', the statement of what we know as the First Law of Motion contains the expression 'vis insita', which also occurs in the set of 'Definitiones' in one of the manuscripts, as follows:

Def. 2. Et vim corporis seu corpori insitam [appello] qua id conatur perseverare in motu suo secundum lineam rectam.

I shall not discuss here Newton's use of the word force or 'vis' in relation to a body persevering in its motion along a right line, a topic often discussed. Clearly this is not a force in the sense of Newton's Law 2, which causes a *change* in a body's state of motion (or of rest), as Newton later pointed out himself. What may be primarily interesting is that Newton used 'vis' in these two very distinct senses and, in particular, the meaning that he gave to 'vis insita' or 'vis insita materiae'.

Throughout this Introduction, and in other writings, I have translated Newton's 'vis insita' by 'inherent force' and 'vis insita materiae' by 'inherent force of matter' or 'force inherent in matter'. But the literature abounds with a different rendering, 'innate force'. We may learn what Newton's intention was by examining the intermediate usage in *De motu corporum*, prior to *LL*, *M*, or the printed editions.

'Vis insita' or 'vis insita materiae' is a difficult concept to render into English. The root meaning of 'insita' is 'implanted' or 'inserted'; it would usually be

[8] In U.L.C. MS Add. 3972, f. 26, 'orbiterrarum' appears as a single word. Possibly Humphrey could have misread 'orbiterrarum' as 'orbiterram'. See note 4 to §7 below.

rendered by 'ingrafted'. But Newton was not referring specifically to an act of putting this 'power' into matter; rather he was using the derived meaning of 'naturally inborn' and hence 'innate' or even 'natural', as commonly used even in classical Latin. Thus Newton's 'vis insita' is necessarily present in a body from time past to time future, infinitely, and so is almost an 'immanent force'.

The translation of 'insita' as 'innate' appears to have originated with William Whiston, who wrote (in the English version of his own work, *Sir Isaac Newton's mathematick philosophy more easily demonstrated*), 'The innate Force of Matter is a Power of...'—a translation of 'Vis insita materiae est...'.[1] Some few years later, this same translation occurs in Andrew Motte's English version of the *Principia* of 1729. There we find:

DEFINITION III.

The Vis Insita, *or Innate Force* of *Matter, is...*

This force is... Upon which account, this *Vis insita*, may, by a most significant name, be called *Vis Inertiae*, or Force of Inactivity...[2]

Motte was evidently aware of the difficulty, and so, although using 'innate force' as a translation of 'vis insita', in the definition itself he gave the Latin original along with a possible English equivalent, as he does—in the explanatory paragraph—for 'vis inertiae'.[3] We do not know whether Motte merely took this phrase from Whiston or conceived it independently. Since Whiston's book constitutes the first translation of a major part of the *Principia*, no doubt it greatly influenced Motte. But all Newtonians did not follow Whiston in his translation of 'vis insita'; John Clarke, for one, used 'inherent force', which was also used at least once by Newton.

The key is given, I believe, in the manuscript *De motu corporum*, which shows how valuable it is to trace Newton's concepts and technical expressions through various states. The copyist or amanuensis at first wrote out a Def. 12 that is much like Def. 2 of *De Motu*:

De Motu, Def. 2. Et vim corporis seu corpori insitam [appello] qua id conatur perseverare in motu suo secundum lineam rectam.

De motu corporum, Def. 12 [state 1]. Vis corporis seu corpori insita et innata est potentia qua id conatur perseverare in statu suo quiescendi vel movendi uniformiter in linea recta, estque corporis quantitati proportionalis.

A comparison of the two shows the changes in Newton's formulation, not because he has altered the grammatical form of an object clause to an independent declarative sentence, but because he has added many new concepts. Thus this 'vis' is no longer merely 'qua id conatur perseverare in motu suo' but has become 'potentia

[1] Whiston (1716), *Newton's mathematick philosophy*, p. 41.

[2] Newton (1729), *Principia* [Motte], vol. 1, p. 2.

[3] But Cajori, in his modernization of Motte, changed this to read more like a work of the twentieth century: 'Upon which account, this *vis insita* may, by a most significant name, be called inertia (*vis inertiae*) or force of inactivity'; see Newton (1934), *Principia* [Motte–Cajori], p. 2. The Marquise du Chastellet rendered this Definition as follows: 'La force qui réside dans la matière (*vis insita*) est le pouvoir qu'elle a de résister.'

qua id conatur'; furthermore, 'conatur perseverare in motu suo' has become 'conatur perseverare in statu suo' and, in fact, 'in statu suo quiescendi vel movendi'. Not only has Newton introduced the concept of 'status'; he has also included the 'status' of resting (or being at rest) as well as of moving. Finally, the moving is not merely 'secundum lineam rectam' but 'uniformly' so.

This list of changes does not include all the novelties introduced; there is in *De motu corporum* a supplement (not found in *De Motu*) to the definition that this 'vis' is proportional to the 'quantity' of the body, which is what Newton also calls mass.

Newton has later altered this statement of Def. 12 in a number of ways in his own hand. He has added a supplement almost as long as the definition itself. He has also changed 'conatur perseverare' to the simple 'perseverat'. And, in the present context, most important of all, he has changed the beginning from

> Vis corporis seu corpori insita et innata est...

to

> Corporis vis insita innata et essentialis est...

At first Newton was intending that 'vis...insita et innata' be another expression for 'the force of a body' or 'vis corporis'. Then he deleted 'insita et innata' and wrote in the three adjectives 'insita innata et essentialis'. Surely this is the same phrase used by Newton a few years later in a letter to Bentley, when he referred to the impossibility that 'Gravity should be innate, inherent and essential to Matter'.[4]

6. DATING NEWTON'S COMPOSITION OF THE 'PRINCIPIA'

According to some statements made by Newton, seventeen or eighteen months elapsed from the time he started in earnest to compose the *Principia* to the end of the job. If we assume that Newton began in November or December 1684, that is, almost immediately following Halley's second visit, then—according to this estimate of time—he would have been finished by April 1686. We know that in this month he sent on to the Royal Society the manuscript of Book I, together with the preliminary Definitions and Laws of Motion, but not the Preface. Among such statements of Newton's, one that was printed by Rigaud, from a draft in the Macclesfield Collection, reads as follows:

> The book of Principles was writ in about 17 or 18 months, whereof about two months were taken up with journeys, & the MS was sent to y[e] R.S. in spring 1686; & the shortness of the time, in which I wrote it, makes me not ashamed of ⌞having⌟ committed [*changed from* committing] some faults & omitting some things. These things I men⟨tion...⟩[1]

[4] [Newton] (1959–), *Correspondence*, vol. 3, p. 254; [Newton] (1958), *Papers and letters* [Cohen and Schofield], p. 302.

[1] Rigaud (1838), *Essay*, p. 92. Quoted from a preliminary English draft of a letter to Varignon in the autumn of 1719. A later Latin version occurs in the Portsmouth Collection (U.L.C. MS Add. 3968, §42, ff. 596[v], 615[r]) and was revised again in a manuscript in the Macclesfield Collection; see Rigaud (1841), *Correspondence*, vol. 2, pp. 436–7. Yet another revision may be found in King's College Library, Keynes MS 142.

Another, a memorandum first published by Brewster and now in the Portsmouth Collection,[2] contains a reference by Newton to an error in E_1 in Prop. X, Book II, in which the 'Tangent of the Arch GH [was drawn] from the wrong end of the Arch which caused an error in the conclusion; but in the second Edition I rectified the mistake'.[3] Then Newton went on:

And there may have been some other mistakes occasioned by the shortness of the time in which the book was written & by its being copied by an Emanuensis who understood not what he copied; besides the press-faults. ffor I wrote it in 17 or 18 months, beginning in the end of December 1684 & sending it to ye R. Society in May 1686; excepting that about ten or twelve of the Propositions were composed before, vizt the 1st and 11th in December 1679, the 6th 7th 8th 9th 10th 12th, 13th 17th Lib. I & the 1, 2, 3 & 4 Lib. II, in June & July 1684.[4]

The first of these quotations proves to be taken from a preliminary draft in English of a letter intended for Varignon, composed in the autumn of 1719. Hence it cannot be considered as an objective memorandum for us to take at its face value. We must see it rather as an instance of Newton's attempting to convince Varignon of a preferred chronology. Similarly, the apparent objectivity of the second memorandum is weakened by the fact that it too was part of a special campaign, and was drafted as part of a preface intended for an abortive revision of the *Principia* some time in 1715 or so; there are related versions or drafts written out for a preface to *De quadratura*, in which there are quotations from the second edition of the *Principia* (1713). Hence any such extracts[5] present two major problems to the critical scholar wishing to make an exact chronology of the events following Halley's visit: first, in what sense could the *Principia* have been completed by April 1686? second, can we accept as accurate the dates assigned to the events in the second memorandum?

Assuming that Newton began the actual writing of the *Principia* (as opposed to *De Motu* or *De motu corporum*[6]) in November 1684, let us see what documentary evidence there is as to the time when he finished writing the *Principia*. In part we face a purely semantic problem: the meaning of the words 'finished writing'. If we have in mind a completed fair copy, with all the figures drawn for the engraver or wood-cutter, and actually delivered to Halley to be turned over to the printer, then the dates are: Book I (and the Definitions and Laws of Motion), April 1686; Book II, March 1687; Book III, March 1687.[7] In this case, Newton's famous statement that the writing of the *Principia* had taken some 17 or 18 months (of which two had been occupied by travelling) is strictly true only of Book I, formally presented to the Royal Society on 28 April 1686.

[2] U.L.C. MS Add. 3968, f. 106.

[3] See the Apparatus Criticus at {252.7 ff.} and {254.16–25, fig.}. The replacement of the original Prop. X of Book II is discussed below in Chapter IX, §4.

[4] This document has been mistakenly said to be lost in Herivel (1960*b*), 'Newton's tract *De motu*', p. 68, n. 10; in Hall and Hall (1962), *Unpublished papers*, p. 239; and in Herivel (1965), *Background*, p. 96. A reference to this manuscript occurs in Whiteside (1964*a*), 'Newton's early thoughts', p. 118, n. 6.

[5] A selection of autobiographical statements of Newton's is printed below in Supplement I.

[6] See above, §§3, 4.

[7] [Newton] (1959–), *Correspondence*, vol. 2, p. 431; see p. 137, below.

But, according to Newton's letter to Halley of 20 June 1686, Book II had been completed in the summer of 1685, and wanted only to be transcribed and to have the diagrams drawn 'fairly' for making the cuts.[8] In fact Newton did not 'make ready' the final fair copy of Book II until the autumn of 1686, guessing 'by ye rate of ye presse' during the summer of 1686 that Halley would be ready for it only by 'about November or December' of 1686.[9] Halley acknowledged its arrival in a letter of 7 March 1686/7.[10] Since the manuscript draft of Book II that was copied by Humphrey Newton has disappeared,[11] we cannot know whether Newton made any alterations or additions after writing to Halley. In the absence of any evidence to the contrary, I do not see why we should not take Newton at his word and assume that—save for minor matters of detail—Book II had also been completed by April 1686, within Newton's time-span of 17 or 18 months.[12]

As to Book III, we are on far less certain ground. In that same letter to Halley of June 1686 referring to Book II as finished and waiting to be transcribed, Newton said that he had designed the *Principia* to consist of three books, the third to be on the system of the world, and that this Book III 'wants the theory of comets'. If the remainder of Book III was indeed finished, more or less as we know it in M and E_1, then—since the section on comets occupies only about one-third of Book III[13]—the *Principia* would have been substantially completed by April 1686, when Halley received the manuscript of Book I.[14] In the absence of dated documents, there is not much that can be said with certainty about how much additional time Newton took to write up the theory of comets.

Another problem arises in the chronology of discovery given by Newton in the second extract quoted at the beginning of this chapter. What are we to make of the explicit statement of propositions found 'in June & July 1684'? W. W. Rouse Ball supposed Newton to have been mistaken about the work done in June and July, 'since we cannot suppose that these propositions were established till after Halley's

[8] *Ibid.* p. 437. [9] *Ibid.* p. 464. [10] *Ibid.* p. 472.

[11] I have not been able to determine the subsequent history of the draft of Book II, once the *Principia* had been printed. Until now, the very existence of this draft has been known only through the reference in Newton's letter to Halley of 20 June 1686. But new evidence (presented below in Chapter IV, §8 and in Supplement VII) not only gives independent confirmation that there was such a draft, but at last enables us to say something about it.

[12] Of course, as all writers know only too well, a manuscript not actually turned over to the printer continually tempts the author to make revisions.

[13] In E_1, thirty-seven pages out of a total in Book III of 110 are devoted to comets.

[14] Since we do not have an actual draft of Book III which was transcribed by Humphrey, we cannot really say much with certainty about the state of composition of this part of the *Principia* at any given time. To indicate how great our uncertainty is, recall that Book III of the *Principia* as we know it was preceded by a somewhat different discussion of the same topic (the system of the world) called *De motu corporum liber secundus* (for which see Chapter IV, §6). Halley seems to have first learned that Newton had expanded his concept of the *Principia* from two books to three in that very letter of 20 June 1686, in which Newton said that the new Book II (on motion in resisting media) had been completed in the previous summer. But when Newton said that the third book lacked the theory of comets, did he have in mind the old Book II, which was later discarded although completed, and which he might still have intended to be a part of the *Principia*, changing only its title from 'liber secundus' to 'liber tertius'? Or had he in mind the new Book III? I would guess the latter, chiefly on the grounds that in his manuscript copy Newton did not alter the title of the old Book II to 'liber tertius'.

visit in August'. Another interpretation, by John Herivel, would use this document and certain others[15] to support the contention that Halley came to see Newton in the spring of 1684, not in August, a suggestion that seems most unlikely and for which there is no direct support and strong counter-evidence.

In the light of the circumstances surrounding the composition of this second quoted memorandum, and especially in the absence of any dated supporting documents, we cannot accept as absolutely accurate the fine details as to which propositions were discovered in any particular month. Newton did not refer to a notebook with dated entries concerning his research, nor to correspondence, nor even to early drafts with dates (such as occur for other topics in the 'Waste Book'). Wholly apart from his bias toward a particular chronology which he wished to implant in the minds of others (and of which he may very likely have even convinced himself), can we really have much confidence in the details of Newton's recollections from the distant past? That is, without any document to help him, could he really have remembered in 1715–19 exactly what had happened in one month or another of 1684, more than thirty years earlier? Furthermore, the second memorandum—for all its detail as to discovery—contains at least one error of fact: Book I of the *Principia* was not sent to the Royal Society in May, but rather late in April.[16] I do not believe, therefore, that we need accept Newton's attribution of any of the propositions to the two months of June or July, say by supposing that he had actually begun to work again on problems of dynamics in early summer 1684, just prior to Halley's visit.

There is, in fact, an additional source of evidence that both supports the August date of Halley's visit and would appear to contradict any suggestion that Newton had been occupied with topics in celestial dynamics during June and July 1684. On 9 June 1684, David Gregory sent Newton from Edinburgh a copy of his *Exercitatio geometrica*,[17] which presumably would have been received some time after mid-June. Practically at once Newton began to write out drafts of two longish tracts on infinite series, the *Matheseos universalis specimina*, revised as *De computo serierum*, which together are several times as long as *De Motu*. If Halley's visit occurred in August, a six-weeks' interval 'is about right for this sudden surge of mathematical activity'; but if Halley had come earlier, where did Newton 'get either time or free creative energy'[18] for both his researches on infinite series and

[15] In the autobiographical statements I have assembled in Supplement I, it may be seen that Newton did refer more than once to a visit by Halley in the spring. But these documents have many mutual inconsistencies and downright errors of date. It must be remembered that they were tentative, often cancelled, and not edited (or even approved) for publication or even public display. D. T. Whiteside (1966*a*, p. 115, note 3) has given strong evidence in his review of Herivel's book that Halley would hardly have been in the mood to go to Cambridge to see Newton in May 1684, since his father and his brother had just died.

[16] Possibly Newton was thinking of the date when—to quote from yet another memorandum of his—the Royal Society 'ordered it to be printed May 19 1686 as I find by a minute in their Jornal book'. (See Supplement I.)

[17] [Newton] (1959–), *Correspondence*, vol. 2, p. 396.

[18] D. T. Whiteside (personal communication). It may be added that the same considerations obviously apply to the possibility of Newton's having been working on dynamics during June and July prior to Halley's visit in August. Newton's *Matheseos universalis specimen* and *De computo serierum* are included in [Newton] (1967–), *Mathematical papers*, vol. 4.

those on dynamics? There is, in fact, no ground for supposing that Newton had engaged in research on dynamics between his correspondence with Hooke in 1679 (on the path of a heavy object let fall down to a rotating earth[19]) and Halley's visit in 1684.[20]

But even though we must reject the assignment of particular propositions to one month or another, we may nevertheless consider this document a statement by Newton of stages of discovery, as his recollection of the group of propositions he had found immediately following Halley's visit. I believe that this kind of information may be more reliable than the assignment of a particular month; that Newton could perfectly well have remembered which propositions he had found in that first great outburst of creative energy following Halley's stimulus. In order to explore the possible consequences of such an assumption, I have constructed a table showing the order of discovery. The first column (Table I) shows the numbering of the lemmas, theorems, corollaries, and problems, as they occur in *De Motu*, while the second gives the corresponding designation and number in the *Principia*. Hence Table I displays the state of Newton's knowledge or formulation (although not necessarily in its entirety) as of November 1684.[21]

In Table I, I have juxtaposed the formally enunciated lemmas, theorems, problems, and corollaries of *De Motu* and their equivalents in *M* and E_1. As we have already seen, and shall see further below, however, these all went through at least one intermediate stage, documented for us in the Lucasian Lectures (*LL*). In *LL*, we may see how certain portions of *De Motu* were first incorporated into a draft of the *Principia*, and then later either moved about or even cancelled. An example occurs in the Scholium to Prob. 4 of *De Motu*, which we find in LL_α in the Scholium to Prop. XX, but which vanished in the final revision. Furthermore, the proofs differ considerably as Newton progressed from *De Motu* to the *Principia*, as may be seen by comparing the entirely different proofs of Theor. 4 of *De Motu* and Prop. XV of the *Principia*. In Table II, I have listed all of the propositions in Secs. I and II of Book I, together with one additional proposition from Sec. VII.

[19] See Whiteside (1964*a*), 'Newton on planetary motion'.

[20] There is a very curious problem about Newton's direct concern with celestial dynamics. In the correspondence with Hooke in 1679, Newton said explicitly that he had no interest in following up Hooke's suggestion of explaining the heavenly motions by 'attractions' from 'direct paths'. Whether we believe Newton or not, the fact remains that we do not have evidence of his having worked seriously at celestial dynamics again prior to Halley's visit. One of the rare occasions between 1679 and 1684 when Newton seems to have exhibited genuine concern for a general view of celestial dynamics was in 1681 (see his letter to Crompton, April 1681, printed in [Newton] (1959–), *Correspondence*, vol. 2, p. 361; and U.L.C. MS Add. 3965, f. 613^r). At this time Newton was invoking a Borellian '*vis centrifuga*...overpow'ring the attraction' and a set of Cartesian principles, e.g. 'Materiam coelorum fluidam esse' and 'Materiam coelorum circa centrum systematis cosmici secundum cursum Planetarum gyrare'—which clearly asserts a deferent vortex. In 1681–3, Newton was working on comets, in testimony whereof we have his correspondence with Flamsteed and others. (See [Newton] (1959–), *Correspondence*, vol. 2, pp. 340–7, 358–67.) But these researches are marked by an apparently complete absence of any attempt by Newton to give a dynamical explanation of cometary motion.

[21] Of course, in each case the statement in the *Principia* is apt to be more precise, or fuller. But I have considered as equivalent in these tables any two that deal *grosso modo* with the same topic, even though not in the identical fashion.

I have then checked these off into four groups, of which the first contains Prop. IV of Book I, discovered long before 1679, in fact in the 1660s, and the last contains those present in *De Motu*. The intermediate groups comprise the two propositions (I and XI) said by Newton to have been discovered in December 1679 and those attributed by him to June/July 1684. I remind the reader that I believe that Newton's 'June or July' 1684 can mean no earlier than 'August or September' (or possibly October); that at best this would be a group of propositions to be assigned to the period between Halley's visit and the writing of *De Motu*.

TABLE I. *Correspondence of propositions in 'De Motu' and 'Principia'*

De Motu	*Principia* (M and E_1)
Lemma 1	Corol. 1 to Leges Motus
Lemma 2	Lemma X, Sec. I, Book I
Lemma 3	Lemma I, Sec. I, Book II
Lemma 4	Lemma XII, Sec. II, Book I
	Book I
Theor. 1	Prop. I
Theor. 2	Prop. IV
Corol. 1–5	*Corol.* 1–2, 4–6
Theor. 3	Prop. VI
Corol.	*Corol.*
Prob. 1	Prop. VII
Prob. 2	Prop. X
Prob. 3	Prop. XI
Theor. 4	Prop. XV
Prob. 4	Prop. XVII
Prob. 5	Prop. XXXII
	Book II
Prob. 6	Prop. II
Prob. 7	Prop. III

Even within this limitation Table II may tell a most interesting story. For it indicates four propositions (Props. VIII, IX, XII, XIII) which Newton would have known by the time of *De Motu*, but which he did not use in *De Motu*. This tract, in other words, would not have represented the limits of his knowledge of dynamics. Even more significant is the pair of propositions (Props. XV, XXXII) which appear in *De Motu*, but which would not have been found by June or July 1684, and so would have been discovered only after Halley's visit, presumed to have been made in the traditionally accepted month of August.

These two propositions may command our special attention, not only because they would be the fruits of research following Halley's visit but because of their content. Proposition XV shows us Newton's true command of the principles of celestial dynamics, because it deals with Kepler's third (or harmonic) law for ellipses, not for the simple case of circular orbits (as in the famous Corol. 6 to Prop. IV, Book I, which is Corol. 5 to Theor. 2 of *De Motu*). In Prop. XV

TABLE II. *Dates of discovery of propositions in 'De Motu' and the 'Principia'*

M and E_1	Known to have been discovered early	Said by Newton to have been discovered in— Dec. 1679	June/July 1684	Present in *De Motu*
Liber I, Sec. II:				
I	...	*	...	*
II				
III				
IV	*	...	...	*
V				
VI	...	...	*	*
VII	...	...	*	*
VIII	...	...	*	...
IX	...	...	*	...
X	...	...	*	*
Sec. III				
XI	...	*	...	*
XII	...	...	*	...
XIII	...	...	*	...
XIV				
XV	...	...	...	*
XVI				
XVII	...	...	*	*
Sec. VII				
XXXII	...	...	...	*

In this table I have listed in column 1 all the numbers of propositions in Secs. II and III of the *Principia*, and the first proposition only of Sec. VII. In the last column I have indicated by an asterisk which of these occur in *De Motu*. In the second column only Prop. IV, discovered by Newton in the 1660s, is marked with an asterisk, while in the third and fourth columns the propositions said to have been found in December 1679 and in June or July 1684 are similarly marked. I have used a leader whenever the proposition in question appears in at least one column.

Proposition IX is partially stated in a scholium of *De Motu*, and Props. XII and XIII are subsumed.

This table could be extended to Props. I–IV of Sec. I of Book II of the *Principia*. For each of these propositions an asterisk would appear in the last two columns. Proposition I is assumed in Prob. 6 of *De Motu* ('ut notum est'); Prop. II is a more precise version of Prob. 6; Prop. III is equivalent to Prob. 7; and Prop. IV is contained in the Scholium following Prob. 7.

(corresponding to Theor. 4 of *De Motu*), Newton shows that the proportionality of 'the squares of the periodic times' to 'the cubes of the transverse axes' necessarily follows from the condition 'that the centripetal force is reciprocally proportional to the square of the distance from the centre'. In the *Principia*, the proof is far shorter and more elegant, invoking the corollary to Prop. XIV (which, as our Table II suggests, was not discovered until after *De Motu* had been written). In *De Motu*, Newton proves that 'the revolutions are performed in the same time in ellipses and in circles whose diameters are equal to the transverse axes of the ellipses'. Then he applies Corol. 5 to Theor. 2 (that, under the given conditions of an inverse-square centripetal force, 'the squares of the periodic times in circles are as the cubes of the diameters') so as to conclude that 'hence they are [also] in ellipses. Q.E.D.'. But in Prop. XV, Book I, of the *Principia*, Newton proves his result directly (that is, without reference to Corol. 6 of Prop. IV), and then states as a corollary that 'the periodic times in ellipses are the same as in circles whose diameters are equal to the major axes of the ellipses'.

This last result, implying also that the periodic times are the same in all ellipses having the same major axis, is the basis on which rests the second of the new propositions seemingly discovered by Newton after Halley's visit. I would call attention to this Prop. XXXII, since it has always seemed to me to be a most interesting idiosyncratic expression of Newton's personal style. In this proposition Newton deals with the problem of how far a body may fall in a given time, starting from rest, toward a centre of force to which it is attracted, the force varying—as gravity does—inversely as the square of the distance. This is a difficult problem, usually requiring the integral calculus (or a protocalculus) for its solution, and is solved most easily by the application of 'energy' considerations.[22] Newton applied to this problem a rather ingenious combination of physical intuition and geometrical insight of which in all probability only he was capable. He begins by considering elliptic motion about a centre of force located in a focus, in relation to a circle of reference drawn so that the major axis of the ellipse would be a diameter. Taking advantage of the fact that under these conditions the ellipse is the result of an affine transformation of the circle,[23] he projects the area-conservation law of the ellipse on to the circle. Since this property holds for any ellipse, it is valid for a whole sequence of ellipses, all with their major axes along the same line. The ellipses in this sequence have a constantly diminishing minor axis; but the same periodic time will be found in every one. In the limit, the ellipse becomes a straight line, in fact the diameter of the circle or the former major axis; and the conditions become those of the original problem, with the centre of force—from which the areas conserved are reckoned—now located at the intersection of the former major axis and the circle.[24] This was not, however, the first time that anyone had been able to go beyond the simple Galilean conditions of constant force (or, more accurately, uniform acceleration), so as to deal with the conditions of motion requisite for an adequate celestial dynamics.[25]

[22] See Cohen (1969*d*), 'Galileo, Newton, and the divine order'.

[23] Newton, of course, was not the discoverer of this property of the transformation of a circle. Possibly one of a select group of contemporaries might have devised such a proof: for instance, Huygens, Leibniz, or Bernoulli, each of whom might have used a similar protocalculus, or primitive equivalent of the calculus.

[24] There is in this proof a major flaw in that, when the falling object reaches S, it will continue on past S along a straight line, until it has reached a point as far away from S as that from which it was originally dropped. Then it will again ascend to S, and so oscillate back and forth. But in Newton's model, at the limit of the ellipse, we would expect the object to turn around at S, and then ascend at once to the point from which it had been dropped. The flaw does not lie in the proof as such, but in the geometrical model. At S the direction of motion is undefined; the differential equation $d^2s/dt^2 \propto 1/s^2$ does not hold at $s=0$.

[25] According to Tables I and II, this proposition is one of those which Newton found after Halley's first visit. It is a sure sign of the advance made by Newton on the subject of dynamics that he was able to deal with motion under a varying force (and hence with a varying acceleration). Galileo, and most of the writers between

Newton's proof may be subject to various types of criticism, and I do not wish to suggest that it is the best way of solving the problem he posed. But I believe there can be no question of the fact that it combines geometric insights with dynamical considerations in a rather unexpected and original fashion. It is difficult to think of any other of his contemporaries who would have had the particular ingenuity that Newton here displays. Certainly, he was outstandingly adept in geometrico-dynamical thinking. And the presentation of motion of descent in a uniformly changing force-field that we find in the *Principia* was probably the first widely circulated general exposition of this most important topic.

An additional result of interest that may be derived from Table II is that Props. II, III, and V (completing Sec. II of Book I of the *Principia*) were apparently not discovered in 1679 or used in *De Motu*. Proposition II is not merely the converse of Prop. I, for it states that the area law implies a centripetal force directed to the point about which equal areas are swept out—whether that point is at rest or moving uniformly forward in a right line. Proposition III deals with the case in which that centre may be accelerated. As to Prop. V, it contains a geometric construction to find the common centre of forces causing a body to trace out a given figure, the velocity being known at three points.

Yet another interpretation is possible: that in this detailed memorandum Newton may only have been referring to the contents of *De Motu* under a date that is too early. In this case the four propositions (VIII, IX, XII, XIII) mentioned by Newton in his memorandum which are not formally present (as explicitly stated theorems or problems) in *De Motu* could have been held by him to be present implicitly. That is, Prop. VIII (of E_1) can be said to have been contained loosely in *De Motu*, Theor. 3, Corol., while Prop. IX appears essentially in *De Motu*, Prob. 1, Scholium. It could also be said that Propositions XII and XIII are invoked in the last paragraph of *De Motu*, Prob. 4. But the memorandum certainly does not mention Prop. XXXII (*De Motu*, Prob. 5).

It may seem that I have dealt at too great length with this memorandum. But my intention was not merely to explore the consequences of one of the most detailed presentations by Newton of the dating of his discoveries in dynamics. I also wished to convey to the reader the great uncertainties in our knowledge concerning the first steps in the creation of the *Principia* and the almost overwhelming difficulties in finding a unique and reliable interpretation of Newton's own statements about his own discoveries.

Galileo and Newton, had been able to work in a satisfactory way only with situations in which the force is constant. An exception of significance is Huygens's analysis of circular motion, in which the direction of the force varies, but—in contrast to Newton's proposition—the magnitude remains constant. Yet another exception is James Gregory's *Tentamina* (1672), which had introduced the problem of projectile motion in which the resultant acceleration downward is not uniform. In *De Motu*, but not in the *Principia*, Newton uses this proposition (Prob. 7) to find the motion of projectiles in our air, supposing gravity to act uniformly 'in parallel straight lines' and the air to be 'homogeneous'. Possibly, therefore, Newton took Gregory's tract as his starting point. Apparently as early as *c.* 1605, Thomas Harriot was also dealing with projectiles in a resisting medium.

7. THE MANUSCRIPT OF THE 'PRINCIPIA': NEWTON'S METHOD OF COMPOSITION

The manuscript of the *Principia*, sent to London for the printer, and currently to be found in the library of the Royal Society, occupies some 460 leaves. It is almost entirely written in the hand of Humphrey Newton, on the recto sides only, with corrections and additions made by both Humphrey and Newton, plus some by Halley.[1]

This manuscript (which we have designated *M*) is a clean 'final' copy and not a rough draft or a first essay with continual revising between the lines or in the margins. Indeed, the corrections and alterations, as may be seen from the Apparatus Criticus, are minor, save for the occasional replacement of one paragraph or scholium by another.[2] Chiefly, Newton's alterations of this manuscript consist of the insertion of cross-references. So perfect a manuscript must have been a fair copy of a prior draft, which would have been worked over, emended, improved, and rewritten much in the manner to be seen in the manuscript of *De motu corporum*. But no such prior draft exists. To be sure, the manuscript deposited as Newton's Lucasian Lectures[3] is a prior draft of much of Book I, together with the Definitions and Laws, but this manuscript itself does not have the appearance of a first version.

Could this manuscript (*M*) have been the product of a first dictation? Almost certainly not; it lacks the fits and starts and cancellations of any dictated work and is singularly free of the errors that always occur in the transcriptions of a dictated text, especially one as difficult as the *Principia*. Of course, some bits here and there may have been dictated, and Humphrey might have been copying an earlier version which itself had been dictated and then corrected by the author.[4] Newton might very well have been standing over Humphrey's shoulder as the latter made his copy, dictating certain further changes which he wished to have introduced.

But even if anyone should suppose that the *Principia* had been dictated (if not in full, then at least in major part), are we to believe that there would not have been an outline or a series of notes? Can anyone seriously advance the hypothesis that even a Newton could keep in his mind—without committing a line to paper—the

[1] The physical characteristics of this manuscript are discussed in the Guide to the Text of this edition. The actual changes made by Halley on the text pages of the manuscript may be identified in the Apparatus Criticus, where each one is followed by the initials EH within square brackets. In addition, Halley has done some checking on the following verso pages: 90ᵛ, 95ᵛ, 120ᵛ, 125ᵛ, 126ᵛ, 200ᵛ, 210ᵛ, 245ᵛ, 268ᵛ, 329ᵛ, 340ᵛ, 408ᵛ, 416ᵛ, 435ᵛ, 436ᵛ. Primarily, these are drawings of figures, or the checking of mathematics, for example, working out a series of proportions (the equivalent to the more modern use of equations).

[2] Some examples are discussed in Chapter IV.

[3] See Chapter IV, §2.

[4] There are, however, one or two possible signs of dictation in *LL* and *M*. For instance, in *LL* Humphrey wrote 'qui essentia' for 'quiescentia' {9.9}, and in *M* he wrote 'limitum' for 'lemmatum' {37.22}. But such occurrences are rare. Such scribal errors could, of course, have resulted from a misreading of a manuscript being copied; they do not prove that the text was being dictated. The important point is only that neither *M* nor *LL* could have been wholly a first draft or the result of a first dictation. See also the final paragraph of §4 above.

whole structure of this complex treatise in three books, and also work out each proof wholly in his mind without pen or pencil and paper? Could the *Principia* then be dictated in so perfect a fashion from memory (or *ad hoc*) that each corollary would have found its proper place at once? Anyone who has studied Newton's manuscripts is aware that he tended to work with pen in hand, making version after version, constantly improving each successive presentation. To assume any other procedure in the making of the *Principia* is thus to put the *Principia* apart from practically every other known creative effort of Newton's—or of anyone else's!

I have said that the actual physical appearance of *M* belies its having possibly been a first dictation. There is another reason why *M* could not have been wholly dictated: the ability of the amanuensis. In several manuscripts Newton complains of Humphrey, saying that the *Principia* was 'copied by an Emanuensis who understood not what he copied'.[5] Can we believe that such an ignorant amanuensis could have written out so perfectly the mathematical proofs in which the *Principia* abounds?

In the statement just quoted, and in others,[6] Newton says specifically that the *Principia* was 'copied', and that Humphrey did not understand 'what he copied'. This sounds as if Newton had worked out a version, perhaps in part dictated, which he then gave to Humphrey to copy. In a letter to Halley of 20 June 1686, Newton said that the second book of the *Principia* 'wants transcribing & drawing the cuts fairly',[7] which reads as if there were a prior version for Humphrey to copy and some rough sketches of the diagrams that needed to be drawn 'fairly' so that cuts could be made. Both the rough sketches and this prior version have long since disappeared from view.

Newton's habit was to write out his works in drafts, often in an abbreviated form, which he would then expand into a more readable copy. Alterations and additions would be made either in the margins or between the lines, and often a little mark would indicate that a paragraph on a separate sheet was to be inserted. Or Newton would rewrite the page (or proposition), indicating by dashes some part of the old copy to be used at that point. To be sure, eighteen months is rather a short interval in which to compose so difficult a work as the *Principia*, and no doubt Newton did not revise his text as much as was his habit when not under such pressure.

If I seem to be dwelling overly long on a single point, the reason is that the paucity of actual work-sheets in Newton's hand is utterly baffling to anyone who wishes to find out how Newton wrote the *Principia*. This want of manuscripts is made all the more striking by the presence of well over 1,000 manuscript pages in the Portsmouth Collection relating to the improvement of the *Principia* after it had been printed. These include long-hand tables, computations, notes, and drafts, which—together with *M* and the annotations in E_1i, E_1a, E_2i, and E_2a—enable us

[5] See Supplement I.
[6] Chiefly in U.L.C. MS Add. 3968.
[7] [Newton] (1959–), *Correspondence*, vol. 2, p. 437.

to trace in considerable detail the evolution of the *Principia* from the final manuscript sent by Newton to Halley in 1686–7 for printing E_1, up to the editing and printing of the third and final edition in 1726.

It must be kept in mind that the Portsmouth Collection contains well over 1,000 leaves of long-hand drafts and versions of the *Commercium epistolicum*[8] and Newton's anonymous review, written for the *Philosophical Transactions*.[9] There are extant an overwhelming collection of notes and extracts on alchemy,[10] and even whole treatises copied out in Newton's hand. We have student notebooks recording intimate details of Newton's personal life,[11] notebooks recording his reading and his first essays in mathematics and science,[12] and page after page of astronomical observations and computations.[13] But among this vast collection of papers there is to be found no series of manuscripts enabling us to trace the growth of the *Principia*, section by section, proposition by proposition. The *Principia*, to use Fontenelle's rather charming expression concerning Newton himself,[14] is like the River Nile: it reveals itself to us only in full stream, its source remaining forever hidden from our view.

The absence of the first rough drafts, the pages of trial and error, of all three books of the *Principia* is especially noteworthy because of Newton's oft-repeated statement that they were very different in form from the *Principia* itself. Newton made a public declaration that he had found the major part of the propositions of the *Principia* by using a method of discovery not at all like the one adopted for presenting (and proving) these propositions in print. In the unsigned 'Recensio libri' which Newton wrote in English (*Phil. Trans.* no. 342, Jan.–Feb. 1715) for the *Commercium epistolicum*, and which was added (in a revised version in Latin) to the next printing of the *Commercium*, along with Newton's anonymous 'Ad lectorem', Newton's words are:

> By the help of the *New Analysis* Mr. Newton found out most of the Propositions in his *Principia Philosophiae*: but because the Ancients for making things certain admitted nothing into Geometry before it was demonstrated synthetically, he demonstrated the Propositions synthetically, that the Systems of the Heavens might be founded upon good Geometry. And this makes it now difficult for unskilful men to see the Analysis by which those Propositions were found out.[15]

[8] The fluxion dispute may be a rather exact parallel to the example of the *Principia*. Very few of the manuscripts remaining in the Portsmouth Collection are prior to December 1712 or January 1713; that is, prior to the *Commercium epistolicum*. There are some rare exceptions, as in the case of the *Principia*: some draft sheets for an introductory preface and a 'History' in English 'which bears the same relation to the published *Commercium* as *De Motu* does to the *Principia*'.

[9] [Collins] (1856), *Commercium epistolicum*; [Newton] (1715), 'Recensio libri'; U.L.C. MS Add. 3968.

[10] A great many of these are now in the Keynes Collection, King's College Library; see Munby (1952*a*) and [Sotheby and Co.] (1936), *Newton papers*.

[11] U.L.C. MS Add. 3996, MS Add. 4000; others are in the Fitzwilliam Museum, Cambridge and the Morgan Library, New York.

[12] See Smith (1927), 'Two unpublished documents'; Hall (1948), 'Newton's notebook'; Westfall (1962), 'Newton's philosophy of nature'; and Westfall (1963*a*), 'Short writing'.

[13] Chiefly U.L.C. MS Add. 3969.

[14] See Fontenelle's 'Eloge' of Newton, reprinted in [Newton] (1958), *Papers and letters* [Cohen and Schofield], p. 446; and Gillispie (1958).

[15] Newton (1715), 'Recensio libri', p. 206.

Writing at the height of the controversy with Leibniz over the discovery of the method of fluxions, Newton was in effect replying to any critic who might ask why he had not composed the *Principia* in the language of fluxions if he had indeed discovered the new mathematics—as he alleged—prior to 1685.

This statement has two distinct meanings. First, Newton is saying that his 'analysis' differed from his 'synthesis' in the *Principia*: that is to say, his mode of discovery was not the same as his logical exposition in print. But, secondly, Newton refers specifically to 'the help of the *New Analysis*', by which he means the whole armoury of fluxional methods (although not, of course, the dotted notation).

As a matter of fact, the superficial ('synthetic') form of the *Principia* is apt to be very misleading. For instance, it is often said that the *Principia* is written in the style of Apollonius, and has the appearance of a treatise on Greek geometry. But Sec. I of Book I (on limits) and the very opening propositions of Book I (Props. I, II, III) show the marks of 'the *New Analysis*'; for in a fashion wholly foreign to Apollonius, Newton uses proofs that involve relations which are true only in the limits. Furthermore, there are a great many applications of obviously infinitesimal methods, notably in Props. XLI and LXXXI of Book I (together with the propositions leading up to them), and in Lemma II of Book II.

Newton even went farther than making such general statements about the *Principia*; he actually said that certain specific theorems had been proved by analysis rather than by the synthetic or geometric methods by means of which they are displayed in the *Principia*. For instance, in a manuscript never printed by Newton, we find:

> By the inverse Method of fluxions I found in the year 1677 [should be 1679/80] the demonstration of Kepler's Astronomical Proposition viz. that the Planets move in Ellipses, wch is the eleventh Proposition of the first book of the Principles.[16]

Newton also claimed specifically that he had used the methods of his tracts *De quadratura*[17] in proving the principal theorems in the *Principia*. For this reason he intended to publish *De quadratura* as an appendix to the second edition of the *Principia*, a plan never realized, and he later thought to publish a version of that tract, originally called *Geometriae liber secundus*, either in a reprint of the second edition of the *Principia* or possibly as a supplement to a projected third edition.[18]

The *Principia* contains several examples of the use of expansions in infinite series, and in his review of the *Principia* in the *Philosophical transactions*[19] Halley

[16] See the collection of autobiographical statements of Newton's in Supplement I.

[17] D. T. Whiteside (private communication) observes that *De quadratura* 'exists in five major versions and was revised continuously over the years 1670–1714+ . . . The tracts *De quadratura* Newton intended to add to the *Principia* in 1686, *c.* 1692–3 and 1714 are three distinct tracts.' It will be very helpful indeed for all students of Newton when these tracts and the other mathematical writings of Newton's will have been made available in the current edition (in progress) of Newton's mathematical papers.

[18] Some of Newton's own statements about the use of fluxions, and analytic methods generally, in discovering and first proving Theorems of the *Principia* are to be found in Supplement I; on *De quadratura* in relation to the *Principia*, see Supplement VII.

[19] Reprinted in [Newton] (1958), *Papers and letters* [Cohen and Schofield].

called the reader's attention particularly to the use of infinite series as a notable feature of Newton's treatise. We are not at all astonished, furthermore, that Newton should claim that the methods of his tracts *De quadratura* should be basic to at least some parts of the *Principia*, since there are propositions (for example, Props. XLVI, LIII, LIV, Book I) in which the premise contains such a phrase as 'Granting the quadratures of curvilinear figures' or 'and the quadratures of curvilinear figures being allowed'. But it is unwarranted to assume (at least in the absence of any further evidence) that Newton proved such propositions as those of Secs. II and III of Book I in the way to be found in today's treatises on celestial mechanics, only using dotted letters in place of our differential notation. Not only had Newton not then invented dotted or 'pricked' letters, but there is no evidence that before Halley's visit he had as yet come into possession of a truly analytic proof that an inverse-square centripetal force may be a necessary condition for motion in a conic section, of which the elliptical motion of planets would be a special case.

On the other hand, Prop. XLI, Book I, of the *Principia*, does show how to find 'the trajectories in which bodies will move' and 'the times of their motions in the trajectories found', on the supposition of 'a centripetal force of any kind'. The resulting semi-analytic expression may easily be restated as a purely analytic expression, using the tools of the calculus. Furthermore, by taking the general condition as the special case of the inverse-square of the distance, the equation for a general conic follows at once.[20]

What ever happened to the work-sheets of the *Principia*? Do they still exist in some obscure private or public collection? Was this particular set of manuscripts —alone out of all of Newton's papers—lost or mislaid, either when the Portsmouth Collection was still in Hurstbourne Castle or during the actual transfer to the University Library in Cambridge? Did such work-sheets still exist among Newton's papers at the time of his death? Or were they lost or destroyed—either by chance or design—during Newton's own lifetime? We may possibly never be certain of the answers to these questions.

[20] Cf. Whiteside (1962), 'Newtonian research', p. 20; Lohne (1960), 'Hooke vs Newton'; Herivel (1965), *Background*, p. 34 [note to p. 17].

CHAPTER IV

WRITING THE 'PRINCIPIA'

1. THE DOCUMENTS: FROM 'DE MOTU' TO THE PRINTER'S MANUSCRIPT (*M*)

THE history of the *Principia* begins when Newton wrote out *De Motu* and however much he actually finished of *De motu corporum*, and then went on to compose his *magnum opus*. In this chapter I shall examine the documents that enable us to trace the development of the *Principia*, following the two tracts just mentioned, up to the completion of the manuscript (*M*) prepared for the printer. I shall give as complete an inventory as possible of all the manuscripts I have found that may be either states of some part of the *Principia* prior to *M*, or materials intended for the *Principia* but then replaced or discarded. In some cases I shall explore the degree to which we can discuss the contents of yet other documents that no longer survive, but whose existence may be inferred. Since my aim here is to show how Newton actually wrote the *Principia*, and not to trace the history of his concepts, methods, and demonstrations,[1] I shall not be concerned with either *De Motu* or *De motu corporum* (for which see Chapter III, §§3 and 4), nor with the antecedent essays and bits and pieces going back to the 1660s, including student notes, the critique of Cartesian physics (recently given the title *De gravitatione et aequipondio fluidorum*), the presentation of dynamics and the solutions of dynamical problems in the Waste Book, the so-called vellum manuscript, the early tract on 'The Lawes of Motion', the untitled manuscript recently given the name *Gravia in trochoide descendentia* and others, now available for study in the volumes of Newton's *Correspondence* and the collections published by A. R. and M. B. Hall and by J. W. Herivel.[2] Portions of these preliminary writings eventually became embodied in the Definitions, Laws of Motion, and some propositions and lemmas of the *Principia*, thus accounting for possibly one-fiftieth of the whole *Principia*. To some degree these may be regarded as work-sheets, particularly the earlier ones, but the versions of *De Motu* are in the nature of terminal stages toward the *Principia*, or finished preliminary states, and do not show us how Newton initially discovered these propositions, nor how he first proved them. *De Motu* and the antecedent texts on dynamics have been discussed by various scholars, with somewhat differing conclusions, and I shall not say anything more about them here. Instead, my goal

[1] A task reserved for the Commentary volume.

[2] To this list must be added vol. 3 of [Newton] (1967–), *Mathematical papers*, especially Part 2, introduction (appendix) and §3 ('Harmonic motion in a cycloidal arc').

will be to deal exclusively with the stages whereby Newton advanced from fragments or short versions to a complete manuscript in three books, ready for the printer.

The documents that are later than the tracts *De Motu* and *De motu corporum* and yet prior to *M* may be grouped as follows:

(*a*) A version of the Definitions, Laws of Motion, and a considerable part of Book I: primarily the text (*LL*) deposited by Newton in the University Library as his Lucasian Lectures for 1684 and 1685, to which I have found a supplement among the manuscripts in the Portsmouth Collection.

(*b*) A page of definitions.

(*c*) A tract on projective conics, of which a portion was used in Sec. V, Book I.

(*d*) A few work-sheets for Book II.

(*e*) A form of Book III that was discarded and replaced by the Book III that we know in the *Principia*. The early form is very different indeed and, in the manuscript, is entitled *De motu corporum, liber secundus*. I have called it *De mundi systemate* (*liber secundus*) to distinguish it from both the *De mundi systemate, liber tertius* of the *Principia*, which replaced it, and the *De motu corporum, liber secundus* of the *Principia*. Whole paragraphs of *De mundi systemate* (*lib. secundus*) were taken over practically *verbatim et litteratim* in *De mundi systemate* (*lib. tertius*).

(*f*) Some work-sheets for Book III.

(*g*) A critique of a manuscript draft of the *Principia* that consisted of Definitions, Laws of Motion, all or at least a major part of Book I, and parts of Books II and III. This critique enables us to infer the existence of a pre-*M* version of Books II and III.

(*h*) A few odd sheets of *M* that were discarded and replaced.

With the exception of the work-sheets for Book II, all of the extant pages in the foregoing list are more or less finished copies, almost all apparently copied out by Humphrey from some preliminary document or documents and hence of no real value in indicating Newton's method of discovering or proving the propositions in the *Principia*. I shall discuss them in the order of their appearance in the foregoing list.

2. THE LUCASIAN LECTURES (*LL*): A DRAFT OF BOOK I

As Lucasian Professor, Newton was required to deliver 'public lectures' for the students and to deposit texts corresponding to them in the University Library. He was appointed Professor on 29 October 1669 and gave his first set of lectures on optics, the next on arithmetic and algebra, the third on the motion of bodies, and a fourth on the system of the world. Manuscripts allegedly corresponding to these four sets of lectures were deposited by Newton in the University Library.[1] A description of the four manuscripts, accompanied by a rough collation of each with the first printed edition, may be found in a supplement to Edleston's 'Synoptic view of Newton's life' in his edition of the Newton–Cotes correspondence.[2]

[1] The Lucasian Lectures, corresponding to part of Book I of the *Principia*, are to be found in U.L.C. MS Dd. 9.46.

[2] Edleston (1850), pp. xci–xcviii.

I have discussed these manuscripts as a group in Supplement III, and shall observe here only that there are reasons to doubt that these deposited manuscripts are necessarily accurate records of what Newton knew or had written out at any given time. Hence, I do not believe that the specific text of the first 'lecture' on the motion of bodies in *LL* was necessarily written in or before October 1684, the date given by Newton in the margin, although he may very well have been lecturing on this general topic at that time. The most that can be said with any certainty is that this set of lectures is an early version of Book I of the *Principia*, together with the Definitions and Laws of Motion, and that it must therefore have been composed between Halley's visit, in August 1684, and April 1686, when the manuscript of the first book was exhibited at the Royal Society.[3]

The Lucasian Lectures (*LL*) consist of a text in Humphrey Newton's hand, made up of parts of two manuscripts (or states) which I have designated LL_α and LL_β. The first state (LL_α) is represented in *LL* by a series of leaves numbered originally by Newton from 9 to 32 in the upper right-hand corner of each recto page. There are additional pages of LL_α which I have found and identified among Newton's manuscripts, and I have been able to make conjectures as to the contents of other portions of LL_α now missing. Hence, LL_α may be divided as follows:

(1) $\langle LL_\alpha^{1-8}\rangle$, now missing;
(2) LL_α, leaves 9–32, now in *LL*, and numbered by Newton;
(3) Further leaves, now in the Portsmouth Collection.

All the existing leaves have a text written by Humphrey on the recto side. Then, there are revisions, corrections, and additions, chiefly in Newton's hand, which spill over to the otherwise blank verso pages. (For two pages from *LL* see Plates 7 and 8.)

When Newton discarded the first eight pages (or leaves) of LL_α, he replaced them by a new beginning, comprising twenty pages (or leaves), all unnumbered save for one (p. 9). I have designated these twenty new pages $LL_\beta(1)$. In the state LL_β, they are followed by leaves 9–24 of LL_α, heavily revised. Then comes a set of fifty-six pages (or leaves) of LL_β, which completes *LL*. But there must have been other pages of LL_β not now part of *LL*. In form, the pages of LL_β resemble those of LL_α. The contents of LL_β may be listed as follows:

(1) $LL_\beta(1)$, twenty leaves, of which only leaf 9 is numbered;
(2) LL_α, leaves 9–24, revised;
(3) LL_β;
(4) Further pages not now in *LL*.

[3] I have referred in Chapter III, §§2 and 6, to the problem that would arise if we had to take the dated portions of *LL* as texts actually completed and read to students at the indicated times. For, allegedly, Newton was lecturing in October 1684 from a manuscript that is almost identically the opening part of the *Principia*, whereas in the following November he sent Halley via Paget a version of the much earlier and more primitive tract *De Motu*; he also sent *De Motu* to be registered at the Royal Society in the following February, after the lectures for 1684 had been finished.

From these descriptions, it is seen that *LL* is made up of the following parts:

(1) $LL_\beta(1)$;
(2) LL_α, leaves 9–24, revised;
(3) LL_α, leaves 25–32;
(4) LL_β, 56 leaves.

When Newton replaced $\langle LL_\alpha^{1-8}\rangle$ by $LL_\beta(1)$ he changed the '9' on the original ninth leaf of LL_α to a '21' and added a second number in a new sequence on each eighth leaf. Thus, the old '17' is also the new '29' and the old '25' is also the new '37'. The whole manuscript bears the title

DE MOTU CORPORUM
LIBER PRIMUS

but on the penultimate unnumbered page of $LL_\beta(1)$ (leaf 22 in the librarian's sequence) this title is repeated; hence we must not too hastily conclude that Newton actually considered the Definitions and Laws of Motion as parts of Book I rather than as preliminaries to the whole treatise. The assignment of dates and division into Lectures is done in Newton's hand in the margin.

Although the whole of *LL* is written out by Humphrey, the handwriting of the first twenty leaves, or $LL_\beta(1)$, while undoubtedly Humphrey's, has a different appearance from that of LL_α, being somewhat smaller and tighter; for example, it has some forty-seven lines per page (on the librarian's page 20) rather than forty (on leaf 17/29). Newton has made changes and additions on both the recto pages and the otherwise blank verso pages.

There were originally ten definitions in $LL_\beta(1)$:

I. Quantitas materiae; II. Axis materiae; III. Centrum materiae; IV. Quantitas motus; V. Materiae vis insita; VI. Vis impressa; VII. Vis centripeta; VIII. Vis centripetae quantitas absoluta; IX. Vis centripetae quantitas acceleratrix; X. Vis centripetae quantitas motrix.

Newton has crossed out II and III and changed the subsequent numbers. Definition I is considerably worked over and a new version is then written out on the verso side; this in turn is worked over.[4] The result is, save for matters of detail, the text of *M* and E_1. These definitions are followed by the scholium on time and space as in *M* and E_1. Then come the 'Axiomata sive Leges Motus', as in *M* and E_1, followed by the corollaries. A replacement of the final three paragraphs of Corol. 2 makes it read as in *M* and E_1; this is in Humphrey's hand on the verso page. It is amusing to see that the final letter of 'plana' comes at the fold of the sheet, so that it cannot be read. The word might thus appear to be 'plano'. This accounts for the error in *M*, E_1, and E_2 {17.1}.

Following the Laws of Motion, the corollaries, and the scholium, *LL* has the text of Sec. I (Lemmas I–XI), then Sec. II (Props. I–IX, Lemma XII, Prop. X), then Sec. III (Props. XI–XII, Lemmas XIII–XIV, Props. XIII–XVII). Proposition

[4] Herivel (1965), *Background*, pp. 321; 325, n. 2.

XVII is mistakenly numbered XVI (the second occurrence of this number). Then *LL* goes on to Prop. XVIII (called mistakenly XVII),[5] which in *M* and E_1 is (following Lemma XV) the beginning of Sec. IV, Prop. XIX (called XVIII), Lemma XVI (called XV),[6] Prop. XXI (called XIX), Prop. XXX (called XX) and Props. XXXII–XXXV (called XXI–XXIV), the latter group beginning Sec. VII in *M* and E_1. Thus, of the Sec. VI of *M* and E_1, 'De inventione motuum in Orbibus datis', containing the discussion of 'Kepler's Problem', *LL* has only Prop. XXX without the corollaries {104.16 ff.} and has a scholium embodying a part only (and that quite different) of the scholium with which Sec. VI concludes in *M* and E_1. It is significant that the whole of the purely geometrical Sec. V of *M* and E_1 is absent, as is much of Sec. IV.[7]

One or two other interesting aspects of LL_α may be noted. The text of Book I is divided into Articles rather than Sections, in which feature it resembles LL_β and *M*; the change was made by Halley with Newton's permission (see Chapter V, §3). The two corollaries following Lemma X {34.1–10} were added by Newton after the manuscript had been finished, as were the final paragraph at the end of Sec. I {38.8 ff.}, the end of the proof of Prop. I {39.29 ff.}, the two corollaries to Prop. I {40.1 ff.}, and Prop. V. Proposition I, Theor. I was apparently at first just 'Theorema I', and Newton then added in his own hand 'Prop. 1' to make the designation contain both a proposition and a theorem number.[8] The same occurred for Humphrey's 'Theorema 2' and 'Theor 3', which were altered by Newton to become 'Prop. II. Theorema II' and 'Prop. III. Theor III'. Although Prop. IV was originally merely 'Theor. 4', there is written above it, centred on the page, in Humphrey Newton's hand (and then cancelled),

Theor. 2. Corporibus in circumferentiis circulorum

Here in LL_α, Prop. IV originally read word for word as in Theor. 2 of *De Motu*:

Corporibus in circumferentiis circulorum uniformitèr gyrantibus, vires centripetas esse ut arcuum simul descriptorum quadrata applicata ad radios circulorum.

That when bodies are revolving uniformly in the circumferences of circles, the centripetal forces are as the squares of the arcs described at the same time divided by [*lit.* applied to] the radii of the circles.

This was then altered by Newton, in several stages, to read:

Corporibus circulos aequabili motu describentibus, tendere vires centripetas ad centra eorundem circulorum et esse inter se ut arcuum simul descriptorum quadrata applicata ad circulorum radios.

That when bodies are describing circles with equable motion, the centripetal forces tend toward the centres of the same circles and are to each other as the squares of the arcs described in the same time divided by the radii of the circles.

[5] When Newton added Prop. V, he increased by one the numbers of the succeeding propositions, but forgot to change the old XVI to XVII, introducing an error in numeration continuing throughout the remainder of the manuscript. [6] Lemma XV of *M* and E_1 is not present in LL_α. [7] See §4 below.

[8] At first this 'Theorema 1' read: 'Gyrantia omnia, radiis ad centrum ductis, areas temporibus proportionales describere', which may be translated, 'That all revolving bodies, by radii drawn to the centre, describe areas proportional to the times'.

Finally, Newton wrote out a new text on the verso page:

Corporum quae diversos circulos aequabili motu describunt, vires centripetas ad centra eorundem circulorum tendere, et esse inter se ut arcuum simul descriptorum quadrata applicata ad circulorum radios.

The centripetal forces of bodies which by equable motion describe different circles tend to the centres of the same circles; and are to each other as the squares of the arcs described in the same time divided by the radii of the circles.

Considering the importance of the Scholium to Prop. IV,[9] it is fascinating to see that in LL_α this scholium is but a short paragraph, reading at first:

Casus Corollarii quinti obtinet in corporibus Coelestibus. Illorum tempora periodica sunt in sesquiplicata ratione distantiarum a centris; et propterea quae spectant ad vim centripetam decrescentem in duplicata ratione distantiarum a centris, decrevi fusius in sequentibus exponere.

The case of the fifth corollary obtains in the celestial bodies. Their periodic times are in the sesquiplicate ratio of their distances from the centres; and therefore I have determined to set forth more fully in the following the things which relate to centripetal force decreasing in the duplicate ratio of the distances from the centres.[10]

In an alteration in Newton's hand, the partial sentence 'Illorum. . . centris' is replaced by 'Planetarum tempora periodica sunt in sesquiplicata ratione distantiarum ab orbium suorum [centro communi *del.*] centro seu nodo communi' ('The periodic times of the planets are in the sesquiplicate ratio of the distances from the common centre or node of their orbits'), after which the whole altered passage is cancelled.

These alterations of *LL* and differences between *LL* and *M* and E_1 show its value in indicating stages of growth and revision of the *Principia*. In the differences between LL_α and *M* we may discern the last additions and changes just before, and while, a manuscript was written out for printing. We do not know how these additions were indicated to Humphrey; they may have been dictated to him as he went along (either in whole or in part) or they may have been written out for him on separate sheets or slips of paper. From Newton's alterations written out on LL_α we may learn what Newton changed and added after Humphrey had copied out the preliminary first version. But *LL* does not show us how any of the propositions had been discovered, nor does it reveal any difference between methods of discovery (and proof) and methods of presentation. On that fundamental topic, we remain wholly in the dark.

The second part of the Lucasian Lectures (LL_β) is far less interesting than LL_α, because it is more nearly identical to *M*. It begins with Corol. 6 to Prop. XVI and ends in the middle of the fourth sentence of the proof of Prop. LIV. The fact that in *LL* both LL_α and LL_β end in the middle of sentences is a major reason for not considering this manuscript (*LL*) a wholly accurate and careful record of the texts of lectures actually delivered at the times specified. There is no question of the

[9] See Chapter V, § 1.

[10] Only the first sentence is identical with a sentence (also the first) of the scholium as found in *De Motu*. But the sense of the first half of the second sentence (up to the semicolon) agrees with that of the second sentence of the scholium in *De Motu*.

fact that LL_β is a later document than LL_α, in part because LL_β is so close to *M* that the propositions and lemmas follow in the same succession and with the same numbering. There is a portion of LL_α that has been expanded and incorporated into LL_β, containing propositions numbered XVI–XXIV; these correspond to the later version in LL_β (and in *M* and E_1) as follows:

LL_α	LL_β (and *M* and E_1)
Prop. XVI	Prop. XVI (Corols. 6–9)
Prop. XVI [*bis*]	Prop. XVII
Props. XVII–XVIII	Props. XVIII–XIX
Lemma XV	Lemma XVI
Prop. XIX	Prop. XXI
Prop. XX	Prop. XXX
Props. XXI–XXIV	Props. XXXII–XXXV

The leaves of LL_β, like those of LL_α, are written on recto pages in Humphrey's hand with corrections and additions made by Newton on these same pages and on occasional otherwise blank verso pages. The leaves of LL_β are numbered by Newton, every eighth leaf bearing a number in the upper right-hand corner: 37, 55, 63, 71, 79, 87, 95; evidently '55' was an error for 45, which was then continued. It is to be observed that, if LL_α had ended just where LL_β begins, then the final eight leaves of LL_α would have been discarded. The first of these, beginning with Corol. 6 (just as does the first page of LL_β), bears Newton's old number '25' and the new number '37'; hence we see why LL_β begins with a page '37'. When Newton combined LL_α and LL_β to deposit *LL* in the University Library, he simply neglected to take out or throw away the final gathering of eight leaves in LL_α.

The overlapping portions of LL_α and LL_β enable us to see that LL_β is a later state than LL_α. Most of the alterations made by Newton in hand in LL_α are incorporated in the text of LL_β. Thus in Corol. 7 to Prop. XVI the word 'revolventis' {62.27} was 'gyrantis' in LL_α, and was then changed in Newton's hand; but in LL_β the text line contains only the final form, 'revolventis'. The last sentence, 'Hinc etiam...in Hyperbola major' {62.33–35}, is added in LL_α in Newton's hand, but in LL_β it is incorporated in the text by Humphrey. And so on.

Our interest in LL_β lies chiefly in the changes and additions in Newton's hand. Thus, the division at 'Artic. IV.' (later to become 'Sect. IV.') was an afterthought, and Lemma XV {66.4–18} was added by Newton in hand, also as an afterthought, to begin this Sec. IV. Likewise, the division at 'Artic. V. Inventio Orbium ubi umbilicus neuter datur' was a later addition in Newton's hand replacing an unnumbered subdivision entitled 'Solutio Problematis Veterum de Loco Solido'. Lemma XXIII {91.15 ff.} was also an addition, appearing on a verso page, but, except for the title, it is written out in Humphrey's hand, not Newton's. Other additions, deletions, and emendations enable us to trace changes in detail, but not in structure, made by Newton after Humphrey had copied out LL_β. Interestingly enough, the divisions at Arts. VI, VII, VIII, and IX (mistakenly written VIII) are all added in Newton's hand after the manuscript had been completed. But Humphrey himself has written out boldly in large letters

Artic. X.

De motu corporum in superficiebus obliquis.

which was altered by Newton to become

De motibus corporum in planis excent⟨r⟩icis et lineis datis.

and then altered once again to become

De motibus corporum in superficiebus datis deque motibus reciprocis funipendulorum.

LL_β shows the discovering mind of Newton even less than LL_α. We may conclude only that since the original versions of LL_α and LL_β made by Humphrey are rather good, representing a semifinal state, there must have been antecedent drafts which have now disappeared. These may have been in whole or in part in Newton's hand. Presumably they were closer to the work-sheets than those pages making up LL_α and LL_β.

It has been remarked above that both LL_α and LL_β end abruptly, in the middle of sentences; the pages that must have followed both LL_α and LL_β have not as yet been found and identified. I have, however, discovered a set of eight leaves in the Portsmouth Collection[11] which can be established as having once formed part of LL_α. They have the physical appearance of the rest of *LL*, being written in Humphrey's hand, on recto sides of the page, with corrections in Newton's hand. The water-mark on the last sheet of LL_α is the same as that on the first sheet of this group of eight leaves, which I have called MS_x; furthermore, the first page of MS_x is numbered 41 in Newton's hand, in a manner similar to the numbering of LL_α. Since the last leaf of LL_α is numbered 32, there must have been an intermediate set of eight leaves, now missing, corresponding to recto page numbers 33–40.

The fragment MS_x begins in the middle of a proof of what must have been a Prop. XXXIV, since it is followed by a Prop. XXXV. Hence, the missing eight leaves (pp. 33–40) must have contained propositions numbered XXV–XXXIV, filling the gap between the final Prop. XXIV of LL_α and the opening Props. XXXIV–XXXV of MS_x. These missing propositions may be identified as follows. First I shall list the contents of MS_x, placing alongside each proposition its designation in *M* and E_1.

MS_x	*M* and E_1
[Prop. XXXIV,]	Prop. LXIV, Prob. XL
Prop. XXXV, Theor. XVII	Prop. LXVI, Theor. XXVI
Prop. XXXVI, Theor. XVIII	Prop. LXVII, Theor. XXVII
Prop. XXXVII, Theor. XIX	Prop. LXVIII, Theor. XXVIII
Prop. XXXVIII, Theor. XX	
Prop. XXXIX, Theor. XXI	Prop. LXX, Theor. XXX
Prop. XL, Theor. XXII	Prop. LXXI, Theor. XXXI
Prop. XLI, Theor. XXIII	Prop. LXXII, Theor. XXXII
Prop. XLII, Theor. XXIV	Prop. LXXIII, Theor. XXXIII
Prop. XLIII, Theor. XXV	Prop. LXXIV, Theor. XXXIV

[11] MS Add. 3965, ff. 7–14.

The gap to be filled in is:

[end of LL_α]	M and E_1
Prop. XXIV, Theor. XI	Prop. XXXV, Theor. XI
↓ ?	↓ ?
Prop. XXXIV, Prob. [?]	Prop. LXIV, Prob. XL
[start of MS_x]	

Missing are Props. XXV–XXXIII, which could have been any nine out of Props. XXXVI–LXIII of M and E_1. But the choice may be narrowed down by a reference in the proof of the Prop. XXXV, Theor. XVII in MS_x (=Prop. LXVI, Theor. XVI of M and E_1) to a 'Prop. XXVIII' which is a 'Theor. XIII'. This turns out to be Prop. LVIII, Theor. XXI of M and E_1. Let us put it into the gap between MS_x and LL_α:

[end of LL_α]	M and E_1
Prop. XXIV, Theor. XI	Prop. XXXV, Theor. XI
↓ ?	↓ ?
Prop. XXVIII, Theor. XIII	Prop. LVIII, Theor. XXI
↓ ?	↓ ?
Prop. XXXIV, Prob. [?]	Prop. LXIV, Prob. XL

It may be seen that the bottom gap on both sides comprises five propositions; undoubtedly, then, the missing Props. XXIX–XXXIII of MS_x would correspond to Props. LIX–LXIII of M and E_1. If we assume that each theorem or problem of M and E_1 was also similarly a theorem or problem in MS_x, then the lower part of the table above may be completed at once as follows:

MS_x	M and E_1
Prop. XXVIII, Theor. XIII	Prop. LVIII, Theor. XXI
Prop. XXIX, Theor. XIV	Prop. LIX, Theor. XXII
Prop. XXX, Theor. XV	Prop. LX, Theor. XXIII
Prop. XXXI, Theor. XVI	Prop. LXI, Theor. XXIV
Prop. XXXII, Prob. [?]	Prop. LXII, Prob. XXXVIII
Prop. XXXIII, Prob. [?]	Prop. LXIII, Prob. XXXIX
Prop. XXXIV, Prob. [?]	Prop. LXIV, Prob. XL

This reconstruction appears reasonable to the extent that it requires three of the propositions to be theorems, numbered XIV, XV, XVI; and indeed the first numbered proposition in MS_x is Prop. XXXV, which is Theorem XVII.

There still remains a gap, however, between the last Proposition of LL_α, Prop. XXIV, Theor. XI (=Prop. XXXV, Theor. XI of M and E_1) and the above-mentioned Prop. XXVIII, Theor. XIII (=Prop. LVIII, Theor. XXI of M and E_1). This gap must contain propositions numbered XXV, XXVI, XXVII, of which one would also be a Theor. XII; hence two of the propositions would have to have been problems. It is possible to find out which of the Theorems XII–XX of E_1 must have been 'Theor. XII' in the missing part, and likewise which two of the problems would have belonged here. If MS_x is a continuation of LL_α, then the two problems must have borne the numbers XIV and XV, since the last problem in LL_α is 'Prop. XXI. Prob. XIII.'. In this case, the foregoing Props. XXXII–XXXIV, listed without the problem numbers, would have been numbered Prob. XVI, Prob. XVII, and Prob. XVIII.[12]

Of these Prop. XXVII must have been also Theor. XII, corresponding to Prop. LVII (Theor. XX) of *M* and E_1. Furthermore, Prop. XXV (Prob. XIV) and Prop. XXVI (Prob. XV) would have corresponded, respectively, to Prop. XXXVI (Prob. XXV) and Prop. XXXVII (Prob. XXVI) of *M* and E_1. (For details, see Supplement IV.)

The agreement between MS_x and LL_α makes it plausible that the two belong together, although the continuation of LL_α may have contained a revised copy of MS_x rather than MS_x itself. Yet the numbering of the pages indicates, at the very least, that at some time MS_x and a prior set of eight leaves must have been part of LL_α. Further information concerning MS_x in relation to LL_α will be presented below in Supplement IV.

We are now in a position to say something definite about how far the manuscripts called LL_α and LL_β may actually have gone, although—in the absence of the last pages of both—we cannot, of course, speak positively. In *M* and E_1, Book I contains fourteen Sections, ending with a Scholium to Prop. XCVIII, Prob. XLVIII. LL_α, to judge by ff. 7–14 of MS Add. 3965, went at least as far as a Prop. XLIII, Theor. XXV, which in *M* and E_1 is Prop. LXXIV, Theor. XXXIV; in *M* and E_1, this Prop. LXXIV occurs as the fifth proposition in Sec. XII of Book I. This much of LL_α, with some omissions, would account for some 197 pages of the 235 printed pages making up the Definitions, Laws of Motion, and Book I in E_1. But LL_α must also have contained some of the propositions later assigned to the new Book II, as shown in Supplement IV.

Since LL_β is practically identical with *M*, and seems to have included at least everything in LL_α, it is a reasonable conclusion that LL_β must similarly have gone at least to the end of Sec. XII of Book I. There is, however, independent evidence that LL_β contained both Sec. XIII (Prop. LXXXV, Theor. XLII through the Scholium following Prop. XCIII, Theor. XLVII) and Sec. XIV (Prop. XCIV, Theor. XLVIII through the Scholium following Prop. XCVIII, Prob. XLVIII). At one time, furthermore, Newton apparently planned to have in LL_β some portions of what we know as Book II in the *Principia*. The conclusion of LL_β is discussed below in §10, and in Supplement IV.

[12] In MS_x, a postil indicates that Fig. 40 goes with Prop. XXXV (=Prop. LXVI [E_1]). In LL_α, a postil similarly indicates that Figs. 31 and 32 go with Prop. XXIV (=Prop. XXXV [E_1]), leaving a gap of seven figures to be accounted for. The seven propositions that almost certainly appeared in the missing pages just before MS_x (Props. LVIII–LXIV of E_1) account for three figures, and there is one figure each for Props. XXXVI and XXXVII, a total of five, which is two short. These occur as a double figure in Prop. LVIII (Theor. XXI) of the *Principia*. But as Prop. XXVIII (Theor. XIII) of LL_α, these two might well have been separate (and separately numbered) figures. Commonly Newton would use a double figure as an illustration in early drafts, and then would combine them into a single composite diagram. Another possibility, however, is that there could have been one or more intervening propositions that were cancelled or discarded and whose past existence may no longer be determinable.

When Newton sent Halley the manuscript of Book I of the *Principia*, he apparently retained *LL* for record purposes, as may be seen by examining *LL* in the light of Newton's correspondence with Halley. For, on 14 October 1686, Halley, in the throes of overseeing the printing of Book I, wrote to Newton about the Scholium following Prop. XXXI, Book I {109.12 ff.}, referring 'to the 63 figure'. In reply, on 18 October 1686, Newton wrote that 'the words *vel Hyperbolae* in y^{e} 3^{d} line are to be struck out, & in y^{e} 5^{t} or 6^{t} line the words *quae sit ad GK* should be quae sit ad $\frac{1}{2}$GK'. In *LL*, this scholium begins on the top of the page; the reference to 'Fig. 63' occurs in the margin. In the third line of the text, the words 'vel Hyperbolae' have been struck out by Newton; and in the sixth line a '$\frac{1}{2}$' has been inserted before 'GK'. Clearly, then, Newton, in answering Halley, merely referred to his own copy, *LL*, making in it the very corrections he asked Halley to make in *M*.[13]

If *LL* served as the record copy while E_1 was being printed, what has happened to the rest of it? Did Newton destroy or lose it? We simply do not know. Of course, once E_1 was in print, there was no need for anyone to keep either *M* or *LL*.[14] Indeed, the only useful purpose that the manuscript sheets of *LL* might possibly have served would have been to provide pages to be handed over to the University Library as the purported text of the lectures Newton was required to deliver and then to deposit under the terms of the Lucasian Professorship.[15]

3. 'DE MOTU CORPORUM: DEFINITIONES'

The foregoing title occurs at the head of a single page in Newton's hand (MS Add. 3965, f. 21^{r}), giving five definitions, as follows:

1. Quantitas materiae
2. Quantitas motus
3. Materiae vis insita
4. Vis impressa
5. Vis centripeta

Definitions 3 and 4 are revised and written out again on the verso side of the page.

These five definitions are anterior to $LL_{\beta}(1)$, as may be seen in one or two brief examples. 'Quantitas materiae' is defined here as follows: 'est quae oritur ex ⌊illius⌋ [*changed from* ipsius] densitate et magnitudine conjunctim'.[1] In *LL* we find 'est [copia seu *del.*] mensura ejusdem orta ex illius densitate et magnitudine conjunctim'. Again, Definition 4 appears here as: '*Vis impressa* est quae corpus ex statu suo...deturbare nititur'; it is then altered by Newton to 'est actio in corpus exercita ad mutandum statum ejus', which is the form written out by Humphrey

[13] In this case Halley did not need to insert the two corrections into *M*, since Newton sent him a replacement of the beginning of the scholium that included the portion containing these corrections. See the Apparatus Criticus {109.13 ff.}.

[14] I have not been able to discover how it was that *M* survived.

[15] In the absence of *M*, Newton could have used *LL* to determine whether a given error in E_1 might have originated with himself, but this would hardly have warranted the preservation of the manuscript.

[1] The corner brackets around 'illius' indicate that this word is a later insertion.

in *LL*, save that 'ejus' precedes rather than follows 'statum'. The sentence explaining Definition '3. *Materiae vis insita*' occurs here as: 'Estque corpori suo proportionalis', following which Newton has inserted 'neque differt ⌊quicquam⌋ ab *inertia* massae nisi in modo conceptus nostri'. This is the form in which Humphrey has written out the version in *LL*, except that 'concipiendi' replaces 'conceptus nostri'. In *LL*, Newton has later altered the beginning of Humphrey's sentence so as to make it read 'Haec semper proportionalis est suo corpori, neque differt', as in *M* and E_1.

This manuscript is of interest in showing the stages of development of the Definitions that appear at the beginning of the *Principia*. There are a number of distinct states, illustrating for us the way in which Newton clarified his thought and expression, in a manner reminding us of the development of the tracts *De Motu* and *De motu corporum*. Here, on a single page, we have a draft of the five fundamental definitions of the *Principia*; lacking as separate entities, but included among the comments, are the three aspects of centripetal force, the 'absolute quantity', the 'accelerative quantity', and the 'motive quantity' (these later became Defs. VI, VII, and VIII). Definition 3 is the most extensively rewritten; at one time Newton would have had the text begin: '*Materiae vis insita* est ⌊inertia sive⌋ potentia resistendi...', but then cancelled his insertion. This set of definitions has been printed by the Halls[2] (but without showing the alterations or giving the cancelled passages), and again by Herivel[3] (who indicates the existence of, but largely does not print, the cancelled passages). The Halls considered this page to be complete in and of itself, but Herivel has suggested that it is continued to another page (f. 25ᵛ). Here we find:

6. Densitas corporis est quantitas ⌊seu copia⌋ materiae collata cum quantitate [spatii quod *del.*] occupati spatii.
7. Per pondus intelligo...
8 Locus
9 Quies
10 Motus
11 Velocitas
12 Quantitas motus est quae oritur...
Def 14 [*This Definition has been cancelled*] Corporis vis exercita est...

I do not believe that this page is a continuation of the group of definitions under discussion. First of all, 'Def 14' occurs toward the bottom of the page, which is itself the verso of a page of definitions of another manuscript, the one I have called *De motu corporum*; thus this 'Def 14' occurs just in face of a gap in a renumbered sequence of definitions of *De motu corporum*, between a 'Def. 13' and a 'Def. 15' (formerly 'Def. 12' and 'Def. 13', changed to 'Def. 13' and 'Def. 14' and then the latter changed to 'Def. 15'). There can be no doubt that this 'Def 14' thus does not belong to the page of 'Definitiones' but rather to *De motu corporum*.[4]

But what of those numbered 6–12? It would seem very doubtful that Newton would have continued his 'Definitiones' in between the pages of *De motu corporum*

[2] Hall and Hall (1962), *Unpublished papers*, pp. 239–42.

[3] Herivel (1965), *Background*, pp. 315–20.

[4] See Chapter III, §4.

(which begins on ff. 25^r and 26^r) on f. 25^v, although it is possible that he could have done so in order to indicate that he wished to copy out from the facing page (f. 26^r) those definitions of which he gave only the titles: Locus, Quies, Motus, Velocitas. More precise information may be obtained by examining closely what Newton actually defined at each stage. I shall start with *De motu corporum* (f. 25^r):

Def. 1. Tempus absolutum est...
Def. 2. Tempus relative ⌊spectatum⌋ est [*changed from* Tempus relativum est]...
Def. 3. Spatium absolutum est [*changed from* Spatium absolute dictum est]...
Def. 4. Spatium relativum est...
Def. 5. Corpora ⌊in sensus ⌊omnium⌋ incurrunt ut⌋...
Def. 6. Centrum corporis cujusque est...[*This Definition is a later insertion by Newton; it has subsequently been cancelled*]
Def. 7. [*originally* 6, *changed to* 7 *by IN*] Locus corporis est...
Def. 8. [*originally* 7] Quies corporis est...

Then, on f. 26^r, the list continues:

Def. 9. [*originally* 8] Motus corporis est...
Def. 10. [*originally* 9] Velocitas est [*changed from* Celeritas motus est]...
Def. 11. [*originally* 10] Quantitas motus est...
Def. 12. [*originally* 11] Vis corporis seu corpori insita et innata est *changed by degrees to* Corporis vis insita innata et essentialis est...
Def. 13. [*originally* 12; *afterwards the whole Definition has been cancelled*] Vis motus seu corpori ex motu ⌊suo⌋ adventitia est...
Def. 15. [*originally* 13, *then altered to* 14, *and finally altered to* 15] Vis corpori illata et impressa est [*originally* Vis impulsus est]...
Def. 16. [*originally* 14, *then altered to* 15, *and finally altered to* 16] Vim centripetam appello...
Def. 17. [*originally* 15, *then altered to* 16, *and finally altered to* 17; *afterwards the whole Definition has been cancelled*] Resistentia est *changed to* Per ⌊medii⌋ resistentiam ⌊in sequentibus⌋ intelligo...

The third and final group of definitions appears on f. 23^r:

Def. 16. [*This definition has been cancelled*] Momenta quantitatum sunt...
Def. 18. [*originally* 17] Exponentes temporum spatiorum motuum celeritatum et virium sunt...

Hence, we see that at first the sequence ran from 1 to 17, that the introduction of a new Def. 6 increased the succeeding numbers by 1 up to Def. 12 (which became 13). Then the new Def. 14 was added by Newton on f. 25^v, so that from then on the numbers were increased by 2. In this sequence the original Defs. 13, 14, and 15 became respectively 15, 16, and 17. The old Def. 16 was cancelled, so that the final Def. 17 became 18 rather than 19 (which would have been the case had the old Def. 16 remained and been renumbered 18). In this sequence, the following definitions and numbers remain:

1. Tempus absolutum	7. Locus corporis	13. Vis motus
2. Tempus relative spectatum	8. Quies corporis	14. Corporis vis exercita
3. Spatium absolutum	9. Motus corporis	15. Vis corpori illata
4. Spatium relativum	10. Velocitas	16. Vis centripeta
5. Corpora	11. Quantitas motus	17. Medii resistentia
6. Centrum corporis	12. Corporis vis insita	18. Exponentes...

Now, let us suppose that the additions on the top of f. 25^v belong to the facing page 26^r (and, in part, the bottom of the preceding page). Then, in place of Def. 6 ('Centrum corporis'), Newton would have us write in a new '6 Densitas corporis est...'. He has also written out a new '7. Per pondus intelligo...'. Then, he lists '8 Locus', '9 Quies', '10 Motus', '11 Velocitas', just giving the single word and, evidently, thus implying that he is not changing the definitions he has already given of these quantities, but merely assigning new numbers so that they now will follow the new Definitions 6 and 7. But then he has written out a new Definition 12, 'Quantitas motus est...'. Hence, the defined quantities now are:

1.–5. Tempus absolutum, Tempus relative spectatum, Spatium absolutum, Spatium relativum, Corpora
6. Densitas corporis
7. Pondus
8.–11. Locus, Quies, Motus, Velocitas
12. Quantitas motus.

The last would replace the old Def. 11 of 'Quantitas motus', to be followed by

12. Corporis vis insita

which Newton should have renumbered 13. Then comes the new

14. Corporis vis exercita

in turn followed by

15.–17. Vis corpori illata, Vis centripeta, Medii resistentia [*renumbered from previous sequence*]
18. Exponentes [*renumbered from* 17].

When, finally, Newton cancelled Def. 14, 'Corporis vis exercita', he did not renumber the succeeding Defs. 15–18; no doubt he had already moved on to the next formulation of a system of definitions for dynamics.

If we do not accept the foregoing reconstruction, and rather suppose that the new Definitions 6–12 plus 14 (cancelled) go with the single page of definitions (f. 21), then we end up with a set in which there is a 'Quantitas Motus' appearing as no. 2 on f. 21 and reappearing as no. 12 on f. 25^v, an absurdity which vanishes if we assign the definitions on f. 25^v to the tract *De motu corporum* and to the pages, ff. 25/26, where they are found.[5]

The set of 'Definitiones' on f. 21 seem to be later than those of both *De Motu* and *De motu corporum*. We have seen that textually they precede *LL*. In *LL*, however, the definitions at first contained two (2, 'Axis materiae'; 3, 'Centrum materiae') which are lacking in the 'Definitiones' on f. 21. But these have later been cancelled. The final version in *LL* thus contains the same eight definitions as *M* and E_1, although there are textual differences. I have referred to the fact that the last three

[5] There are other difficulties that arise from assuming that Definitions 6–12 and 14 (f. 25^v) belong to the set found on f. 21. For instance, Def. 6 ('densitas') and Def. 7 ('pondus' and 'copia materiae') would already have been included in Def. 1 ('quantitas materiae'); Def. 10 ('motus') would have been in Def. 2 ('quantitas motus'); while Def. 14 ('corporis vis exercita') is an alternative formulation of Def. 4 ('vis impressa').

(on the absolute, accelerative, and motive measures of centripetal forces) are not present as separate 'Definitiones' on f. 21; they occur, however, within the paragraph of commentary on the final Definition 5, 'Vis centripeta'.

4. A PRIOR DRAFT OF SECTION V, BOOK I

Sections IV and V of Book I deal with a set of classical geometric problems, which have been described by H. W. Turnbull as follows:

> Newton found elegant constructions and proofs for describing a conic to satisfy five conditions of the type either to pass through a given point or else to touch a given line: and he dealt with all the six cases so arising. His treatment included a broad survey of the ancient Apollonian problem, the locus *Ad tres et quattuor lineas* (a problem which had led Descartes to his discovery of analytical geometry), wherein a conic is the locus of a point, the product of whose distances from two given lines is proportional to the [square of the] distance from a third line, or else to the product of its distances from a third and a fourth given line.[1]

It is at once obvious to anyone who reads the *Principia*, even superficially, that very little—if indeed any at all—of Sec. V is of real use to the astronomer in the practical determination of planetary or cometary orbits. Nowhere in the *Principia* does Newton ever refer to a lemma or proposition in this Sec. V of Book I. But it would be incorrect to characterize the other geometrical section (Sec. IV) in the same terms. For instance, in Prop. XLI, Book III, there is a reference to Prop. XIX (Sec. IV of Book I); while in Prop. XLII, Book III, there is a similar reference to Prop. XXI (also in Sec. IV of Book I). Possibly, therefore, Rouse Ball was overstating his case when he lumped Sec. IV and Sec. V together as 'a digression on pure geometry' that has 'but little connection with the rest of the Principia'.[2]

We may assume that Sec. IV was considered by Newton as a review of geometric principles that would be useful in the construction of planetary orbits, and particularly of cometary orbits. In this case, Sec. IV would have been written out especially for the *Principia* some time after *De Motu* and in time to be included in *LL*, that is, late in 1684 or (more probably) early in 1685. I have found no prior drafts of Sec. IV, nor any evidence that it may have been composed independently of the *Principia*.[3]

But the situation is quite different for Sec. V, which does appear to be rather digressive. One is led to suspect that Newton had by him a small tract which he incorporated into the *Principia* as an easy mode of publication. Or, possibly, he had been thinking independently of the aspects of conic sections treated in Sec. V, or he was stimulated into such thinking because Sec. III deals with motion along conic sections, as does also a part of Sec. II, and he began to write a tract or at

[1] Turnbull (1945), *Mathematical discoveries*, pp. 54–5; cf. Huxley (1959), 'Newton and Greek geometry'.

[2] Ball (1893), *Essay*, p. 81.

[3] D. T. Whiteside writes (in a private communication): 'I am sure that Section IV (which rings variants on the focus-directrix property of conics) was specially composed in 1685 for the *Principia*: there are certainly no comparable former drafts of it that I know of.'

least to write out the propositions in Sec. V and then introduced into the *Principia* whatever it was he had in hand. One can, albeit weakly, find some justification for the presence of Sec. V in the fact that Newton had just dealt with motion along conic sections, and was about to present yet further aspects of such motion; hence, any discussion of the geometric properties of these curves would not be wholly out of place.

I have said above that I have not been able to find an original draft of Sec. IV, the contents of which appear in *LL* and *M*.[4] But in the case of Sec. V, the early version, written out as the opening part of a separate tract, is readily identified. I do not know when Newton wrote out (or copied) the portion of this tract that became Sec. V of Book I. But it does not appear in LL_α, where the Scholium to what is Prop. XXI in *M* and E_1 is followed directly (without any subdivision) by what is Prop. XXX in *M* and E_1, although in LL_β, *M*, and E_1 this scholium marks the end of Art. IV/Sec. IV.

In LL_β, Prop. XVII (Prob. IX) was originally followed by Prop. XVIII (Prob. X) directly, the two being denoted respectively as 'Prob. VIII' and 'Prob. IX'. Newton altered these numbers; he then placed before Prop. XVIII the subdivision 'Artic. IV.' and followed it by a new Lemma XV, which is just the way the text appears in *M*, altered in E_1 only by the change from 'Artic. IV.' to 'Sect. IV.'. In *M* and E_1, this Articulus or Sectio ends with a short Scholium to Prop. XXI, but in LL_β there are two unnumbered lemmas which were later crossed out. Then, as mentioned earlier (§2), there was a subheading

Solutio Problematis Veterum
de Loco solido.

crossed out by Newton, who has put in its place

Artic. V.
Inventio Orbium ubi umbilicus neuter datur.

This first subheading enables us to identify positively the manuscript from which this portion of LL_β (essentially Sec. V of Book I) was taken. For among Newton's manuscripts there is a short treatise with this exact title; the title is even written out on two separate lines exactly as in LL_β (U.L.C. MS Add. 3963, §13, fol. 152^r). Since this fragment was an independent entity in itself when written, it began with 'Prop. 1'. (Humphrey copied this form into LL_β; later 'Prop.' is crossed out and 'Lemma XVII' is written in Newton's hand.)

This work exists in several states. The beginning of it occurs, written out entirely in Newton's hand, on a single page,[5] which is a revised copy made by Newton of an earlier version (MS Add. 3963, §12, f. 137^r) entitled:

De Compositione Locorum solidorum.

[4] Newton's habit of copying, revising, and recopying makes it now difficult always to recognize the early states of his writings. It would take a bold man indeed to say that an early version of any given text does not exist in any form within the vast reaches of the Portsmouth Collection.

[5] I do not believe there was any more since this f. 152^r is the only page of a folded sheet (f. 152^v and ff. 153^r and 153^v are blank) to have any writing. Possibly, Newton merely rewrote the first page of the earlier version for Humphrey to copy.

That the latter is a prior state may be seen in certain changes. For instance, in these two versions, Prop. 1 (=Lemma XVII in *M* and the printed editions) begins:

f. 137ʳ	f. 152ʳ
Si a Conicae Sectionis puncto Aliquo indefinite spectato P, ad Trapezii alicujus ABCD in Conica ista Sectione inscripti latera quatuor, si opus est infinite producta, AB, CD, AC, DB...	Si a datae conicae sectionis puncto quovis P, ad Trapezii alicujus ABCD in Conica ista sectione inscripti, latera quatuor [si opus est *del.*] infinite producta, AB, CD, AC, DB...

This second version[6] is very close indeed to LL_β, *M*, and E_1, even to capitalization and punctuation (save for the omission of the comma after 'producta' and the introduction in E_1 of a comma after 'ABCD'). But this text itself was redone by Newton in yet another version (MS Add. 3963, §12, f. 127ʳ), and there was one further revision leading to the final draft (MS Add. 3963, §12, ff. 145–6) which Humphrey copied out into LL_β.

It appears that a complete tract, perhaps corresponding to about half of Sec. V, may be found among the pages of MS Add. 3963, together with prior work-sheets, much written over. These occur not only in §12 of MS Add. 3963 but elsewhere (§6, now §8, ff. 63–6 and most of ff. 127–46). Possibly the parts were separated by the cataloguers when the Portsmouth Collection was presented to the University (see Chapter I, §5).[7]

For the student of the *Principia*, there are a number of fascinating aspects of this tract. Its history actually proves that it was not conceived and written as an integral part of the *Principia*, but was a separate and independent creation. Newton had it ready and merely directed Humphrey to copy the beginning of it into LL_β, changing the designations of propositions to lemmas, or corollaries to lemmas, and perhaps adding further material. Humphrey copied out this tract exactly, even reproducing the title line for line, so that at first, before Newton crossed out the title and introduced the subdivision of 'Artic. V', the pages containing the tract actually had the physical appearance of being extraneous: a digression, or a separate tract with its own title. This example is particularly interesting in showing us how Humphrey worked; that is, the way in which at times he actually copied out into *LL*, as presumably into *M* and any other pre-*M* versions or drafts, certain textual material written out by Newton.

[6] This prior version consists of a single folded sheet comprising ff. 137–8, on which Newton has filled f. 137ʳ and written a few lines on f. 137ᵛ, containing a Prop. 1, followed by 'Cas. 1', 'Cas. 2', and the first four words of 'Cas. 3. Si denique lineae quatuor', corresponding to Lemma XVII in *M* and the printed editions {73.11–74.24}. This too is a revised partial draft.

[7] D. T. Whiteside (private communication) believes this tract to be 'pre-1680 (as are all the sheets I gathered a few years ago in 3963.8). If you check variants, you will find that f. 152ʳ is *revised* on f. 145ʳ, itself *revised* by f. 127ʳ (the extended tract "De Compositione Locorum Solidorum").' See, further, [Newton] (1967–), *Mathematical papers*, vol. 4.

5. SOME WORK-SHEETS FOR BOOK II

For Book II there exists neither a draft comparable to *LL* (for Book I), nor an alternative earlier version (as for Book III), although the tract *De Motu* does contain two of the propositions that appear in Book II.[1] Furthermore, *De motu corporum* obviously was planned to deal with motions in at least one sort of resisting fluid; how far Newton ever got with this plan we cannot tell from the obviously incomplete fragment preserved in the Portsmouth Collection.

Newton's first concept of the *Principia* did not include a Book II as we know it; and we have seen that the original Book II (*De motu corporum, Liber Secundus*) was an early form of what was to become the *Liber Tertius* of the *Principia*, there entitled *De Mundi Systemate*. The first announcement to Halley of the existence of a Book II on motion in resisting media was evidently made in Newton's letter of 20 June 1686, stating that there were to have been three books, of which the second had been 'finished last summer'.[2] This is the letter in which Newton said specifically that Book II 'only wants transcribing'. The manuscript from which this transcription would have been made apparently does not now exist in any known collection, and we have therefore no way of knowing whether it was in Newton's hand or in Humphrey's. But that it did really exist is proved by the comments made on it by a contemporary critic (see §7 and Supplement VII).

The situation with respect to Book II differs greatly from that of Book I and Book III in that the Portsmouth Collection does contain a small set of actual work-sheets giving us the barest glimpse of how Book II evolved. These are especially precious because they help us to see Newton's method of composition. It is therefore a source of intense regret that these pages are so very few in number.

One set of preliminary work-sheets for Book II is related to Sec. VII, 'Of the motion of fluids, and the resistance made to projected bodies'. These are to be found in the University Library, Cambridge (in MS Add. 3965, §10). The cataloguer first described this set of manuscripts as 'Revision of Principia. Resistance of fluids', but then changed his mind; the words 'Revision of' are crossed out. A later note reads:

> These Prop^{ns} do not agree with the numbering of the Principia & there is no reference to the pages Hence they probably belong to an earlier stage of the work, than the printing of the 1st Ed^{n}

The contents of §10 of MS Add. 3965 are foliated from 100 to 146. Even the most superficial examination shows that much of this §10 must be of a later date than *M*, leading one to wonder how much sorting and rearranging was done prior to the deposit in the University Library. For instance, the very first pages (ff. 100–2) contain a report by F. Hawksbee on experiments made in 1710, in which balls were dropped from the cupola of 'St Pauls Church'.[3] The presence of this document

[1] See Chapter III, §3.
[2] This is the letter announcing that he had decided 'to suppress' the third book.
[3] See Chapter XI, §4, and Book II, Exp. 13 of Scholium following Prop. XL {351.29 ff.}.

no doubt explains the final form of the description of §10 in the published *Catalogue*:

10. On the Resistance of fluids; account of Hawksbee's experiments, with Newton's deductions; chiefly rough notes.[4]

Before analysing the pre-*M* portion of §10, let me say a few words about some of the later material that is mixed with the early work-sheets.

Following the first pages (ff. 100–2), just mentioned, there is a f. 103 which contains some solutions of problems of motion in fluids, using dotted letters for fluxions, and hence must be later than *M*. Another page (f. 109^r) contains in the middle a note that 'The Electors of Brandenburg & Saxony & the house of Lunenburg have agreed for ye convenience of trade to coin their moneys of equal value though not of equall alloy', and so this page must date from Newton's time in the Mint in the 1700s. Further studies using dotted letters occur in ff. 110–11, along with several pages covered with computations, while ff. 112, 118–19 contain notes on experiments with falling bladders made in 1719 by Desaguliers and others.[5] The manuscript notes on falling globes (including a set of six corollaries) and other problems, on ff. 113–14, must be later than *M* since they are written on the blank parts of a letter to Newton from Jos: Lily, dated '24th No. 1707'. Similarly f. 116, containing notes for revision of a part of Book II, Prop. XXXVII, referring to 'cas. 5 & Corol. 6. Prop. XXXVI', must be later than *M* since the same scrap (on verso) mentions 'Ann. 1705'. Newton's notes on falling bodies on ff. 120–1 may be dated by their occurrence on a letter to Newton mentioning the Abbé Conti. Folios 122–6 deal with experiments on the oscillations of pendulums in air and water. Then, on f. 139, there are some first versions of propositions for the new part of Sec. VII, which was almost entirely rewritten for E_2; these are followed on ff. 141–2 by the text and discussion (with the table) of the new Prop. XL for E_2 and also the new scholium with which Sec. VII terminates in E_2. The final folios, 143–6, are almost entirely covered with calculations; that these are much later than *M* may be seen from the handwriting and also from extraneous matter such as a draft of a letter 'To the R^t Honble the Lords Comm^{ers} of his Ma^{ts} Treãry' (which is visible under the calculations) and some notes referring to the fluxion controversy, together with two drafts of a letter to Levino Vincent (F.R.S. 1715). Some further pages of Sec. 10 (of MS Add. 3965) are discussed at the end of the present §5, and in Supplement V.

The only pre-*M* sheets in §10 of MS Add. 3965 appear to be ff. 104–5, 106–7, and 134. Of these the most significant may be ff. 104–5 containing drafts in Newton's hand of a corollary to a Prop. XVIII, cancelled after having been revised and then replaced by a set of Corols. 1–3 (of which the first two have been extensively rewritten) and a scholium. This is not the Prop. XVIII of Book II as we know it, which appears in Sec. IV, but must have been part of what we know as

[4] [Portsmouth Collection] (1888), *Catalogue*, p. 4.

[5] See Chapter XI, §4.

Sec. VII, since it deals with pendulums and resistances and has a scholium that is apparently an early form of the Scholium Generale with which Sec. VII concludes in *M* and E_1, but which was transferred to Sec. VI in E_2 and E_3. The scholium reads (in its final version[6]):

Schol. Tentari possunt haec omnia in motu Pendulorum. Nam quamvis motus iste non sit uniformis, sunt tamen proportionales motus qui in oscillationum diversarum partibus proportionalibus fiunt, adeoque si ex data proportione velocitatum datur ratio resistentiarum rationem illam patefacient. ffeci igitur ut pendulum pedum fere duodecim oscillando describeret arcum duorum digitorum longitudinis inferne: et oscillationibus 37 amissa est quarta pars motûs, pendulo jam describente arcum digiti unius cum semisse. ffeci deinde ut idem oscillaretur in arcu digitorum quatuor et oscillationibus 34 amissa est quarta pars motus pendulo jam eunte et redeunte in arcu digitorum trium. Feci postea ut idem successivè oscillaretur in arcubus digitorum octo, sexdecim, triginta duo, sexaginta octo, centum viginti et octo, et oscillationibus $30\frac{1}{2}$, $20\frac{2}{3}$, $13\frac{1}{4}$, $7\frac{3}{4}$, 4 respectivè, amissa est pars quarta motus totius. Singulis experimentis duplicabatur et velocitas seu motus penduli et motus amissus. Quare cum resistentiae sint ut motus amissi 1, 2, 4, 8, 16, 32, 64 et tempora numeris oscillationum portionalia [*sic*] 36, 34, $30\frac{1}{2}$, $20\frac{2}{3}$, $13\frac{1}{4}$, $7\frac{3}{4}$, 4: applicentur hi motus ad tempora et orientur resistentiae, sunt igitur resistentiae illae $\frac{1}{37}$ in experimento primo, $\frac{2}{34}$ in secundo, $\frac{4}{30\frac{1}{2}}$ in tertio, $\frac{8}{20\frac{2}{3}}$ in quarto, $\frac{16}{13\frac{1}{4}}$ in quinto, $\frac{32}{7\frac{3}{4}}$ in sexto et $\frac{64}{4}$ in septimo. In primis duobus experimentis ubi motus erant tardissimi proportio resistentiarum $\frac{1}{37}$ ad $\frac{2}{34}$ seu 17 ad 37 est paulo major quam velocitatum 1 ad 2. In ultimis duobus ubi motus erant velocissimi proportio illa $\frac{32}{7\frac{3}{4}}$ ad $\frac{64}{4}$ seu 8 ad 31 est duplicata proportionis velocitatum quampoxime [*sic*]. Ponamus resistentiam esse ut summa quantitum [*sic*]

Scholium. These things can all be tested in the motion of pendulums. For although that motion is not uniform, there are nevertheless proportional motions that occur in the proportional parts of diverse oscillations, and therefore if from a given proportion of velocities there is given the ratio of resistances, they will reveal that ratio. Therefore, I made a pendulum of about 12 feet describe by oscillating an arc of two inches in length at the bottom; and a fourth of the motion was lost by 37 oscillations, the pendulum now describing an arc of $1\frac{1}{2}$ inches. Then I made the same pendulum oscillate in an arc of 4 inches and a quarter of the motion was lost by 34 oscillations, the pendulum now going and returning in an arc of 3 inches. Afterwards I made the same pendulum oscillate successively in arcs of 8, 16, 32, 68, 124 inches, and a quarter of the total motion was lost by $30\frac{1}{2}$, $20\frac{2}{3}$, $13\frac{1}{4}$, $7\frac{3}{4}$, 4 oscillations respectively. In individual experiments both the velocity or motion of the pendulum and the lost motion was doubled. Therefore since the resistances are as the lost motions 1, 2, 4, 8, 16, 32, 64 and the times proportional to the numbers of oscillations 37, 34, $30\frac{1}{2}$, $20\frac{2}{3}$, $13\frac{1}{4}$, $7\frac{3}{4}$, 4; let these motions be divided by the times and the resistances will arise; therefore those resistances are $\frac{1}{37}$ in the first experiment, $\frac{2}{34}$ in the second, $\frac{4}{30\frac{1}{2}}$ in the third, $\frac{8}{20\frac{2}{3}}$ in the fourth, $\frac{16}{13\frac{1}{4}}$ in the fifth, $\frac{32}{7\frac{3}{4}}$ in the sixth and $\frac{64}{4}$ in the seventh. In the first two experiments where the motions were slowest the proportion of resistances $\frac{1}{37}$ to $\frac{2}{34}$ or 17 to 37 is a little larger than that of the velocities 1 to 2. In the last two where the motions were swiftest, that proportion $\frac{32}{7\frac{3}{4}}$ to $\frac{64}{4}$ or 8 to 31 is the duplicate of the proportion of the velocities as closely as possible. Let us set the resistance to be as the

[6] I have not corrected errors: 'sexaginta octo' for 'sexaginta quatuor' in lines 17–18, and '36' for '37' in line 24 (correctly given three lines later in the fraction '1/37') in the Latin original.

$AV+BV^2$ ubi V velocitatem designet et AB quantitates datas. Et in experimento secundo scribendo 2 pro V prodibit resistentia $2A+4B$ aequalis $\frac{2}{34}$. In experimento autem sexto scribendo 32 pro V prodibit resistentia $32A+1024B$ aequalis $\frac{32}{7\frac{3}{4}}$. Ex his aequationibus elicietur A aequalis $\frac{1}{44}$ et B aequalis $\frac{1}{30}$, adeoque si Corollarium secundum recte se habet debebit resistentia esse ut $\frac{1}{44}V+\frac{1}{30}V^2$. Igitur pro V in experimentis illis septem subtituo [*sic*] velocitates 1, 2, 4, 8, 16, 32, 64 successive pro V et prodeunt resistentiae ut in Tabula sequente.

sum of the quantitities $AV+BV^2$ where V designates the velocity and A,B designate given quantities. And in the second experiment by writing 2 for V the resistance $2A+4B$ will come out equal to $\frac{2}{34}$. And in the sixth experiment by writing 32 for V the resistance $32A+1024B$ will come out equal to $\frac{32}{7\frac{3}{4}}$. From these equations A will be ascertained equal to $\frac{1}{44}$ and B equal to $\frac{1}{30}$, and therefore if the second Corollary is correct the resistance ought to be as $\frac{1}{44}V+\frac{1}{30}V^2$. Therefore for V in those seven experiments I substitute the velocities 1, 2, 4, 8, 16, 32, 64 successively for V and the resistances come out as in the following table:

	Resistentiae		Tempora	
Velocitates	ex calculo	ex Observat.	ex calc.	ex Obs.
1	$\frac{1}{38}$	$\frac{1}{37}$	38	37
2	$\frac{2}{34}$	$\frac{2}{34}$	34	34
4	$\frac{4}{30\frac{1}{2}}$	$\frac{4}{30\frac{1}{2}}$	$30\frac{1}{2}$	$30\frac{1}{2}$
8	$\frac{8}{20\frac{1}{3}}$	$\frac{8}{20\frac{2}{3}}$	$20\frac{1}{3}$	$20\frac{2}{3}$
16	$\frac{16}{13\frac{1}{6}}$	$\frac{16}{13\frac{1}{4}}$	$13\frac{1}{6}$	$13\frac{1}{4}$
32	$\frac{32}{7\frac{3}{4}}$	$\frac{32}{7\frac{3}{4}}$	$7\frac{3}{4}$	$7\frac{3}{4}$
64	$\frac{64}{4\frac{1}{5}}$	$\frac{64}{4}$	$4\frac{1}{5}$	4

Quod tempus in ultima observatione prodiit paulò majus per calculum quam per observationem, exinde evenisse puto, quod oscillationes non fierent in Cycloide adeoque non essent satis isochronae. Nam pendulis aequalibus tempora oscillationum in circulo paulo majora sunt quàm in Cycloide idque excessu quodam qui est in duplicata circiter ratione arcûs a pendulo descripti. Inde factum est quod oscillationes in experimentis primis propemodum isochronae fuerint, in ultimo verò ob magnitudinem arcûs excessus temporis evaserit notabilis.

That the time in the last observation came out a little greater through calculation than through observation I think arose from this, that the oscillations were not being made in a cycloid and therefore were not sufficiently isochronous. For in equal pendulums the times of oscillations in a circle are a little greater than in a cycloid and that with a certain excess which is in about the duplicate ratio of the arc described by the pendulum. Thence it happened that the oscillations in the first experiments were almost isochronous, but in the last on account of the magnitude of the arc the excess of time came out noticeable.

Hisce similia quaedam tentavi in aqua sed minus commodè. Globum ferreum cujus diameter erat digiti unius, pondus vero granorum

I tried certain things like these in water, but less conveniently. I suspended by an iron wire an iron globe whose diameter was one

972 suspendebam a filo ferreo. Longitudo penduli inter punctum suspensionis et centrum globi erat 27 digitorum. Regulam in digitos et partes digiti divisos [*for* divisam?] collocabam horizontaliter ultra filum ferreum proxime supra superficiem aquae sic, ut oscillantis penduli filum ferreum propemodum tangeret a latere, essetque plano in quo motus fili peragebatur omninò parallela. Hoc pacto per partes digitorum in Regula notatas observavi longitudines arcuum quos fili pars inferior e regione Regulae describebat. Si arcus ille in oscillatione prima erat duorum digitorum pendulum oscillationibus 13 amisit quartam partem motus sui. Sin arcus ille erat digitorum quatuor vel octo, pars quarta motus totius oscillationibus $6\frac{3}{5}$ vel $3\frac{1}{3}$ amissa fuit. Partes $\frac{3}{8}$ oscillationis integri colligebam notando differentiam inter oscillationes 6 et 7 e regione Regulae et hanc differentiam secando in puncto ubi arcus primus digitorum duorum octava sui parte parte [*sic*] minuendus esset. Nam partibus duabus octavis hinc inde subductis motus reddebatur quarta sui parte minor. Hoc pacto colligebam partes oscillationum in alii [*sic*] casibus.

inch, and its weight 972 grains. The length of the pendulum between the point of suspension and the centre of the globe was 27 inches. I placed a ruler divided into inches and parts of an inch horizontally beyond the iron wire immediately above the surface of the water in such a way that it almost touched from the side the iron wire of the oscillating pendulum, and was completely parallel to the plane in which the wire's motion was going on. In this way through the parts of inches notated on the ruler I observed the lengths of the arcs which the lower part of the wire in the line of the ruler was describing. If that arc in the first oscillation was two inches the pendulum lost a quarter of its motion in 13 oscillations. But if that arc was four or eight inches a quarter of the total motion was lost by $6\frac{3}{5}$ or $3\frac{1}{3}$ oscillations. I inferred $\frac{3}{8}$ of a whole oscillation by noting the difference between 6 and 7 oscillations in the line of the ruler and by cutting this difference in the point where the first arc of two inches was to be lessened by an eighth part of itself. For with $\frac{2}{8}$ parts subtracted from both sides the motion was made less by one quarter part of itself. In this way I determined the parts of oscillations in other cases.

On comparison of this scholium with the Scholium Generale at the end of Sec. VI in E_3 {307.13 ff.}, it is at once apparent to what degree Newton improved on both his apparatus and his results. The method of analysing his data became more profound too; for example, in the Scholium Generale {308.14 ff.} he deals more accurately and more fully with the difference between oscillations along a circular and a cycloidal arc. Also the discussion {308.12 ff.} of the difference in arcs where the resistance is set at $AV+BV^{3/2}+CV^2$ is lacking here. The 'hydrostatical experiment' {309.31 ff.} was new in the Scholium Generale; furthermore, in the scholium printed above only one pendulum bob was used in air (and its size is not given), whereas in the Scholium Generale we read that Newton used a 'wooden ball' {307.16}, a 'leaden globe' {310.20}, and a 'globe whose diameter was $18\frac{3}{4}$ inches' {311.27–29}. Additionally, the experiments with a pendulum partly immersed in water are greatly refined and expanded in the Scholium Generale {313.10 ff.}, and Newton also added an experiment with a pendulum oscillating in quicksilver.

This document has thus a special significance in that we may see both an early and a late form of an experimental attack on the very same problem. Unfortunately, this scholium does not contain anything relating to the final subject of the Scholium Generale {316.4 ff.}: an investigation, by means of an experiment with pendulums, of the possible existence of a certain 'aethereal medium extremely rare and subtile, which freely pervades the pores of all bodies'. Professor R. S. Westfall has called our attention to the significance of this experiment in relation to the decline of Newton's adherence to a belief in the aether.[7] How fortunate we would have been had this manuscript provided information concerning the experiment itself, which, Newton says in the *Principia* {317.23–24}, 'is related by memory, the paper being lost in which I had described it'.

A scrap (f. 133^{v}) enables us to see some further aspects of the growth and change of this scholium after it became the Scholium Generale. This page of manuscript indicates that portions of the text of the Scholium Generale were at one time written out by Newton in the discussion of a 'Pr. XXXII. Resistentias ffluidorum captis [factis *written above* captis, *but the latter not cancelled*] experimentis determinare' ('To determine the resistances of fluids by $\left\{\begin{matrix}\text{making}\\ \text{taking}\end{matrix}\right.$ experiments'). This scrap of manuscript reads:

> Exper. 1. Globum ligneum pondere unciarum Romanarum $57\frac{7}{22}$ diametro digitorum Londinensium $6\frac{7}{8}$ fabricatum - - - partes digiti respective. [Et hae differentiae oriuntur ex resistentia globi in una oscillatione mediocri &] Et per has differentias Resistentiae Globorum in oscillatione mediocri exponi possunt.
>
> Caeterum cum velocitates - - - Haec ratio in casu primò [*sic*] est [*space*] ad [*space*] in secunto [*sic*] est 6283 ad 6275, in tertio est [*space*] ad [*space*] in quarto est 12566 ad 12533 in quinto est [*space*] ad [*space*] & in sexto 25132 ad 24869. Et inde Resistentiarum exponentes $\frac{1}{656}$ $\frac{1}{242}$ $\frac{1}{}$ [*space*] evadent [*space*] id est in numeris decimalibus [*space*] respective.

The square brackets in the first paragraph are Newton's, probably indicating deletion; the dashes are likewise Newton's.

As may be seen from the Apparatus Criticus (and Appendix I), this Scholium Generale appears at the end of Sec. VII, Book II, in M and E_1; but when Newton undertook the extensive revision of Sec. VII for E_2, he transferred the Scholium Generale from Sec. VII to Sec. VI. In the new position, the Scholium Generale follows Prop. XXXI rather than Prop. XL. Newton's plan, as shown on f. 133^{v}, to incorporate some or all of the Scholium Generale into a Prop. XXXII, must therefore have been made after he had decided to jettison most of the original Sec. VII, and to have the Scholium Generale (or its contents, now placed in a Prop. XXXII) follow Prop. XXXI. Hence, the manuscript fragment in question cannot be part of the preparation for M or E_1, but must represent some stage of contemplated revision of either E_1 or E_2.

Here we have a somewhat typical manuscript puzzle. It can be resolved, however, by a close inspection of the actual text, fragmentary as it is. The first paragraph contains a gap, indicated by three hyphens, corresponding to that part of the Scholium Generale at {307.18–308.6}, which is not notably different among

[7] In a paper read at the Newton conference in Austin, Texas, in November 1966, and again at the annual meeting of the History of Science Society in Washington, D.C., in December 1966. See Westfall (1967), and Guerlac (1967).

the three editions. But the gap in the second paragraph is of a different sort altogether. It begins at {308.14}, corresponding to line 5 of page 341 of E_1. But, as the Apparatus Criticus reveals, in E_3 (as in E_2) the sentence in question reads 'Cum velocitates...' and is the second sentence of the paragraph, whereas in E_1 it reads 'Caeterum cum velocitates...' and opens a paragraph—thus corresponding exactly with the manuscript fragment on f. 133^v. Furthermore the conclusion of the manuscript paragraph revises the printed text of E_1, which, as may be seen in the Apparatus Criticus, differs at this point from that of E_2 and E_3. There can thus be little doubt that this manuscript fragment represents a contemplated revision of E_1, undertaken when Newton had already decided to transfer the Scholium Generale (or its contents) from the end of Sec. VII, following Prop. XL, to the end of Sec. VI, following Prop. XXXI, presumably around 1710. This example shows how valuable it is to have available an edition with variant readings for identifying a manuscript and determining to what stage of revision of the *Principia* it should be assigned.

In the text of M and E_1 numerical results are given only for the second, fourth, and sixth cases, where in this f. 133^v there are blanks left for cases one, three, and five. On f. 132^v some raw data and computations are given for this Scholium Generale, as well as the tables {310.16–19, 27–32}.

The next part of §10 (MS Add. 3965) is of interest to us less for its intellectual content than as an exhibit of how Newton actually wrote out at least a part of the *Principia* prior to its having been copied out in a final (or near-final) draft in M by Humphrey. This material begins on f. 106, with a statement of Prop. XXXVI, Prob. VIII, Book II, exactly as in M and E_1 (save for capitalization), but with blank spaces left for the proposition and problem numbers. Then comes the first paragraph of the demonstration (bottom of p. 327 of E_1, reproduced in the present edition in Appendix I to the text). This is succeeded in turn by the demonstration of Prop. XXXV, Theor. XXVIII, which uses the same figure (as on page 325 of E_1, reproduced in this edition). These demonstrations of two propositions of M and E_1, here combined in one, are identical to the later divided versions save for minor details.[8] It would thus seem as if, some time after Newton had written out the proposition and proof on f. 106, he decided to have two separate propositions and then inserted a scholium (pp. 326–7 of E_1) between them. On the verso of f. 106 there is the long second paragraph of the proof of Prop. XXXVI as we find it in M and E_1 (p. 328). It is followed, on that same f. 106^v, by the five corollaries, as they appear in M and E_1 (p. 329). These corollaries are much worked over by Newton and show us that Corol. 5 was originally Corol. 4 and was changed to Corol. 5 when the Corol. 4 of M and E_1 was added.

[8] Following 'Q.E.D.' at the end of the proof (corresponding to line 12, p. 326, of E_1), the manuscript has an additional incomplete sentence reading, 'Ideoque cùm resistentia in Medio hic posito et resistentia in Medio elastico de quo Propositio est, si &c.' In the parenthetical reference (line 27, p. 324, E_1), there is a space for a proposition number to be added later on.

Folios 106 and 107 are two leaves of the same sheet, folded once; on f. 107^v Newton has written out the text of Prop. XXXV, Theor. XXVIII, but again without numbers. It is followed by the scholium that intervenes between Prop. XXXV and XXXVI in *M* and E_1. One cannot readily tell whether or not Newton may have first written out the proposition (XXXVI in *M* and E_1) and proof on f. 106, and then decided to have two propositions rather than one—in which case he would have separated the two paragraphs of proof, written out the text of the new proposition (XXXV in *M* and E_1), the intervening scholium, and then written the remainder of the proof of the first proposition (XXXVI) and the five corollaries. It is also possible that these parts were merely improvements of rough scraps and that the order in which they were entered on this double sheet has no particular significance.

That the former is more likely the true situation may be seen in yet another sheet, f. 134. On the verso side Newton has stated Prop. XXXVI (again leaving blank the number of the proposition and of the problem), followed by the first paragraph of proof (p. 327 of E_1), and the single word 'Nam' which could either designate the first word of the next paragraph in *M* and E_1 (p. 328) or the first word of the proof of the preceding proposition (XXXV in *M* and E_1). On f. 134^r, however, this same proposition is stated (again without a number), but with the proof that appears in *M* and E_1 for Prop. XXXV (pp. 324 ff.). Newton has written out both of these statements of the proposition itself without the word 'raro' (which appears in the final form in *M* and E_1), but he has inserted this word on a caret on f. 134. That f. 134 is prior to ff. 106–7 may be seen in the fact that this and other insertions on f. 134 are incorporated into the text on ff. 106–7. For instance, the first sentence of the proof of Prop. XXXV on f. 106^r contains in the text three insertions made on f. 134^r.

It would thus appear that Newton had initially considered (f. 134) both the first paragraph of the proof of Prop. XXXVI (in *M* and E_1) and the proof of Prop. XXXV (in *M* and E_1) as separate proofs to go with this Prop. XXXVI. Then he combined the two (ff. 106–7) and then added further material so as to produce two propositions and proofs, plus five corollaries and a scholium.

Though the statements of these two propositions, their proofs, and the corollaries do not differ notably from the printed versions, there are minor variations. For instance, the second paragraph of the proof of Prop. XXXVI in E_1 (p. 328, line 1) begins 'Nam (per motuum Legem tertiam) motus quem cylindrus...', whereas on f. 106^v the manuscript at first began 'Nam spatium' altered to 'Motus quem cylindrus...' altered to 'Nam (per Legem tertiam) Motus quem cylindrus ...'. In *M*, this sentence is written out in Humphrey's hand exactly as on f. 106^v (except with 'motus' for 'Motus') and the word 'motuum' was inserted later by Newton. This very minor change shows us how closely *M* followed ff. 106–7 for these two propositions. But the situation is quite different for the scholium. The first two paragraphs and the beginning of the third are practically identical in *M* and in ff. 106–7, but the figures differ. In f. 107^v, the line QR is absent, and hence the texts of M and E_1 and f. 107^v begin to disagree; the final sentences are also different, as may be seen by comparing the two:

f. 107^v [final version]	M and E_1
Quod si...et ducatur recta GP quae parallela sit rectae figuram tangenti in N, et axem productum secet in P, fuerit MN ad GP ut GP^{cub} ad $4BP \times GB^q$ solidum quod figurae hujus revolutione circa axem AB describitur resistetur minimè omnium ejusdem longitudinis & latitudinis.	Quod si...& a puncto dato G ducatur recta GR quae parallela sit rectae figuram tangenti in N, & axem productum secet in R, fuerit MN ad GR ut GR *cub.* ad $4BR \times GBq$: Solidum quod figurae hujus revolutione circa axem AB facta describitur, in Medio raro & Elastico ab A versus B velocissime movendo, minus resistetur quam aliud quodvis eadem longitudine & latitudine descriptum Solidum circulare.

We may conclude that between the state of ff. 106–7 and of M there probably existed an intermediate copy, either in Humphrey's hand or in Newton's, in which numbers were assigned in all the blank spaces (for the propositions themselves and the references), in which the parts were rearranged in their proper sequence, and in which the final paragraph of the scholium was rewritten.

Regrettably, these variations, and the alterations on ff. 106–7 in Newton's hand, are not of very great consequence in showing the growth of methods or of major concepts. But they have an importance of a different sort: as further witnesses that Newton actually wrote out rough drafts of propositions of the *Principia*, and then rewrote and corrected them, to make them ready for Humphrey to copy. In this case, Newton did so before the structure of Book II had achieved its final form, that is, before he even knew what numbers he would assign to these propositions. Certainly, the text from which Humphrey 'transcribed' Book II must have been made up of sheets like these (some cruder, others with the parts in proper sequence), in Newton's hand.

Another small group of sheets in §10 (MS Add. 3965) attracts our attention by differing rather notably from the text of the *Principia*. Not only are these sheets of great intrinsic interest in displaying the growth of Book II of the *Principia* from stages otherwise unknown to us, but some of them are of consequence because they take us back of Sec. II, and its Prop. X, a paramount example of Newton's use of 'moments' in the *Principia*.

We shall see below (in Chapter IX, §4) that an error in this proposition became a significant focus of criticism of the *Principia*, and thus it came to play a special role in the controversy between Newton and the Continental mathematicians. Hence these particular work-sheets are of primary interest for the student of Newton's mathematical methods. They refer at times to yet other manuscripts, or printed texts, as by the use of dashes which here (as elsewhere in Newton's manuscripts) indicate a section to be taken from another manuscript or on another page.

Thus on f. 135^r Newton writes:

Est autem resistentia ut Medii densitas - - - proportionalis $\frac{Cf-CF-fh}{FG}$ - - - ut $\frac{Cf-CF-fh}{FG}$ directe et $\frac{CF^q}{FG}$ inverse, id est ut $\frac{Cf-CF-fh}{CF^q}$. Q.E.I.

Corol. 1. Et hinc colligitur - - - fiet Medii densitas ut $\frac{FG-kl-fh}{CF \text{ in } \overline{FG+kl}}$.

The Apparatus Criticus shows that in E_2 (and E_3) the concluding paragraph of the demonstration differs entirely from that just quoted from f. 135^r, as does Corol. 1; but the gaps may be readily filled from the text of *M* or E_1.[9] Thus, once again, the advantage may be seen of having available an edition with variant readings, permitting an immediate identification of the text being revised in an otherwise undated manuscript: the document in question is undoubtedly a stage in Newton's last-minute revision of Prop. X, Book II, undertaken in the autumn of 1712 after the pages in question had been printed off. (On this topic, see below, Chapter IX, §4.) The verso side of this f. 135 contains material relating to the Scholium to Prop. X.

These pages of Sec. II differ from those that have been described above in relation to Sec. VII in that they do not contain almost finished or finished versions that all but correspond to *M* and E_1. Rather these pages of Sec. II are documents of true trial and error, the actual work-sheets, showing Newton attempting to forge an improved text of Prop. X and the Scholium, evidently just before he decided to recast this material altogether.

From yet another group of sheets in §10, we learn that at some time in the course of preparing or revising Book II, Newton had planned to have a somewhat different kind of Sec. VII, as may be seen in a fragment (on f. 129^v) as follows:

Sect. VII	Section VII
De resistentia [Projectilium *del.*] corporum in fluidis.	On the resistance of bodies in fluids.
Prop.	Prop.
In Medio quocunque fluidissimo pars illa resistentiae Globorum aequivelocium quae oritur ex attritione partium fluidi est ut diameter Globi, pars autem altera resistentiae quae oritur ex inertia materiae fluidi est in duplicata ratione diametri.	In any very fluid medium, that part of the resistance of globes with equal velocities that arises from the attrition of the parts of the fluid is as the diameter of the globe, while the other part of the resistance that arises from the inertia of the matter of the fluid is in the duplicate ratio of the diameter.

This is followed by a 'Cas. 1.' written out in full and then a heading 'Cas. 2.' without text. Instead, following a short horizontal rule, we find (in English!): 'The Peripateticks placed the 4 Elements of earth water air & fire under y^e orb

[9] This occurs in the Apparatus Criticus at p. 252 of E_3, in the final paragraph before Corol. 1 in the text of E_1, and in the first Corollary.

of y[e] Moon & made all the heavens above them to be filled wth a Quintessence, pretending y[t] Nature abhorred a vacuum that is that matter was necessarily in all places or that matter was self existing. The Cartesians tell us that matter is space it self & therefore must be every where. But what [*here the text breaks off*].'

By referring to the Apparatus Criticus (and Appendix I), we can readily see that in this manuscript fragment of a Sec. VII Newton was revising Prop. XXXVI of Book II of E_1 (p. 327), but had not yet reached the final state of revision achieved in the new Prop. XXXV which replaced the original Prop. XXXVI when Newton recast most of Sec. VII for E_2. Presumably, therefore, the text of Sec. VII just quoted should be assigned a date of 1710 or soon after.

In another group of pages in this §10 of MS Add. 3965, Newton discusses fluid inertia and fusiform bodies (which are not mentioned as such in the *Principia* at all). Fusiform bodies are spindle-shaped (from 'fusus', the Latin word for spindle). They are solids formed by revolving an arc about its chord as axis. The topics presented here (f. 128[v]; see Plate 9) follow an opening statement which reads:

Corporum Sphaericorum in Mediis quibuscunque fluidissimis resistentiam ex inertia [*written* iniertia] materiae fluidi oriundam determinare.

To determine the resistance of spherical bodies in any very fluid mediums arising from the inertia of matter of the fluid.

Only the first short paragraph deals with so general a topic, Newton proceeding at once to the properties of spindle-shaped bodies moving in a resisting medium. But on the next facing page (f. 129[r]) Newton generalizes the statements about spindle-shaped bodies to the properties of bodies in general. These documents are printed below in Supplement V. They do not appear in any edition of the *Principia* in the state in which we find them here, and may be part of the revision of Book II undertaken about 1710 or soon after, when Newton rewrote most of Sec. VII. But they are not work-sheets for *M* or E_1, and thus are not of direct concern to us now.

6. THE FIRST 'DE MOTU CORPORUM (LIBER SECUNDUS)' AND BOOK III OF THE 'PRINCIPIA'

The history of Book III marks it off in two major respects from Books I and II. Book III is presented to us among the Newton manuscripts only as a finished product and it replaced a somewhat different 'Liber' on the identical subject. Unlike Book I, which exists in major part in a manuscript prior to *M* (namely *LL*), Book III appears to us only in *M*. I have found no prior draft from which Book III was copied into *M*, nor many true work-sheets containing trials and errors or rough drafts that were improved, polished, rewritten, and which then—only after several tries—achieved the final form of *M*.[1] And even though some paragraphs are taken

[1] I have, however, found in MS Add. 3966, ff. 102–7, some work-sheets that may have been preliminary to Book III; they are discussed below in §7. These, however, are not preliminary pages to actual versions occurring in *M* and E_1.

directly from another work (a second differentiating feature of Book III), that work too exists only in a semi-final draft, complete, without the anterior work-sheets. Happily, that draft is somewhat worked over by Newton, so that we can see some stages in its evolution. I should add that the Portsmouth Collection does contain hundreds of pages of astronomical tables and calculations, some of which may indeed prove to be sources of statements and of numerical results in both of these presentations by Newton of the System of the World; here is an intriguing task for some mathematically minded puzzle-solver.[2]

The first version of *De mundi systemate* is the tract published (after Newton's death) under the title 'De mundi systemate liber Isaaci Newtoni'. This work was published in 1728[3] from a manuscript now in the Portsmouth Collection (U.L.C. MS Add. 3990) written largely in Humphrey's hand with corrections by Newton and entitled 'De motu corporum, liber secundus'.[4]

Since LL_α is called 'De motu corporum, liber primus', and is written in the same physical format and presentation as this 'De motu corporum, liber secundus', the two together must have comprised a first stage of the *Principia*, consisting of a Book I on the general principles of motion and a Book II on the system of the world, exemplifying these principles in the physical universe.[5] It has been generally supposed that this tract or 'Liber secundus' must have been that prior version of Book III of the *Principia* to which Newton refers in the beginning of Book III {386.5 ff.}:

> In the preceding books I have laid down the principles of philosophy; principles, not philosophical, but mathematical...It remains, that from the same principles, I now demonstrate the frame of the System of the World. Upon this subject, I had indeed compos'd the third book in a popular method, that it might be read by many. But afterwards considering that such as had not sufficiently enter'd into the principles, could not easily discern the strength of the consequences, nor lay aside the prejudices to which they had been many years accustomed; therefore to prevent the disputes which might be rais'd upon such accounts, I chose to reduce the substance of that book into the form of propositions (in the mathematical way) which should be read by those only, who had first made themselves masters of the principles establish'd in the preceding books.

I am now able to state categorically that there need be no more doubt that this *was* the earlier rejected version, since I have found that fourteen paragraphs of Book III were taken almost verbatim from this manuscript; examples are displayed in Supplement VI. There are some instances of corrections made by Newton to Humphrey's copy in this manuscript; in *M* these corrections have been incorporated into the text of Book III.

[2] I must repeat that among the many pages of Newton's manuscripts there may be some work-sheets or preliminary versions of either of these two 'Books' *De mundi systemate* that I have missed and that may happily reveal themselves to the fresh eyes of some other investigator.

[3] Newton (1728*a*), *De mundi systemate.*

[4] The evidence that this was the copy used in printing consists of the compositor's marks corresponding to the end of each printed page, and occasionally also the abbreviated form of the compositor's name, as well as proof marks.

[5] The references to propositions by number indicate that this 'De motu corporum, liber secundus' was originally composed as a companion to LL_α, but was later altered (in part only) to conform to LL_β. For details, see Supplement VI.

In 1728, when *De motu corporum* (*liber secundus*) was being published posthumously, the need was felt for a more appropriate title. It was meaningless in two different senses to print this work as entitled by Newton: first, there was now no 'Liber primus' to go with it, and secondly, it was not a tract 'De motu corporum' —save in the special sense of 'De motu corporum coelestium'. Hence, the more attractive and appropriate title of *De mundi systemate* was used instead of Newton's original.[6] But as a result, it has been difficult ever since to find a way of referring to this 'De mundi systemate' that would easily distinguish it from Book III of the *Principia*, which is entitled on page 401 of E_1, 'DE/Mundi Systemate/LIBER TERTIUS'.[7] Since the version put aside by Newton was called by him 'Liber secundus', I shall designate it here by *De mundi systemate* (*liber secundus*), which not only marks it off from 'Liber Tertius' of the *Principia*, which I shall refer to as *De mundi systemate* (*liber tertius*), but happily also designates the order of composition (II→III).

De mundi systemate (*lib. 2us*) exists in a number of states and copies. First of all, there is a copy of the first part, made by Humphrey, which Newton then deposited in the University Library as a set of 'lectures' (MS Dd. 4.18); it is divided into five 'lectures', of which the first is designated 'Sept 29 Praelect. 1'. Cotes made a copy of this version which is at present in the Trinity College Library (MS R. 16.39). Another manuscript copy of this portion is to be found in the Clare College Library (MS Kk. 5.8), and is part of a volume of Newtoniana belonging to Charles Morgan, who got the text of the first five 'lectures' from 'Mr Robt Smith ffellow of Trin Coll & Plumian Professor of Astron. &c. in Camb.', and so presumably from Cotes's copy. Morgan obtained the remainder of *De mundi systemate* (*lib. 2us*) from the Secretary of the Royal Society, Martin Folkes. Edleston[8] conjectured that Smith in turn borrowed Morgan's copy of the remainder and had it copied out to go with Cotes's transcript.[9] Newton's own manuscript, now in the Portsmouth

[6] According to Stukeley, the Latin version of this manuscript was 'edited' for publication by Conduitt. Stukeley's own copy (item 111 in Catalogue 140, Dawsons of Pall Mall, London, *Makers of modern science*; see Stukeley (1936), *Newton's life, 1752*, p. 25) is 'inscribed on the board endpaper "W. Stukeley d.d. Johēs Conduit ar 1728". On the title-page in the hand of the presentation inscription is "edidit Johēs Conduit Ar"; the handwriting is Stukeley's.' Of course, Stukeley was not using the word 'edit' in the present sense of the word. Possibly it was Conduitt who introduced the new title; hence the appearance of a separate tract. But in this case he was only rendering into Latin the title of the previously published English version, *Newton's system of the world*.

On the history of Newton's manuscripts after his death, see D. T. Whiteside's General Introduction (pp. xvii ff.) to [Newton] (1967–), *Mathematical papers*, vol. 1.

[7] In Florian Cajori's revision of Motte's translation of the *Principia*, the English translation of this *De mundi systemate* is printed immediately following Book III. Since both are translated as *The system of the world*, Cajori has added to the title of Book III the qualifying phrase, 'in mathematical treatment', as if Book III had been so designated by the author! See Newton (1934), *Principia* [Motte–Cajori], p. 397.

[8] Edleston (1850), *Correspondence of Newton and Cotes*, p. xcviii.

[9] I have not been able to identify the manuscript that Martin Folkes made available to Charles Morgan. Possibly it was Newton's complete manuscript, from which Conduitt printed the Latin version of the tract *De mundi systemate*. Only the final portion was transcribed by (or for) Morgan, since he had obtained the first part from Cotes's copy of the text deposited in the University Library as Newton's lectures. A rough comparison of this final portion in Charles Morgan's volume (in Clare College Library) with Newton's own manuscript (in the Portsmouth Collection, U.L.C.) has disclosed no notable differences. But the possibility does exist that the manuscript that Folkes in part made available to Morgan was not the one now catalogued as MS Add. 3990

Collection (MS Add. 3990), from which Latin original the edition of 1728 was printed, is largely written out in the hand of Humphrey, with many passages cancelled or replaced and other corrections and alterations by Newton. In the margin Newton has entered summary postils in his own hand, of which only the first twenty-seven are numbered.

In MS Dd. 4.18 paragraph XVII is missing, paragraph XVI being followed by XVIII, and hence this is true also of Cotes's and Morgan's copies. The reason for the gap is that in his own full version (MS Add. 3990), Newton rather completely rewrote par. XIV, cancelled par. XV, rewrote the old par. XVI and quite properly renumbered it XV, and then assigned to the old par. XVII the new number XVI; but he did not go on with the renumbering, with the result that the succeeding paragraph is XVIII, and so on. None of the early Latin editions or the printed English versions have paragraph numbers, though they do have the postils. In the Cajori version, the numbers are given in correct sequence so that from par. XVIII onward the Cajori numbers are one less than those in Newton's own manuscript.[10]

I do not propose to discuss this significant document here (reserving that pleasure for another occasion), save in relation to the *Principia* as such.[11] There is one aspect of the manuscript itself which is a key to Newton's method of composition. I have said that the manuscript in the Portsmouth Collection (MS Add. 3990) is largely written in Humphrey's hand; for the most part the text is on the recto side only. Newton has added paragraph numbers and descriptive or summary postils, and has entered many corrections, alterations, and additions. Even more interesting, however, is the fact that at times—for example, the bottom five lines of f. 27, most of f. 28 (beginning after 'quo in casu' in line 4), the first line and bottom twelve lines of f. 29, and the first four lines of f. 30—the copy is written out by Newton himself. Was he dictating and going too fast for Humphrey, or was Humphrey getting tired? Or was Humphrey copying out an earlier draft that had been revised with some corrections being dictated, when Newton suddenly decided to add to his text? The latter may be more probable. In this event, Newton would not surprisingly have picked up the pen in order to write out the additional material right on the page, rather than either putting it on a scrap of paper for Humphrey to copy or dictating it. That this may have been the case appears likely from a close examination of f. 29. Humphrey had written out:

His in locis aestus ascendit ad pedes 40 vel 50 et ultra. Alibi ascensus ut plurimum est pedum quatuor sex vel octo et raro superat pedes decem vel duodecim.	In these places the tide ascends to 40 or 50 feet and beyond. Elsewhere the ascent at the most is of four, six, or eight feet and rarely exceeds ten or twelve feet.

(U.L.C.). It may very well have been yet another manuscript (in Newton's hand or in Humphrey's), which either no longer exists or is in private hands and not available for study, or lies unidentified or 'unknown' in some institutional library. The reason for making such a supposition is that the English version, published in 1728, does have a notable difference from all the manuscripts mentioned above. This variation indicates either that great liberties were taken by the translator or that he had a text which differed in several respects from the manuscripts we know. See Supplement VI. [10] Newton (1934), *Principia* [Motte–Cajori], pp. 567 ff.

[11] For the treatment of Hooke in this manuscript see the end of this §6, below.

This sentence is crossed out and followed by a passage in Newton's hand beginning:

His in locis mare magna cum velocitate accedendo et recedendo litora nunc inundat nunc arida relinquit ad multa milliaria. Neque impetus [accessus & recessus *del.*] accedendi vel recedendi prius frangi potest quàm aqua attollitur vel deprimitur ad pedes 40, vel 50 et amplius.

In these places the sea by approaching and receding with great velocity sometimes inundates the shores, sometimes leaves them dry for many miles. Nor can the impetus of approaching or receding be broken before the water is raised or depressed to 40 feet, or 50 or more.

Newton appears to have changed his mind while Humphrey was making his copy.

While some paragraphs of *De mundi systemate* (*lib. 2us*) found their way into *De mundi systemate* (*lib. 3ius*), these account for only a very small part of the latter. The remainder certainly was copied out by Humphrey from a prior manuscript.[12] But of this version I have found not so much as a single page, and so I cannot say whether it would have been in Humphrey's hand (like *LL*) or in Newton's.

In any event, I believe we may safely assume that Newton must have had a more or less complete manuscript version of Book III of the *Principia* (as well as of the Definitions, the Axioms or Laws of Motion, and Books I and II) from which, with further final emendations and improvements, *M* was transcribed. As I have said earlier, this would explain the otherwise unaccountable fact that *M* appears to us as a fair copy. Furthermore, the existence of such a complete manuscript would answer another very puzzling question: would Newton have entrusted his only copy to be sent on to Halley for printing, without keeping another text, however imperfect, as insurance against loss or fire?[13] This question was long a disturbing one for me until I found evidence of the existence of a duplicate copy of all parts of the *Principia*, which no longer exists save for *LL* (more exactly, LL_β).

We may assign a probable date to *De mundi systemate* (*lib. 2us*) by reference to Newton's letter to Halley of 20 June 1686, the one in which he told Halley that he had designed the *Principia* in three books.[14] Recall that he said that the 'third wants y^e^ Theory of Comets. In Autumn last I spent two months in calculations to no purpose for want of a good method, wch made me afterwards return to ye first Book & enlarge it wth divers Propositions some relating to Comets others to other things found out last Winter.' A comparison of the summary account of comets in *De mundi systemate* (*lib. 2us*) and the more extended treatment of comets and their orbits in *De mundi systemate* (*lib. 3ius*) will show at a glance the kind of progress obviously made by Newton following the autumn of 1685. Since Newton's letter to Halley seems to refer explicitly to *De mundi systemate* (*lib. 3ius*), we may presumably date *De mundi systemate* (*lib. 2us*) at some time prior to the autumn of 1685.

12 The arguments for the existence of a pre-*M* version of all three books of the *Principia* have been given earlier. Further evidence for the existence of just such a pre-*M* version of Book III is given below in §7 and Supplement VII.

13 From the Newton–Halley correspondence during the course of the printing of E_1, we learn of the losses that were apt to occur. On 13 February 1686/7 Newton wrote to Halley, 'I have sent you ye sheet you want.' Halley's reply, 24 February 1686/7, explains what happened: 'I return you most hearty thanks for the copy you sent me of the sheet which was lost by the printers Negligence'; see [Newton] (1959–), *Correspondence*, vol. 2, pp. 464, 469.

14 *Ibid.* p. 437.

Newton's manuscript copy of *De mundi systemate* (*lib.* 2us) is of special interest in relation to a letter of his about Hooke's claim to 'the invention of ye rule of the decrease of Gravity, being reciprocally as the squares of the distances from the Center', to quote from Halley's letter about this issue to Newton on 22 May 1686. Halley warned that Hooke 'seems to expect you should make some mention of him, in the preface'.[15] In his reply on 27 May 1686,[16] Newton said rather explicitly, 'In the papers in your hands there is noe one proposition to which he can pretend, & soe I had noe proper occasion of mentioning him there.' And, indeed, the only reference to Hooke in Book I occurs in the Scholium to Prop. IV, which—as we shall see—was not part of the original manuscript at all, but was sent on to Halley later, on 14 July 1686.[17] Then, having said that Hooke had no share in any part of Book I, Newton went on, 'In those [papers] behind where I state the systeme of the world, I mention him and others.' Does he? not really! Hooke's name does not appear even once in Book III of the *Principia* in E_1. He is, however, mentioned in Book III fifteen times in E_2 and twelve in E_3,[18] but only for his observations of comets. Clearly, then, Newton was not referring to the ultimate version[19] of Book III when he said that in 'those behind' he had mentioned Hooke and others. Furthermore, this cannot be a reference to the earlier *De mundi systemate* (*lib.* 2us), for in that work the only reference to Hooke in relation to the system of the world as such occurs in the following passage (quoted from the contemporaneous English version, published in 1728):

we do not know in what manner the ancients explained the question, how the planets came to be retained within certain bounds in these free spaces, and to be drawn off from the rectilinear courses, which, left to themselves, they should have pursued, into regular revolutions in curvilinear orbits...

The later philosophers pretend to account for it either by the action of certain vortices, as *Kepler* and *Descartes*; or by some other principle of impulse or attraction, as *Borelli*, *Hooke*, and others of our nation; for, from the laws of motion, it is most certain that these effects must proceed from the action of some force or other.[20]

I hardly think that this can be what Newton had in mind. Surely, such a statement in print would have inflamed the feelings of Hooke rather than have served as balm.

It is interesting, furthermore, that in Newton's manuscript copy of *De mundi systemate* (*lib.* 2us), the reference to Hooke did not originally contain his name at all. The original text, in Humphrey's hand, reads:

Philosophi recentiores aut vortices esse volunt aut aliud aliquod sive impulsus sive attractionis principium...

[15] *Ibid.* p. 431. [16] *Ibid.* p. 433. [17] *Ibid.* p. 445.

[18] {504.21, 30, 33, 37; 505.1, 12, 19; 521.14, 21, 30, 32, 36}. The three additional occurrences in E_2 are at {505.30} (twice) and {520 tab. 17–20}.

[19] Of course, Hooke may have been originally mentioned in a prior version of Book III, from which *M* was copied, and which exists no longer. In this event, his name would have been eliminated some time after 27 May 1686.

[20] Newton (1728), *System of the world*, p. 4; revised and reprinted in Newton (1934), *Principia* [Motte–Cajori], p. 550.

Then Newton made two additions in his own hand: 'ut Keplerus et Cartesius' as an insert after 'esse volunt' and 'ut Borellius et ex nostratibus alii [*changed from* aliqui]' as an insert at the end of the sentence. Next, and only then, apparently as a final emendation, did he add 'Hookius' to the second insert so as to follow 'Borellius'. Even this small acknowledgement of Hooke in relation to 'impulse or attraction' was only an afterthought.[21] It is possible, of course, that Newton, when writing to Halley, had in mind an earlier draft, not now available or possibly no longer extant, in which there had been quite a different sort of reference to Hooke.

7. SOME WORK-SHEETS FOR BOOK III

Any student of the *Principia* who leafs through the set of manuscripts classified as U.L.C. MS Add. 3966 will be struck by a familiar sight when he reaches §12 (consisting of ff. 102–13). At the front of this §12, there is a set of folio sheets, each with a single fold so as to make two leaves; each leaf contains a vertical rule on the recto side, placed an inch or so from the right-hand margin. These folded sheets form a gathering: that is, the sheet comprising ff. 104–5 is enclosed within that comprising ff. 103–6, and this pair is then enclosed within the folded sheet comprising ff. 102–7. The handwriting is without doubt Humphrey Newton's; and there are additions, corrections, and emendations by Newton. Hence the general appearance is similar to the pages of *LL*, of *M*, and of *De mundi systemate* (*lib. 2us*). The subject-matter is the motion of the Moon's apogee, and is developed in two lemmas and two propositions—both without numbers. Clearly these manuscript pages are related to the *Principia*. But are they an independent exercise based on the *Principia*, a pre-draft of Book III, or a revision essayed after the completion of *M* or E_1? To find the answer, let us turn to the contents of these lemmas and propositions.

The first page of §12 begins (f. 102):

Prop. Prob.

Invenire motum horarium Apogaei Lunae.

Designet ABCD orbem Lunae, A & B syzygias, C & D quadraturas, S Terram, P Lunam & PI & 2IS vires Solis ad perturbandos motus Lunares ut supra; quarum virium altera PI trahat Lunam versus I, altera 2IS distrahat Lunam a linea CD...

This particular proposition (or problem) does not appear in *M* or in any of the printed editions of the *Principia*, but from its content it appears to be related to the family of propositions about the motion of the Moon (Book III: Props. XXV–XXXV).

Since the handwriting is Humphrey's, either these sheets must have been written almost immediately after the completion or the publication (partial or complete)

[21] Later on, Newton apparently decided that 'Hookius' should be cancelled; Hooke's name is crossed out in the copy made by Humphrey and deposited in the University Library, with the result that his name does not even appear at this point in either the Cotes or the Morgan transcriptions. But it is not crossed out in the complete manuscript copy that Newton retained.

of the *Principia*—while Humphrey was still in Newton's service—or they must have been preliminary essays undertaken before the final version was made ready for the press. The lack of assignment of numbers to designate this proposition (and problem), and the others in the sequence, does not help us to decide whether these are work-sheets towards the first edition (or *M*) or are an early form of revision or extension of Book III. If this proposition, and the others that go with it, were destined for the *Principia*, as they certainly seem to have been, then they must have been written for Book III of the *Principia* as we know it: that is, for *De mundi systemate* (*lib. 3ius*) rather than for *De mundi systemate* (*lib. 2us*). The reason is that throughout this group of manuscripts (§12) we find blank spaces left for numbers in the designation of propositions, not only in the headings but also within the body of the proofs. The proof of the foregoing proposition on 'the horary motion of the Apogee of the Moon', for instance, contains a reference (f. 104) to an as yet unnumbered 'Corol. Prop. ', two blank spaces having been left for future numbers, and there is a similar reference in blank to 'vi V in Lemmate & propterea (per Lemma illud)'. Since Book III of the *Principia*—or *De mundi systemate* (*lib. 3ius*)—is made up of numbered propositions (each of which is a numbered problem or theorem), whereas *De mundi systemate* (*lib. 2us*) is not, there can be no doubt that these propositions are related to Book III of the *Principia*.

The sheet comprising ff. 102–7 is, however, folded 'inside out', and the correct order of these leaves should be as follows: ff. 105r–106r, 107r–102r, 103r–104r, 110r–111r, forming a continuously written sequence.[1] This is a typical example of the lack of order that may occur within the Newton manuscripts in the Portsmouth Collection, even among papers that are grouped together, and that may often cause scholars much wasted time and energy before they discover how the fragments are to be put together. In this same §12 of MS Add. 3966 there are two other sequences, one occurring on the recto and the verso sides of f. 113, the other on the recto sides only of ff. 108–9, and the recto and verso sides of f. 112. As we shall see in a moment, ff. 108–9, 112 contain an earlier version or pre-draft of the main sequence under discussion, whereas f. 113 contains material of a quite different sort: the text of Prop. XXXIV, Book III, and the four corollaries, which actually happens to be of the highest interest.

In order to show what these sheets represent, I shall begin with the main sequence in its final draft. First there are two lemmas, which I shall denote as Lemma [*a*] and Lemma [*b*]. Lemma [*a*] begins at the head of f. 105r and is completed in the top half of f. 106r. Lemma [*b*] begins in the middle of f. 106r and is completed on f. 107r. On the verso of f. 106 there occurs a figure lettered in Newton's hand (and presumably drawn by him) for these two lemmas and for the Proposition [*i*] immediately following. Lemma [*a*] reads as follows:

[1] I am grateful to D. T. Whiteside for bringing to my attention this difference between the true sequence of these pages and the order in which they are to be found in §12.

Si Luna P in orbe Elliptico QPR axem QR umbilicos S, F habente, revolvatur circa Terram S et interea vi aliqua V a pondere suo in Terram diversa continuò impellatur versus Terram; sit autem umbilicorum distantia SF infinitè parva: erit motus [Aphelii *del.*] ⌊Apogaei⌋ ab impulsibus illis oriundus ad motum medium Lunae circa Terram in ratione composita ex ratione ⌊duplae⌋ vis V ad Lunae pondus mediocre P, et ratione [⌊duplae⌋ *del.*] lineae SE ⌊quae centro Terrae et⌋ perpendiculo PE ⌊interjacet⌋ [abscissae *del.*] ad umbilicorum [*illegible word del.*] distantiam SF.

The proof of Lemma [*a*] consists of a 'Cas. 1.' only, without a 'Cas. 2.'. It is followed by:

Corol. Valet Propositio quamproximè ubi excentricitas finitae est magnitudinis, si modo parva sit.

The second lemma[2] also has only a 'Cas. 1.'. It too has a corollary, stating that 'Obtinet etiam Propositio quamproximè...', even when the eccentricity of the elliptical orbit is not infinitely small.[3] When these two lemmas, and the 'Cas. 1.' and 'Corol.' to each, were printed as Appendix III to the Preface of the *Catalogue of the Portsmouth Collection*, the editors (or compilers) gave them the correctly descriptive title, 'On the Motion of the Apogee in an Elliptic Orbit of very small Eccentricity'.[4]

Lemma [*a*] contains two references to the *Principia*:

Et per ea quae in Prop.   Lib. 1 ostensa sunt...

Nam latus rectum quod sit (per Prop.   Lib. 1 Princip.) in duplicata ratione ⟨areae⟩ quam Luna radio...describat...

[2] The full text of both Lemma [*a*] and Lemma [*b*], but not of the two accompanying propositions, is printed in [Portsmouth Collection] (1888), *Catalogue*, Appendix III to the Preface, pp. xxvi–xxx. The order of these two lemmas (dictated by the fact that Lemma [*a*] occupies f. 105 and the top half of f. 106, while Lemma [*b*] occupies the bottom half of f. 106 and all of f. 107) is confirmed by the occurrence of two references to 'the above (or foregoing) Lemma'. The first appears in line 4 of f. 107, and reads: 'Est autem (ut in Lemmate superiore)...' The second appears in a sentence half-way down f. 107 beginning: 'Jam vero in Lemmate superiore...' Both refer to Lemma [*a*].

[3] Since it is almost certain that these lemmas and the two propositions based upon them were composed after the *Principia* (or at least after Book I), I shall not enter here into the fascinating question of their intellectual purpose. As we shall see below, these lemmas and propositions, so far as we can now tell, never even attained such a degree of commitment on Newton's part that he assigned numbers to them as belonging to a revised Book III, much less entered them into one or the other of his personal annotated copies of E_1. It has been pointed out that these lemmas deal with 'the motion of the apogee in an elliptic orbit of very small eccentricity due to given small disturbing forces acting, (1) in the direction of the radius vector, and (2) in the direction perpendicular to it'. This sentence, presumably by John Couch Adams, is quoted [Portsmouth Collection] (1888), *Catalogue*, Preface, p. xii, where a two-page discussion is given of the relation of these lemmas (and of the two propositions accompanying them) to the presentation in the *Principia* of the motion of the Moon and in particular the problem of deducing 'the horary motion of the moon's apogee for any given position of the apogee with respect to the sun'. These conclusions of Adams's are repeated almost verbatim and partially supplemented in Rouse Ball's *Essay* (1893), p. 109, and also p. 85.

[4] No indication is given, however, of where this manuscript is to be found within the Portsmouth Collection. Nor is anything said of the propositions that accompany these lemmas, nor of the prior draft. The editors did not explore the possibility of whether these might be part of a revision of the *Principia* or of an earlier version than *M*, and merely said (p. xxvi): 'From a somewhat mutilated MS. which seems to have been prepared for the press.' Of course, these lemmas were not necessarily 'prepared' for the press; the form of presentation is merely that which was customarily used by Humphrey in his copies. In the body of the Preface (p. xii) it is said, 'These lemmas are carefully written out, as if in preparation for the press, and they were probably at first intended to form part of the *Principia*.' These two statements are less extravagant than a note in the *Catalogue* proper (p. 6: IX, 12): 'The Propositions are not numbered, and therefore they were perhaps intended to be worked up for the 1st Edition.' This hypothesis, as will be seen, cannot be sustained.

From the context it is apparent that the first blank should be filled with XVII and the second with XIV. Two puzzles thus are presented to us. First, why should Newton refer to 'Princip.' if these lemmas were intended for Book III of the *Principia* itself? Secondly, why are the numbers left blank? In Lemma [*b*], there are also two references:

...(per Prop. XIV Lib. I. Princip.)...
...(per Lem. Lib. II Princip.)...

Since Newton has filled in the number (XIV) in the first of these, whereas he has not done so in Lemma [*a*], the reason for the blanks in Lemma [*a*] cannot be that these lemmas were written out before the propositions in Book I had had numbers assigned to them. Corroborating evidence is that Newton refers to 'Lib. II Princip.'. As we have seen (Chapter III, § 6), by the time Newton had decided to have three books and to have a Book II as we know it (that is, by the summer of 1685), Book I had been essentially completed and the propositions numbered. Incidentally, this reference to 'Lib. II Princip.' also confirms the statement made in the beginning of this section that these manuscripts were written in relation to *De mundi systemate* (*lib. 3^{ius}*) rather than *De mundi systemate* (*lib. 2us*), since the 'Lem. Lib. II Princip.' is Lemma II of Book II of the *Principia* and not a lemma from *De mundi systemate* (*lib. 2us*).

The sequence goes on (following ff. 105^r–106^r, 107^r containing Lemma [*a*] and Lemma [*b*] in their entirety) to the first of the propositions at the head of f. 102^r. There is no doubt that this proposition was intended to follow immediately after the two lemmas, since f. 102^r (on which it is found) must come next in succession after f. 107; the evidence lies in the fact that an insert for f. 102^r occurs on f. 107^v at the very part of the page alongside the line where it is to go on f. 102^r. This proposition (already quoted at the beginning of this § 7) reads:

Prop. Prob.
Invenire motum horarium Apogaei Lunae.

Newton has not assigned numbers to either the proposition or the problem. For convenience, I shall refer to it as Prop. [*i*], Prob. [*y*].

In developing this Prob. [*y*], Newton refers explicitly to 'Lem. 2. Lib. II Princip.',[5] whereas in Lemma [*b*] he has only written 'Lem. Lib. II Princip.'. But he has also left some references blank, as in the example cited above (f. 103):

...vi V in Lemmate & propterea (per Lemma illud) motus Apogaei...

which occurs twice; the first reference is to Lemma [*a*] and the second to Lemma [*b*]. In the corollary (f. 103^r, bottom; f. 104^r), there is a blank reference to 'Corol. Prop. '. This can be identified as Corol. 3, Prop. XXXIV, Book III.

[5] The argument here is that if t is the 'momentary increment' or 'moment' of the time T (and hence infinitely small with respect to T), then $[(T+t)/T]^2=(T+2t)/T$ because $(T+t)^2=T^2+2tT$. The latter follows from the cited Lemma II, Book II, where it is shown that if a and b are 'moments' respectively of A and B, then $(A+a)\ (B+b)=AB+aB+bA$.

The second proposition (problem) in the sequence is stated on the bottom of f. 104^r and developed on ff. 110^r and 111^r. It reads:

Prop. Prob.

⟨Pos⟩ito quod excentricitas orbis Lunaris sit infinite parva, ⟨invenire mo⟩tum medium Apogaei.[6]

For convenience, I shall refer to this as Prop. [*ii*], Prob. [*z*].

I have stated earlier that these manuscript pages are written out in Humphrey's hand, with later corrections, alterations, and additions by Newton. But on f. 111, the end of the sequence, the handwriting ceases to be Humphrey's about a fifth of the way down the page and becomes Newton's. There is no apparent way of telling whether or not the concluding portion in Newton's hand was added at a later date. A portion of the text in Newton's hand is cancelled; a replacement, also in Newton's hand, appears on the facing page (f. 110^v). The concluding lines (at the bottom of f. 111) are also cancelled. There is no indication on these sheets (or on any others that I know) whether Newton had ever written out any further propositions in this sequence; nor can one readily determine whether there had ever been any additional textual material in Newton's hand for this Prop. [*ii*], Prob. [*z*], on a page or pages now missing.

In our efforts to determine Newton's intentions in composing this pair of lemmas and propositions, we are frustrated by the lack of external information; they are not mentioned in correspondence, nor does there appear to be an explicit mention of them in any other manuscripts or in memoranda. Because of the references to Lemma II, Book II, of the *Principia*, they must have been written out after the summer of 1685, by which time—according to Newton's letter of 20 June 1686 to Halley—Newton had completed the second book. Furthermore, since these two lemmas constitute a generalization or extension of Prop. XLV, Book I, it may be presumed that by the time Newton had worked them out it was too late to rewrite Prop. XLV so as to incorporate the new results.[7] If Newton had still had the manuscript of Book I by him, would he not then have altered his treatment of this most important proposition? Such considerations would enable us to date the sequence under discussion at some time after February or March 1687, when the printing of the first edition had progressed beyond Prop. XLV.[8]

[6] The portions in angle brackets have been restored.

[7] This point was suggested to me by D. T. Whiteside.

[8] In E_1, Prop. XLV occurs on pp. 137–42 (first six lines), while Corols. 1 and 2 occupy pp. 142 (remainder of page) to 144; these eight pages comprise the whole of a single sheet (of eight pages, or four leaves), signature T. Reckoning from B (containing pp. 1–8), since A was—as usual—reserved for the front-matter and composed at the end, and counting I but not J, Prop. XLV occurs in its entirety within the eighteenth sheet of E_1. On 13 February 1686/7 Newton wrote to Halley: 'I have eleven sheets already, that is to M. When you have seven more printed off, I desire you would send them.' Hence we may assume that, if Newton had still wanted to make a change in the sheets yet to come, he could have. On 14 March 1686/7 Halley wrote to Newton that he was sending him this eighteenth sheet, printed. Most probably, therefore, the manuscript pages under discussion here must have been written after February or March 1687. But we cannot entirely dismiss the possibility that they date from late 1686 (or even January or February 1687), and that Newton—for whatever reason—did not wish to revise the pages containing Prop. XLV which were either at the printer's

These lemmas and propositions deal explicitly with the motion of the Moon's apogee, a topic that occurs briefly in E_1 in the short Scholium to Prop. XXXV, Book III, which was replaced in E_2 by a much longer scholium in which Newton applies the theory of gravity to various inequalities in the motion of the Moon.[9] We know that Newton was never fully satisfied with the results of his studies of the motion of the Moon and returned to them again and again. In E_3, as well as in E_2, Newton added new material on this topic, and not very long after the first edition had been published, in about 1694, he wrote out a list of propositions on the motion of the Moon as a plan evidently made for a revised edition.[10] The present documents are earlier, since—as mentioned above—they are written out by Humphrey. Possibly they are to be dated in 1687 or 1688, while Newton was still being carried forward by the momentum of his achievement; and perhaps they are even a kind of first response to Halley's letter to him of 5 July 1687, announcing the completion of the printing of the *Principia* and exhorting him to 'resume those contemplations, wherin you have had so good success, and attempt the perfection of the Lunar Theory'.[11]

If this were a separate exercise in supplement to the *Principia*, rather than a proper revision of the *Principia*, then—understandably—Newton would mention his *Principia* by name. That Newton does in fact refer to 'Lem. Lib. II Princip.' would reinforce the conclusion that, as written, these lemmas and propositions were an amplification or extension of the finished *Principia*, rather than work-sheets in preparation for Book III. This does not, of course, rule out the possibility that at any time Newton might have contemplated introducing them into the *Principia* as part of some stage of revision, and then assigning numbers.

There seems to be no doubt that Newton had written these lemmas and propositions after composing Book III, and possibly even after Book III had been printed off. But, since there is no way of telling whether they date from just before or just after publication of the *Principia*, we may conclude only that they are of 1686–7–8.

The remainder of §12 of MS Add. 3966 consists of two parts. Folios 108–9 and 112 contain earlier versions of these two lemmas and two propositions, and are discussed in Supplement IX. Pages 113^r–113^v are of a different kind altogether, true work-sheets that are part of the preparation of *M*. The reason why they are grouped together with ff. 108–9, 112 and ff. 102–7, 110–11 is that ff. 112–13 comprise a single folio sheet, folded once down the centre to give two leaves. But

or just about to be taken by Halley to the printer's. When Halley sent Newton the eighteenth sheet, he wrote 'of the extraordinary trouble of the last sheet, which was the reason that it could not be finished time enough to send it you the last week: I have not been wanting to endeavour the clearing it of Errata, but am sensible that notwithstanding all my care some have crept in, but I hope none of consequence'.

[9] See note 3 above.

[10] See Rouse Ball (1893), *Essay*, p. 110; this list of propositions is reprinted by him on pp. 126–8 of his *Essay* from the text given in [Portsmouth Collection] (1888), *Catalogue*, Appendix II to the Preface, pp. xxiii–xxvi.

[11] See Chapter V, §3.

we must not conclude that the text of pages 112^r–112^v is necessarily of the same date as that of pages 113^r–113^v, merely because ff. 112–13 are part of the same sheet. It was Newton's habit to use any blank leaf or even a part of any handy page as so much 'waste' paper for scribbling. Thus we find Newton manuscripts written out on otherwise blank parts of pages of letters sent to him, of official documents (as of the Mint), and of his own writings. In this case it seems that Newton had used both sides of one leaf (f. 113) for a version of a proposition and its corollaries in the preparation of *M*, and then turned the sheet inside out, so that the first leaf was now blank, and used this blank leaf as scrap-paper. Thus, when the cataloguers encountered this sheet, they numbered the top leaf 112, and the bottom one 113, without regard to the fact that the text of leaf 112 was written later than that of 113.

Folio 113^r contains (at the head) the text of Prop. XXXIV, Prob. XIV, Book III, written out in Humphrey Newton's hand:

Prop. Prob.
Invenire variationem inclinationis Orbis Lunaris ad planum Eclipticae.

The development of this problem, also written out in Humphrey's hand, is identical (save for capitalization and punctuation) with the text—unchanged in all editions—as printed in the *Principia* {454.24–455.7}. The only difference is that in the manuscript there is no number assigned to this proposition (and problem), and the reference {455.4} 'per Prop. ' is also blank. But the paragraph following the proof ('Haec ita se habent ex hypothesi... & propterea idem manebit atque prius') is written out in Newton's hand immediately after Humphrey's 'Q.E.I.'.[12] This work-sheet does not show us the actual steps of invention, since Humphrey's text and Newton's are presented to us as finished products, ready for copying out in *M*; nevertheless, we may learn from it that the paragraph in Newton's hand is apparently of a later date, an afterthought.

The rest of the page contains the text in Humphrey's hand of Corol. 1 and Corol. 2. A large blank space has been left between the proposition and Corol. 1, and between Corol. 1 and Corol. 2, presumably for any additions or a rewriting that Newton might care to make. On the top of f. 113^v there is another version of Corol. 2 in Humphrey's hand, with some proportions worked out by Newton. This version has been cancelled. In the centre of the page Humphrey has written out Corol. 3, which has been revised by Newton; at the bottom of the page, Newton himself has written Corol. 4, which is much worked over.

Pages 113^r–113^v have the physical appearance of work-sheets. Unlike ff. 105–6, 101–2, 103–4, 110–11, containing the revised versions of Lemma [*a*] and Lemma [*b*] and of Prop. [*i*] and Prop. [*ii*], these pages do not have the customary right-hand rule used by Humphrey in his more-or-less final copy. Furthermore, the great gaps of open space for Newton to fill with revisions testify to the tentative

[12] This paragraph differs in *M* and E_1 in that the final sentence (as it appears in E_2 and E_3 {455.12}) ends with 'Sinus IT' rather than 'inclinationis variatio' and there is an additional sentence. The manuscript under discussion here corresponds with *M* and E_1.

nature of the texts given on pages 113^r–113^v. There is no way of dating these two pages, but the absence of numbers to designate the proposition itself (and the proposition to which Newton refers) may indicate that these versions were made before the rest of Book III had taken final form.

8. A CRITIQUE OF A PRE-*M* VERSION OF THE 'PRINCIPIA'

I have mentioned again and again in this Introduction the growth of my conviction that there must have been a tolerably complete draft of all three books of the *Principia*, which was the basis of Humphrey's final copy (*M*), and of *LL* and *De mundi systemate* (*lib. 2us*). I believe that any scholar who studies Newton's manuscripts and steeps himself in Newton's methods of work will come to the conclusion that such drafts had necessarily to exist, and have either perished or been deliberately destroyed. The likelihood seems small indeed that they have merely been lost and may yet turn up, in separate parts or in their entirety.

The reader may well imagine, therefore, how exciting it was to locate among Newton's manuscripts (U.L.C. MS Add. 3965, ff. 94–9; see Plates 10 and 11) a set of fragments, not by Newton, which I was able to identify as a set of all-too-brief comments, in part on an earlier version of the *Principia* than *M* or E_1. Some of the suggestions of this critic were actually later incorporated into the final text; others were apparently rejected. This critic saw a version of the preliminary Definitions and Laws of Motion, and substantial portions of Books I, II, and III. Since his observations precede the final draft (*M*), we can now be absolutely certain that there once existed a prior draft of Books II and III, as well as of Book I, together with the Definitions and the Laws of Motion. I believe there can be little doubt that this critic was Edmond Halley, although we have no documentary evidence other than the existence of these sheets that Halley might have exercised so critical an influence at this early stage of composition of the *Principia*.

Further information about the contents of this manuscript critique is given in Supplement VII. Here I shall content myself with a bare summary of my findings. First of all, since I have identified the individual comments, there is no doubt that the critic saw the actual sheets of LL_α that survive. Not only does he refer to LL_α by page and line, but his suggestions are sometimes adopted by Newton and may be seen written out in LL_α in Newton's hand. There are other comments which may be identified as references to a continuation of LL_α, no longer extant save for a set of eight pages.[1] Only once does this critic appear to have made a comment directly on LL_β; this occurs separately from the others, and may be presumed to have been made at a later date; but it does show that the critic saw both LL_α and LL_β. The fragments show that the critic saw a manuscript version of Book I that in *LL* went up to Props. LXXI and LXXII of *M* and E_1, well into what is Sec. XII (of the fourteen sections) in the final version of Book I.

[1] See Supplement IV, and §2 above.

The references to Book II encompass only about a third of the final version, including the first nineteen propositions. Again, since some of the critic's suggestions have been incorporated into *M*, we may be sure that he had seen a pre-*M* version rather than *M* itself. The same holds true for Book III. We may be certain, moreover, that it was in fact *De mundi systemate* (*lib. 3ius*) that the critic saw, and not the earlier *De mundi systemate* (*lib. 2us*), since the references include proposition numbers, which are absent in the prior work. The critic saw Book III through Prop. XXIV, about one-third of the whole, or one-half of Book III without the material on the comets. We have no way of telling how complete the existing fragments of critical comment may be. There is simply no way of telling how much more of either Book II or Book III the critic actually may have seen.

Who could this critic have been? I have not been able to find any documentary evidence of any friend or associate of Newton's ever having seen such a preliminary version. I believe that this is the first time that the very existence of these comments has been brought to light, giving us at last some specific information about versions of the *Principia* other than *LL*.

In attempting to identify this critic, we must keep in mind that he must not only have been an intimate of Newton's but also a person of such stature that Newton would have been to some degree guided by his comments. Candidates whose names come to mind are Paget, Babbington, Vincent, and Halley. We have no manuscript of Paget's that I know of, and there is no reason to suppose that Vincent was more than a messenger who took the manuscript of Book I to the Royal Society. As to Humphrey Babbington, a tutor in Trinity, we are on better ground. He was uncle to Arthur Storer, who sent to Babbington for Newton (and to Newton directly) various astronomical observations from Maryland.[2] Babbington, later Vice-Master of Trinity, was uncle to the famous 'Miss Storey' (no doubt a Miss Storer) of Stukeley's account of Newton,[3] and like Newton himself had been educated at Grantham Grammar School and Trinity. But I have found no evidence that Babbington knew enough science or mathematics to have made such a critique of the *Principia*.

In terms of competence, Halley and Paget would seem to be the leading contenders. Paget may be eliminated, however, since the handwriting of this critique differs so markedly from that of his signature in the Trinity College exit-book of this period. Halley may be more likely, since there are marked similarities between his hand and this. But there are some differences too, which could possibly be due to his having jotted down these notes in haste rather than having written them out at leisure in a more usual and formal hand.

At this stage of the investigation, my quest for a positive identification of the critic was considerably sharpened by my recognition of the significance of the very

[2] [Newton] (1959–), *Correspondence*, vol. 2, pp. 269, 272, 275, 280, 368, 387.

[3] Stukeley (1936), *Newton's life, 1752*, pp. 45–6; see also More (1934), *Newton*, pp. 15, 19, 28, 30, 48; Brewster (1855), *Newton*, vol. 2, pp. 91, 101, 412, 546. I accept the supposition that 'Miss Storey' was really 'Miss Storer'.

existence of these comments; that is, that they show the extent of an earlier and now (largely) missing draft of the *Principia*, as well as the fact that Newton had sought and received critical comments at an early stage of composition. Of all the known associates of Newton, Halley was by all odds the likeliest candidate, but identification by handwriting alone is always a chancy business, especially when —as in this case—the result of writing in haste may have produced some differences between this document and others written by Halley more carefully and at leisure.

I therefore asked D. T. Whiteside, who strongly identified the writer as Halley. Later, in support of Halley's candidature, he called my attention to two specific pieces of evidence: (1) on f. 95 there are figures drawn by Newton which are related to a letter of his to Halley, written on 14 July 1686;[4] (2) on f. 96^v there are calculations in Newton's hand, in which he is surely checking on Halley's query of 14 October 1686, answered by Newton four days later.[5] The fact that Newton on two separate occasions worked out on these sheets his replies to Halley's questions certainly so increases the probability of Halley's authorship that I think we may consider him to have been the critic in question. The result is that we must now be cognizant of a greater degree of intimacy between Newton and Halley than is demanded by other documentary evidence, such as correspondence. This example shows why it is valuable for the scholar to go through Newton's manuscripts again and again, to be certain that odd scraps and fragments of comments do not continue to hide their secrets.

9. SOME ADDITIONAL PAGES OF *M*

In addition to the materials described in the foregoing sections (§§4, 5, 7) there are among the manuscripts in the Portsmouth Collection a few more pre-*M* pages, which appear to be rejected pages of *M*. Among these are three sheets, consisting of two leaves each, written on one side of the page only in Humphrey's hand (U.L.C. MS Add. 3965, ff. 15–16, 17–18, 19–20). These sheets were originally written out for Book I in *M* and correspond to the sheets (two leaves each) there foliated 7, 44, 53. They contain, respectively, the end of Corol. 2 and the beginning of Corol. 3 to the Laws of Motion (f. 7); the end of the proof of Prop. XXXI and the beginning of the scholium (f. 44); and the end of the proof of Prop. XLI, the whole of Prop. XLII, and the statement of Prop. XLIII and the beginning of the proof (f. 53). Save for the usual occasional alteration in Newton's hand, these pages are so very much like the others composing *M* that on first examination the reader may very well wonder why they were replaced.

The reason for the rejection of these three sheets becomes clear from a close comparison of their contents with both the final versions in *M* and the corresponding texts of *LL* (the deposited 'lectures' *De motu corporum*: U.L.C. MS Dd. 9.46). The rejected ff. 15–16, which contain the end of Corol. 2 and the beginning of

[4] [Newton] (1959–), *Correspondence*, vol. 2, p. 444. [5] *Ibid.* pp. 452–3. See also Chapter V, §1.

Corol. 3 to the Laws of Motion, are based on the versions given on the recto pages of *LL*, with some revisions. But Newton later changed his mind and rewrote the end of Corol. 2, cancelling the last two paragraphs altogether and most of the antepenultimate paragraph (on f. 15[r], in the modern librarian's numbering, of *LL*). On the otherwise blank facing page (f. 14[v]), there is written out—in Humphrey's hand—the new conclusion to Corol. 2, as it appears in the final copy of *M* and is printed in E_1. Incidentally, the fact that Humphrey has written out the new text on f. 14[v] shows once again that Newton was using *LL* as his copy for recording changes before the printing of E_1.

With respect to the rejected ff. 17–18, containing the beginning part of the scholium to Prop. XXXI, Book I, the problem of following Newton's steps is vastly more complex. First, as will be seen from the Apparatus Criticus, the text of this portion of the scholium differs greatly in E_1 and in E_2 and E_3. Secondly, the manuscript *M* contains two stages of composition of this very part of the scholium. A major difference between the two is that the original state of *M* as sent to Halley is shorter; it contains the same opening paragraph ('Caeterum...quamproxime') as E_1 (p. 109), omits altogether the next two paragraphs ('Nam...sequentem' and 'Per...reddet' as on pp. 109–11 of E_1), then has the fourth paragraph but for the final sentence ('Si Ellipseos...Hyperbola'), and continues with the next paragraph ('Si quando locus ille P...') as in E_1 (p. 111). In *M*, the two paragraphs at the beginning of the scholium ('Caeterum...quamproxime' and 'Si Ellipseos...Hyperbola') are cancelled and are replaced by a sheet folded so as to make two pages of replacement, corresponding to the first four paragraphs of the scholium as printed in E_1. The original version in *M* is referred to in the Apparatus Criticus {109.13–14, 14} as M_i and the replacement pages as M_{ii}. There are further differences between M_i and M_{ii}, but they need not detain us at this point since the absence in M_i of the second and third paragraphs as found in M_{ii} and E_1 shows that the rejected ff. 17–18 correspond to M_i (that is, are pre-M_i) and were therefore not cast aside because (and when) Newton decided to rewrite and enlarge the scholium, which he did in M_{ii}.

A careful collation of the scholium, to the extent that it is found on ff. 17–18, shows no difference whatsoever from M_i, save for punctuation, capitalization, and the spelling out of 'et' in place of the ampersand. Clearly, then, Newton did not reject ff. 17–18 because of a fault in the copy for the scholium. Incidentally, it can readily be shown that ff. 17–18 are prior to M_i, the corresponding sheet in *M*, because certain minor emendations made by Newton himself on ff. 17–18 are incorporated into the text line in *M*. These are six in number: (1) 'principale' in the first sentence of the paragraph beginning 'Si Ellipseos latus transversum principale...'; (2) 'ea' later on in the same sentence ('...et in ea ratione ad GK quam habet area AVPS ad...'); (3) the alteration of 'ordinata PR' to 'ordinatim applicata PR' in the fourth sentence ('Demissaque ad axem Ellipseos...') of the next paragraph (beginning 'Si quando locus...'); (4) 'angulus' three sentences later ('Sit angulus iste N'); (5) 'et' in the next sentence ('Tum capiatur et angulus D...'); and (6) 'tum' in the next sentence ('Postea capiatur tum angulus...'). I have listed these alterations in full so that it may be seen that their correction would not appear to have required a whole new sheet of two pages. As I mentioned, all of these corrections have been incorporated into the text of M_i; and yet another, not present in f. 18, has been so incorporated in M_i, the second occurrence of 'ordinata' in (3), which reads 'ordinata RQ' on f. 18 (without alteration) and which becomes 'ordinatim applicata RQ' in M_i.

It is interesting to find that, of the foregoing changes introduced by Newton in ff. 17–18, (1) has not been made in *LL*, nor does it appear in M_{ii} or E_1, although like the other five it is found in M_i; (2) has been made in *LL* and is found in M_i, M_{ii}, E_1; (4), (5), (6) have not been made in *LL*, but do occur in M_i, M_{ii}, E_1; while the correction of (3) in *LL* is at the second occurrence of 'ordinata' and not the first, but both are given in the altered form in M_i, M_{ii}, E_1. Newton thus entered some, but not all, of the alterations in *LL*, after these pages had been copied out from ff. 17–18 into M_i. In two instances in the first paragraph—the deletion of 'vel Hyperbolae' in the second sentence ('Ellipseos vel Hyperbolae...') and the insertion of the phrase in parentheses ('qui casus in Theoria Cometarum incidit')—the change has been made only in *LL*. Thus neither *LL* nor ff. 17–18 correspond exactly with M_i or M_{ii}.

I have said that I do not believe these changes in the scholium would have been sufficient to cause Newton to have this sheet wholly recopied. But at the top of f. 17, there is a rather significant change made in the final paragraph of the demonstration of Prop. XXXI. This paragraph is the same in *LL* and in the original version in f. 17 (save for minor and unessential details of presentation). To exhibit the difference between the two versions, I have printed it in parts; those that are centred on the page are common to both (i) the first version in f. 17 (and *LL*) and (ii) *M* (and f. 17's final version); those that are of half-width are different in (i) the first version in f. 17 (and *LL*) and (ii) *M* (and f. 17's final version); the left-hand column contains the first version from f. 17, the right-hand column the final from *M*:

Nam centro O... perpendiculum SR. Area APS est ut area

AQS, id est, ut differentia [inter aream AQ et aream SR *del.*] arearum OQA − OQS seu $\frac{1}{2}$OQ in AQ − SR; hoc est,	inter Sectorem OQA et triangulum OQS, sive ut differentia rectangulorum $\frac{1}{2}$OQ × AQ et $\frac{1}{2}$OQ × SR, hoc est ob datam $\frac{1}{2}$ OQ,

ut differentia inter arcum AQ et rectam SR,

quae est ad sinum arcûs illius ut OS ad AO, adeoque	adeoque (ob aequalitatem rationum SR ad sinum arcûs AQ, OS ad OA, OA ad OG, AQ ad GF, et divisim AQ − SR ad GF − sin. arc. AQ)

ut GK differentia inter arcum GF & sinum arcus AQ. Q.E.D.

No doubt, this revision determined the replacement of ff. 17–18.

The situation with respect to ff. 19–20 is somewhat different. These contain the very last lines of the proof of Prop. XLI, Book I, and its Corols. 1 and 2, the whole of Prop. XLII, and the beginning of Prop. XLIII. In *LL* Prop. XLI has the third corollary, present in *M*, but not in this rejected sheet. This is apparently the reason for the rejection of ff. 19–20.

10. FOUR ADDITIONAL PAGES OF LL_β: THE CONCLUSION OF BOOK I

Among the manuscripts in the Portsmouth Collection devoted to optics (U.L.C. MS Add. 3970) there is a pair of sheets (ff. 615–17) containing the concluding part of Book I, also written out in Humphrey's hand, which have the physical appearance of the sheets in *M*. These three leaves[1] contain the end of the Scholium following Prop. XCVI, Book I, in a form that differs from the version in *M* and E_1; also Prop. XCVII plus its two corollaries, Prop. XCVIII, and the concluding Scholium to Sec. XIV; that is, they contain the concluding pages at the very end of Book I. Unlike the final pages of *M*, however, these three do not contain the numbers in the margin referring to figures, of which there are in E_1 two for Prop. XCVII and one for Prop. XCVIII (a repeat of one of the previous pair). Hence, these are not rejected pages of *M*, but must have been part of a version of Book I prior to *M*. Another page that belongs to this set may be found elsewhere in MS Add. 3970, on f. 428b^r.

These pages contain many minor corrections, alterations, and improvements,[2] but also certain major changes. One such occurs in the final sentence of the Scholium following Prop. XCVI in E_1:

Igitur ob analogiam quae est inter propagationem radiorum lucis & progressum corporum, visum est Propositiones sequentes in usus opticos subjungere; interea de natura radiorum (utrum sint corpora necne) nihil omnino disputans, sed trajectorias corporum trajectoriis radiorum persimiles solummodo determinans.

Therefore on account of the analogy that exists between the propagation of the rays of light and the motion of bodies, it seemed best to subjoin the following propositions for optical uses; meanwhile considering not at all, the nature of the rays (whether they are bodies or not), but only determining the trajectories of bodies which are similar to the trajectories of the rays.

The original version, on ff. 428*b* and 615, reads as follows:

Refractio igitur quia incipit in aere non fit per resistentiam vitri, et quia perseverat in vitro non fit per resistentiam aeris, sed attractioni jam expositae similior est. Si radii in medio quovis

Refraction, therefore, since it begins in the air, does not take place through the resistance of the glass; and since it perseveres in the glass does not take place through the resistance of the air, but is more like the attraction already set forth. If rays in any medium experienced resist-

[1] These leaves are part of a set of four on two sheets, each sheet containing two leaves, on only three of which there is writing, the fourth being blank.

[2] For example, in the statement of Prop. XCVII, Newton himself has later written in 'juxta' {227.4} for 'ad', 'brevissimo' {227.4} for 'minimo', 'determinare' {227.5} for 'requiritur', and so on.

resistentiam sentirent hi (sive motus sint sive mota corpuscula) perpetuo retardarentur et redderentur debiliores, omninò contra experientiam. Utrum vero reflexio et refractio fiant per attractiones disputet qui volet. Malim Propositione una et altera de motu, inventionem figurarum usibus opticis inservientium docere; ⌞interea de natura lucis nihil definiens sed [ex analogia tantum inter propagationem⌞radiorum⌟ lucis et motum corporum motubus corporum *del.*[3]] ⌞trajectorias corporum trajectoriis radiorum⌟ lucis persimiles [definiens *del.*] determinans.

ance, these (whether they be motions or moving small bodies) would perpetually be retarded and made weaker, totally contrary to experience. But whether reflection and refraction come about through attractions let anyone dispute who wishes to. I should prefer with a couple of propositions on motion to elucidate the finding of figures serving for optical uses; meanwhile defining nothing about the nature of light but [only by the analogy between the propagation of the rays of light and the motion of bodies... to [?] the motions of bodies *cancelled*] determining the trajectories of bodies which are very similar to the trajectories of the rays of light.

When Humphrey recopied these pages into *M*, he included the new conclusion to the scholium and other changes. The conclusion of Prop. XCVI, and the beginning part of the scholium which is continued on f. 615, occur on f. $428b^r$. This is one half of a folded sheet having originally two leaves (like ff. 615–16, 617–18, and the sheets in *LL* and *M* generally); the other half is not presently in MS Add. 3970. Nor have I been able to locate (or to identify) any other pages of the final Sec. XIV in this pre-*M* state.

Folio 616 has only minor corrections by Newton, but the final sheet has a major change; the latter has only eight lines of text (only the first page, f. 617^r, has any writing on it). On this page (f. 617), the material in Humphrey's hand ends with the penultimate sentence of *M* and E_1, '...vel alias quascunque perfici possit' {229.22–5}, followed by a cancelled sentence in a new paragraph: 'Hactenus exposui motus corporum in spatiis liberis.' Newton has added another incomplete sentence in his own hand, which he has then crossed out, to conclude with the final sentence as in *M* and E_1, 'Nisi corrigi possunt errores...imperite collocabitur' {229.25–6}, in which 'imperite' was a replacement (probably) of 'frustra'. The cancelled unfinished sentence, in its final version, reads:

Errores inde oriundi sunt (in Telescopiis) longè majores quàm qui ex figuris minus aptis oriri solent: et vitio vertendum est siquis ignorata errorum causa principali

The errors thence arising are (in telescopes) far greater than those which usually arise from less suitable figures; and it must be reckoned a fault if anyone, not knowing the principal cause of the errors...

From the foregoing analysis, it appears that the above-mentioned four pages (ff. 615–17, 428*b*) are immediately pre-*M*, and hence must have been part of LL_{β}, the manuscript from which Humphrey copied *M*. These four pages are not today among the papers relating to the *Principia* (chiefly U.L.C. MS Add. 3965), but

[3] Possibly Newton first meant to continue the cancelled passage with the words 'lucis persimilem', that is, '...the analogy between the propagation of the rays of light and the motion of bodies which is very similar to the motion of bodies of light', but decided not to lean toward either side of the question whether rays of light consisted of 'bodies' or not.

rather among the optical papers (U.L.C. MS Add. 3970), possibly placed there by the cataloguers who may have rearranged and reclassified the *disjecta membra*. It is also possible, of course, that Newton himself may have placed these four pages of geometrical optics together with the mass of his other optical manuscripts, once he had completed the text of the *Principia*. But in either case, these four pages are now separated from the other manuscript drafts of the *Principia* and the other manuscripts relating to the *Principia*, and the four pages in question are in two wholly different parts of the optical manuscripts: f. 428*b* is in Sec. 3 of MS Add. 3970, while ff. 615–17 are in Sec. 9. This melancholy example shows the necessity of looking at every conceivable group of manuscripts, even those classified as dealing with subjects other than the researcher's special interest. Above all, it shows us why great caution must always be exercised when any scholar (myself included, of course) claims to have made an inventory of all of Newton's manuscripts on a particular topic; for example, all manuscripts containing texts of the *Principia* prior to *M*.

What stage of writing the *Principia* can these pages represent? First of all, they must be later than LL_α, since the proposition numbers (XCV–XCVIII) agree with those in M, E_1, E_2, and E_3. Thus they are post-LL_α and pre-M, and hence by definition an early state of LL_β.

CHAPTER V

THE COMPLETION AND PRINTING OF THE 'PRINCIPIA'

1. BOOK I

IN Chapter IV we have seen the varieties of documents that show the stages whereby Newton advanced from the tract *De Motu* to the finished manuscript of the *Principia*. From other sources we may find the nature of the subject he was working on. For instance, during the last months of 1684 and all of 1685, following the completion of *De Motu*, we know from Newton's continued correspondence with Flamsteed[1] that he was working away at astronomical problems such as appear in Book III of the *Principia*. But we cannot, on the basis of presently known documents or letters, establish a chronology of the writing of the *Principia*, prior to Halley's acknowledgement of the receipt of the manuscript of Book I of the *Principia* in April 1686.[2] The situation becomes markedly different once the manuscript of Book I had been received, for we can then trace out in considerable detail the stages leading up to the start of printing, beginning with the official presentation of the manuscript to the Royal Society on 28 April 1686:

> Dr. Vincent presented to the Society a manuscript treatise intitled, *Philosophiae Naturalis principia mathematica*, and dedicated to the Society by Mr. Isaac Newton, wherein he gives a mathematical demonstration of the Copernican hypothesis as proposed by Kepler, and makes out all the phaenomena of the celestial motions by the only supposition of a gravitation towards the centre of the sun decreasing as the squares of the distances therefrom reciprocally.
>
> It was ordered, that a letter of thanks be written to Mr. Newton; and that the printing of his book be referred to the consideration of the council; and that in the mean time the book be put into the hands of Mr. Halley, to make a report thereof to the council.[3]

This description is especially interesting since Propositions I–XI deal with all three of Kepler's laws of planetary motion—the law of areas (Props. I–III), the harmonic law (Corol. 6 to Prop. IV), and the law of elliptical orbits (Prop. XI)—so that Newton's achievement can be quite properly described as having been to provide a dynamical foundation (or 'mathematical demonstration') of the Keplerian version of the Copernican hypothesis.[4] Yet nowhere in Book I of the

[1] [Newton] (1959–), *Correspondence*, vol. 2, *passim*; cf. Baily (1835), *Flamsteed.*

[2] [Newton] (1959–), *Correspondence*, vol. 2, p. 431.

[3] Birch (1756–7), *History*, vol. 4, pp. 479–80; [Newton] (1958), *Papers and letters* [Cohen and Schofield], pp. 489–90.

[4] The word 'hypothesis' was customarily used in those days to designate a system of the world.

Principia does Newton acknowledge Kepler's contribution, or even once mention his name.[5]

On 16 May, Halley wrote about the *Principia* to a German correspondent, Johann Chr. Sturm, as follows:

> Gravity, although a force far dissimilar [to the magnetic force] and equally spread everywhere, doubtless draws its origin from causes not much different; in fact, a most keen mathematician and philosopher Mr Newton of Cambridge has already brilliantly investigated the causes and effects of gravity, and his books about this are now in press. He demonstrates that the force of gravity is greatest on the surface of the earth and decreases upward in the duplicate proportion of the distance from the centre, and downward in the simple proportion; that is, the gravities of bodies placed a semidiameter of the earth upwards from the surface decrease to a fourth part of their own weight, and below, in the mid-point toward the centre bodies are pressed toward the centre by only half their weight; and in the very centre weights disappear; and he demonstrates that such a force is present extensively but is far stronger in the Sun, as also in Jupiter; and hence he easily deduces that bodies moving by any impulse—provided that they be acted on by such gravitation—necessarily describe either circles, or ellipses or parabolas or hyperbolas according to the degree of velocity impressed; and among the celestial phenomena not a single one is found that does not agree exactly with this hypothesis, or rather demonstration. Moreover, he shows that such a force is composed of the conjoint forces of innumerable smallest particles composing the bodies of the earth, sun, &c., by which they seek one another, and from all sides act together toward union, as one can see in the smallest particles of fluid bodies, namely drops of quicksilver and rain, which as long as they are very small take on a spherical shape of their own accord.[6]

On 19 May 1686, the records of the Royal Society show the order that 'Mr. Newton's *Philosophiae naturalis principia mathematica* be printed forthwith in quarto in a fair letter; and that a letter be written to him to signify the Society's resolution, and to desire his opinion as to the print, volume, cuts, &c.'.[7] Three days later, on 22 May 1686, Halley wrote to Newton about the reception accorded to the *Principia*:

> Your Incomparable treatise intituled *Philosophiae Naturalis Principia Mathematica*, was by Dr Vincent presented to the R. Society on the 28th past, and they were so very sensible of the Great Honour you do them by your Dedication, that they immediately ordered you their most hearty thanks, and that a Councell should be summon'd to consider about the printing therof; but by reason of the Presidents attendance upon the King, and the absence of our Vice-presidents, whom the good weather had drawn out of Town there has not since been any Authentick Councell to resolve what to do in the matter; so that on Wednesday last the Society in their meeting, judging that so excellent a work ought not to have its publication any longer delayd, resolved to print it at their own charge, in a large Quarto, of a fair l⟨ett⟩re; and that this their resolution should be signified to you and your opinion therin be desired, that so it might be gone about with all speed. I am intrusted to look after the printing it, and will take care that it shall be performed as well as possible, only I would first have your directions in what you shall think necessary for the embellishing therof,

[5] I have discussed elsewhere (Cohen (1971*a*), *Newton and Kepler*) Newton's refusal to give Kepler credit for the laws of plantetary motion in Book I. Be it noted here, however, that in the tract *De Motu* (see Chapter III, §§3 and 5) Newton does attribute the law of elliptical orbits and the law of areas to Kepler. In Book III of the *Principia*, Newton attributes the third (or harmonic) law to Kepler but not the law of areas, or the law of elliptical orbits.

[6] [Halley] (1932), *Correspondence* [MacPike], p. 63, translated from the Latin.

[7] Birch (1756–7), *History*, vol. 4, p. 484.

and particularly whether you think it not better, that the Schemes should be enlarged, which is the opinion of some here; but what you signifie as your desire shall be punctually observed.

...When I shall have received your directions, the printing shall be pushed on with all expedition, which therfore I entreat you to send me, as soon as may be.[8]

In the same letter Halley also told Newton of Hooke's 'pretensions upon the invention of ye rule of the decrease of Gravity'. We have seen that in this letter Halley told Newton that Hooke seemed 'to expect you should make some mention of him, in a preface, which, it is possible, you may see reason to praefix'. I shall not enter here into the question of Hooke's possible claims to a share in the discovery that gravitation decreases inversely as the square of the distance—a question that has been discussed again and again[9] and that hinges largely on the philosophical problem of the meaning of the word 'discovery'. Newton's reply (27 May 1686) was an attempt to specify just what his possible indebtedness to Hooke could have been.[10]

On 2 June, at the Royal Society,

It was ordered, that Mr. Newton's book be printed, and that Mr. Halley undertake the business of looking after it, and printing it at his own charge; which he engaged to do.'[11]

Within a week, on 7 June 1686, Halley was able to send Newton 'a proof of the first sheet of your Book, which we think to print on this paper, and in this Character; if you have any objection, it shall be altered: and if you approve it, wee will proceed'.[12] Halley must have seen Newton, or have heard from him either by a letter now lost or by a third party, for he now told Newton that 'care shall be taken that it shall not be published before the End of Michaelmass term, since you

[8] [Newton] (1959–), *Correspondence*, vol. 2, p. 431.

[9] Cf. Koyré (1952), 'Letter of Hooke to Newton', reprinted in Koyré (1965), *Newtonian studies*, chap. V. See also Whiteside (1964*a*), 'Newton on planetary motion', pp. 131 ff., and studies on Hooke by Andrade (1950), 'Espinasse (1956), and Patterson (1949, 1950). Especially valuable are Lohne (1960), 'Hooke v. Newton', and Westfall (1967*a*), 'Hooke and gravitation'.

[10] 'The summe of w^t^ past between Mr Hooke & me (to the best of my remembrance) was this. He soliciting me for some philosophicall communications or other I sent him this notion. That a falling body ought by reason of the earth's diurnall motion to advance eastward and not fall to the west as the vulgar opinion is. and in the scheme wherein I explained this I carelesly described the Descent of the falling body in a spirall to the center of the earth: which is true in a resisting medium such as our air is. Mr Hooke replyd it would not descend to the center, but at a certaine limit returne upwards againe. I then took the simplest case for computation, which was that of Gravity uniform in a medium Resisting—Imagining he had learnd the Limit from some computation, and for that end had considerd the simplest case first, and in this case I granted what he contended for, and stated the Limit as nearly as I could. he replyed that gravity was not uniform, but increased in descent to the center in A Reciprocall Duplicate proportion of the distance from it. and thus the Limit would be otherwise then I had stated it, namely at the end of every intire Revolution, and added that according to this Duplicate proportion, the motions of the planets might be explaind, and their orbs defined. This is the summe of w^t^ I remember. If there was any thing more materiall or any thing otherwise, I desire Mr Hooke would help my memory. Further that I remember about 9 years since, Sir Christopher Wren upon a visit Dr Done and I gave him at his Lodgings, discoursd of this Problem of Determining the Hevenly motions upon philosophicall principles. This was about a year or two before I received Mr Hooks letters. You are acquainted wth Sr Christopher. Pray know whence & whence he first Learnt the decrease of the force in a Duplicate Ratio of the Distance, from the Center.' [Newton] (1959–), *Correspondence*, vol. 2, pp. 433–4. Halley's reply has been discussed in Chapter III, §1.

[11] Birch (1756–7), *History*, vol. 4, p. 486.

[12] [Newton] (1959–), *Correspondence*, vol. 2, p. 434.

desire it'.[13] This fascinating letter not only deals with the technical questions of printing the book,[14] but refers to the application of the mathematical results of Book I 'to the System of the world', in what Halley calls 'the second part', which he hoped he would receive soon. This 'second part' or 'Liber Secundus', he said, would be important for making the 'Mathematical part...acceptable to all Naturalists, as well as Mathematiciens'. But that would not be all. Halley, who now had to print the *Principia* at his own expense, a heroic act since he was far from wealthy, could not help adding that the 'System of the world' will 'much advance the sale of ye book'. Halley's reference to Newton's 'System of the world' as the 'second part' of the whole treatise agrees with the fact that the manuscript which I have called *De mundi systemate* (*liber 2us*), of which a part was deposited by Newton as the text of his Lucasian Lectures for 1687, is entitled 'Liber Secundus'.[15]

Evidently Halley did not yet know that Newton had changed his mind about having only two 'books'; accordingly, we find Newton informing Halley in reply (on 20 June 1686) that he had intended to have the *Principia* 'consist of three books', not two. That is, the new plan was to have three books as we know them, of which the presentation of the system of the world was to occupy Book III. Newton, however, in this very same letter, announced a further change of plan, as follows:

> The third I now designe to suppress. Philosophy is such an impertinently litigious Lady that a man had as good be engaged in Law suits as have to do with [as have to do with *replaces* as come neare] her. I found it so formerly & now I no sooner come near her again but she gives me warning. The two first books without the third will not so well beare ye title of Philosophiae naturalis Principia Mathematica & therefore I had altered it to this De motu corporum libri duo: but upon second thoughts I retain ye former title. Twill help ye sale of ye book wch I ought not to diminish now tis yo^{rs}.[16]

Newton then answered an inquiry of Halley's, presumably on the proof sheets which Halley had sent on with his letter to Newton of 7 June 1686,[17] and concerning which Newton said, 'The Proof you sent me I like very well.' The inquiry was about calling the divisions of Book I 'Articles' or 'Sections'. Newton explained that 'Articles are wth ye largest to be called by that name', but added, 'If you please you may change ye word to *sections*, thô it be not material.' And, indeed, by turning to our Apparatus Criticus [{28.4}, {38.26}, {54.21}, {66.1}, ...] the reader may see how Halley consistently altered Newton's 'Artic.' to 'Sectio'. Newton also told Halley, 'In ye first page I have struck out ye words *uti posthac docebitur* [{1.17} 'as will be shown below'] as referring to ye third book.' But Halley, who was able

13 Newton's 'order' that 'the press proceed...very slowly' is mentioned in Flamsteed's letter to Towneley of 4 November 1686, quoted below at the end of §1.

14 [Newton] (1959–), *Correspondence*, vol. 2, pp. 435–6: 'Pray please to revise this proof, and send it me up with your Answer, I have already corrected it, but cannot say I have spied all the faults. when it has past your eye, I doubt not but it will be clear from errata. The printer begs your excuse of the Diphthongs, which are of a Character a little bigger, but he has some a casting of the just size. This sheet being a proof is not so clear as it ought to be; but the letter is new, and I have seen a book of a very fair character, which was the last thing printed from this set of Letter; so that I hope the Edition may in that particular be to your satisfaction.'

15 See Chapter IV, §6. 16 [Newton] (1959–), *Correspondence*, vol. 2, p. 437. 17 *Ibid.*

to persuade Newton to allow Book III to remain in the *Principia*, did not delete these three words.

The remainder of Newton's letter is a lengthy recounting of the grounds for maintaining that Hooke's claim, if any, was minimal. I shall not discuss this part of the letter, since the question of Hooke *v.* Newton deals with matters antedating the *Principia* as such and its elaboration would take us too far afield.

Halley proved to be a master of diplomacy and his saving of Book III may just possibly have been his most significant contribution to the *Principia*. Halley wrote (on 29 June 1686):

> I am heartily sorry, that in this matter wherin all mankind ought to acknowledg their obligations to you, you should meet with any thing that should give you disquiet, or that any disgust should make you think of desisting in your pretensions to a Lady, whose favours you have so much reason to boast of. Tis not shee but your Rivalls enviing your happiness that endeavour to disturb your quiet enjoyment, which when you consider, I hope you will see cause to alter your former Resolution of suppressing your third Book, there being nothing which you can have compiled therin, which the learned world will not be concerned to have concealed; Those Gentlemen of the Society to whom I have communicated it, are very much troubled at it, and that this unlucky business should have hapned to give you trouble, having a just sentiment of the Author therof.
>
> ...I am sure that the Society have a very great satisfaction in the honour you do them, by your dedication of so worthy a Treatise.[18]

Then, following a discussion of Hooke's claims, in which Halley recounted how he himself had come on the inverse-square law and how he had come to Cambridge to see Newton,[19] he again begged Newton 'not to let your resentments run so high, as to deprive us of your third book, wherin the application of your Mathematicall doctrine to the Theory of Comets, and severall curious Experiments, which, as I guess by what you write, ought to compose it'. Halley was understandably still concerned lest the suppression of Book III should restrict readers to mathematicians. He once again told Newton that it was this Book III which would make the *Principia* acceptable to 'those that will call themselves philosophers without Mathematicks, which are by much the greater number'.[20] Newton's response (14 July 1686) was in keeping with Halley's noble sentiments:

> I am very sensible of ye great kindness of ye Gentlemen of your Society to me, far beyond w^t^ I could ever expect or deserve & know how to distinguish between their favour & anothers humour. Now I understand he was in some respects misrepresented to me I wish I had spared ye Postscript in my last...And now having sincerely told you ye case between Mr Hook & me I hope I shall be free for ye future from ye prejudice of his Letters. I have considered how best to compose ye present dispute & I think it may be donc by ye inclosed Scholium to ye fourth Proposition.[21]

[18] *Ibid.* pp. 441, 443.

[19] See Chapter III, §1.

[20] The remainder of the letter dealt with the illustrations: 'Now you approve of the Character and Paper, I will push on the Edition Vigorously. I have sometimes had thoughts of having the Cutts neatly done in Wood, so as to stand in the page, with the demonstrations, it will be more convenient, and not much more charge, if it please you to have it so, I will trie how well it can be done. otherwise I will have them in somewhat a larger size than those you have sent up.' Newton replied on 14 July 1686: 'I have considered yor proposal about wooden cuts & beleive it will be much convenienter for ye Reader & may be sufficiently handsome but I leave it to your determination. If you go this way, then I desire you would divide ye first figure into these two. I crouded y^m^ into one to save ye trouble of altering ye numbers in ye schemes you have.' [Newton] (1959–), *Correspondence*, vol. 2, p. 443.

[21] *Ibid.* pp. 444–5.

The new scholium, as may be seen from the Apparatus Criticus {45.5 ff.}, differs greatly from the original, which comprised but a single sentence reading (in English translation):

SCHOLIUM

The case of the sixth corollary obtains in the celestial bodies; and therefore in what follows, I intend to treat more at large of those things which relate to centripetal force decreasing as the squares of the distances from the centres.

The new scholium began with a paragraph identically like the original one but for the addition of a parenthetical statement so as to read: 'obtains in the celestial bodies (as Wren, Halley, and Hooke of our nation have severally observed)'. Then Newton went on to discuss the fact that from circular motion it was easy to get to an inverse-square force, particularly once Huygens had published 'in his excellent book *De horologio oscillatorio*' a comparison of 'the force of gravity with the centrifugal forces of revolving bodies'. I believe that Newton must have added this paragraph as a way of saying to the *cognoscenti* that the credit to be given to Hooke (and for that matter, to Wren and Halley) was not really very great, and that if anyone deserved a pat on the back it was Huygens and not Hooke.

Then, as Newton told Halley, 'In turning over some old papers I met with another demonstration of that Proposition, wch I have added at ye end of this Scholium.' Herivel has recently suggested that this additional proof is similar to an early one in the Waste Book, dating from the 1660s.[22] We may hazard a guess that Newton added this final paragraph to the scholium as a means of asserting his proper priority over Hooke, and also over Wren and Halley; they would not have made their guesses until the publication of Huygens's *Horologium oscillatorium* in 1673, whereas he (Newton) had independently discovered the law of centripetal force in uniform circular motion long before Huygens's book had been published.

The Apparatus Criticus contains an amusing variant reading in this scholium. Newton's words, in the manuscript text sent to Halley, are: 'ut seorsum collegerunt etiam nostrates Wrennus, Halleus et Hookius' ('as observed severally by Wren, Halley and Hooke of our nation'). But as printed in E_1, and in E_2 and E_3 {45.6}, the order of names is 'Wrennus, Hookius & Hallaeus [Halleus E_1]'. I do not know by whose authority Hooke's name was put ahead of Halley's, but the alteration was evidently made in proof (and so presumably by Halley), since the addition to M as sent by Newton to Halley and by Halley to the printer is unaltered. My guess is that Halley put Hooke's name ahead of his own because he knew he would have to face Hooke after publication.[23]

As the printing of Book I got under way, Halley wrote (on 19 July 1686) the

[22] Herivel (1960*a*), 'Newton's discovery of the law'; also (1965), *Background*, chap. I and pp. 129–30.

[23] The value of an edition with variant readings—even for such apparently trivial questions as word order —may be seen by a statement by S. P. Rigaud (1838), *Essay*, p. 68. Commenting on the request by Halley that Newton make some mention of Hooke, Rigaud says: 'Newton therefore here added a scholium in which he says, that this property "obtinet in corporibus coelestibus (ut seorsum collegerunt etiam nostrates Wrennus, Hookius et Hallaeus)". It may be observed that he arranges the names in the order of the time in which he was persuaded the individuals had come to a knowledge of the truth.'

following account of its contents to a Dr Salomon Reisel [or Reiselius], 'Leibarzt d. Herzogs von Würtemberg in Stuttgart', who was the author of two communications published in Latin (in 1685 and in 1686) in the *Philosophical Transactions* of the Royal Society of London:[24]

> And now a truly outstanding book is in press, whose title is the Mathematical Principles of Natural Philosophy, by Isaac Newton, professor of Mathematics at Cambridge, of all geometers who have ever existed perhaps the greatest. By this example it will be proved how far the human mind properly instructed can avail in seeking truth. For with firm and certain demonstrations he shows that the orbital motion of the planets is composed of an impressed force and a force gravitating toward the centre of the sun, and that those forces decrease in the duplicate ratio reciprocally of the distances from the sun; in what way the orbs of all revolving bodies, whose velocities do not exceed a given limit, become ellipses; while those attaining that given velocity describe a parabolic line, & those surpassing it a hyperbolic line, the sun being set as a centre in the focus of them all, with the law of motion that the areas between the curve and the radii are proportional to the times, with many other extremely worthwhile discoveries.[25]

News concerning the progress of the *Principia* was disseminated by word of mouth and by correspondence. On 4 November 1686, the Astronomer Royal, John Flamsteed, wrote a letter to Richard Towneley which is notable for a statement that the *Principia* will be useful in helping him in 'the reforming of the planetary motions'. Furthermore, Flamsteed gives credit to Kepler for the law of areas and the harmonic law (omitting the law of elliptical orbits), and saying explicitly that neither had been explained by Kepler. Indeed, there was no adequate system of dynamics on which to base an explanation of these three laws before Newton. Flamsteed correctly reported that Newton had 'overthrown' the Cartesian philosophy, and had replaced it by 'demonstration and demonstrated principles'. This letter (Roy. Astron. Soc. MS 243 (Flamsteed), No. 68) reads in part as follows:

> Mr Newton's Treatise of Motion is in the press. Mr Halley takes care of it and 13 sheets (he tells me) are wrought off. He [Mr Newton] lays by the notion of a vortex which in a letter to me he says would disturb the planets motion and render them more irregular than they are. He solves the Recess of the Equinoctiall Points and Nodes of the Moon's orbit without one, and has determined the motion of the nodes of the orbit of Jupiter's satellites. He proves that, according to the Laws of Motion, the cube roots of their periods squared are, and ought to be, proportional to their distances from the Sun, and that the radius vector proceeding from the planet to the Sun sweeps over equal areas in equal time, which Kepler found out first but could assign no reason for but what was drawn from experiment. The Cartesian philosophy in this point will be overthrown, but we shall have demonstration and demonstrated principles in the room of it.
>
> I have lost a cause I assigned for the Recess of the Equinoctiall Points, but I shall gain infinitely by it in the assistance those discoveries will lend me in the reforming of the planetary motions, so that, in the room of mourning I congratulate my own business. He has fresh and strong arguments for the Earth's motion, but that I look upon as so well settled that they will only exercise the wits of those who have ignorance enough to deny it. But enough of this. We shall have the piece in good time, though the press proceed by Mr Newtons order very slowly.

Since the *Principia* is printed in quarto, the first '13 sheets' comprise pages 1–104 (gatherings B–I, K–O), including the first five sections of Book I.

[24] See J. C. Poggendorf, *Biographisch-litterarisches Handwörterbuch* (Leipzig: Johann Ambrosius Barth, 1863), col. 600.

[25] [Halley] (1932), *Correspondence* [MacPike], p. 68, translated from the Latin.

2. BOOK II AND BOOK III

On 1 March 1686/7 Newton announced to Halley that he was sending Book II 'by ye Coach...to be left wth Mr Hunt at Gresham Coll.' and begging 'ye favour of a line or two to know of ye receipt'. He admitted that, as far as he personally was concerned, he 'could be as well satisfied' to let the edition 'rest a year or two longer', but 'because of peoples expectation' he was obliged to Halley 'for pushing on'. Six days later, on 7 March, Halley acknowledged receiving the letter 'and according to it, your Second Book'. Halley had found a second printer to begin composing Book II, and who agreed to have the job done in seven weeks. He reckoned that Book II would make about 20 sheets, so that Book I (about 30 sheets) would be 'finished much about the same time'.[1] He would send Newton within the week 'the 18th sheet according to your direction'.[2] Halley asked Newton about Book III. If it were ready and not too long, he would 'endeavour by a third hand to get it all done togather', if Newton would see fit to send it on.[3]

Correcting the proof errors in such a work as the *Principia* was a difficult task. Sending on sheet T on 14 March, Halley remarked on 'the extraordinary trouble of the last sheet'[4] [S? or T? or even R?] which had been printed off. Newton was to check it carefully for any 'mistakes' that might be 'noted at the end' of the book, in the printed errata. If any errors should prove to be 'very materiall' the whole sheet would have to be done over, as had been the case for sheet D, and half of P (where, on page 112, the figure had been printed upside down).[5]

On 5 April Halley reported to Newton that on the previous day he had received Book III, 'the last part of your divine Treatise'.[6] It was presented to the Royal Society on the next day.[7] Halley could not expect that the 'first part' (Book I) of the *Principia* would be finished in three weeks' time. Hence,

> considering the shortness of the third [book] over the second, the same press that did the first will get it done so soon as the second can be finished by another press; but I find some difficulty to match the letter Justly.

In the event, the first printer did Books I and II, and another did Book III.

[1] [Newton] (1959–), *Correspondence*, vol. 2, pp. 470–3.

[2] Since the *Principia* was being printed in quarto, each sheet corresponds to eight pages. Book I, ending on p. 235 (Gg2), begins on p. 26, hence occupying 210 pages, somewhat less than 30 sheets (which in a quarto book, eight pages to a sheet, make 240 pages); but if Halley were including the 'Definitiones' and 'Axiomata sive Leges Motus' preliminary to 'De Motu Corporum, Liber Primus', and beginning on p. 1 (sheet B), he was just a bit short in his estimate. I gather that Halley meant by 'the 18th sheet' the sheet T (reckoning B as the first sheet), and thus containing pp. 137–44 (Prop. XLV), concluding Sec. IX.

[3] Because of the guessing about the eventual page-numbers, there is a hiatus in the numeration in E_1. Book III begins on the first page of the sheet bearing the signature 'Aaa' and is numbered '401'. The previous signatures are 'Zz' and '${}_{*}{*}{}_{*}$'. The last page of signature '${}_{*}{*}{}_{*}$', facing the first page of the next signature 'Aaa', is numbered '400'; but if it had been numbered in the sequence of the previous signatures, it would have borne the number '384'. Since the page numbers had to be put on the sheets as they were printed off, a guess had to be made as to what number the first page of Book III would have. The number '401' was a pretty good guess, not off by much.

[4] [Newton] (1959–), *Correspondence*, vol. 2, p. 473.

[5] A few copies have the original p. 112, rather than the cancel.

[6] [Newton] (1959–), *Correspondence*, vol. 2, p. 473.

[7] Birch (1756–7), *History*, vol. 4, p. 529.

Halley concluded on a somewhat pessimistic note:

> I find I shall not get the whole compleat before Trinity term, when I hope to have it published; when the world will not be more instructed by the demonstrative doctrine therof, than it will pride it self to have a subject capable of penetrating so far into the abstrusest secrets of Nature, and exalting humane Reason to so sublime a pitch by this utmost effort of the mind.[8]

No further correspondence is recorded between Newton and Halley until,[9] on 5 July 1687, Halley wrote to Newton that he had 'at length brought your book to an end, and hope it will please you. the last errata came just in time to be inserted'.[10]

3. THE COMPLETION OF THE JOB: HALLEY'S CONTRIBUTION TO THE 'PRINCIPIA'

No one knows for certain just how large the edition of the *Principia* was. Rouse Ball reckoned it to be some 250 copies,[1] but A. N. L. Munby held it to be more nearly 300–400.[2] Some copies have an imprint reading 'Prostat apud plures Bibliopolas', whereas others have a cancel title-page carrying the imprint 'Prostant Venales apud Sam. Smith...aliosque nonnullos Bibliopolas'.[3] Usually, those copies with 'Sam. Smith' prove to have been intended for Continental distribution, which has been suggested to imply that Halley made an agreement with Smith for foreign sales only after the printing had been completed. Although the two forms of the title-page give rise to the bibliographical distinction between a 'first issue' and a 'second issue', the texts themselves are not the result of separate printings.

When Halley announced to Newton on 5 July 1687 that the job had been brought to a conclusion, he also informed him that he would 'present from you the books

[8] Some additional information is contained in a letter from Halley to Wallis of 9 April 1687, reading in part: 'I have lately been very intent upon the publication of Mr Newtons book, which has made me forget my duty in regard of the Societies correspondants; but that book when published will I presume make you a sufficient amends for this neglect. I have now lately received the last Book of that Treatise wch is entituled de Systemate Mundi; wherin is shown the principle by which all the Celestiall Motions are regulated, together with the reasons of the several inequalities of the Moons Motion and the cause and quantity of the progression of the Apogeon and retrocession of the Nodes...I hope to get the whole finished by Trinity term, but the correction of the press costs me a great deal of time and paines.' [Halley] (1932), *Correspondence* [MacPike], pp. 80–1.

[9] On 24 June 1687, Fatio wrote an account of the *Principia* to Huygens, saying, 'Quelques uns de ces Messieurs qui la composent sont extremement prévenus en faveur d'un livre de Mons. Newton qui s'imprime presentement et qui se debitera dans trois semaines d'ici'; Huygens (1888–), *Oeuvres*, vol. 9, pp. 167–8. On 25 June, Halley said in a letter to Wallis, 'I was willing to make you a more valuable present: I mean Mr. Newton's book of Mathematicall Philosophie, which is now near finished: but I must entreat your patience yet ten days more, by which time it will be compleat. To hasten the edition of this book, I have been obliged to attend 2 presses, which has so farr taken me up, that for that reason, and for want of such communications as might be worthy of you, I have so long forborn to write to you, but the book, which needs must please you, will I hope obtain my pardon'; [Halley] (1932), *Correspondence* [MacPike], p. 85.

[10] [Newton] (1959–), *Correspondence*, vol. 2, p. 481. The printed errata occupy a single (recto) side of an unnumbered page immediately following p. 510 of E_1, the verso side being blank. This is the seventh page of the final sheet, bearing the signature 'Ooo'.

[1] Ball (1893), *Essay*, p. 67; see Macomber (1953), 'Principia census', p. 269.

[2] Munby (1952), 'Distribution of first edition', p. 37.

[3] See the account of the editions of the *Principia* in Appendix VIII to the text.

you desire to the R. Society, Mr Boyle, Mr Pagit, Mr Flamsteed and...any [one] elce in town that you design to gratifie that way'.[4] He entreated Newton 'to accept' twenty copies which he had sent him 'to bestow on your friends in the University'[5] and told Newton that there were 40 more in 'the same parcell' which he hoped the author would 'put into the hands of one or more of your ablest Booksellers to dispose of them'.[6] The price to the trade was to be 6 shillings in quires (or sheets) (reduced to 5 shillings if payment was in cash or 'at some short time'), but 9 shillings 'bound in Calves leather and lettred'.[7] Halley concluded by expressing the hope,

> after you shall have a little diverted your self with other studies, that you will resume those contemplations, wherin you have had so good success, and attempt the perfection of the Lunar Theory, which will be of prodigious use in Navigation, as well as of profound and subtile speculation.

There is no further correspondence known to exist between Newton and Halley until 1695, but there must always remain the possibility that such letters may still turn up. For instance, would not Newton have written to Halley about receiving 60 copies of the *Principia* 'by waggon'? (Of course, he could have sent a message to Halley by a third party, or possibly he might have deemed his signature in the carrier's book to have been sufficient.)

It would surely be pleasant to read his remarks (if there were any) concerning the Latin ode to Newton which Halley wrote to go before the Definitions, and also any letter of appreciation for Halley's truly heroic efforts on his behalf. But if we still do not know any private expression of Newton's gratitude to Halley, the world at large learned from the Preface what were Newton's public sentiments:

> *In the publication of this work the most acute and universally learned* Mr. Edmond Halley *not only assisted me with his pains in correcting the press and taking care of the schemes* [i.e., *preparing the geometrical figures*], *but it was to his solicitations that its becoming publick is owing. For when he had obtained of me my demonstrations of the figure of the celestial orbits, he continually pressed me to communicate the same to the* Royal Society; *who afterwards by their kind encouragement and entreaties, engaged me to think of publishing them.*

In E_1 no date is given for this Preface, but in E_2 and E_3 it is dated

Dabam *Cantabrigiae*, e Collegio
S. Trinitatis, Maii 8. 1686.

The accuracy of this date is open to much suspicion.[8]

Apart from Halley's moral encouragement, and his practical services in dealing with printers and booksellers and in correcting the proofs, what did he actually

[4] Flamsteed's annotated copy is now in the Library of the Royal Society.

[5] The Library of Emmanuel College has a presentation copy, in which someone (not Newton) has written: 'Authoris sum͠è docti Julii 13tio 1687.'

[6] [Newton] (1959–), *Correspondence*, vol. 2, p. 481.

[7] 'I entend the price of them bound in Calves leather and lettred to be 9 shill⟨ings⟩ here, those I send you I value in Quires at 6:shill to take my money as they are sold, or at 5sh a price certain for ready or elce at some short time; for I am satisfied that there is no dealing in books without interesting the Booksellers, and I am contented to lett them go halves with me, rather than have your excellent Work smothered by their combinations.'

[8] See Edleston (1850), pp. lviii (n. 89), lxxi (n. 147).

contribute as an editor? We may judge the extent of his intellectual share in the final product by reading his correspondence and chiefly by closely studying the variant readings of *M* in the Apparatus Criticus to the present edition. On a purely stylistic level, we have seen Halley suggest that each book be divided into Sections rather than Articles; accordingly, this change was made in *M* in Halley's hand, as may be seen in the Apparatus Criticus. Again it was Halley who thought of 'having the Cutts neatly done in Wood, so as to stand in the page with the demonstrations'.[9] It would be 'more convenient' than to have them on pull-out pages, from engravings, and of 'not much more charge', though to a man of Halley's moderate means any charge was considerable. We must keep in mind that the method adopted required much more work on the part of the editor to see that each cut was in its proper place.[10] Newton agreed in a letter to Halley dated 14 July 1686.[11] On the back of this letter, Halley noted that he had received a communication from Newton on 20 August 1686.[12] Since Newton wrote on 13 February 1686/7 about an earlier 'letter wth two Corollaries [which] I sent you in Autumn', Edleston conjectured that the missing letter must have contained Corols. 2 and 3 to Prop. XCI of Book I.[13] In these corollaries Newton discusses the attraction of a 'round solid' or spheroid on either (Corol. 2) 'any body [corpuscle] P situate externally in its axis AB' or (Corol. 3) a 'corpuscle...placed within the spheroid and in its axis', though the demonstration holds—as Newton says—'whether the particle be in the axis or in any other given diameter'. Edleston gave as the basis for his conjecture 'the fact that the corollaries above-mentioned are not found in Newton's MS', a statement repeated by the editors of Newton's *Correspondence*.[14] But these two corollaries *are* to be found in *M* {217.4 ff.}, although they are not part of the first manuscript; they appear on a separate sheet presently bound in at the end of Book I.

On 14 October Halley asked Newton about Problem XXIII, the second proposition (that is, Prop. XXXI) in Sec. VI, Book I.[15] This is Kepler's problem, 'To find the place of a body moving in a given elliptic trajectory at any assigned time'. Halley thought there must 'be some mistake committed' and entreated Newton to send, 'revised by your self, those few lines that relate therto', that is, his approximate 'effection' in the scholium following the proposition. As may be seen in the Apparatus Criticus {109.13–14}, the original manuscript sheet communicated to Halley lacked a factor of $\frac{1}{2}$.

Halley also pointed to the lack of a demonstration, and hoped Newton might 'be prevailed upon to subjoyn something of the Demonstration'. Newton's reply[16] does not mention Prob. XXIII, and, as may be seen in the Apparatus Criticus, no changes were made in *M*. But as to the scholium, Newton sent Halley a

[9] [Newton] (1959–), *Correspondence*, vol. 2, p. 443.

[10] As mentioned, one got upside down and part of the sheet had to be reprinted.

[11] [Newton] (1959–), *Correspondence*, vol. 2, p. 444.

[12] *Ibid.* p. 445, n. 5.

[13] Edleston (1850), pp. xxx, lvii.

[14] *Ibid.*

[15] [Newton] (1959–), *Correspondence*, vol. 2, p. 452.

[16] *Ibid.* pp. 453–4

revised statement (in a letter of 18 October 1686) which was substituted for the original one in *M*. In the Apparatus Criticus, therefore, the reader will find printed for the first time {109.13 ff.} the original manuscript version of this scholium as sent by Newton to Halley in *M*, as well as the first printed version. Newton thanked Halley 'heartily for giving me notice that it [this scholium] was amiss'. Newton also explained to Halley in English the sense of the Latin lines giving the 'ground of ye transmutation of a trapezium into a parallelogram' (Lemma XXII).[17] From the fact that Newton wrote the explanation in English, we may be sure that he was giving Halley a little private explanation, and certainly did not intend to alter the text of *M*. Had Newton wished to emend what he had written, rather than merely instruct Halley, he would have sent on a new version or an addition in Latin. And indeed, in the Apparatus Criticus we find no basic variation between *M* and E_1.

On 13 February 1686/7 Newton wrote:

> I think I have y^e^ solution of your Problem about y^e^ Suns Parallax, but through other occasions shall scarce have time to think further on these things & besides I want something of observation. ffor if my notion be right, the Sun draws y^e^ Moon in y^e^ Quadratures, so that there needs an equation of about 4 or 4½ minutes to be subducted from her motion in ye first Quarter & added in ye last.[18]

In reply, Halley wrote:

> Your demonstration of the Parallax of the Sunn from the Inequalitys of the Moons Motions, is what the Societie has commanded me to request of you, it being the best means of determining the dimensions of the Planetary Systeme, which all other ways are deficient in; and they entreat you not to desist, when you are come so near the Solution of so noble a Probleme. This done there remains nothing more to be enquired in this matter, and you will do your self the honour of perfecting scientifically what all past ages have but blindly groped after.[19]

To which Newton in turn replied:

> The deduction of ye Suns Parallax from ye Moons variation I cannot promise now to consider. When Astronomers have examined whether there be such an inequality of her motion in ye Quadratures as I mentioned in my last & determined ye quantity thereof, I may take some occasion perhaps to tell them ye reason.[20]

A glance through the Apparatus Criticus shows the extent of Halley's care in preparing *M* for the press, since all alterations made by him are indicated in the Apparatus by the initials 'EH' within square brackets. Halley certainly sent on to Newton the sheets as they were printed off (M, the '11th sheet', had been received by 13 February 1686/7, T by mid-March[21]), but he almost certainly did not send Newton every sheet for proof-reading. Hence, it is probable that many of the alterations made in proof (appearing in the Apparatus Criticus as variant readings of *M*, not in E_1) that are not corrections of printers' errors are due to Halley. (Of course, some variation between *M* and E_1 is due to the printers.) Furthermore, we do not know the full extent of Halley's contact with Newton during the printing

17 *Ibid.* pp. 454–5. 18 *Ibid.* p. 464. 19 *Ibid.* pp. 469–70.
20 *Ibid.* p. 471. 21 *Ibid.* p. 464.

of the *Principia*. There are certain letters now wanting. I have referred to a missing letter of Newton's to Halley (20 August 1686). Another, written 'in Sommer' of 1686, is mentioned in Newton's letter to Halley of 13 February 1686/7, apparently telling Halley that the 'second book...should come out wth ye first', but it is missing from the sequence.[22] We know, moreover, that Halley paid at least one visit to Newton during the printing of the *Principia* (though we are not sure of the exact date), since in a letter to Flamsteed of 3 September 1686, Newton mentioned 'Mr Halley (who was lately here)'.[23] Newton also used Paget as an intermediary (on his stay with Newton, Paget found some errata in the first printed sheets) and at least once Newton 'desired my intimate friend Mr C. Montague to enquire of Mr Paget how things were'.[24]

Halley's final services for Newton in the completion of E_1 were the preparation of a book review that he wrote and published in the *Philosophical Transactions* of the Royal Society, of which he was the editor,[25] and a popular account of the book, concentrating on the theory of tides, prepared for James II and later reprinted in the *Philosophical Transactions*.[26] Halley also printed a preliminary statement about the *Principia*[27] before the actual publication of the book, in an article on gunnery in the *Philosophical Transactions*.[28]

In any such summary, we must now include Halley's critique of an early version of the *Principia* (see Chapter IV, §8, and Supplement VIII).

[22] *Ibid.* There are other letters missing from the sequence; for example, on 13 February 1686/7 Newton sent Halley a sheet 'lost by the printer's Negligence', but we have no letter in which Halley had requested it (*ibid.* pp. 464, 469). Again, Halley agreed that the book 'should not be published before the End of Michaelmass term' (*ibid.* p. 434), but we have no letter requesting this.

[23] *Ibid.* p. 448.

[24] *Ibid.* p. 464.

[25] Other reviews were published in the *Bibliothèque Universelle*, the *Journal des Sçavans*, and the *Acta Eruditorum*. See Chapter VI, §1.

[26] Reprinted in [Newton] (1958), *Papers and letters* [Cohen and Schofield], pp. 405 ff.; cf. Munby (1952).

[27] Halley (1686), 'Discourse concerning gravity', p. 6. This article presents a summary account of the 'Affections or Properties of *Gravity*, and its manner of acting on *Bodies falling*...discovered, and...made out by *Mathematical demonstration*...lately by our worthy Country-man Mr. *Isaac Newton*, (who has an incomparable *Treatise* of *Motion* almost ready for the *Press*)'.

[28] There exist two copies of E_1 with manuscript corrections by Halley; one is in the library of the University of Texas, while the other belongs to Miss M. Norman, of Cremorne, New South Wales, Australia. For details, see Chapter VIII, §2, and Appendix IV to the text.

PART THREE

REVISING THE ‘PRINCIPIA’

CHAPTER VI

THE FIRST CRITICAL EVALUATIONS OF THE 'PRINCIPIA'

1. THE REVIEW IN THE 'BIBLIOTHÈQUE UNIVERSELLE'

THE first impact of the *Principia*[1] upon Newton's contemporaries may be traced through scientific correspondence (notably the letters of Huygens and Leibniz[2]) and through four book reviews: two in French, one in English, and one in Latin. The French reviews, in the *Journal des Sçavans*[3] and the *Bibliothèque Universelle*,[4] were both anonymous, but currently there is a move to attribute the one in the *Bibliothèque Universelle* to John Locke.[5] The English review, in the *Philosophical Transactions* of the Royal Society,[6] was written by Halley, and is the only one of the four whose author is known to us by name.[7] The Latin review, by far the most extensive, appeared in the *Acta Eruditorum*[8] published in Leipzig.

I shall not discuss here the stages whereby the new science of the *Principia* was gradually admitted into the discourse of physics and astronomy, nor shall I concern myself with the opposition aroused by Newton's concepts, these subjects being more properly part of a Commentary on the *Principia* than of an Introduction to this edition. But there are one or two effects of these reviews that must be considered here, since they are embodied in subsequent revisions of the text of the *Principia* by Newton.

The presentation of the *Principia* in the *Bibliothèque Universelle* (it is hardly a 'review') consists primarily of a translation into French of the Latin headings of the successive sections of Books I and II, followed by the reviewer's own summary

[1] The *Principia* was published by 5 July 1687, when Halley announced to Newton that the printing had been completed and that copies were being distributed; see [Newton] (1959–), *Correspondence*, vol. 2, p. 481.

[2] See [Huygens] (1888–), *Oeuvres*, vols. 9 and 10.

[3] *Journal des Sçavans* (2 August 1688), p. 128.

[4] *Bibliothèque Universelle*, vol. 8 (March 1688), pp. 436–50.

[5] First suggested by Rosalie Colie (1960), this thesis has now been advanced even more forcibly in Axtell (1965, 1968).

[6] *Phil. Trans.* no. 186 (January, February, March 1687), pp. 291–7, reprinted in [Newton] (1958), *Papers and letters* [Cohen and Schofield], pp. 405–11.

[7] See Schofield (1958), *ibid.* pp. 397–404.

[8] *Acta Eruditorum* (June 1688), pp. 303–15.

of Book III, together with an introductory paragraph.[9] Indeed, to such an extent has Locke (or whoever was the author) limited himself to a summary or outline in Newton's own words that he apologizes for the fact that Newton has not divided up Book III into sections (as in Books I and II); since Newton has not provided headings to translate, the reviewer has had to fall back on his own meagre resources.[10] The result is curious indeed. A great part of this presentation of Book III, far out of proportion to either its novelty or its importance to the reader, is taken up by a summary of what we know as the 'Phaenomena',[11] even reproducing the tables of the periodic times and mean distances of the six major planets and the satellites of Jupiter. And although it is said that

> Il y a de la pesanteur en tous les corps, selon la quantité de matiere qui est en chacun d'eux, & l'effort que fait la pesanteur sur toutes les particules égales d'un corps est reciproquement comme le quarré de la distance des lieux de ces particules,

and that

> Le flux & le reflux de la Mer procedent de l'impression du Soleil & de la Lune,

I doubt whether any reader would gain from such a presentation any real insight into the way in which Newton had shown how the gravitational force binds the members of the solar system together and also regulates and determines their motions. For instance, such a reader would come away with data concerning the semi-diameters of the Sun and planets and their relative densities, and with bits and pieces of other information, such as that Newton had shown the Earth to be an oblate spheroid. But this same reader would not have learned that Newton had introduced into physics a new technical expression, centripetal (as opposed to centrifugal) force; nor that Newton had shown that Kepler's Law of Areas is both a necessary and a sufficient condition for a deviating force to be central (Props. I–III), and had actually demonstrated that elliptical orbits imply an inverse-square force directed toward a focus. Even the Axioms, or Laws of Motion,

[9] According to Rosalie Colie, it is this introductory paragraph, appearing in Locke's manuscript outline of the *Principia*, that points to Locke's authorship. This paragraph reads: 'Si ceux qui travaillent dans les Méchaniques entendoient parfaitement les regles de la Géometrie, ou qu'ils fussent tout à fait maîtres de leur matiere, ils ne manqueroient jamais leur but, & ils pourroient donner à leurs Ouvrages toute l'exactitude & la perfection que les Mathematiciens sont capables d'imaginer. C'est pourquoi les Philosophes & principalement les modernes se sont imaginé que Dieu s'est prescrit de semblables Loix, pour la formation & la conservation de ces Ouvrages, & ont tâché d'expliquer par là divers effets de la Nature. Mr. *Newton* se propose le même but, & prend la même voie dans ce Traité, expliquant dans les deux premiers Livres les regles génerales des Mechaniques naturelles, c'est à dire les effets, les causes & les degrez de la pesanteur, de la legereté, de la force élastique, de la résistance des fluides, & des vertus qu'on appelle attractives & impulsives. Il entreprend, dans le III. Livre d'expliquer le Systeme du monde, les degrez de pesanteur, qui portent les corps vers le Soleil, ou vers quelque Planete, & qui étant connus lui servent à rendre raison du mouvement des Planetes, des Cometes, de la Lune & de la Mer.' See Colie (1960); see also Axtell (1965, 1965*a*).

[10] 'M. Newton se sert dans le troisième Livre des veritez qu'il a prouvées, pour former le systeme du Monde; & continuë à proposer ses raisons selon la méthode des Géometres; mais c'est sans diviser sa matiére en Sections. C'est pourquoi il faudra se contenter d'extraire quelques unes de ses observations les plus curieuses.'

[11] They were called 'Phaenomena' in E_2 and E_3, but in E_1 they were still a part of the 'Hypotheses'; see Chapter I, §1, and Chapter II, §2.

are not given, but only referred to *en passant*, in a statement that the author has followed the method of the geometers, 'posant avant qu'entrer en matieres plusiers définitions & axiomes touchant le mouvement'.

2. JOHN LOCKE AND THE 'PRINCIPIA'

My presentation of the review of the *Principia* in the *Bibliothèque Universelle* may appear overly harsh toward Locke, if he was the author. There are excuses aplenty for his having given so superficial a summary, not the least of which is the shortness of the time allotted for the preparation. The *Principia* was, and always has been, a difficult book—all the more so on its first appearance and before there had been much discussion of it. Locke was not, furthermore, trained in mathematics.[1] Later on, Newton wrote out for him in English a simplified version of the proof that elliptical orbits require an inverse-square force.[2]

Among the members of Newton's circle there was circulated a story that Locke simply could not understand the mathematics of the *Principia*, but was content to limit himself to the physical principles. We have a contemporaneous statement to this effect by J. T. Desaguliers, a physicist in his own right, and an intimate member of the Newtonian circle:

> But to return to the *Newtonian Philosophy*: Tho' its Truth is supported by Mathematicks, yet its Physical Discoveries may be communicated without. The great Mr. *Locke* was the first who became a *Newtonian Philosopher* without the help of Geometry; for having asked Mr. *Huygens*, whether all the mathematical *Propositions* in Sir *Isaac's Principia* were true, and being told he might depend upon their Certainty; he took them for granted, and carefully examined the Reasonings and *Corollaries* drawn from them, became Master of all the Physics, and was fully convinc'd of the great Discoveries contained in that Book.[3]

Desaguliers states unequivocally that the foregoing information had been told to him 'several times by Sir *Isaac Newton* himself'.

The example of John Locke may be related to an amusing emendation which Newton once contemplated making to Book III. It occurs in the introductory paragraph to this book, in which Newton has been dealing with the topic of readers who are well trained in mathematics, contrasted by implication with those who are not. This passage, unaltered in all three editions, is embedded in a discussion of the relation of this third book, 'De Mundi Systemate', to the first two, in which—he says—he has laid down 'principles of philosophy' which are mathematical, that is, 'laws and conditions of motions and forces'. Now, in Book III,

[1] On Locke's mathematical training, see Axtell (1968), pp. 84–5.

[2] This essay, printed by King, Rouse Ball, the Halls, and Herivel, is dated (in Locke's copy) Mar. 1689/90, but was very likely a copy of an earlier (that is, a pre-*Principia*) text. See Herivel (1965), *Background*, pp. 108 ff., and Hall and Hall (1963), 'The date'.

[3] Desaguliers (1763), *Experimental philosophy*, vol. 1, p. viii. This statement appears in the first edition (1734). Earlier still, Conduitt wrote out an equivalent version, crediting 'Moivre' as the source of the story: 'Lock took his prop^s^ for granted on hearing Hugens say that he had proved them' (King's College Library, Keynes MS 130).

he will, 'from the same principles...demonstrate the constitution of the system of the world'. He would not advise all readers to study 'every proposition' of the first two books before turning to the third, since it would 'cost too much time, even to readers of good mathematical learning' ('lectoribus etiam mathematice doctis'). It is rather enough to concentrate on the Definitions, the Laws of Motion, and the first three sections of Book I.[4]

In E_2i, Newton added an additional concluding sentence {386.25}:

Propositiones etiam legere possunt qui mathematice docti non sunt, & mathematicos de veritate Demonstrationum consulere.[5]	Those who are not mathematically learned can read the Propositions also, and can consult mathematicians concerning the truth of the Demonstrations.

Certainly, this statement (with its explicit contrast of those who are *not* mathematically learned ['mathematice docti'] to the previous mention of readers who *are*) may suggest that Newton had in mind a real person who had actually so consulted a mathematician. The existence of this annotation gives added plausibility to Desaguliers's story about Locke and Huygens.

3. EDMOND HALLEY'S REVIEW IN THE 'PHILOSOPHICAL TRANSACTIONS'

Halley's review in the *Philosophical Transactions*, like the anonymous one in the *Acta Eruditorum*, is an admirable example of skill in presenting clearly and succinctly the main achievements of a difficult and complex treatise. Any reader of either of these accounts of the *Principia* would at once become aware of the form and structure of Newton's great book; he would learn of Newton's method of limits and his physical concepts; and he would gain some knowledge of Newton's main results concerning force and motion in general, and their specific application to the solar system, including comets.

A notable feature of Halley's presentation is his statement that Newton's

> great skill in the old and new Geometry, helped by his own improvements of the latter, (I mean his method of *infinite Series*) has enabled him to master those Problems, which for their difficulty would have still lain unresolved, had one less qualified than himself attempted them.[1]

Surely, Halley was being critical (however mildly) when, after describing the first eleven propositions, in which Kepler's three laws of planetary motion are analysed from the point of view of dynamics, he wrote:

> All which being found to agree with the *Phenomena* of the Celestial Motions, as discovered by the great Sagacity and Diligence of *Kepler*, our Author extends himself upon the consequences of this sort of *Vis centripeta*...

We have seen that nowhere in Book I does Newton mention Kepler's name, while in Book III he gives Kepler credit only for the third (or harmonic) law,

[4] Such a reader could also 'consult such of the remaining Propositions of the first two Books, as the references in this, and his occasions, shall require'.

[5] The final form of this sentence, as may be seen in the Apparatus Criticus, represents a third draft.

[1] [Newton] (1958), *Papers and letters* [Cohen and Schofield], p. 405.

and not for the first two![2] Hence Halley's statement may be viewed as a critical comment.

As a typical example of Halley's lucid exposition, let me quote his description of Secs. II and III of Book I, following an account of the Definitions and Laws of Motion, together with their Corollaries, the 'necessary *Praecognita*':

our Author proceeds to consider the Curves generated by the composition of a direct impressed motion with a gravitation or tendency towards a Center: and having demonstrated that in all cases the Areas at the Center, described by a revolving Body, are proportional to the Times; he shews how from the Curve described, to find the Law or Rule of the decrease or increase of the Tendency or Centripetal forces (as he calls it) in differing distances from the Center. Of this there are several examples: as if the Curve described be a Circle passing through the Center of tendency; then the force or tendency towards that Center is in all points as the fifth power or squared-cube of the distance therefrom reciprocally. If in the proportional Spiral, reciprocally as the cube of the distance. If in an Ellipse about the Center thereof directly as the distance. If in any of the *Conick* Sections about the *Focus* thereof; then he demonstrates that the *Vis Centripeta*, or tendency towards that *Focus*, is in all places reciprocally as the square of the distance therefrom; and that according to the Velocity of the impressed Motion, the Curve described is an *Hyperbola*; if the Body moved be swift to a certain degree then a *Parabola*; if slower an *Ellipse* or Circle in one case. From this sort of tendency or gravitation it follows likewise that the squares of the Times of the periodical Revolutions are as the Cubes of the *Radii* or *transverse Axes* of the *Ellipses*.

Of course, Halley was in a better position to write such a review than anyone else except the author, having seen the work through the press.

On the final page of the review there is an advertisement explaining why 'the Publication of these *Transactions* has for some Months...been interrupted'; it reads in part:

The Reader is desired to take notice that the care of the Edition of this Book of Mr. *Newton* having lain wholly upon the Publisher (wherein he conceives he hath been more serviceable to the

[2] Although in some of his earlier writings Newton attributed all three laws to Kepler, he withheld credit for the first two in the *Principia*. In a letter to Halley of 20 June 1686, Newton referred to Kepler and the law of elliptical orbits in comparison with Hooke's claims to the law of the inverse square: 'But grant I received it afterwards from Mr Hook, yet have I as great a right to it as to ye Ellipsis. ffor as Kepler knew ye Orb to be not circular but oval & guest it to be Elliptical, so Mr Hook without knowing what I have found out since his letters to me, can know no more but that ye proportion was duplicate quam proximè at great distances from ye center, & only guest it to be so accurately & guest amiss in extending y^t proportion down to ye very center, whereas Kepler guest right at ye Ellipsis. And so Mr Hook found less of ye Proportion then Kepler of ye Ellipsis. There is so strong an objection against ye accurateness of this proportion, y^t without my Demonstrations, to wch Mr Hook is yet a stranger, it cannot be beleived by a judicious Philosopher to be any where accurate. And so in stating this business I do pretend to have done as much for ye proportion as for ye Ellipsis & to have as much right to ye one from Mr Hook & all men as to ye other from Kepler.' [Newton] (1959–), *Correspondence*, vol. 2, pp. 436–7.

In the tract *De Motu*, however (for which see Chapter III, §§3 and 4), Newton included a Scholium to Theorem 3, reading: 'Therefore the major planets revolve in ellipses having a focus in the centre of the Sun; and the radii-vectores to the Sun describe areas proportional to the times, exactly as Kepler supposed.' Again, in a paper of 'Phaenomena', Newton wrote, 'The Planets Saturn Jupiter Mars Venus & Mercury move in Ovals about the Sun placed in the inferior node of the Oval, & every Planet with a right line drawn from it to the center of the Sun describes equal areas in equal times. Kepler by an elaborate discourse has proved this in the Planet Mars & Astronomers find that it holds true in all the primary Planets.' These extracts come from Hall and Hall (1962), *Unpublished papers*; the first (pp. 253, 277) is taken from the Halls' translation, and the second (p. 385) comes from a manuscript of Newton's in English. Newton's knowledge and use of Kepler's work in physics and in astronomy are the subject of a separate study; see Cohen (1971*a*), *Newton and Kepler*; see also Whiteside (1964*a*), 'Newton on planetary motion'.

Commonwealth of Learning) and for some other pressing reasons, they could not be got ready in due time; but now they will again be continued as formerly, and come out regularly...

Halley was giving public notice of his services to Newton. His admiration, expressed in the 'advertisement' just quoted, also appears in the concluding sentiment, that 'so many and so Valuable *Philosophical Truths*, as are herein discovered and put past Dispute, were never yet owing to the Capacity and Industry of any one Man'.

Immediately below the 'advertisement' there appears an imprint, reading:

LONDON,
Printed by *J. Streater*, and are to be sold by *Samuel Smith*
at the *Princes Arms* in St. *Paul's* Church-yard.

It will be recalled that there are two states of the title-page of the first edition of the *Principia*: the original one contains Streater's name ('*LONDINI*, Jussu *Societatis Regiae* ac Typis *Josephi Streater*. Prostat apud plures Bibliopolas. *Anno* MDCLXXXVII'), but the second, a cancel pasted on the stub of the original, bears both Streater's and Smith's names; the sentence 'Prostat apud plures Bibliopolas' is replaced by 'Prostant Venales apud *Sam. Smith* ad insignia Principis *Walliae* in Coemiterio D. *Pauli*, aliosq; nonnullos Bibliopolas'.[3]

4. THE REVIEW IN THE 'ACTA ERUDITORUM'

The review in the *Acta Eruditorum* of Leipzig was surely the most important of the four. These *Acta* were widely read and respected, and the account of the *Principia* no doubt spread news about Newton's work to intellectual circles beyond the range of the *Philosophical Transactions*. Furthermore, this review or summary is a full-scale presentation, running to eighteen pages. When its author had progressed to page 35 of the *Principia*, he apologized to his readers as follows:

As for what pertains furthermore to the arguments of the first two books of the work, we had indeed decided to exhibit them to the reader set forth in order [*presumably, in the order of their appearance*], with as much abridgement as could be made; but in fact...our writing grew to such a bulk that it far exceeded the measure of our intention.[1]

Hence the reviewer hastened on and treated the main parts of Books I and II with less thoroughness than he had done in presenting the contents of the Preface, the Definitions, the Laws of Motion, and Sec. I of Book I. However, he makes it perfectly clear that in the *Principia*

motions of bodies of every kind are discussed, [bodies] spherical and not spherical, ascending and descending, projected, pendulous, fluid, and agitated by whatever forces, motions in a straight line, in a curved line, motions circular, spiral, in conic sections, concentric with the centre of forces and eccentric, with movable or immovable orbits, progressive motions, motions propagated through fluids; likewise, the centripetal, absolute, accelerative forces of motions; times, velocities, and the increases and decreases of the latter, centres, areas, places, apsides, spaces, mediums, and the

[3] For bibliographical information concerning the editions of the *Principia*, see Appendix VIII to the text.

[1] *Acta Eruditorum* (June 1688), p. 307, here translated from the Latin.

densities and resistances of mediums, and that they are unfolded with a penetration worthy of so great a mathematician. Here and there among the propositions lemmas have been scattered, to no small degree perfecting geometry, especially of conics; and there are added from time to time corollaries showing the fullness of things demonstrated; and lest the learning should possibly seem sterile, scholia illustrating philosophy.

This is the tone in which the reviewer epitomizes the main achievements of the *Principia*, in pure geometry and other parts of mathematics, in dynamics and general physics, and in celestial mechanics.

Newton's critique of Cartesian vortices at the end of Book II is adequately summarized, and a few lines are quoted {385.25–27} about the irreconcilability of celestial phenomena and the concept of vortices. Then it is pointed out that in Book III there are hypotheses, 'partly relying upon astronomical observations'. The account continues:

From these and the pronouncements of the preceding books he demonstrates: that the forces by which the circumjovial planets, the primary planets, and the moon, perpetually drawn off from rectilinear motions, are kept in their orbits, are from their gravitation toward Jupiter, the sun, and the earth and are reciprocally as the squares of the distances from the centre of Jupiter, of the sun, and of the earth; that all bodies gravitate toward individual planets, and that their weights toward any one [and the same] planet, at equal distances from the centre of the planet, are proportional to the quantity of matter in each one (whence he further infers: that the weights of bodies do not depend upon their forms and textures; that a vacuum is necessarily given...that gravity is of a different nature from magnetic force); that gravity occurs in all bodies universally and is proportional to the quantity of matter in each individual one; that if there has been posited the matter of two globes gravitating toward each other, homogeneous on all sides in regions equidistant from the centre, the weight of either globe on the other is reciprocally as the square of the distance between their centres...that gravity by proceeding downwards from the surfaces of the planets decreases as nearly as possible in the ratio of the distances from the centre; that the motions of the planets in the heavens can be conserved for a very long time; that the common centre of gravity of the earth, sun, and all the planets is at rest; that the sun is agitated by perpetual motion, but never recedes far from the common centre of gravity of all the planets, which accordingly he contends should be considered as the centre of the world; that the planets move in ellipses, having their focus in the centre of the sun, and that by radii drawn to that centre they describe areas proportional to the times (where [he speaks] concerning the perturbations of motions, from the mutual gravity of the planets); that the aphelia and nodes of the orbits are at rest (where from the imperceptible parallax of the fixed stars he asserts that they cannot produce perceptible effects about us); that the diurnal motions of the planets are uniform and that the libration of the moon arises from its diurnal motion; that the axes of the planets are less than the diameters drawn perpendicularly to the same [axes] (whence he gathers that the diameter of the earth through the poles is to the diameter along the equator as 689 to 692, and so with the earth's mean semi-diameter of 19615800 Parisian feet, taken according to the recent measurement of the French, the earth is here 85200 feet higher or 17 miles ...)...that the equinoctial points regress, and that the axis of the earth by oscillating twice in individual revolutions is inclined toward the ecliptic, and twice is restored to its former position; that all the lunar motions, and all the inequalities of the motions follow from the principles laid down; that the flux and reflux of the sea arise from the actions of the sun and moon, the actions of the luminaries in conjunction and opposition conspiring toward the greatest effect of the motion, in quadratures impeding each other and bringing about the smallest tide...

Concerning comets: that they are higher than the moon (the witness being the lack of diurnal parallax) and that they are in the region of the planets (their annual parallax proving this)...he infers that comets shine with light reflected from the sun, and that in the circumsolar region many more are seen than in the region opposite to the sun. Further, from their motion, in every way very free, preserved for a very long time even contrary to the courses of the planets, he declares that the aether has no resistance, and conjectures that they are a kind of planets, returning by perpetual motion into orbit, and that they move in conic sections (ellipses, but so like a parabola that the latter can be used in place of the former without perceptible error) having their focus in the centre of the sun, and by radii drawn to the sun describe areas proportional to the times; and from an analogy with the planets, that those are smaller which revolve in orbits [that are] smaller and nearer to the sun.

The reviewer included numerical data and physical observations throughout the account of the *Principia*. No reader could come away from this account without a true feeling for the nature and magnitude of Newton's achievement.[2]

5. LEIBNIZ AND THE REVIEW IN THE 'ACTA ERUDITORUM': THE MENTION OF GOD IN THE 'PRINCIPIA'

I have indicated above how interesting the assignment would be to trace the spread of the influence of the *Principia* through the readers of these four book reviews; I regret that such an investigation lies outside the scope of this Introduction. A single major example must suffice, though this particular example is of more than ordinary interest since the reader in question was Leibniz.

In 1689 Leibniz published three articles in the *Acta Eruditorum*, one 'On the resistance of a medium and the motion of heavy projectiles in a resisting medium', another 'On dioptric and other curves', and a third, 'An essay on the causes of the motions of the heavenly bodies' (or 'Tentamen de motuum coelestium causis').[1] As to the last Leibniz wrote that it was composed after he had 'come across an account of the celebrated Isaac Newton's Mathematical Principles of Nature' in the June 1688 issue of the *Acta*, which he had read in Rome.[2] Although the subject was 'far removed from my present line of thought', Leibniz wrote, it tended to recall to his mind some earlier results of his:

conclusions about the resistance of the medium, which I have put on a special sheet, I had reached to a considerable extent twelve years ago in Paris, and I communicated some of them to the famous Royal Academy [of Sciences]. Then, when I too had chanced to reflect on the physical cause of celestial motions, which have every appearance of truth, I thought it worth while to bring before the public some of these ideas in a hasty extemporization of my own, although I had decided to suppress them until I had the chance to make a more careful comparison of the geometrical laws with the most recent observations of the astronomers. But (apart from the fact that I am tied by occupations of quite another sort) Newton's work stimulated me to allow these notes, for what they are worth, to appear, so that sparks of truth should be struck out by the clash and sifting of arguments, and that we should have the penetration of a very talented man to assist us.[3]

[2] As to the authorship of the review, see §5, note 10.

[1] Leibniz (1689, 1689*a*, 1689*b*). These essays are reprinted in Leibniz (1849–), *Mathematische Schriften*, vols. 6 and 7.

[2] Cf. [Newton] (1959–), *Correspondence*, vol. 3, p. 3.

[3] *Ibid.* p. 5, where an English translation is given.

This essay of Leibniz's on the cause of celestial motions is of special fascination to the historian of the exact sciences, not only as a specimen of Leibniz's thinking about celestial mechanics, but also because it is a superb example of the tendency of the thinking men in that age to base their concepts on an all-pervading medium rather than on simple attraction between bodies; that is, on a force acting at a distance.[4] Thus Leibniz postulates that 'the planets are moved by their aethers'[5] and hence depart from rectilinear paths. He defines two kinds of motion: an 'harmonic circulation', in which 'the velocities which a given circling body has are inversely proportional to the radii or distances from the centre of circulation', and a 'paracentric motion', either directly toward or away from the centre. Leibniz allegedly proves that, whenever 'a mobile is transported by a harmonic circulation (whatever its paracentric motion may be)', the areas 'swept out by radii drawn from the centre of circulation to the mobile will be proportional to the elapsed times'; he also presents his proof of the converse. He then concludes:

> It follows therefore that the planets move in a harmonic circulation, the primary [planets] around the Sun as centre, the satellites around their primary [planet]. For the radii drawn from the centre of circulation describe areas proportional to their times (as is observed).

Throughout his life Newton was convinced that Leibniz had not been wholly honest in stating that he had found the propositions expounded in his three papers (published in the *Acta* in 1689) prior to—and thus independently of—Newton's *Principia*. Indeed, in an anonymous review of the *Commercium epistolicum* Newton stated his views strongly as follows (referring to himself in the third person):

> when Mr. *Newton's Principia Philosophiae* came abroad, he [Leibniz] improved his Knowledge in these Matters, by trying to extend this [Differential] Method to the Principal Propositions in that Book, and by this means composed the said three Tracts. For the Propositions contained in them (Errors and Trifles excepted) are Mr. *Newton's* (or easy Corollaries from them...)[6]

In another document, a draft of a letter, possibly intended for publication, Newton (writing presumably *c.* 1712) referred to the account of his book as 'an epitome of it'.[7] Leibniz, Newton said, having read this 'epitome', then composed the three

[4] Another outstanding example is the view of Huygens, who was convinced by Newton that the hypothesis of Cartesian vortices is untenable. Nevertheless, Huygens did not accept the concept of force acting at a distance, but rather sought (for example, in his *La cause de la pesanteur*) a different kind of ether which might produce the effects attributed to central forces. Cf. Koyré (1965), *Newtonian studies*, pp. 116–23 ff.; Dijksterhuis (1961), *Mechanization*, III, F (a, b); Hesse (1961), *Forces*.

[5] This and the following extracts from Leibniz's *Tentamen* are taken from an English version prepared by E. J. Collins for my Harvard course, History of Science 110.

[6] Newton (1715), 'Recensio libri', p. 209.

[7] Rouse Ball, praising Halley's review of the *Principia*, disparaged the account in the *Acta* by describing it as something which 'is, and purports to be, little more than a synopsis of the contents'; Ball (1893), *Essay*, p. 68. Considering the scope and difficulty of the *Principia*, the preparation of such a synopsis is no mean achievement. Indeed, I know of no other 'synopsis' of the *Principia* save the one made by Rouse Ball himself.

essays 'as if he himself also had discovered the principal propositions of Newton concerning these matters, and had done so by a different method, and had not yet seen Newton's book'. Newton declared his sentiments as follows:

> If such unrestrained licence is allowed, any author can easily be robbed of his discoveries... Leibniz had seen the epitome in the Leipzig *Acta*. Through the wide exchange of letters which he had with learned men, he could have learned the principal propositions contained in that book [that is, the *Principia*] and indeed have obtained the book itself. But even if he had not seen the book itself, he ought nevertheless to have seen it before he published his own thoughts concerning these same matters, and this so that he might not err through haste in a new and difficult subject, or by stealing unjustly from Newton what he had discovered, or by annoyingly repeating what Newton had already said before.[8]

I certainly do not wish to enter into the complexities of Newton's feud with Leibniz. Very likely, the presentation in the *Acta* was sufficiently comprehensive and so sound that it could have served to give Leibniz the essential features of Newtonian science he needed. In order to write the *Tentamen* without actually having seen the *Principia*, all he really needed to be aware of was the absolute primacy of Kepler's law of areas; and this, as newly generalized by Newton, appears clearly in the review in the *Acta*. But Leibniz may even have recognized the significance of Kepler's second law on his own. We must not forget that the details of Leibniz's mathematical argument are wholly original and completely independent of the *Principia*.

Leibniz's copy of the first edition of Newton's *Principia* has just turned up and is being studied by Dr E. A. Fellmann of Basel, who has very kindly sent me information concerning Leibniz's annotations. There are no notes in the Leges Motus, and but one in the Definitiones ('Def. VI. Vis centripetae quantitas absoluta...'), although there is an underscoring of the final two words ('vim inertiae') in the penultimate sentence of the discussion of Def. IV. There are no annotations or other marks in the section of 'Hypotheses' in Book III. Dr Fellmann informs me that annotations occur on the following pages (of E_1): 3, 26, 31, 33, 37, 41, 42, 44, 45, 48, 50, 105, 107, 246, 249, 254, 257, 327, 373, 405, 410, 411, 412, 416, 417; and there are underscorings on pages 3, 23, 40, 105, 122, 139, 141, 142, 264. We eagerly await Dr Fellmann's studies of this important document.

To the degree that the review in the *Acta* was truly an epitome, rather than a critical commentary, it did not produce a series of revisions in the second edition; there was nothing of a derogatory or a challenging nature to reply to. But one subsequent alteration of the *Principia* of great interest may very well have been caused by this review. This alteration may attract our attention because it has led to a misunderstanding about Newton ever since.

[8] Edleston (1850), pp. 308–9. Translation by E. J. Collins. On this subject see the penetrating analysis by E. J. Aiton (1962), 'Celestial mechanics of Leibniz', showing convincingly that the *Tentamen* must have been written even before Leibniz had seen the review of the *Principia*. In most respects it follows more closely on Borelli's researches of 1666 than on Newton's and it is certainly an ingenious and original piece of research.

The author of the review, referring to page 415 of E_1, wrote:

Ex quo Deum concludit collocasse Planetas in diversis a Sole distantiis, ut pro ratione densitatum suarum a Sole calorem reciperent.	From which he concludes that God placed the Planets at different distances from the Sun, so that they would receive heat from the Sun according to the proportion of their densities.[9]

A glance at the Apparatus Criticus {405.31–6} shows that this reference to God is indeed present in E_1, but that it has disappeared in E_2, and indeed has been deleted in the alterations made by Newton in E_1i. It seems likely that Newton had not originally intended to make quite so pronounced a statement about God in the midst of the propositions of Book III; in any event, he regretted having done so and accordingly altered the text in E_1i and E_2. The prominence given in the *Acta* to this aside of Newton's could well have been a determining factor.[10]

Newton's views on this subject, as expressed in the first edition of the *Principia*, were also picked up by Richard Bentley, who used them in his penultimate sermon in the inaugural set of Boyle Lectures, published under the title, *A Confutation of Atheism from the Origin and Frame of the World*, Part II (preached at St Martin-in-the-Fields on 7 November 1692, and published in London in 1693). Bentley (see Chapter VIII, §2) cited Newton's *Principia* as his source for the following conclusions:

Was it mere Chance then, or Divine Counsel and Choice, that constituted the Earth in its present Situation? To know this; we will enquire, if this particular Distance from the Sun be better for our Earth and its Creatures, than a greater or less would have been. We may be mathematically certain, That the Heat of the Sun is according to the density of the Sun-beams, and is reciprocally proportional to the square of the distance from the Body of the Sun. Now by this Calculation, suppose the Earth should be removed and placed nearer to the Sun, and revolve for instance in the Orbit of *Mercury*; there the whole Ocean would even boil with extremity of Heat, and be all exhaled into Vapors; all Plants and Animals would be scorched and consumed in that fiery Furnace. But

[9] *Acta Eruditorum* (June 1688), p. 312.

[10] We do not know for sure whether Newton ever read the review of the *Principia* in the *Acta Eruditorum*. His personal library contained a set of fifty-five volumes of the *Acta* in quarto, beginning in 1682, and the *Supplementum* (1718) in octavo; see De Villamil (1931), p. 63. From two letters written in 1693 by Newton to Otto Mencke, the editor of the *Acta Eruditorum*, we may say with certainty that by 30 May 1693 he had received the whole set, 'from the year 1682, when the publication started', through the issue for August 1692, and thus had in his possession the issue containing the review of the *Principia*. In this letter of 30 May 1693, Newton refers to a copy of the *Principia* sent to Mencke in Newton's name by Halley as a 'complimentary' copy ([Newton] (1959–), *Correspondence*, vol. 3, pp. 270–1; the letter is there given both in the original Latin and in English translation). But from a later letter, 22 November 1693, we learn of Newton's amazement at Mencke's kindness 'in sending me in exchange for one copy of my book complete copies of your *Acta*, and in still continuing to send them until I discovered my mistake from your letter. For when Mr Halley, who saw my book through the press, sometimes sent copies of it to my friends even without my knowledge, I thought that he had sent a copy to you. But now I gather from your letter that you have had several copies not on my account but on the strength of an agreement made with Mr Halley by Mr Detlev Cluver's agency, namely that a number of copies of my book and your *Acta* should be exchanged' (*ibid.* pp. 291–2). In this letter, too, Newton mentioned his having 'received all those copies of your *Acta* that you sent me, that is, from the year 1682 right up to August 1692 inclusive, along with twelve sections of supplements'. Since the review of the *Principia* was published in the issue for June 1688, there can be no doubt that a copy was in Newton's hands at the very latest within a few years of publication. Possibly Mencke himself was the author of the review of the *Principia*, although his talents ran more to philosophy and theology than to mathematical physics.

suppose the Earth should be carried to the great Distance of *Saturn*; there the whole Globe would be one *Frigid Zone*, the deepest Seas under the very Equator would be frozen to the bottom; there would be no Life, no Germination; nor any thing that comes now under our knowledge or senses. It was much better therefore, that the Earth should move where it does, than in a much greater or less Interval from the body of the Sun. And if you place it at any other Distance, either less or more than *Saturn* or *Mercury*; you will still alter it for the worse proportionally to the Change. It was situated therefore where it is, by the Wisdom of some voluntary Agent; and not by the blind motions of Fortune or Fate. If any one shall think with himself, How then can any thing live in *Mercury* and *Saturn* in such intense degrees of Heat and Cold? Let him only consider, that the Matter of each Planet may have a different density and texture and form, which will dispose and qualifie it to be acted on by greater or less degrees of Heat according to their several Situations; and that the Laws of Vegetation and Life and Sustenance and Propagation are the arbitrary pleasure of God, and may vary in all Planets according to the Divine Appointment and the Exigencies of Things, in manners incomprehensible to our Imaginations. 'Tis enough for our purpose, to discern the tokens of Wisdom in the placing of our Earth; if its present constitution would be spoil'd and destroy'd, if we could not wear Flesh and Blood, if we could not have Human Nature at those different Distances.

The result of this alteration has been that almost all commentators on Newton have erroneously assumed that Newton mentioned God in the *Principia* only in the later editions. They have had in mind that the concluding Scholium Generale, with its reference to this 'most beautiful System of the Sun, Planets, and Comets, [which] could only proceed from the counsel and dominion of an intelligent and powerful being', and its lengthy discussion of God (his names, his duration, his omnipresence, his attributes), was not present in E_1 but was added in E_2. But these same writers have been ignorant of the fact that in a discussion of God's providence in Book III an explicit mention of God is present even in E_1, but was eliminated before the Scholium Generale was written. This example shows how valuable it is to have an edition of the *Principia* with variant readings of the several editions. We may now reject wholly the view that Newton's introduction of God into the *Principia* was a result of senility, intellectual decline, or even a later development after E_1, as Laplace and J.-B. Biot would have had us believe.[11]

6. THE REVIEW IN THE 'JOURNAL DES SÇAVANS': A POSSIBLE CAUSE OF THE ELIMINATION OF 'HYPOTHESES' IN THE BEGINNING OF BOOK III

I shall say only a word or two concerning the fourth review, in the *Journal des Sçavans*.[1] It was highly critical, though peppering its generally derisory comments with occasional bits of extravagant praise. The review was printed anonymously, and the author has never been identified.

The reviewer had no doubt that the *Principia* was 'une Mécanique la plus parfaite qu'on puisse imaginer'. But he nevertheless felt the need of adding the

[11] This topic is explored further in Cohen (1969), 'Newton's *Principia* and divine providence'.

[1] *Journal des Sçavans*, vol. 16 (2 August 1688), pp. 237–8.

stricture that 'on ne peut regarder ces démonstrations que comme mécaniques, puisque l'auteur reconnoit lui-même...qu'il n'a pas considéré leurs principes en Physicien, mais en simple Géométre'. Newton, furthermore, 'avoue la même chose au commencement du 3. livre où il tâche neanmoins d'expliquer le systême du monde':

> Mais ce n'est que par des hypothéses qui sont la plupart arbitraires, & qui par conséquent ne peuvent servir de fondement qu'à un traité de pure Mécanique.

The only mention of any specific contents of the book occurs in an example to illustrate the foregoing point: the flux and reflux of the sea. The explanation, we are told, is founded 'sur ce principe *que toutes les planetes pésent réciproquement les unes sur les autres*'. But this is an 'arbitrary' supposition, not having been proved. Hence, the demonstration depending on it 'ne peut être que mécanique'.

In conclusion the reviewer recommends to Newton that he now compose 'un ouvrage le plus parfait qu'il est possible'. To do so, Newton needs only to give us 'une Physique aussi exacte qu'est la Mécanique'. He will have given us such a work, finally, when he will have substituted 'de vrais mouvemens en la place de ceux qu'il a supposés'.

I shall not discuss here either the bias of the reviewer or the justifiability of the distinction between a work that is only a 'Mécanique' and one that is truly a 'Physique'. But Newton would have been furious to find that his presentation of the System of the World in Book III was described as having been based on a set of introductory 'hypotheses' that are for the most part arbitrary and hence not to be considered as the basis of a true physics. Newton had surely not intended that any reader should suppose that he had dealt with imagined motions and not the true motions of the heavenly bodies. Indeed, that these were not in any general sense 'hypothetical', or purely imagined, should have been apparent to anyone who saw the tables of data following Hypoth. V and Hypoth. VII (Phaen. I and Phaen. IV in E_2 and E_3), or who read carefully the account of observations to support Hypoth. VI, VIII, and IX (Phaen. III, V, and VI). We have no evidence, say from correspondence or manuscript notes, that Newton ever reacted directly to this review,[2] but it may be that it was the allegation of his having dealt with 'hypothetical' rather than real motions that caused him to alter the Hypotheses at the opening of Book III to Regulae Philosophandi and Phaenomena[3] and to remove to a later place the one that remained an Hypothesis in E_2 and E_3.

[2] At least, I have been unable to find any such direct evidence. Newton's library contained a set of seventy-five volumes of the *Journal des Sçavans* in octavo beginning in 1669 (cf. De Villamil (1931), p. 82), but we do not know when he obtained them.

[3] Of course, even in E_1 these motions were not really hypothetical, save to a particularly perverse reader or critic, since Newton gave numerical data or observational details to support his every statement in these 'Hypotheses' concerning the motions of planets and their satellites. On the change from Hypotheses to Phaenomena, see Chapter I, §1; Chapter II, §2.

7. GILBERT CLERKE'S COMMENTS (1687)

I have referred earlier[1] to the fact that Newton had begun to revise or to correct the printed sheets of the *Principia* as they were sent on to him, and that the first fruits of this re-reading appear in the printed sheet of errata bound into copies of E_1. Once the book was printed and disseminated, further corrections of errata and improvements were introduced by the author himself or brought to his attention by readers who communicated their suggestions either in person and orally or in correspondence. Among the latter was Gilbert Clerke, a former mathematician of Sidney Sussex College, Cambridge (Fellow, 1648–55), who had been ordained a Presbyterian minister in 1651 and who in 1687, at the age of 61, was living in retirement. Clerke was the author of several mathematical and theological works, including a commentary on the *Clavis mathematicae* (1631) of the English mathematician William Oughtred.[2] He ended one of his letters to Newton by telling something about himself:

> I have long livd in an obscure village, in worldly businesse & field-recreations & have not been acquainted wth ye brave notions of Galileus, Hugenius &c. & so despaire of understanding your booke well, but I would willingly know as much as would satisfie my selfe about ye Tydes & some other phaenomena; you must give me leave to talke though it be not very good sense, for I am one of your forefathers. I will be bold to say y^t^ Dr Barrow & I contributed neare 40 yeares since, as much or more than any two others, (to speake modestly) *in diebus illis*, to bring these things into place in ye university.[3]

Clerke was evidently the first critic to propose any changes in the *Principia* and as such he merits our special attention. He was so obviously a sincere old man, who really was puzzled by Newton's ambiguities, that we may well understand Newton's courteous reply, beginning with a reassurance, 'I doe not wonder that in reading a hard Book you meet wth some scruples.' Newton hoped that the 'removal of those you propound may help you to understand it more easily'.[4] Clerke was concerned with two major puzzles, although he noted that 'there are one or two things more in wch I am not wel satisfied'. As to the first, he could not work out the algebra, and in reply Newton wrote out for him the solution in full.[5]

[1] Chapter V, §3, especially note 24.

[2] According to John Collins (in a letter to Vernon, printed by Rigaud), Clerke's *Oughtredus explicatus* (1682) was not a major work: 'Clerke...hath likewise writ a Comment on the said Clavis; but I have no appetite to recommend either the one or the other to the press, knowing what Mr. [John] Kersey hath writ to be much better than either'; Rigaud (1841), *Macclesfield collection*, vol. 1, pp. 153–4.

[3] [Newton] (1959–), *Correspondence*, vol. 2, p. 493.

[4] *Ibid.* p. 487.

[5] In Newton's explanation to Clerke, it is intended to reduce the equation

$$\mathrm{SP}^q - 2\mathrm{KPH} + \mathrm{PH}^q = \mathrm{SP}^q + 2\mathrm{SPH} + \mathrm{PH}^q - \mathrm{L} \times \overline{\mathrm{SP} + \mathrm{PH}}$$

to

$$\mathrm{L} \times \overline{\mathrm{SP} + \mathrm{PH}} = 2\mathrm{SPH} + 2\mathrm{KPH},$$

which involves (for us) merely cancelling the terms SP^q and PH^q which appear on both sides of the equation and transposing the terms $-2\mathrm{KPH}$ and $-\mathrm{L} \times \overline{\mathrm{SP} + \mathrm{PH}}$. Newton's method, however, is to add to both sides of the equation the expression

$$2\mathrm{KPH} + \mathrm{L} \times \overline{\mathrm{SP} + \mathrm{PH}} - \mathrm{SP}^q - \mathrm{PH}^q,$$

But as to the second, Clerke was perfectly right in claiming to have found ambiguity in Newton's mathematical language. He was concerned with a sentence {60.30–61.2} in the proof of Prop. XV:

atque adeo rectangulum sub axibus est in ratione composita ex dimidiata ratione lateris recti & sesquiplicata ratione axis transversi.

Now, Clerke says,

by *sesquiplic*: it appeares from Cor. 2. lem: 11. y^t you meane y^e same with *triplicata* so $\frac{Ac}{1}$ is y^e triplicate of $\frac{A}{1}$, or sessquipl. of $\frac{Aq}{1}$, and by *dimidiata* you meane y^e root, as p. 57. l. 3. & certainly by *integra ratio* you mean $\frac{A}{1}$. if I mistake your meaneing, pray let me know in a word or two.[6]

In this letter we may observe the disadvantage of describing ratios by using words such as 'duplicate' ($A^2:1$), 'triplicate' ($A^3:1$), 'subduplicate' ($A^{\frac{1}{2}}:1$), 'dimidiate' (*also* $A^{\frac{1}{2}}:1$), 'sesquiplicate' ($A^{\frac{3}{2}}:1$), and 'sesquialter' (*also* $A^{\frac{3}{2}}:1$), which are possibly ambiguous, rather than the algebraic expressions which are unambiguous.

In his reply[7] Newton explains that by 'sesquiplicata' he does not mean *triplicata* (the ratio $A^3:1$) but rather '*sesquialtera,* a ratio & an half or y^e root of y^e *ratio triplicata*'. That is, since the prefix 'sesqui-' means 'one-half more', the sesquiplicate or sesquialter ratio must be $A^{\frac{3}{2}}:1$, or $\surd(A)^3:1$.[8] But with respect to the expression 'dimidiate', there can be real confusion. The word means to divide into halves, or to reduce to the half, as to cut into two equal parts. Newton therefore assumed that the dimidiate ratio was unambiguously $A^{\frac{1}{2}}:1$. But Clerke called Newton's attention to a sentence by Oughtred, 'Si dimidiandum sit $\surd c$ 32, vel dividendum per 2 pro 2 sumatur $\surd c$ 8', which means 'if $\sqrt[3]{32}$ is to be halved, or divided by 2, put $\sqrt[3]{8}$ for 2'. Clavius, too, had used 'dimidiata' in this sense. So, Clerke wrote, 'I confesse I did not very well approve of your calling y^e root by y^e *dimidiata ratio*, for *dimidiare* is properly to divide by 2.' He concluded by advising Newton 'that for most men's understanding, this ratio had been better called *sub-duplicata*'.[9] Newton evidently agreed, and in order to avoid any further confusion, he altered

so that the left-hand side becomes

$$SP^q - 2KPH + PH^q + 2KPH + L \times \overline{SP + PH} - SP^q - PH^q$$

while the right-hand side becomes similarly enlarged. Then, 'by striking out y^e terms ($SP^q - SP^q$ & y^e rest) wch destroy one another', there 'will remain only' $L \times \overline{SP + PH} = 2SPH + 2KPH$.

[6] In those days the second and third powers were apt to be expressed by letters, as A*q*, A*c*, or A^q, A^c, the former being the style used by Oughtred and the latter the one often used by Newton. In the printed editions of the *Principia*, the style varies, encompassing A*q*, A*q*., A *quad.*, A^q, A^2.

[7] [Newton] (1959–), *Correspondence*, vol. 2, p. 487.

[8] The ratios $A^{1/2}:1$ and $A^{1/3}:1$ are respectively the subduplicate and subtriplicate ratios, since $A^2:1$ and $A^3:1$ are the duplicate and triplicate ratios. Hence the sense of sesquiplicate or sesquialter ($1+1/2$, or $3/2$) follows at once. Clerke was correct in stating that A*c*:1 (or $A^3:1$) is the triplicate of A:1, but wrong in ignoring that *Ac*:1 is the sesquiplicate of A*q*:1 (or $A^2:1$) though not, of course, of A:1.

[9] [Newton] (1959–), *Correspondence*, vol. 2, p. 490.

'dimidiata' to 'subduplicata' in both E_1i and E_1a, and the change is made throughout E_2 and E_3. Newton concluded his letter with thanks to Clerke for

signifying your doubts to me in these things because they might have proved my mistakes. If there be any thing els you think material for me to know or stick much at in reading y^e Book, pray do me y^e favour of another letter, or two.[10]

This correspondence may have given the occasion for Newton's making an addition to E_1i and E_1a, reading:

Rationem vero sesquiplicatam voco quae ex triplicata & subduplicata componitur, quamque alias [*originally* aliqui *in* E_1a] sesquialteram dicunt.	Indeed I call sesquiplicate a ratio which is compounded out of the triplicate and the subduplicate, and which elsewhere they call sesquialter.

The final printed version of this sentence in E_2 and E_3 {36.6–8} differs somewhat.

Clerke took advantage of Newton's offer and wrote three further letters of questions and comments. Newton's replies are lost, but portions are quoted by Clerke in his discussions of them. Clerke was responsible for one further alteration in the *Principia*. In the proof of Prop. XI in Book I, the basic proposition that Keplerian motion in an ellipse implies a force varying inversely as the square of the distance, the first edition used a notation that is confusing even to a mathematically trained reader (at least at first), and we may agree with Clerke's statement that 'your booke is hard enough',[11] and that Newton should 'make it as easie as you can'.[12] To see the force and justice of Clerke's criticism, until at last he was able to 'perceive' that Newton meant multiplication and not addition, the reader has only to turn to the Apparatus Criticus {55.21–22}, where the alteration of the text of E_1 will show Newton's response to Clerke's criticism. In Newton's statement,

Et conjunctis his omnibus rationibus, $L\times QR$ fit ad QT *quad.* ut AC ad $PC+L$ ad $Gv+CPq$ ad $CDq+CDq.$ ad CBq id est ut $AC\times L$ (seu $2CBq.$) $\times CPq.$ ad $PC\times Gv\times CBq.$ sive ut $2PC$ ad Gv.

'$PC+L$' does not mean 'PC plus L', nor are the other two occurrences of '$+$' to be read 'plus'. Newton's statement means that, when all the ratios are joined together,

$L\times QR$ is to QT *quad.* as AC to PC [conjoined with] L to Gv, and also [conjoined with] CPq to CDq, and also [conjoined with] $CDq.$ to CBq, that is as $AC\times L$ (or $2CBq.$) $\times CPq.$ to $PC\times Gv\times CBq$, or as $2PC$ to Gv,

which is hardly immediately apparent from the Latin text.[13] Clerke suggested that Newton would save his readers much effort if he placed these ratios 'in rank & file

[10] *Ibid.* p. 487.

[11] *Ibid.* p. 492.

[12] He went on, 'but you masters doe not consider y^e infirmities of your readers, except you intended to write only to professours or intended to have your bookes lie, moulding in libraries or other men to gett y^e credit of your inventions'.

[13] D. T. Whiteside points out (in a private communication) that Clerke was ignorant of the notation used by Newton for the 'addition' of ratios '$\alpha:\beta+\gamma:\delta$', and that a sign of Clerke's low level of comprehension is his confusion of 'sesquialtera' with 'triplicata'. In the 'addition of ratios', furthermore, 'the notation is Barrow's and was widely used in English mathematical texts between, say, 1660 and 1730. Since always brackets can be put in uniquely, Barrow's notation is always *consistent*. Like any other notation, it is extremely convenient to use once you gain the hang of it, and it avoids confusion between the letter X and the multiplication sign $\times$.'

under one another', so as to show 'easily...how to multiply & divide', in other words,

$$\begin{array}{rcrcrcl} L\times QR & : & L\times Pv & :: & AC & : & PC \\ L\times Pv & : & Gv\times Pv & :: & L & : & Gv \\ Gv\times Pv & : & Qv^2 & :: & PC^2 & : & CD^2 \\ Qv^2 & : & QT^2 & :: & CD^2 & : & CB^2 \end{array}$$

At once, it is seen that the quantities '$L\times Pv$' and '$Gv\times Pv$' and 'Qv^2' which occur in successive lines in the left-hand part of the array may be cancelled. Similarly, 'CD^2' and 'PC' may be cancelled in the right-hand part, leaving

$$L\times QR:QT^2::AC\times L\times PC:Gv\times CB^2$$

which may be further simplified to

$$L\times QR:QT^2::2PC:Gv$$

(since $L=2CB^2/AC$). Newton did not wish to so simplify his proof, and in E_2 he merely improved the confusing text of E_1 by writing:

Et conjunctis his omnibus rationibus, $L\times QR$ fit ad QT *quad.* ut $AC\times L\times PCq.\times CDq.$ seu $2CBq.\times PCq.\times CDq.$ ad $PC\times Gv\times CDq.\times CBq.$ sive ut $2PC$ ad Gv.

And compounding all these ratios together, we shall have $L\times QR$ to QT^2 as $AC\times L\times PC^2\times CD^2$, or $2CB^2\times PC^2\times CD^2$, to $PC\times Gv\times CD^2\times CB^2$, or as $2PC$ to Gv.

It is to be observed that in stating the result of multiplying the four ratios, Newton does the cancelling on the left-hand side[14] to get

$$L\times QR:QT^2$$

but at first merely multiplies the terms together on the right-hand side to get

$$AC\times L\times PC^2\times CD^2:PC\times Gv\times CD^2\times CB^2$$

or

$$2CB^2\times PC^2\times CD^2:PC\times Gv\times CD^2\times CB^2$$

since

$$AC\times L=2CB^2.$$

It is clear that, on cancellation, the final ratio does become $2PC:Gv$, as Newton says. Although the change introduced in E_2 may go back ultimately to Clerke's criticism, the actual alteration of the text was apparently not made by Newton until much later,[15] possibly coming to mind as he was making the final revision on the eve of sending the printer's copy for E_2 to Cotes.

[14] Perhaps because of the easy 'diagonal' cancellation on the left-hand side.

[15] Thus it was not recorded in E_1i or E_1a, nor is the correction in Gregory's list or Fatio's; and hence it is not printed by Groening.

CHAPTER VII

REVISIONS, CHIEFLY OF THE 1690s, AND PLANS FOR A SECOND EDITION

1. REVISING THE 'PRINCIPIA'

THE revisions of the *Principia* essayed by Newton during the two decades after publication were stimulated by a variety of circumstances. Some arose from the mere fact that, once pushed by Halley into writing the *Principia*, Newton was carried forward by the momentum of his own intellectual activity, which did not cease at once with the printing of the volume. Other revisions were made because Newton was aware that he had not fully solved certain problems, notably the trajectories of comets, the path of the moon, 'Kepler's problem', and others. Further emendations were the result of criticism, or of suggestions by members of his circle: Halley, Fatio de Duillier, Gregory. There were also new data of observation to be used, such as Cassini's reports on the satellites of Saturn, and Flamsteed's tables which Newton needed so desperately that he was responsible for a pirated edition of them. We shall see below that the *Principia* very soon went out of print, which is hardly surprising in view of the small number of copies that had been printed. One result of the pressure being applied to Newton to produce a new edition, or at least to sanction a corrected reprint, must have been to make him consider what further alterations should be made before he released the text to a publisher.

Much of the information concerning these revisions, both those actually made by Newton and those that never got beyond being merely projected, is recorded for us in successive stages by Fatio and Gregory, both of whom had hoped to be associated with a new edition in some fashion. We may also learn of Newton's plans for the *Principia* during this period from his correspondence, especially with Flamsteed, and from the latter's memoranda.

Many of the changes made or contemplated by Newton soon after the publication of the *Principia* are recorded in the Apparatus Criticus, especially the *variae lectiones* arising from E_1a and E_1i. There are also a considerable number of attempted revisions which, for one reason or another, Newton never entered into E_1a or E_1i, and which are to be found chiefly among the loose sheets comprising MS Add. 3965 in the Portsmouth Collection in the University Library, Cambridge. These have not been included in the Apparatus Criticus to the present edition,[1] and they are far too voluminous to describe in full here,[2] but I shall present one or two samples which may be of interest in exhibiting the intensity of Newton's activity in polishing and improving his book once it had been printed.

2. REVISIONS OF THE SECOND LAW OF MOTION

In the years following publication of the *Principia*, Newton became dissatisfied with the Second Law of Motion as he had stated it. Thus we find him writing out a number of new versions which may be dated at 1692–3. One sequence[1] begins:

Vis omnis [in corpus liberum *add. and del.*[2]] impressa motum sibi proportionalem a loco quem corpus alias occuparet in plagam propriam generat.	Every force impressed [upon a free body *add. and del.*] generates a motion proportional to itself from the place the body would otherwise occupy into its own region [that is, in its own direction].

Then Newton writes:

Vis omnis impressa motum sibi proportionalem g	Every impressed force g[enerates] a motion proportional to itself

We may observe that Newton has introduced three conceptual variations: (1) he writes of a motion that is generated rather than a change of motion, (2) he no longer speaks of the new motion as merely being along the right line in which the force acts, and (3) he casts the sentence so as to read that every force generates a motion rather than that every change in motion is proportional to a force.

A little later on in this sequence[3] Newton tries again:

Lex II. Motum [in spatio vel immobili vel mobili *del.*] genitum proportionalem esse vi motrici impressae & fieri secundum lineam rectam qua vis illa imprimitur.[4]	Law II. That a motion arising [in a space either immobile or mobile *del.*] is proportional to the motive force impressed and occurs along the right line in which that force is impressed.

This 'law' is followed by a commentary, in which Newton says in part:

Et hic motus si corpus ante vim impressam quiesceret computandus est in spatio immobili juxta determinationem vis impressae sin corpus antea movebatur computandus est in spatio suo mobili in quo corpus absque vi impressa relative quiesceret.[5]	And this motion if the body was at rest before the impressed force must be computed in an immobile space according to the direction of the impressed force, but if the body was moving before must be computed in its own mobile space in which the body without the impressed force would be relatively at rest.

[1] See Chapter II, §5.

[2] A discussion of these manuscripts will be found in the Commentary volume, to accompany this edition of the *Principia*.

[1] U.L.C. MS Add. 3965, f. 274ᵛ.

[2] I omit certain other deletions and alterations.

[3] I have omitted the many intermediate versions, which I have published elsewhere; Cohen (1967*a*), 'Newton's second law'.

[4] U.L.C. MS Add. 3965, f. 274.

[5] *Ibid.*

This is a particularly fascinating way of conceiving Law II. In the *Principia*, in all printed versions, a body that receives a blow (or 'impulsive' force) may be either at rest or in uniform motion, presumably because it has received such a blow at some anterior time. Then the new motion is to be added to the pre-existing motion; or, if two blows are given simultaneously, the two motions must be added together. Examples are to be found in Corol. 1 to the Laws of Motion {14.21 ff.} and Prop. 1, Book I {38.28 ff.}.

But now Newton asks us to think in a new way about the situation in which a body in motion may receive a blow and gain a new component of motion to be added to the original motion. The body, says Newton, is to be considered at rest within a uniformly moving space. Then it gains a motion from the blow according to the principles of bodies at rest; that is, the ordinary form of Law II. To an observer at true rest, the motion of the body is then composed of the motion within the space of reference, resulting from the blow, and the motion of the space of reference itself. This way of looking at the Second Law does not add (or require) any new physical concept or principle, since Newton had already established that physical events (notably the actions of forces and the motions they produce) occur in exactly the same manner in a system of reference at rest and in a system of reference in uniform rectilinear motion. But what is of major interest about this new way of casting the Second Law is that the eventual trajectory shown by Newton in a diagram is a curve looking much like a parabola, as if it were the product of a continuously acting force, whereas the context of the Second Law puts it beyond any doubt that Newton was conceiving of an impulsive (quasi-instantaneous) force and the change in motion (or change in momentum) it may produce. We would have supposed that the effect of such a 'force' (impulse) in a uniformly moving body would be to produce a linear path according to the diagonal construction, and not a parabola-like curve.

In these manuscripts, as in the *Principia*, it is particularly clear that Newton had in mind that any given 'force' (or impulse) will produce the same change in momentum (i.e. in the speed and direction of motion) whether applied 'simul & semel' or 'gradatim & successive', i.e. 'altogether and at once' or 'by degrees and successively' {13.19}. These words, taken from the discussion of Law II in the *Principia*, are obviously true for the impulse-momentum form of the law. For if three impulses, $\boldsymbol{\Phi}_1$, $\boldsymbol{\Phi}_2$, $\boldsymbol{\Phi}_3$, produce changes in momentum $\Delta m\mathbf{V}_1$, $\Delta m\mathbf{V}_2$, $\Delta m\mathbf{V}_3$, respectively, a single impulse $\boldsymbol{\Phi}$, where

$$\boldsymbol{\Phi} = \boldsymbol{\Phi}_1 + \boldsymbol{\Phi}_2 + \boldsymbol{\Phi}_3,$$

will produce a change in momentum $\Delta m\mathbf{V}$, where

$$\Delta m\mathbf{V} = \Delta m\mathbf{V}_1 + \Delta m\mathbf{V}_2 + \Delta m\mathbf{V}_3.$$

If there is a succession of impulses ($\boldsymbol{\Phi}_1$, $\boldsymbol{\Phi}_2$, ...) 'gradatim & successive', the trajectory will be a polygon,[6] but the final direction and magnitude of motion will

[6] If all the blows are struck in the same direction as the original motion, the polygon will be a straight line.

be the same as if there had been one blow (by an impulse $\mathbf{\Phi}=\mathbf{\Phi}_1+\mathbf{\Phi}_2+\ldots$) rather than many. In the limit, as the time between successive blows becomes zero (or 'infinitesimally small'), the polygon will approach the limiting smooth curve shown in Newton's diagram. This would also be true if there were a sequence of infinitesimal impulses, whose sum were a finite 'force' $\mathbf{\Phi}$.

In the *Principia*, Newton evidently regarded the impulse-momentum form of Law II as fundamental; at least, this is the form in which the law is stated in the 'Axiomata sive Leges Motus', and this is the form in which the same effect is produced whether the 'force' act (or 'be impressed') 'simul & semel' or 'gradatim & successive'. In Prop. I, and in Def. VIII (par. 3) {5.27 ff.}, it seems that Newton considered the continuous form of the law to be derivable, in the limit, from a series of blows, i.e. an action applied 'gradatim & successive'. Hence, in Newtonian practice, there need be no real distinction made between the forms of the Second Law which we would write today as

$$(1)\ \mathbf{\Phi}\propto\Delta\mathbf{V},\quad (2)\ \mathbf{f}\propto d\mathbf{V},\quad (3)\ \mathbf{f}\propto\frac{d\mathbf{V}}{dt},\quad (4)\ \mathbf{f}\,dt\propto d\mathbf{V}$$

(where $\mathbf{f}$ stands for a continuously acting rather than an instantaneous force), because, in Newton's fluxional conception of force, time is always a uniformly fluent variable. In other words, the increments are always constant and so need not be specified by Newton. Thus, Newton would not conceive of equation (1) and its limit, equation (2), as separate and distinct from equation (3) and equation (4). We today are concerned that $\mathbf{\Phi}$ in equation (1) and $\mathbf{f}$ in equations (2), (3), and (4) have a different dimensionality, so that the continuous 'force' $\mathbf{f}$ cannot strictly be the limit of a series of impulsive 'forces' $\mathbf{\Phi}$. Hence, we make a distinction between force and impulse, and define an impulse ($\mathbf{\Phi}$) in terms of a 'force' ($\mathbf{F}$),[7]

$$\mathbf{\Phi}=\mathbf{F}\cdot\Delta t, \tag{5}$$

so that we can make explicit the time Δt in which the 'force' acts; but we do so because $\mathbf{\Phi}$ is not for us a force at all, but a different type of quantity. We now substitute $\mathbf{F}\cdot\Delta t$ for $\mathbf{\Phi}$ in equation (1) to get

$$\mathbf{F}\cdot\Delta t\propto\Delta\mathbf{V}, \tag{6}$$

and then take the limit either directly,

$$\lim_{\Delta t\to 0}\mathbf{F}\ \Delta t=\mathbf{f}\,dt\propto\lim_{\Delta t\to 0}\Delta\mathbf{V}=d\mathbf{V}, \tag{7}$$

or after dividing equation (6) by Δt,

$$\mathbf{F}\propto\frac{\Delta\mathbf{V}}{\Delta t}, \tag{8}$$

$$\lim_{\Delta t\to 0}\mathbf{F}=\mathbf{f}\propto\lim_{\Delta t\to 0}\frac{\Delta\mathbf{V}}{\Delta t}=\frac{d\mathbf{V}}{dt}. \tag{9}$$

[7] A distinction is made between $\mathbf{F}$ and its limiting value $\mathbf{f}$, because if the force should vary during the time interval Δt, $\mathbf{F}$ is the average value during Δt.

But such a procedure of ours is *not* Newton's. For he did not conceive a continuous force as primary, and an impulse as a derived concept obtained by multiplying the force by the time interval in which it acts, whether finite or infinitesimal. Clearly, in Newton's view, the same constant increment of uniformly flowing time was built into the concept of $\mathbf{\Phi}$ in equation (1), and again in the limit form, when $\mathbf{\Phi}\to\mathbf{f}$, in equation (2). In equations (3) and (4) this factor, dt, is made explicit, and of course the proportionality constant cannot be the same in equations (2) and (3). Newton's undifferentiated use of 'force' for what I have called $\mathbf{\Phi}$ in equation (1), $\mathbf{f}$ in equation (2), and $\mathbf{f}$ in equations (3) and (4) is the source of many difficulties to today's reader.

The problem of understanding Newton's production of a curved trajectory thus depends on keeping distinct the two ways in which a 'force' may act: 'simul & semel', in which the orbit is constructed as the result of a sequence of infinitesimal discrete force-impulses; and 'gradatim & successive', in which 'a series of infinitesimal arcs is generated by a continuous force (of second-order infinitesimal discrete force-impulses)'.[8] Both of course lead to exactly the same theory of central forces.[9] Confusion on this point, and failure to read the text of the *Principia* (apart from the Definitions and Laws) has caused many writers to assert incorrectly that Newton did not 'know' the continuous form of the Second Law. Of course, he did not ever write

$$\text{`}\mathbf{F}=km\mathbf{A}\text{'} \quad \text{or} \quad \text{`}\mathbf{F}=km\frac{d\mathbf{V}}{dt}=km\frac{d^2\mathbf{s}}{dt^2}\text{'},$$

as we would; nor did he ever write out a fluxional equation such as '$\mathbf{F}=km\ddot{\mathbf{s}}=km\dot{\mathbf{V}}$'. But anyone who studies carefully Prop. XLI of Book I (or many others, such as Prop. XXIV, Theor. XIX, Book II) will see at once that the mere absence of an equation or the differential or fluxional algorithm cannot mask the fact that Newton was perfectly aware of the Second Law in the form in which we know it.

[8] Quoted from D. T. Whiteside (in a private communication), who would call attention to the fact that the building up of an orbit in terms of these 'infinitesimal discrete force-impulses' was favoured by Leibniz, whereas the second method (in which we have 'a series of infinitesimal arcs generated by a continuous force') was favoured by Newton. The relations between Newton's applications of the Second Law and his fluxional concepts are developed in Whiteside (1966*a*), 'Newtonian dynamics'. Newton's revisions of the Second Law are discussed at greater length, together with an analysis of Newton's use of this law in the *Principia*, in Cohen (1967*a*), 'Newton's second law'.

[9] In revising E_2, Newton planned to introduce arguments similar to those I have been discussing, showing how the Second Law could be applied directly to continuous forces, and once again showing how a curved (parabola-like) trajectory might be produced. He never introduced these revisions into the third edition, but he did add in E_3 some new material {21.22–22.11} to explain his statement (in E_1 and E_2) that Galileo had used the first two Laws of Motion and the first two corollaries to discover 'that the descent of bodies varied as the square of the time and that the motion of projectiles was in the curve of a parabola' {21.18–20}. On Newton and Galileo, see Herivel (1965), *Background*, chap. 2; Herivel (1964), 'Galileo's influence'; Cohen (1967), 'Newton's attribution to Galileo'.

3. A NEW PROP. I AND PROP. II FOR SEC. II OF BOOK I

One rather startling revision proposed by Newton would have a new Prop. I, Theor. I, based on Prop. IV (Book I) of E_1, and, with its corollaries, reading as follows (here only final versions of the texts are given):

Prop. I. Theor. I.

Vim qua corpus uniformi cum motu in circulo quocunque dato, revolvens retrahitur semper a motu rectilineo et in perimetro circuli illius retinetur dirigi ad centrum circuli illius et esse ad vim gravitatis ut quadratum arcus dato quocunque tempore descripti ad rectangulum sub diametro circuli et spatium quod corpus eodem tempore cadendo describere posset.

Corol. 1. Igitur in circulis aequalibus vires centripetae (ob aequalitatem diametrorum) sunt ad invicem ut quadrata arcuum simul descriptorum hoc est in duplicata ratione velocitatum, et propterea reciproce in duplicata ratione temporum periodicorum.

Corol. 2. In circulis autem inaequalibus vires illae sunt ad invicem ut quadrata arcuum simul descriptorum applicata ad[2] circulorum diametros hoc est ut quadrata velocitatum applicata ad easdem diametros vel (ut cum Geometris loquar) in ratione composita ex duplicata ratione velocitatum directe & ratione diametrorum inverse et propterea sunt etiam in ratione composita ex duplicata ratione temporum periodicorum inverse et ratione diametrorum directe.

Corol. 3. Unde si tempora periodica aequantur erunt tum vires centripetae tum velocitates ut Radii circulorum: et contra

Corol. 4. Si quadrata temporum periodicorum sunt ut radii vires centripetae sunt aequales, & velocitates in subduplicata ratione radiorum: et contra.[3]

Prop. I. Theor. I.

That the force by which a body revolving with a uniform motion in any given circle is drawn back always from rectilinear motion and is kept in the perimeter of that circle is directed to the centre of that circle and is to the force of gravity as the square of an arc described in any given time to the rectangle of the diameter of the circle and the space which the body could have described by falling in the same time.[1]

Corol. 1. Therefore in equal circles centripetal forces (on account of the equality of diameters) are to one another as the squares of the arcs described in the same time, that is in the duplicate ratio of the velocities, and therefore reciprocally in the duplicate ratio of the periodic times.

Corol. 2. But in unequal circles those forces are to one another as the squares of the arcs described in the same time applied to [divided by[2]] the diameters of the circles, that is as the squares of the velocities applied to the same diameters or (to speak as Geometricians do) in a ratio composed of the duplicate ratio of the velocities directly and the ratio of the diameters inversely and therefore are also in a ratio composed of the duplicate ratio of the periodic times inversely and the ratio of the diameters directly.

Corol. 3. Whence if the periodic times are equal, both the centripetal forces and the velocities will be as the Radii of the circles: and conversely.

Corol. 4. If the squares of the periodic times are as the radii the centripetal forces are equal, and the velocities in the subduplicate ratio of the radii; and conversely.

These Corols. 3 and 4 differ from the Corols. 3 and 4 of Prop. IV in E_1 only in minor changes in wording.[4]

[1] That is, in the ratio of the square of the said arc to the product of the diameter and the said space.

[2] Here, and in the following extracts, 'applicata ad' ('applied to') means 'divided by'.

[3] This and the next five quotations are from U.L.C. MS Add. 3965, ff. 36^v, 37, and 37^v.

[4] In E_1, Corol. 3 ends with 'ut radii, & vice versa', whereas this Corol. 3 ends 'ut Radii circulorum: et contra'. The Corol. 4 of E_1 differs from this Corol. 4 in having 'dimidiata ratione' for 'subduplicata ratione' and in concluding with ': Et vice versa' rather than ': et contra'.

Then there were to be Corols. 5, 6, and 7, as in E_1, followed by the first and third paragraphs of the scholium, thus *inter alia* omitting the reference to Huygens. To remedy this, Newton has directed:

After y^e Demonstration of Prop 1 add. ∧ [IN's caret] Et hujusmodi Propositionibus Hugenius in eximio suo Tractatu vim gravitatis cum revolventium viribus centrifugis contulit. Cartesius Borellus et alii de conatu revolventium nonnulla scripserunt at Hugenius quantitatem conatus illius primus omnium determinavit.

After the Demonstration of Prop. 1 add 'And by Propositions of this sort Huygens in his outstanding Treatise compared the force of gravity with the centrifugal forces of revolving bodies. Descartes, Borelli, and others have written some things about the *conatus* of revolving bodies but Huygens was the first of all to determine the quantity of that *conatus*.'

This was to be followed by a

Prop. II. Theor. II.

Vires quibus corpora diversos circulos aequabili motu describentia retrahuntur a motibus rectilineis & in circulis retinentur, sunt ad invicem ut arcuum simul descriptorum quadrata applicata ad circulorum radios.

Sunt enim ad gravitatem ut quadrata arcuum eodem tempore descriptorum sunt ad rectangula sub circulorum diametris & spatio quod grave cadendo eodem tempore describit (per Prop. 1.) hoc est ut quadrata arcuum applicata ad diametros sunt ad spatium cadendo descriptum. Et propterea sunt ad invicem ut quadrata illa ad diametros vel (quod perinde est) ad semidiametros applicata. Q.E.D.

Prop. II. Theor. II.

The forces by which bodies describing diverse circles with an equable motion are drawn back from rectilinear motions and are kept in circles, are to one another as the squares of the arcs described in the same time applied to[5] the radii of the circles.

For they are to gravity as the squares of the arcs described in the same time are to the rectangles of the diameters of the circles and the space which the body in falling describes in the same time (by Prop. 1), that is, as the squares of the arcs applied to the diameters are to the space described in falling. And therefore they are to one another as those squares applied to the diameters or (which is the same thing) to the semi-diameters. Q.E.D.

There then follow the seven corollaries and two paragraphs of the scholium originally intended for the above-mentioned version of Prop. I, Theor, I.

In a variation, the foregoing material on Huygens, Descartes, and Borelli was to have been included in a Corol. 2 to a new Prop. V, Theor. IV:

Corol. 2. Si corpus uniformi cum motu in circulo revolvitur erit vis qua retrahitur a motu rectilineo ad ipsius pondus ut quadratum arcus dato quocunque tempore descripti ad rectang. sub diametro circuli et spatium quod grave eodem tempore cadendo describere posset. Nam sagitta arcûs minimo tempore descripti aequalis est quadrato arcûs ejusdem applicato ad diametrum circuli, hoc est spatio quod grave cadendo eodem tempore describit. Et aucto tempore in ratione quacunque augentur aequalia illa in eadem ratione duplicata et propterea

Corol. 2. If a body revolves with uniform motion in a circle the force by which it is drawn back from rectilinear motion will be to the weight of the same body as the square of the arc described in any given time to the rectangle of the diameter of the circle and the space which the body in the same time could have described by falling. For the sagitta of the arc described in the smallest time is equal to the square of the same arc applied to the diameter of the circle, that is to the space which the body describes by falling in the same time. And with the time increased in any ratio those equal quantities are increased in the same ratio squared and there-

[5] See note 2 above.

semper sunt aequalia. Et hujusmodi Propositionibus Hugenius in eximio suo tractatu de Horologio oscillatorio vim gravitatis cum revolventium viribus centrifugis contulit. De his viribus Cartesius, Borellus et alii nonnulla disseruerunt. Sed Hugenius earum quantitatem primus docuit, quanquam ipse etiam eandem seorsum inveni.

fore are always equal. And by this sort of Propositions Huygens in his outstanding treatise on the oscillatory [i.e. pendulum] Clock compared the force of gravity with the centrifugal forces of revolving bodies. Concerning these forces Descartes, Borelli, and others have discoursed to some extent. But Huygens was the first to teach their quantity, although I myself also found the same quantity separately.

In this sequence, Corols. 6–9 correspond to Corols. 3–6 to Prop. IV in E_1, E_2, and E_3, while Corol. 10 embodies a generalization probably suggested by Fatio (see Chapter VIII, §9). It now reads:

Corol. 10. Et universaliter si tempus periodicum sit ut Radii R potestas quaelibet R^n et propterea velocitas reciproce ut R^{n-1} erit vis centripeta reciproce ut R^{2n-1}.

Corol. 10. And universally if the periodic time be as any power R^n of the radius R and therefore the velocity reciprocally as R^{n-1}, the centripetal force will be reciprocally as R^{2n-1}.

This is essentially the new Corol. 7 of E_2, reprinted in E_3 {44.26–28}. This is followed by Corol. 11, 'Eadem omnia...applicata', which is the old Corol. 7 of E_1, without the final sentence added in E_2. This Corol. 11 was originally numbered Corol. 9 and was to have been followed by a Corol. 10, a Corol. 11 (*bis*, the first cancelled), a Corol. 12, and the words 'Corol 13' (but without any text to go with it), as follows:

Corol. 10. Si corpus P in perimetro circuli cujusvis PQT motu quocunque revolvitur et vis centripeta tendit ad centrum aliquod S extra circuli centrum C consistens: erit vis illa ut quadratum velocitatis applicatum ad circuli chordam istam PV quae a corpore per centrum virium ducitur. Nam sagitta PR dato tempore quam minimo descripta aequatur $\frac{PQ^{quad}}{PV}$ & PQ^q est ut quadratum velocitatis.

Corol. 11. Ideoque si area quae radio PS ad centrum virium acto describitur sit ut tempus, et in circuli tangentem demittatur perpendiculum SZ; vis centripeta erit reciproce ut solidum $PV \times SZ^q$. Nam velocitas corporis in hoc casu est ut [*sic*] et propterea reciproce ut SZ.

Corol 12 Iisdem [*sic*] positis vis centripeta erit etiam ut solidum $\frac{PS^q \times PV^c}{PT^q}$ propterea quod $\frac{PS \times PV}{PT}$ aequalis sit ipsi PZ.

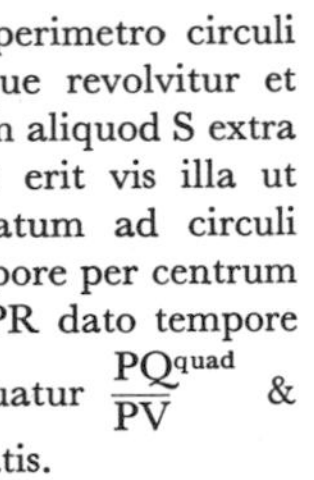

Corol. 10. If a body P revolves with any motion in the perimeter of any circle PQT and the centripetal force tends toward some centre S standing outside the circle's centre C: that force will be as the square of the velocity applied to[6] that chord of the circle PV which is drawn from the body through the centre of forces. For the sagitta PR described in a given minimal time is equal to $\frac{PQ^2}{PV}$ and PQ^2 is as the square of the velocity.

Corol. 11. And therefore if the area which is described by a radius PS drawn to the centre of forces be as the time, and a perpendicular SZ be dropped to the tangent of the circle; the centripetal force will be reciprocally as the solid $PV \times SZ^2$. For the velocity of the body in this case is as [*sic*] and therefore reciprocally as SZ.

Corol. 12. The same things being supposed, the centripetal force will be also as the solid $\frac{PS^2 \times PV^3}{PT^2}$ because $\frac{PS \times PV}{PT}$ is equal to PZ itself.

[6] See note 2 above.

In another redaction of this material, the foregoing Prop. II is now

Prop. V. Theor. IV

Corporum quae diversos circulos aequabili motu describunt, vires centripetas ad centra eorundem tendere et esse inter se ut arcuum simul descriptorum quadrata applicata ad circulorum radios.[7]

Prop. V. Theor. IV

That the centripetal forces of bodies which describe diverse circles with equable motion tend toward the centres of the same [circles] and are among themselves as the squares of the arcs described in the same time applied to the radii of the circles.

There follows a series of corollaries, similar to those following Prop. IV in E_1, but with the new corollary (suggested by Fatio) inserted as Corol. 7, so that the old Corol. 7 becomes Corol. 8. In the notes that follow, some of the final alterations of Props. V–IX are present.[8]

4. THE ACTUAL CHANGES MADE IN SECTIONS II AND III OF BOOK I

Now it is an interesting fact that, despite these many essays at new ways of presenting Secs. II and III of Book I, Newton's actual revisions appearing in E_2 are far less drastic. He did not, in the end, even adopt the alteration found in E_1i and E_1a {43.26 ff.}, by which Prop. V, Prob. I, would have become Prop. IV, Prob. I, while Prop. IV, Theor. IV (together with its seven corollaries and its scholium), would have come next as Prop. V, Theor. IV. In E_2, Sec. III appears much as in E_1, save for the addition of alternative proofs, as for Prop. XI {55.28–56.4} and Prop. XII {57.6–10}.

But in Sec. II, there were some very real changes made in E_2, even though different from those just described. For example, the original two corollaries to Prop. I in E_1 were rewritten and then transferred to Prop. II in E_2 {41.29–42.3}; they are so noted in E_1i and E_1a. The six corollaries to Prop. I, new in E_2 (and E_3), occur first in E_1i and their presence is indicated in E_1a by '&c.'. By the use of the new Corols. 2 and 4 to Prop. I, the proof of Prop. IV is greatly simplified in E_2 {43.31 ff.}. There are also significant changes in the corollaries to Prop. IV, notably the new Corol. 7 {44.26–28}, which may have been suggested by Fatio de Duillier (see Chapter VIII, §9). The second paragraph of the proof of Prop. V {46.14–18} is altered in E_2. The old Prop. VI of E_1—with some revisions—becomes in E_2 a mere Corol. 1 to Prop. VI {47.10–23}, whereas the old unnumbered corollary becomes Corol. 5 and there is a wholly new Prop. VI, with three other corollaries (five in all) {46.27–47.9; 47.24–48.14}. This Prop. VI, Theor. V,

[7] U.L.C. MS Add. 3965, f. 181. The proof begins with a very clear statement that 'the velocity, which a given force can generate in a given matter in a given time [!], is as the force and time directly, and the matter inversely'.

[8] For instance, the new Corol. 5 to Prop. VI {48.6–14}; an early form of the 'idem aliter' proof of Prop. VII {49.9 ff.}, of Prop. IX {52.2–5}, and Prop. X {53.8 ff.}. But the 'idem aliter' proof of Prop. VIII was discarded, and in E_2 Newton merely wrote {51.4} 'Idem facile colligitur etiam ex Propositione praecedente' ('And the same thing is likewise easily inferred from the preceding Proposition'); *ibid.* ff. 181–2.

appears in both E_1i and E_1a, but (as may be seen in the Apparatus Criticus) E_1i has a preliminary Prop. VI, Theor. V. Some alterations occur in the proof of Prop. VII, and an alternative proof is added {49.8–14}, of which (as may be seen in the Apparatus Criticus) some drafts may be found in E_1i; there are also {49.15–50.15} three new corollaries. Alternative proofs are added to Prop. IX {52.15} and to Prop. X {53.7–19}. The final scholium to this Sec. II is considerably expanded {54.10–20}.

5. SOME FURTHER CHANGES PROPOSED FOR BOOK I

No doubt the most interesting aspect of the revisions of the early part of Book I in the 1690s occurs in the lemmas comprising Sec. I.[1] Among Newton's manuscripts in the Portsmouth Collection there is a page (MS Add. 3965, f. 635) beginning with a 'Lem. XII', and marked at the top 'Lemmata Generalia. Lem. I. Lem II &c', of which Lemma XII is essentially the Lemma XII of the *Principia*, occurring not in the general Sec. I of lemmas, but in Sec. II, just before Prop X {52.6–9}. But Lemma XIII and Lemma XIV, given without proof, correspond in part to the material of Prop. XI in E_1, E_2, and E_3:

Lem. XIII

Recta quae per centrum sectionis conicae ducitur & cuilibet ejus tangenti parallela est, de recta quae ducitur a puncto contactus ad umbilicum sectionis, abscindit versus punctum contactûs longitudinem semissi axis principalis aequalem.

Lemma XIII

A right line which is drawn through the centre of a conic section, and is parallel to any tangent of it, cuts off from the right line which is drawn from the point of contact to the focus of the [conic] section (toward the point of contact) a length equal to half of the principal [major] axis.

Lem. XIV

In Ellipsi et Hyperbola axis principalis est ad diametrum quamvis ut diameter illa ad rectam quae per umbilicum ad Curvam utrinque ducitur ac diametro illi parallela est.

Lemma XIV

In the Ellipse and Hyperbola the principal axis is to any diameter as that diameter to the right line which is drawn through the focus to the curve on both sides and is parallel to that diameter.

Apparently what Newton was trying to do here[2] (apart from presenting the general mathematical truths involved in Book I in a collective preliminary to his dynamical text) was to stress the fluxional character of the *Principia*.[3] The first eleven lemmas (Sec. I) deal with limits and quasi-calculus problems. In the sequence now being described Newton eventually got up to Lem. XXII, which is Lemma II of Book II

[1] Sec. I ('On the method of prime and ultimate ratios') is a mathematical introduction to the whole treatise, and Sec. II (comprising Props. I–X) introduces the first propositions dealing with dynamics, the basis for Sec. III ('On the motion of bodies in eccentric conics').

[2] This point was suggested by D. T. Whiteside.

[3] In E_2i, at the beginning of Sec. I, Book I, which deals with the method of first and last ratios, there is a sheet of paper listing those parts of the *Principia* in which the method of fluxions may be found. This document, printed in Appendix III to the text, is entitled *De methodo fluxionum*. It directs attention in particular to the following parts of the *Principia* (E_2): 'Lib. I. Sect. I; Lib. I. sect. XIII Prop. 93. Schol. pag. 202; Lib. II. Lem. II. pag. 224; Lib. II. Prop. 19. pag 251.'

{243.18 ff.} on 'moments' (fluxions) and their applications to general polynomials, and written here by Newton only as

Lem. XXII. XV.

Momentum Genitae etc[4]

Newton has also written out two pages of 'Errata et emendata in Ph. nat. Prin. Math.'[5] and a more elaborate 'In Philosophiae Naturalis Principiis Mathematicis corrigenda et addenda',[6] in which may be found the lengthy addition to Definition V {3.8 ff.}, and the six new corollaries to Prop. I, Book I. Then there occurs the new Prop. VI and its Corols. 1–5 as printed in E_2 {46.26–48.14}, and marked for 'pag 44. lin. 20.'—which is the very place in which the old Prop. VI, Theor. V occurs in E_1. But although at first Newton has indicated that the new proposition is also to be Prop. VI, Theor. V, he has then changed his mind, altering the numbers to Prop. IX, Theor. VIII. Nevertheless it is followed by the two new proofs of Prop. VII, Prob. II (the numbers unchanged) and the new Corol. 1 {49.15–17}.[7]

There can be no doubt that, during the 1690s, Newton was contemplating rather serious changes in the structure of the first part of the *Principia*.

6. REVISING BOOK III: THE MOTION OF THE MOON AND THE QUARREL WITH FLAMSTEED

During the 1690s, Newton was planning substantive improvements for Book III. We have already seen (Chapter II, §2) how he introduced the new Hypothesis III, and then converted the 'Hypotheses' into Phaenomena and Regulae. During 1694–5 the news was circulated that Newton was preparing a new edition, and David Gregory mentioned a visit made by Newton to London in 1694 in connection with Newton's efforts to get the *Principia* back into print.[1] In the course of Newton's exchanges with Flamsteed at this time, it was made clear that Newton's research was being directed toward a new edition; when Newton wrote on 1 November 1694

[4] The double numbering, 'XXII.XV', arises from the fact that in the original sequence the two lemmas (XIII and XIV) just quoted were followed by seven others, Lemmas XV–XXI (five of which are the equivalents of the Lemmas XII–XIV, XXVIII–XXIX of Book I of the *Principia*), so that the next one would be Lemma XXII. Lemmas XIII and XIV were subsequently cancelled and replaced by a new Lemma XIII (on curvature) and a new Lemma XIV (which is equivalent to Lemma V, Book III). The remaining seven Lemmas (XV–XXI) are then wholly cancelled, so that the Lemma XXII became XV.

[5] MS Add. 3965, f. 180.

[6] *Ibid.* f. 635.

[7] Corollaries 2 and 3 {49.18–50.15} are not present on this sheet. Newton has added a short Corol. 2 which he has not used in E_2.

[1] On 1 September 1694 Gregory recorded: 'Mr. Newton visited Flamsteed at Greenwich on 1 September 1694, when he spoke about the new edition of his *Principia*. He believes that the theory of the Moon is within his grasp. To find her position he will need five or six equations. Flamsteed disclosed one, which is the greatest in quadratures: he showed him about fifty positions of the Moon reduced to a synopsis. Newton's equation indicated the correct position where they were near quadratures. Whether it is greater or smaller depends on other physical causes. Observations are not sufficient to complete the theory of the Moon. Physical causes must be considered.' [Newton] (1959–), *Correspondence*, vol. 4, p. 7.

to ask Flamsteed for 'such observations as tend to perfecting the theory of the planets', he referred expressly to the use he would make of them 'in order to a second edition of my book'.[2]

Two of the major problems in Book III that were occupying Newton's attention were planetary perturbations[3] and the related problem of the motion of the Moon.[4] It was during this period that Newton also invented a new means of calculating tables of atmospheric refraction, which later on elicited the highest praise from J.-B. Biot: 'Newton is therefore the creator of the theory of astronomical refractions, as he is of that of the theory of gravity.'[5]

To complete his theory of the Moon's motion, Newton needed Flamsteed's tables of observations of the Moon. Very soon there began the dispute between Newton and Flamsteed that reached its climax about a decade later with respect to the publication of Flamsteed's *Historia coelestis*; it can be compared in intensity with Newton's other scientific quarrels, with Hooke and Leibniz. We may, I believe, agree with Brewster's conclusion that 'Flamsteed gave Newton his best assistance, though perhaps with some unnecessary delay, very often in an irritating form, and accompanied by reflections on Halley—of whom Flamsteed had an unreasonable dislike—which were particularly annoying to Newton'.[6]

The problem of the Moon's motion is most difficult and has occupied some of the best mathematical minds, even in modern times.[7] The history of the attempts to deal with the observed irregularities in the Moon's motion shows how shallow the judgement is of those who say that Newton had been able to demonstrate how the solar system is 'governed' by the law of universal gravitation. It is an interesting matter of record that both Newton and Halley were aware, at the time of publication of the first edition of the *Principia*, that this subject had not yet been treated in a fully satisfactory manner. On 5 July 1687, immediately after publication, Halley urged Newton to continue with his attack on the problem.[8] According to Conduitt, Halley said that when Newton was pressed to go on to complete his theory of the Moon he replied that 'it made his head ache, and kept him awake so often, that he would think of it no more'. Years afterwards, Newton remarked to John Machin that 'his head never ached but when he was studying [the lunar theory]'. Yet he

[2] Baily (1835), p. 138.

[3] In E_2, Prop. XIII (Book III) contains notable revisions of the presentation of the perturbation of Saturn's orbit by Jupiter, and of the Earth's orbit by the Moon.

[4] One of Newton's plans for elaborating his theory of the Moon has been published in the *Catalogue* of the Portsmouth Collection, pp. xii–xvi, and reprinted by Rouse Ball (1893), *Essay*, pp. 126–8; I have discussed it in Chapter IV, §7.

[5] *Journal des Savants* (November, December 1836), pp. 642, 735; also Brewster (1855), vol. 2, pp. 173, 175.

[6] As summarized in Ball (1893), *Essay*, p. 129.

[7] For example, George William Hill, Henri Poincaré, and Ernest William Brown. Hill made celebrated researches on the 'restricted problem of three bodies' for the case in which one of three mutually gravitating bodies may be considered to have a null mass (and so produce no perturbations of the motions of the other two); even this problem has no analytic solution. See Poincaré (1905); [Brown] (1941); Brouwer and Clemence (1961); Marcolongo (1919).

[8] [Newton] (1959–), *Correspondence*, vol. 2, p. 482.

also told Conduitt that if he were to live until Halley could make six years of observation, 'he would have another stroke at the moon'.[9]

Newton had planned in July 1691 to see Flamsteed at Greenwich, in order to arrange for Flamsteed to provide accurate observations to test Newton's equations.[10] But Flamsteed was away; Newton apparently did not establish direct contact with him until 1694.[11] Newton's hopes for his lunar theory are clearly expressed in a letter to Flamsteed of 7 October 1694, formally requesting a 'set of observations' by means of which, he said, 'I believe I could set right the moon's theory this winter'.[12] Writing from Cambridge, to which he had just returned, Newton said that he had been 'comparing' Flamsteed's observations with his own theory, 'and now I have satisfied myself that, by both together, the moon's theory may be reduced to a good degree of exactness: perhaps to the exactness of 2 or 3 minutes'.

In a letter dated 16 February 1694/5 Newton assured Flamsteed that he would not 'communicate' his observations 'to any body, and much less publish them, without your consent', and then expressed his point of view about combining theory and observations in a single publication:

> But if I should perfect the moon's theory, and you should think fit to give me leave to publish your observations with it, you may rest assured that I should make a faithful and honorable acknowledgment of their author, with a just character of their exactness above any others yet extant. In the former edition of my book, you may remember that you communicated some things to me, and I hope the acknowledgments I made of your communications were to your satisfaction: and you may be assured I shall not be less just to you for the future. For all the world knows that I make no observations myself, and therefore I must of necessity acknowledge their author: and if I do not make a handsome acknowledgment, they will reckon me an ungrateful clown.[13]

This statement is particularly striking in the light of Newton's subsequent removal in E_2 of many of the references to Flamsteed that had been present in the original edition.[14] It is also to be noted that Newton is less certain than formerly about either the rapidity or the certainty with which he will 'perfect the moon's theory'.

But this very same letter discloses the fundamental difference between Newton's scientific outlook and Flamsteed's. Newton says to Flamsteed that it will be to Flamsteed's great advantage to publish his observations in conjunction 'with a theory which you ushered into the world, and which by their means has been made exact'. The reason is, he says, that 'such a theory will be a demonstration of

[9] Brewster (1855), vol. 2, pp. 157–8.

[10] [Newton] (1959–), *Correspondence*, vol. 3, p. 164.

[11] But he did send David Gregory to Flamsteed as a kind of emissary, armed with a letter of introduction; Brewster (1855), vol. 2, p. 119; Baily (1835), p. 129.

[12] Baily (1835), p. 133; Brewster (1855), vol. 2, p. 169. Newton had 'paid a visit to the Royal Observatory on the 1st of September 1694. Flamsteed shewed him 150 places of the moon calculated from his own observations, either by himself or "his hired servants", with the differences, in three synopses, between these places and those in the common tables of the moon, "in order to correct the theory of her motions". These observations were given to Newton on two conditions, which he accepted, 1st, that he would not, without Flamsteed's consent, communicate them to anybody, and 2ndly, that he would not, in the first instance, impart the result of what he derived from them to anybody but himself.' Brewster (1855), vol. 2, p. 166.

[13] Baily (1835), p. 151.

[14] These may be found readily from the Index Nominum appended to the text of the present edition.

their exactness, and make you readily acknowledged the exactest observer that has hitherto appeared in the world'. Contrariwise, if Flamsteed were to publish his observations 'without such a theory to recommend them', then they would be only some further raw data to 'be thrown into the heap of the observations of former astronomers', and there they would rest 'till somebody shall arise that, by perfecting the theory of the moon, shall discover your observations to be exacter than the rest'.[15] Newton closed this observation with a warning:

But when that shall be, God knows: I fear not in your life-time, if I should die before it is done. For I find this theory so very intricate, and the theory of gravity so necessary to it, that I am satisfied it will never be perfected but by somebody who understands the theory of gravity as well, or better than I do.

Flamsteed held a diametrically opposite view. He simply could not give assent to the opinion that 'to have the theory of the moon published with my observations, would be a great proof of their accuracy'. Quite the contrary! As he remarked in a comment on Newton's letter, 'theories do not commend observations; but are to be tried by them'. Furthermore, 'theories are then only probable, when they agree with exact and indubitable observations'.[16]

Newton's optimism did not wholly abate. On 23 April 1695 he wrote to Flamsteed that he would 'let the moon's theory alone at present, with a design to set to it again'. When he would have received Flamsteed's 'materials', he boasted, 'I reckon it will prove a work of about three or four months: and when I have done it once I would have done with it forever.'[17] But on 29 June 1695, thanking Flamsteed for 'your solar tables', Newton repeated that 'I want not your calculations, but your observations only'. He went on:

I will therefore once more propose it to you, to send me your naked observations of the moon's right ascensions and meridional altitudes; and leave it to me to get her places calculated from them. If you like this proposal, then pray send me first your observations for the year 1692, and I will get them calculated, and send you a copy of the calculated places. But if you like it not, then I desire you would propose some other practicable method of supplying me with observations; or else let me know plainly that I must be content to lose all the time and pains I have hitherto taken about the moon's theory, and about the table of refractions.

Even under the best of circumstances, these two very difficult men would have had a hard time getting on.[18] When Newton sent Flamsteed his table of atmospheric

[15] In this extract, and the one immediately following, the word 'than' is an incorrect editorial expansion of 'y^n' or an editorial 'correction' from 'then'; Newton almost universally wrote 'then' as the expanded form of 'y^n' and not 'than'. The texts of Newton's letters to Flamsteed, taken from Baily's life, have not been corrected by a comparison with the originals, mostly in the Royal Greenwich Observatory (Herstmonceaux Castle). Volume 4 of Newton's *Correspondence*, containing some of these letters, appeared while this Introduction was in the press.

[16] Baily (1835), p. 152.

[17] Brewster (1855), vol. 2, p. 175; Baily (1835), pp. 154, 157.

[18] And, possibly, as Brewster shrewdly suggested, 'had Halley not been a party, there is reason to believe that no difference would have arisen between Newton and Flamsteed'. Flamsteed charged Halley with 'infidelity and libertinism'; Brewster (1855), vol. 2, p. 165. David Gregory, in his 'Notae' (for which see §§12, 14), recorded a quite different source of Flamsteed's enmity to Halley; it was, to use S. J. Rigaud's phrase, Flamsteed's 'detected act of dishonesty' (*ibid.*); S. J. Rigaud, (1844), *Defence of Halley*. The source is Newton, and

refractions, Flamsteed observed that Newton was 'not pleased to impart the foundations on which you calculated it'; and so Newton hastened to reply that he had not concealed 'as a secret' his method of determining the table, but had merely 'omitted [it] through the haste I was in, when I wrote my last letter'. And so, on 20 December 1694, he set it down in outline.[19] In the summer of 1695 Flamsteed complained in a note on the back of a letter:

> Let the world judge whether Mr. Newton had any cause to complain of want of observations, when all these were imparted to him. I was ill of the headache all the summer, which ended in a fit of the stone: yet I forbore not, as I was able, to serve him without reward, or the prospect of any. I contend it.[20]

He might also have been angered by Newton's imperious tone.[21] But I think the state of Flamsteed's health may have been a conditioning factor of his attitude.

We may conclude this section with an account by Newton, in a letter to Flamsteed of 17 November 1694, of 'my method in determining the moon's motions'. Newton says:

> I have not been about making such corrections as you seem to suppose, but about getting a general notion of all the equations on which her motions depend; and considering how afterwards I shall go to work, with least labor and most exactness, to determine them. For the vulgar way of approaching by degrees is bungling and tedious. The method which I propose to myself, is, first to get a general notion of the equations to be determined, and then by accurate observations to determine them. If I can compass the first part of my design, I do not doubt but to compass the second: and that made me write to you, that I hoped to determine her theory to the exactness of two or three minutes. But I am not yet master of the first work; nor can be, till I have seen something of the moon's motions when her apogee is in the summer signs: and to go about the second work, till I am master of the first, would be injudicious; there being a complication of small equations which can never be determined till one sees the way of distinguishing them, and attributing to each their proper phenomena. Sir, if you can have but a little patience with me till I have satisfied myself about these things, and make the theory fit to be communicated without danger of error, I do intend that you shall be the first man to whom I will communicate it.[22]

---

is therefore, of course, greatly prejudiced. Gregory writes, 'Newton often told me, but especially in December 1698, that these tables [Flamsteed's lunar ones] were first made and computed by Edmund Halley, and communicated to Flamsteed, and published by him without the knowledge of Halley, and that this theft was the origin of the eternal quarrels between Halley and Flamsteed. Newton said that he had seen the handwriting of Halley'; Brewster (1855), vol. 2, pp. 165–6; S. J. Rigaud (1844), *Defence of Halley*, p. 20. This story is to be taken more as a gauge of Newton's feelings concerning Flamsteed than as an historical rendition of what occurred.

[19] Baily (1835), pp. 143, 145–6.

[20] *Ibid.* p. 142; cf. Baily's note on p. 157.

[21] As in the above-quoted line: 'I want not your calculations, but your observations only.' Or, 'I desire that you would send the right ascensions and meridional altitudes of the moon, in your observations of the last six months. You may do it in three columns under these titles—

Tempus apparens Grenovici.	*Lunae ascensio recta observata.*	*Lunae altitudo meridiana apparens.*

And for the trouble you are at in this business, besides the pains you will save of calculating (and that upon an erroneous hypothesis as I must do) the observations you communicate to me, and the satisfaction you will have to see the theory you have ushered into the world brought (as I hope) to competent perfection, and received by astronomers, I do intend to gratify you to your satisfaction: though at present I return you only thanks.' *Ibid.* p. 140.

[22] *Ibid.* pp. 139–40.

Flamsteed's comment on having 'a little patience', written in a note on the letter itself, is typical: 'As much as he pleases: I have waited 5 years for them.'[23]

7. N. FATIO DE DUILLIER AS POSSIBLE EDITOR OF A NEW EDITION

Some time in the 1690s, Nicolas Fatio de Duillier, a Genevese mathematician who had come to England in 1687 and had become a great favourite of Newton's, was ready to bring out a new edition of the *Principia*.[1] Fatio had ambitions to become a major scientist and he was particularly proud of his hypothesis on 'la Cause de la Pesanteur', which he communicated to the Royal Society of London on 27 June 1688, and which at one time or another had won the approbation of Huygens (1691) and of Newton and Halley (19 March 1689/90).[2] Although Newton once held that Fatio had 'found out' the only possible 'Mechanical cause of Gravity', and even went so far as to say so in a projected revision of the *Principia*, he later changed his mind and refused to take this hypothesis seriously (see §§8 and 9).

Fatio interests us primarily because he must have been among the earliest close readers of the *Principia*. By March 1689/90 he had already jotted down quite a few possible improvements to be made in any new edition, and had picked up a considerable number of author's and printers' errors, but he had not by then completed a thorough study of every page of Newton's book.[3] I have not been able to determine either the date or the circumstances of Newton's first meeting with Fatio; Newton said that in 1687 he had given Fatio a copy of the *Principia* for him to send on to Leibniz.[4]

Fatio went to Holland in April 1690, taking with him a list of errata and addenda to the *Principia*, of which he left a copy with Huygens. Writing of Newton to Huygens in September 1691, Fatio asked him to return 'les Errata de son livre [that is, le livre de Newton] que je Vous ai laissez et desquels malheureusement je n'ai point gardé de Copie'.[5] This list had a curious subsequent history, to which reference has already been made,[6] eventually being published by Groening in 1701 as if partly the work of Huygens.

[23] *Ibid.* p. 140. In the first edition of the *Principia*, Flamsteed not only appeared among the observers of Jupiter's satellites, and of the Moon, but was mentioned explicitly for his outstanding observations of comets, the latter being one of the references removed in the second edition; see the Apparatus Criticus {496, tab.}. A manuscript of Newton's containing a reference to Flamsteed's cometary observations may be found in [Newton] (1959–), *Correspondence*, vol. 4, p. 167; see also pp. 169 ff.

[1] See Fatio (1949), *La cause*; Bopp (1929).

[2] Printed in Bopp (1929) and in Fatio (1949), *La cause*.

[3] See S. P. Rigaud (1838), *Essay*, p. 93.

[4] Edleston (1850), p. 309, Newton's 'Ex epistola cujusdam ad amicum'. Fatio had written to Huygens from London on 24 June 1687 describing Newton's book 'qui s'imprime presentement et qui se debitera dans trois semaines d'ici'. Huygens replied on 11 July, 'Je souhaitte de voir le livre de Newton', and concluded, 'Ayons le livre de Newton'; [Huygens] (1888–), *Oeuvres*, vol. 9, pp. 168, 190. It was in this letter that Huygens made his oft-quoted comment that he hoped Newton's work would not be another theory of attraction!

[5] *Ibid.* vol. 10, p. 146.

[6] See also §10.

Our information concerning Fatio's proposed edition of the *Principia* is derived chiefly from his correspondence with Christiaan Huygens and from some memoranda of David Gregory's. We have seen (§ 6) that by 1694 Newton had openly committed himself to produce a new edition.[7] But in December 1691 he had not yet decided to do so, or at least to embark full-scale on a revised printing or new edition without finding someone to act as editor. For in that month Fatio wrote to Huygens that it was useless to ask Newton to undertake a new edition of his *Principia*: 'Je l'ai importuné plusieurs fois sur ce sujet sans l'avoir jamais pu flechir.'[8] But Fatio hastened to add that it was not impossible that he himself would undertake to prepare such an edition. Newton had been busy revising his *Principia* and correcting the first printing, but he may have felt that yet more alterations and additions were needed before the work would be ready to be turned over to a printer.

Fatio was confident of his qualifications to edit the new edition and to see it through the press. In any event, as he assured Huygens, he could easily go down to Cambridge to seek Newton's help on any point that he could not master by himself. He planned a lengthy addition of his own, or a commentary, which—despite its length—could probably be read more quickly than Newton's book. The task, he supposed, would probably take him two or three years, and he would seek subscriptions in advance for a certain number of large-paper copies.[9] Fatio concluded with the observation that the list of errata was becoming ever greater in proportion with his progress in mastering and reading the *Principia*. Huygens's reply contained his opinion that Newton should be pleased by the prospect of the new edition, but he wondered why the English printers would require an advance subscription.[10] In his comments on the margin he jotted down the wish that Fatio should not add

[7] In a letter to Newton of 24 September 1694, David Gregory wrote: 'I am glade to understand that you have been at London, and that your design of the new edition of your book is going on and in forwardness.' [Newton] (1959–), *Correspondence*, vol. 4, p. 20.

[8] [Huygens] (1888–), *Oeuvres*, vol. 10, p. 213.

[9] The part of this letter giving the details of the proposed new edition reads as follows: 'Il est assez inutile de prier Monsieur Newton de faire une nouvelle édition de son livre. Je l'ai importuné plusieurs fois sur ce sujet sans l'avoir jamais pu flechir. Mais il n'est pas impossible que j'entreprenne cette édition; à quoi je me sens d'autant plus porté que je ne croi pas qu'il y ait persone qui entende à fonds une si grande partie de ce livre que moi, graces aux peines que j'ai prises et au temps que j'ai emploié pour en surmonter l'obscurité. D'ailleurs je pourrois facilement aller faire un tour a Cambridge et recevoir de Mr. Newton même l'explication de ce que je n'ai point entendu. Mais la longueur de cet ouvrage m'epouvante, puis que par les differentes choses que j'y voudrois ajouter il feroit un folio assez raisonnable. Ce folio neanmoins se liroit et s'entendroit en beaucoup moins de temps que l'on ne peut lire ou entendre le quarto de Mr. Newton. Voila un dessein Monsieur capable de m'occuper pendant deux ou trois années: et je ne voi point trop comment le reconcilier avec l'état de ma fortune, à moins que je ne me puisse resoudre à rechercher qu'un assez bon nombre de persones s'accordent à faire des souscriptions, comme on les pratique ici, pour s'assurer des exemplaires en papier roial, et cela à un prix qui puisse me mettre l'esprit en repos.'

[10] 'Monsieur Newton serait bien heureux si vous vouliez entreprendre cette seconde Edition de son ouvrage, qui serait tout autre chose avec vos eclaircissements et additions, qu'il n'est a present. Mais il ne faudroit pas que ce travail nuisit a vostre santè. Pour la depense, cela me paroit etrange qu'en ce païs la il n'y a pas d'imprimeurs qui veuillent hasarder a leur frais l'impression des livres de cette importance. On en trouveroit assurement icy. Cette maniere de souscriptions n'est pas aisée dans des ouvrages qui se doivent debiter par toute l'Europe; car l'Angleterre avec ce païs icy n'en fourniroit pas assez.' Huygens to Fatio, 5 February 1692 (N.S.); [Huygens] (1888–), *Oeuvres*, vol. 10, p. 241.

too much of his own and expressed the opinion that '200 Exemplaires suffiront'.[11] News of the proposed edition spread rapidly. Leibniz, for one, thought Newton was fortunate that Fatio was to prepare the new edition. It would thus appear that by 1691, four years after publication, the first edition had gone out of print and that scholars were hard put to find copies.[12]

8. FATIO'S PROPOSED EDITION AND COMMENTARY

I have not been able to find any extensive notes of Fatio's[1] that would enable us to know just what revisions he planned to make of the *Principia*, or even what his commentary would contain. His copies of the first and second editions of the *Principia* were purchased for the Bodleian Library in 1755 from an Oxford bookseller and have been described by Rigaud.[2] The copy of the first edition was presented to Fatio by Halley, not by Newton, according to Fatio's inscription—'N. Fatius Duillierius; ex dono Cl. Hallaei Societati regiae a Secretis. Londini anno 1687'—which says, furthermore:

> Many of those [things], which in this copy I had noted down, Newton inserted into the second and third edition; many in fact he neglected. Furthermore, some things—on the third page and elsewhere perhaps—are written in his hand. I wanted also to add at the end of this volume the corrigenda from Newton's own codex, 13 March 1689/90; but being occupied with other affairs, and shortly about to set out for Holland, I neglected to do it.[3]

There exists independent evidence to support Fatio's statement to Huygens on 18/28 December 1691 concerning the proposed edition that he thought he would

[11] *Ibid.* p. 215. Presumably, although it is not transparently clear, Huygens's figure of 200 refers to the number of advance subscriptions, not to the size of the whole edition.

[12] The Journal Book of the Royal Society of London mentions a letter from Leibniz, which was read on 31 October 1694, in which Leibniz 'recommends to the Society to use their endeavours to induce Mr Newton to publish his further thoughts and emprovements on the subject of his late book Principia Philosophiae Mathematica, and his other Physicall & Mathematicall discoverys, least by his death they should happen to be lost.' See [Newton] (1959–), *Correspondence*, vol. 4, p. 24.

[1] The two major collections of Fatio's manuscripts are in the Library of the University of Geneva and in the Bodleian Library, Oxford.

[2] Since the copy of the third edition contains a note saying that Newton had given it to Fatio in 1713 ('Exemplar hocce Nicolao Facio Duillierio dono dedit clarissimus autor anno 1713'), Rigaud concluded 'that he was also in possession of the second edition, and had by mistake written the memorandum in the wrong volume: nothing, however, is known of what may have become of it.' Rigaud (1838), *Essay*, pp. 89–90.

[3] This extract is translated from the Latin; the latter is printed in Rigaud (1838), *Essay*, p. 90. Rigaud discussed what Fatio may have meant by 'codex'. He was referring either to manuscript additions occurring in one of Newton's own copies of E_1 or to some manuscript corrigenda on loose sheets. Since those emendations that Fatio did copy out before sailing for Holland were taken from the MS Errata entered by Newton on the end sheets (containing the single page of printed errata) of E_1a, perhaps he was stating his wish to have been able to copy out the remaining corrigenda dispersed throughout that copy. I have found no evidence that he ever had access to E_1i. But he certainly did see and transcribe some proposed emendations that do not appear in either E_1i or E_1a and that he must have seen in loose manuscript sheets of Newton's. (These contain a statement of high praise for Fatio's hypothesis on the cause of gravity; see §10.) There are reasons to hypothesize the existence of yet another personal copy of E_1 of Newton's, containing manuscript additions, of which we have no knowledge today. As to the word 'codex' itself, David Gregory used it at least once in reference to the *Principia* as a whole, and thus—as Rigaud correctly assumed—not 'in the technical sense of a manuscript'. See p. 302 of Gregory's 'Notae' (in the Edinburgh copy).

be allowed to supervise. For on 28 December 1691 (presumably before he had met Newton) David Gregory, in London, noted that

> M[r] Fatio designs a new edition of M[r] Newtons book in folio wherin among a great many notes and elucidations, in the preface he will explain gravity acting as M[r] Newton shews it doth, from the rectilinear motion of particles the agregate all which is but a given quantity of matter dispersed in a given space. he says that he hath satisfied Mr Newton, Mr Hugens & Mr Hally in it.[4]

I do not know when Fatio gave up the idea of bringing out the new edition, or was discouraged from doing so by Newton. In fact, on 30 November 1693 Huygens wrote to Fatio raising the question whether he had 'encore dessein de contribuer à une seconde Edition du livre de Monsr Newton'.[5] Perhaps Huygens had already (and quite properly) recognized that Fatio was never going to undertake the task. Actually, Fatio had written to him in the spring of 1692 about his difficulties in being able to understand the whole of the *Principia*, admitting that he had worked through only the first five sections and the ninth (of Book I, presumably) and the 'Treatise on Comets' (apparently the final part of Book III), and concluding: 'Si j'étois prêt à imprimer le livre de Mr Newton je ne m'éloignerois peut etre pas trop de m'attacher à la ville d'Amsterdam pendant quelque temps.'[6]

I have no doubt that Fatio would have been able to do a conscientious job as editor. Of course, he did not have the mathematical ability of Cotes and consequently could not have given Newton either the kind of real help or the intellectual stimulation that Cotes was to give in the years 1708–13 (see Chapter IX). But, after all, there was only one Cotes! Fatio for a time enjoyed an especially close relation with Newton; its rupture evidently had severe repercussions. Most likely, Newton's change of mind with regard to Fatio's hypothesis on the cause of gravity, recorded in Gregory's later paragraph, is a reflection of Newton's change of heart with regard to Fatio himself. Perhaps it was also this change in relations that caused

[4] R. S. Greg. MS 70, 71. In a very different hand, and so presumably at a somewhat later date, Gregory added this comment in a bottom corner of the page: 'M[r] Newton and M[r] Hally laugh at M[r] Fatios manner of explaining gravity.' See [Newton] (1959–), *Correspondence*, vol. 3, p. 191.

[5] [Huygens] (1888–), *Oeuvres*, vol. 10, p. 567.

[6] *Ibid.* vol. 22, pp. 158–9. In this letter to Huygens, dated London, 29 April (=9 May N.S.) 1692, Fatio had written: 'De temps en temps ma paresse ou d'autres Etudes interrompent considerablement mon application aux Mathematiques. Mais je n'y reviens point aprez mes égaremens que je ne sente redoubler mon envie d'imprimer le livre de Mr Newton. La raison en est peutetre que dans ce premier feu je surmonte ordinairement quelques-unes des plus grandes difficultez qui se trouvent dans ce livre: ce qui me fait croire que je l'entendrai parfaitement aprez y avoir deja donné tout le temps necessaire [?] je n'aurois plus besoin que de consulter l'Auteur sur 20 passages au plus. Mais quoi que j'y [aye] perdu bien du temps il y a du plaisir neanmoins d'étudier un livre qui laisse beaucoup à trouver avant qu'on ne puisse l'entendre. Or je ne conte pas que j'entende ce que je n'ai pas parfaitement compris et dont je n'aie pas saisi toute l'evidence. Les Traittez entiers que j'ai étudiez à fonds ne sont encore que les 5 premieres Sections, la neuvieme, et le Traitté des Cometes. Ailleurs je n'ai étudié à fonds que quelques propositions par ci par là. Dans ces Traittez il n'y a rien ou presque rien qui m'ait échappé et dont je ne pusse donner la demonstration dans le besoin. Par exemple la 9e Section qui est peutetre la plus difficile de tout le livre étant corrigée comme je l'ay fait en mon exemplaire dans les endroits ou j'ai trouvé qu'il devoit l'etre ne contient absolument rien ni ne suppose rien que je ne puisse parfaitement demontrer. Si de votre côté M[r]. vous entrepreniez quelqu'une des autres Sections il ne seroit pas difficile de venir bien tot à bout de tout le livre. Et nous pourrions nous rendre conte l'un à l'autre des difficultez que nous aurions rencontrées et nous faciliter reciproquement l'etude d'un livre qui est assurement fort excellent mais en mesme temps fort obscur.'

Newton to abandon any possible intention of allowing Fatio to be the editor of a new edition. In any case, the loss in not having the *Principia* available in print during about two decades from 1694 to 1713 was eventually more than compensated for by the quality of the contributions made to the second edition by Roger Cotes.

9. FATIO'S IMPROVEMENTS

Some of the improvements that Fatio would have introduced into his edition of the *Principia* may be ascertained from the handwritten alterations in his copy of E_1. Evidently he planned to have running heads and postils, of which there are none whatever in the first edition. In the second edition, each verso page has the running head PHILOSOPHIAE NATURALIS, which, with the running head of the next recto page, PRINCIPIA MATHEMATICA, gives the title in full. Furthermore, following the front matter, each page has a postil in the margin: DEFINITIONES, AXIOMATA SIVE LEGES MOTUS (divided so that the first two words appear on the verso and the last two on the recto page), LIBER PRIMUS, LIBER SECUNDUS, LIBER TERTIUS. In the third edition, this was improved somewhat by having the postil on the verso pages in Books I and II read DE MOTU CORPORUM and in Book III DE MUNDI SYSTEMATE. But Fatio's proposed heads would have made the book far easier to use. For in his system, no matter to what page the book would be opened, the reader would see at once whether the text belonged to the *Definitiones*, the *Leges Motus*, or *Lib. I*, *Lib. II*, or *Lib. III*. Furthermore, in Books I and II, there was to be additionally the section number and an abbreviated title of the section, as: *Lib. I. Sect. I. De Methodo rationum primarum*,[1] or *Lib. I. Sect. XIV. De Motu Corpusculorum a magno Corpore attractorum*.[2] In Book III, Newton had made no subdivisions, so that here Fatio had to invent his own. Thus, the pages of Book III would reflect Fatio's analysis by topics as follows: *Lib. III. De Systemate Mundi. Hypotheses*; *Lib. III. De Systemate Mundi. De Gravitate in Jovem, Solem & Terram*. . . ;[3] *Lib. III. De Systemate Mundi. De Corrigendo Cometae Orbe*.

Fatio was also responsible for a neat improvement—the introduction of the letters L and T in Prop. III of Book I, as obvious abbreviations of *Luna* and *Terra*

[1] This abbreviation might have been better had Fatio also included *& ultimarum*, though quite properly omitting *cujus ope sequentia demonstrantur*.

[2] Abbreviated from *Sect. XIV. De motu corporum minimorum, quae viribus centripetis ad singulas magni alicujus corporis partes tendentibus agitantur*.

[3] The remaining subdivisions are: *De Gravitate in Terram Jovem & Solem*; *De Gravitate in Planetas singulos*; *De Gravitate in Corpora universa*; *De Gravitate Globorum duorum in se mutuo*; *Deque Gravitate in Planetas*; *De Densitatibus Planetarum*; *De Resistentia Motui Planetarum*; *Deque eorum communi gravitatis Centro*; *De Planetarum motu in Ellipsibus*; *De Planetarum Excentricitatibus, Apheliis, Motibus diurnis, et figura non sphaerica*; *De Planetarum Figura non Sphaerica*; *De non aequali ubique gravitate in ipsa Superficie Terrae*; *De Motibus Lunae*; *De Motu Satellitum*: *deque Maris Fluxu et Refluxu*; *De Maris Fluxu & Refluxu*; *De Motibus Lunae*; *De Vi Solis ad Mare movendum*; *De Vi Lunae Ad Mare movendum*; *De Corporis Lunae Figura*; *De Actione Solis in Terram non Sphaericam*; *De Praecessione Aequinoctiorum*; *De Cometarum a Terra distantia*; *De Cometarum Orbibus*; *De Inventione loci Cometae ad quodvis tempus intermedium ex observatis aliquot Cometae locis*; *De Cometarum Motu in Orbibus Parabolicis*; *De Inventione Orbis Cometae*; *Observationes Caudae Cometae*; *De Natura Cometarum*.

and thus showing that the application of this Prop. III was to the Earth–Moon system. This improvement, adopted by Newton, brought this proposition into harmony with its neighbours, in which the centre of force had been designated by *S*, in obvious suggestion of *Sol*, and the body on which the force acts had been designated by *P* for *Planeta*.[4]

At Prop. XXXVII, Book II, replaced in E_2 and E_3 in its entirety by the new Prop. XXXVI, on the problem of defining the motion of water flowing out of a vessel through a hole, Fatio wrote, 'Tota haec demonstratio fallax est ejusque conclusio falsa' ('This whole demonstration is fallacious and its conclusion false'). Fatio evidently had difficulty in convincing Newton that he had made an error in stating that the speed of the effluent water would be as half the height *A*; he should have said the whole height *A*, not half. In the margin of his copy of E_1, Fatio wrote:

Newtonum ab erroribus in hac propositione contentis nullatenus liberare potui, quam per Experimentum; constructo scilicet Vase ad hunc ipsum usum destinato.[5]	I could by no means free Newton from the errors contained in this proposition, save by an Experiment; namely after a vessel had been constructed, intended for this very purpose.

In the event, as Fatio recorded, Newton cancelled the whole demonstration and substituted another. Finally, he cast out both the original Prop. XXXVI and its demonstration,[6] and substituted a different Prop. XXXVII on a similar subject, along with the other new propositions that make up most of Sec. VII in E_2 and E_3.

One interesting suggestion is to be found on page 42 of Fatio's copy, a proposed new corollary to Prop. IV, Book I, on the rule for finding the centripetal acceleration and force in uniform circular motion. Fatio's marginal note reads as follows:

Si sint T, t; tempora period: R, r; radii circulorum; erunt $\frac{R}{T}, \frac{r}{t}$; ut velocitates. $\frac{R}{T^2}, \frac{r}{t^2}$; ut vires centrip. Si sit praeterea

$$T^n \cdot t^n :: R^m \cdot r^m$$

erunt $R^{1-\frac{m}{n}}$; $r^{1-\frac{m}{n}}$ ut veloc. $R^{1-\frac{2m}{n}}$; $r^{1-\frac{2m}{n}}$ ut vires Centr.

Itemque erunt $T^{\frac{n}{m}-1}$; $t^{\frac{n}{m}-1}$; ut veloc. $T^{\frac{n}{m}-2}$; $t^{\frac{n}{m}-2}$; ut vires centr. Sunt autem n et m indices potestatum &c.[7]

[4] The English translation kept the letters *L* and *T*, rather than transforming them into *M* and *E*.

[5] Fatio also wrote on a slip of paper, now bound in his copy: 'Newtonum nostrum ab hoc errore vix liberare potui, idque facto demum experimento ope vasis quod conficiendum curavi' ('I could scarcely free our friend Newton from this error, and that only in the end, after making an experiment with the help of a vessel which I arranged to have constructed').

[6] See the Apparatus Criticus {356}, and also Appendix I to the text, note (3) to Prop. XXXVII, Prob. IX of E_1.

[7] A variant form of this proposed addition appears in the set of suggestions included with the list of Newton's corrigenda given by Fatio to Huygens in 1691 (see §8). This version reads: 'Et in universum si sunt T, t tempora periodica; R, r radii circulorum; erunt $\frac{R}{T}, \frac{r}{t}$ ut velocitates; $\frac{R}{Tq}, \frac{r}{tq}$ ut vires centripetae. Si erit praeterea T^n ad t^n ut R^m ad r^m, erunt $R^{1-m/n}$, $r^{1-m/n}$ ut velocitates. $R^{1-2m/n}$, $r^{1-2m/n}$, ut vires centripetae, itemque erunt $T^{n/m-1}$, $t^{n/m-1}$, ut velocitates, $T^{n/m-2}$, $t^{n/m-2}$, ut vires centripetae.' [Huygens] (1888–), *Oeuvres*, vol. 10, p. 149.

That is, Newton had shown (Prop. IV) that in uniform circular motion the centripetal forces are as the squares of the arcs described in equal times divided by the radii of the circles, and hence (Corol. 1) as the squares of the velocities divided by the radii, or (Corol. 2) as the radii divided by the periodic times. Then Newton goes on to explore three conditions: that (Corol. 4) the periodic times be as the square roots of the radii, that (Corol. 5) the periodic times be as the radii, and that (Corol. 6) the periodic times be as the $\frac{3}{2}$ power of the radii. The centripetal forces, accordingly, will be: 'equal one to another' (Corol. 4), or 'inversely as the radii' (Corol. 5), or 'as the square roots of the radii' (Corol. 6), the latter an error in E_1 (corrected in E_2) for 'inversely as the square roots of the radii'. Fatio would state these proportions as letters, T, t being any pair of periodic times corresponding to a pair of radii R, r. Then (since the velocities will be $2\pi R/T$, $2\pi r/t$) the velocities will be as R/T, r/t. According to Prop. IV, then, the centripetal forces will be as R/T^2, r/t^2. The conditions of the above-mentioned three corollaries may be stated by us as follows:[8]

$$T:t::\sqrt{R}:\sqrt{r},\quad T^2:t^2::R:r; \qquad \text{(Corol. 4)}$$
$$T:t::R:r,\quad T^2:t^2::R^2:r^2; \qquad \text{(Corol. 5)}$$
$$T:t::R^{\frac{3}{2}}:r^{\frac{3}{2}},\quad T^2:t^2::R^3:r^3. \qquad \text{(Corol. 6)}$$

This suggests a generalization to:

$$T^2:t^2::R^m:r^n$$

and in fact Fatio proposed including all the results under the most general condition of all, namely, one we may express as:

$$T^n:t^n::R^m:r^m.$$

It follows that the ratio of the speeds (radius÷periodic time) will be as $R^{1-m/n}$: $r^{1-m/n}$ because $T:t::R^{m/n}:r^{m/n}$ and hence the ratio of the forces (proportional to $(\text{speed})^2 \div \text{radius}$) will be as $R^{1-2m/n}:r^{1-2m/n}$ (since $(R^{1-m/n})^2/R = R^{1-2m/n}$). Similarly, the speeds prove to be as $T^{n/m-1}:t^{n/m-1}$ and the forces as $T^{n/m-2}:t^{n/m-2}$. Newton did not accept Fatio's suggestion entirely, but in E_2 he did add a Corol. 7 in which he stated the condition that, in general, 'if the periodic time be as any power R^n of the radius R', so that the velocity will be inversely proportional to R^{n-1}, the centripetal force will be inversely as R^{2n-1}. This is Fatio's result, save that instead of the condition

$$T^n:t^n::R^m:r^m$$

Newton has used

$$T:t::R^n:r^n,$$

which is equivalent to Fatio's if Newton's n is taken equal to Fatio's m/n. Thus the forces, which under Fatio's condition were as $R^{1-m/n}:r^{1-m/n}$, become under Newton's condition as $R^{1-n}:r^{1-n}$ or as $r^{n-1}:R^{n-1}$ or, generally, inversely as R^{n-1}.

[8] In presenting Fatio's arguments and Newton's response, I have used the colon and double colon for expressions of proportion, even though they do not appear in Newton's or Fatio's annotations or in the *Principia* (although they are sometimes used by Newton's contemporaries, e.g. David Gregory); and I have enclosed in parentheses the anachronistic phrase involving the factor 2π.

Incidentally, this example proves the justice of Fatio's remark on the fly-leaf of his copy of E_1 that in E_2 Newton had availed himself of some of Fatio's suggestions.

10. FATIO AS RECORDER AND TRANSMITTER OF NEWTON'S ALTERATIONS: GROENING'S 'HISTORIA CYCLOEIDIS'

In the history of the *Principia*, our interest in Fatio does not derive particularly from any positive contributions he may have made to the ultimate revision, by suggesting improvements to Newton. Rather, we value his role as a recorder of the growth of Newton's views, particularly during the early and middle 1690s. In two major documents of his, we may not only see stages of Newton's revision, but date them. One of these is described as 'Extrait d'une Copie faite par l'Auteur en 8bre 1692 des Corrections et des Additions que Mr. Newton destinoit alors à son Liure des Principes Mathematiques de la Philosophie Naturelle'. These notes are to be found in the library of the University of Basel,[1] along with a copy of Fatio's essay, 'De la Cause de la Pesanteur', both in the hand of the same amanuensis.[2] From this transcript we learn that in October 1692 Newton had abandoned the old Hypoth. III at the beginning of Book III, on the transformation of any kind of matter into matter of any other kind, taking on every intermediate degree of quality, and had replaced it by a new Hypoth. III, similar to the eventual Regula III of E_2 and E_3. There was also a new Hypoth. IV, which Newton never used in E_2 or in E_3.[3] In addition, Fatio had copied out a lengthy supplement planned by Newton to be added to Corol. 4 on page 411 of E_1; that is, to the Corollary to Prop. VI, Book III, which in E_1 was denoted by Corol. 3. In this extraordinary corollary, Newton discusses matter and gravity, and states unequivocally that the only hypothesis by which gravity can be explained mechanically is the one invented by that ingenious geometer Fatio ('eamque Geometra ingeniosissimus D. N. Fatio primus excogitavit').[4] There is also a new Corol. 5 dealing with the vacuum, pores in matter, and so on. In E_2 {402.27–403.2} Newton introduced a very different pair of corollaries (3 and 4) covering these topics, and the old Corol. 4 of E_1 then became (with some changes) Corol. 5.

In a private memorandum (in English) Fatio recorded that he had communicated his views on the 'cause de la pesanteur' to various scientists, and went on:

[1] On deposit from the Library in Schloss Friedenstein in Gotha; they were first described in Bopp (1929).

[2] The library of the University of Geneva has two further versions of this essay in Fatio's hand, with many alterations.

[3] This one is: 'Pro novis rerum generibus habenda non sunt quae cum aliis rerum generibus, quoad omnes suas notas qualitates congruunt'; that is, 'Things which agree as to all their known qualities with other kinds of things are not to be considered as new kinds of things.' This Hypoth. IV also appears on a manuscript sheet on which Newton has written out both the original Hypoth. III of E_1 and the new Hypoth. III, the forerunner of Regula III. See Cohen (1971), *Scientific ideas*, chap. 2, 'From Hypotheses to Rules'.

[4] See Hall and Hall (1962), *Unpublished papers*, pp. 313, 315; Bopp (1929), p. 22.

Sir Isaac Newton's Testimony is of the greatest weight of any. It is contained in some Additions written by himself at the End of his own printed Copy of the first Edition of his Principles, while he was preparing it for a second Edition, And he gave me leave to transcribe that Testimony. There he did not scruple to say *That there is but one possible Mechanical cause of Gravity, to wit that which I had found out*: Thô he would often seem to incline to think that Gravity had its Foundation only in the arbitrary Will of God.[5]

Evidently, if we are to believe Gregory's memorandum (quoted in §8), Newton had undergone a considerable change of mind since writing out the corollary in which Fatio received such high praise. It may be added that all the aspects of the relations between Fatio and Newton are puzzling, and are perhaps grist for the mill of the psychoanalytically oriented scholar rather than the chronicler of the *Principia*.

Fatio's reference to manuscript additions 'at the End' of one of Newton's copies of the *Principia* suggests E_1a, where there are extensive notes on the page containing the printed Errata and the final otherwise blank pages. But in those annotations there is no mention either of Fatio or of his hypothesis. In E_1i there are no such additions in hand at the 'end' of the book. Hence there may have been yet another copy of the *Principia* in which Newton had been entering manuscript corrections, alterations, and additions.[6] It is also possible that, after Newton broke with Fatio, he tore out the page (most likely in E_1a) containing his praise of Fatio. But there can be no doubt of the accuracy of Fatio's report on Newton's warm sentiments concerning Fatio's hypothesis, since there is, among Newton's manuscripts, a draft of the statement praising Fatio in the text of the proposed Corol. 4 to Prop. VI, Book III.[7]

Another possibility is that Fatio had had a lapse of memory and that, writing his memorandum some ten years or more later,[8] he had confused in his own mind the corrigenda that he had transcribed directly from the MS Errata to E_1a and those that he had taken from the separate pages of Newton's manuscript. Such a confusion would be all the more understandable since both this latter set of extracts and the material copied out from the MS Errata to E_1a contain a version of the new Hypoth. III.[9]

[5] Fatio (1949), *La cause*, p. 117; see also Bopp (1929), p. 22.

[6] We shall see in Chapter VIII, §4, that there are a number of other indications of the one-time existence of at least one other copy of the *Principia* (in addition to E_1a and E_1i) containing manuscript alterations by Newton.

[7] This may be found in U.L.C. MS Add. 3965, f. 311; see also Hall and Hall (1962), *Unpublished papers*, pp. 312 ff. This topic is discussed further in Cohen (1971*b*), *Newton's philosophy*. Perhaps Fatio was referring to one or two loose pages at the end of Newton's copy of E_1, which were later removed.

[8] At least after 1701, since this date is mentioned as the time when he had 'communicated the Original itself with several Additions to Mr James Bernoulli Professor of Mathematics at Basil whose letters I have also'. Bopp (1929), p. 20. The date must, indeed, be later than 1705, when Newton was knighted.

[9] In the MS Errata to E_1a this new Hypoth. III begins, 'Leges et proprietates corporum...', as in Fatio's list (and hence as printed in 1701 by Groening). But in the version copied out from the separate manuscript pages of corrigenda, this Hypoth. III begins 'Qualitates corporum...'. (The Basel copy, in the hand of an amanuensis, actually reads 'Quantitates corporum...'—obviously a scribal error.) In Newton's manuscript pages, Hypoth. III (beginning with the word 'Qualitates') occurs together with Newton's proposed new Corol. 4 to Prop. VI (Book III) praising Fatio's hypothesis for explaining gravity mechanically. The presence

The second document had a most interesting subsequent history, since Huygens's copy of it, which, as we have seen, eventually ended up among Leibniz's manuscripts in Hanover, was printed (incompletely and imperfectly) in 1701 in J. Groening's *Historia cycloeidis*. Groening published a set of 'Annotata in Newtonii Philos. Nat. Principia Mathematica', which is a jumble of: two manuscripts sent by Newton to Huygens (in Newton's hand with comments written by Huygens), Huygens's copy (with comments in the margin) of Fatio's transcripts of some of Newton's emendations from the MS Errata to E_1a, and Huygens's copy of Fatio's own conjectures, on which Newton had written comments and additions before Huygens copied the whole thing. Furthermore, although these 'annotata' derive from both Huygens and Fatio, neither name is given in the heading. From the title-page of Groening's book the reader would probably conclude that all these 'annotata' were by Huygens.

The confusion about these 'annotata' begins in the first paragraph. They are said to be 'Annotata in Newtonii Philos. Nat. Principia Mathematica', but the first comment, 'Ad Propos. 2. Neuton', is completely misleading since it does not contain Huygens's remarks on a Prop. II of the *Principia* at all, but rather a Prop. 2 in a manuscript that Newton had sent to Huygens in 1689, on the motion of bodies in a medium with a resistance proportional to the square of the velocity.[10] And, indeed, the contents of this very manuscript are printed following Huygens's comments, with no typographical or editorial separation, as if the two were a single document. Then, as if it were a continuation of what precedes it, there is printed another manuscript of Newton's, containing an additional Proposition 2. This, in turn, is followed by a short comment of Huygens's (this time on the *Principia*), 'Ad propos. 37. Lib. 2. pag. 33.' (the latter an error for 'pag. 330'). Finally, Groening printed some of the material copied out by Fatio from the MS Errata to E_1a, as if it too were part of the single document. The latter begins with Newton's directions with respect to Prop. XXXVII, Book II ('dele omnia'), and the new statement that was to replace it.[11] But, of course, in E_2 a wholly different

of the two forms of Hypoth. III shows us that Fatio had copied out both versions, one from the MS Errata to E_1a and one from the separate manuscript pages. Later on, Fatio would have had the evidence before him that the new Corol. 4 had been written in Newton's hand together with a version of the new Hypoth. III, but he could easily have forgotten which of the two versions (the one with 'Qualitates' and the new Corol. 4 or the one with 'Leges et proprietates' and without the new Corol. 4) had been found in the MS Errata to E_1a and which in the manuscript pages of emendations.

[10] Groening (1701), pp. 105–16; see [Huygens] (1888–), *Oeuvres*, vol. 9, pp. 321–7.

[11] This statement is to be found in Appendix I to the Apparatus Criticus at note 3 to Prop. XXXVII of E_1, taken from the MS Errata to E_1a. The beginning may be translated as follows: 'pages 330–1. Cancel the whole and read [instead]: If a vessel is filled with water and pierced at the bottom so that the water flows out by the hole, it is obvious that the outflowing water will be equal to the water [Huygens, on perusing Fatio's copy, here underlined the words *manifestum est quod aqua defluens aequalis erit aquae*, and wrote in the margin *non apparet mihi quidem* ("it is not at all apparent to me"); see *Oeuvres*, vol. 10, p. 154] that falls in air in the same time in a round column, whose width at the foot is the same as that of the hole, and whose height is the same as that of the water standing in the vessel: and so the speed with which the water issues from the hole is the same as if it had fallen from the top of the water in the vessel, and consequently if that motion is turned upwards, the issuing water will rise to the height of the water in the vessel.'

Prop. XXXVII was introduced, when Newton almost completely rewrote Sec. VII of Book II.

Following the revised Prop. XXXVII, Groening prints from the MS Errata to E_1a: 'p. 402. l. 10. Hypoth. III. Leges & proprietates corporum omnium...' with the discussion ending '...in infinitum dividi possent', followed by 'p. 402. l. 23. Astronomi' and a comment on the final corollary to Prop. VI, Book III, on the magnetic force:

p. 411. l. 22. Haec fallacibus nituntur observationibus...ut ipse expertus sum. Plures afferre possem observationes, quibus constat vim Magnetis juxta superficiem non multo minorem esse quam in ipsa superficie...Sed & minora ferri frustula in magnetem admotum [*printed* ad motum] & adhuc a se distantem prosiliunt, cum tamen a [*printed* in] magnete vi mediocri adhibita... separari possint.

The remainder of the text published by Groening is entitled '*Conjecturae de Sphalmatis Typographicis* In Cl. Nevvtonii Philosoph. Natur. Principia Mathem.', which gives the impression that they had been written by someone other than Newton himself.

In the original manuscript copy, however, there are two parts, of which the first has a title reading (in Huygens's hand):

Conjecturae de sphalmatis typographicis in Cl. Newtoni Philosophiae Principiis Mathematicis.

A note in the margin reads:

Quae inducta sunt Newtonus deleverat in M.S. D. Fatii ex quo haec descripta sunt, quippe frustra aut sine causa reprehensa.

This may be translated as follows:

Conjectures concerning typographical slips in Newton's Mathematical Principles of Philosophy. Newton had cancelled the things which are crossed out in the MS of Mr Fatio from which these things are copied, on the grounds that they were criticized in vain or without reason.

This list thus contains Fatio's proposed emendations, some of which were rejected by Newton, and are accordingly crossed out in Huygens's copy. A final paragraph contains some emendations 'Ex Newtoni Codice'.

Then a second list is entitled 'Alia Errata ex Newtoni mei codice. Londini. 13 Mart. die 1689/90'. This is not a mere assemblage of errata noted by Fatio, but contains some of Newton's own emendations and additions, for example, the new Hypoth. III. The entries are given in this order: p. 3, l. 14 {4.19–20}—p. 231, l. 14; p. 105, l. 20—p. 490, l. 4; p. 488; p. 23, l. ult.; pp. 37, 39, 38, 40, 36, 39; p. 1, l. 9—p. 410, l. 33 {402.14–16}. Every item on this list is taken from the MS Errata to E_1a, which enables us to assign a terminal date to those improvements. Groening, however, combined into a single sequence both Fatio's own suggestions and Newton's alterations.[12]

[12] This list, as printed by Groening, has many mistakes in it. Huygens's original copy, now among the Leibniz manuscripts in Hanover, was published in vol. 10 of Huygens's *Oeuvres* (pp. 147–55), but the editors inadvertently omitted two whole manuscript lines and there are yet other omissions as well as errors in transcription.

II. DAVID GREGORY AND THE 'PRINCIPIA'

A second aspirant to the editorship of a new edition of the *Principia*[1] was David Gregory, the Scots astronomer and mathematician, nephew of James Gregory.[2] David Gregory was appointed professor of mathematics at Edinburgh in 1683 at the age of twenty-two and elected Savilian Professor of Astronomy at Oxford in 1691.

Gregory had come into contact with Newton by 1691 when Newton felt sufficiently sure of Gregory's ability to recommend him for the post of Savilian Professor.[3] Gregory became one of those to whom Newton confided his most intimate thoughts about scientific questions and the mystical knowledge of the ancients.[4] In Gregory's *Astronomiae physicae & geometricae elementa* (1702), there are included two previously unpublished writings of Newton, of which one (pages 332–6), *Lunae theoria Newtoniana*, was also published as a pamphlet.[5] It was certainly a mark of real favour to Gregory thus to have been permitted to include something new of Newton's in the *Astronomiae...elementa*. The second of these unpublished works is presented in a curious fashion, for it appears as the major part of Gregory's own Praefatio; that is, under Gregory's name and as if it had not been written by Newton at all, but had been Gregory's own composition. If this was done with Newton's permission, it would not be the only occasion on which something written by Newton appeared as if it had been the product of another, or of others. His later 'recensio' of the *Commercium epistolicum* and his 'Scala graduum caloris' (both published in the *Philosophical transactions*) bore no author's name at all. The *Commercium epistolicum* appeared as the work of a committee of the Royal Society, appointed to examine the respective rights of Newton and Leibniz in the matter of the discovery of the calculus, though it too was actually edited by Newton. But there is a fundamental difference between these works and Gregory's Praefatio. In the *Commercium episto-*

[1] Gregory, as we shall see below, contemplated an edition of the *Principia* in which all of his extensive notes would be printed either as a running commentary to Newton's text or as a supplement, save that those which served to correct the text (rather than to explain it) would be rendered unnecessary if Newton were to adopt them as textual emendations. But if Newton were not to accept these proposed emendations, then all of Gregory's notes were to be printed as a separate work.

[2] See Turnbull (1939), *James Gregory*, and Stewart (1901).

[3] [Newton] (1959–), *Correspondence*, vol. 3, pp. 154–5.

[4] See J. C. Gregory (1832), 'An autograph MS by Newton'; see also McGuire and Rattansi (1966).

[5] Newton (1702). Gregory introduces 'Sir Isaac Newton's Theory of the Moon' with these words: 'I have thought fit to subjoyn the Theory of the Moon made use of by Sir *Isaac Newton*, by which this incomparable Philosopher has compass'd this extremely difficult Matter, hitherto despair'd of by Astronomers; namely, by Calculation to define the Moon's Place even out of the Syzygies, nay, in the Quadratures themselves so nicely agreeable to its Place in the Heavens (as he has experienc'd it by several of the Moon's Places observ'd by the ingenious Mr. *Flamstead*) as to differ from it (when the difference is the greatest) scarce above two Minutes in her Syzygies, or above three in her Quadratures; but commonly so little that it may well enough be reckon'd only as a Defect of the Observation. In this Calculation, which we give in the Words of the Author, he does not wholly mention all the Inequalities, whose Causes are above explain'd, nor those which are as yet only suspected; but omitting those which he knew wou'd take off one another, and others of less Moment, he only confines those to Aequations and Tables, that have the greatest Force and produce the most sensible Effects.' Quoted from the English translation, Gregory, *Elements of astronomy* (1715), vol. 2, p. 562. Gregory had initially translated Newton's text from English into Latin.

licum, and by implication in the 'recensio' of it, Newton is affirming his claim to discovery. In the Praefatio, the appearance is given that Gregory was denying to Newton the originality of his invention. Had Newton said in his own name that the ancients knew both the concept of gravity and the form of the inverse-square law, his readers could have attributed this statement to an author's modesty. But for Gregory to say it must certainly have given the appearance that he was attempting to show that Newton's work was actually less original than might otherwise have seemed the case.

During the 1690s Newton had been exploring the possible knowledge of ancient sages. He not only wrote out a tentative essay on this topic, but even planned to add to the beginning of Book III of the *Principia* (notably Props. IV–IX) a series of scholia containing references to this supposed ancient wisdom, going back not only to presumably historical figures such as Thales and Pythagoras, but even further to mythical ancient sages.[6] Newton himself wisely refrained from including such material in any of his printed scientific treatises, but faint echoes of these speculations may possibly be detected in the concluding Scholium Generale of the *Principia* and the final Queries of the *Opticks*.[7]

Folded between pages 412 and 413 of the interleaved copy of the first edition ($E_1 i$) there is to be found in Newton's handwriting an essay on the ancient sages. Newton's writings on this subject have recently been analysed in a most intriguing and revealing fashion by J. E. McGuire and P. M. Rattansi. A reference to Newton's plans to incorporate a discussion of this topic into a proposed revision of the *Principia* is recorded by Gregory in a memorandum (printed in §12 below) under the date of 5, 6, 7 May 1694.

12. GREGORY'S 'NOTAE'

Gregory had certainly been one of the first men to make a thorough and detailed study of Newton's *Principia*. Unlike Fatio, Gregory did not read this treatise 'cursorily or partially', as Rigaud pointed out, but 'went regularly through the whole, noting what occurred to him, and examining every step in detail'.[1] His 'Notae in Newtoni Principia Mathematica Philosophiae Naturalis' embody the results of this study. This work exists in four manuscript copies, of which the original—in Gregory's hand—is in the library of the Royal Society of London.[2]

[6] We are reminded that Newton held to a euhemeristic theory in general, and that his chronological investigations took as a point of departure that such figures as Jason were real people and that there had been a good ship *Argo*, the date of whose voyage Newton attempted to determine by applying knowledge of the precession constants.

[7] See Cohen (1964), 'Quantum in se est', J. C. Gregory (1832), and Appendix III to this edition of the *Principia*. A study of these texts of Newton's may be found in McGuire and Rattansi (1967).

[1] Rigaud (1838), p. 98.

[2] The other three are in Aberdeen, Edinburgh, and Christ Church, Oxford. In the title of the copy in Christ Church, Gregory inverted the order of 'Principia Mathematica' and 'Philosophiae Naturalis', but the title of the copy in the Royal Society reads 'Notae in Newtoni Principia Mathematica philosophiae Naturalis', and the title of the Edinburgh copy reads 'Notae in Isaaci Newtoni Principia Philosophiae'. See Wightman (1953, 1955, 1957).

From the dates written in his copy, it appears that he began these notes in Edinburgh in September 1687, within two months of the publication of the *Principia*, and completed them at Oxford in January 1693/4. These notes comprise an explanatory commentary, giving alternative proofs, explaining Newton's meaning, and even translating Newton's verbal proportions into algebraic expressions. Thus he explains Lemma I, Book II,[3] as follows:

Let A be to A − B as B to B − C and C to C − D, &c. & *dividendo* A will be to B as B to C & C to D &c. Q.E.D.

For if A:A − B::B:B − C, it will so be the case that AB − AC = AB − BB. Therefore AC = BB. Therefore A:B::B:C.

Sometimes Gregory's notes merely give a reference to Euclid or Apollonius or are explanatory, as in Prop. XI:

[Aequentur ES, EI] Per 2.6^{ti}.
[Angulos aequales IPR, HPZ] Per 48. lib: 3. Apollonii.
. . . .
[Ut QR ad Pv] Nam QR:Pv::(Px:Pv::) PE:PC, ob xv‖PR‖EC.
[Ut CP*q* ad CD*q*] ex natura Ellipseos.

Sometimes, as following Prop. XI, Book I, Gregory inserts something 'indicated' to him 'by the renowned Author' ('Mihi methodum a Clariss: Auctore indicatam'): in this case a Lemma, followed by a Theorema, and four Corollaria. At other times, notably in Book III, Gregory gives examples, or verbal explanations. There can be no doubt that by 1693, when these notes had been substantially completed, Gregory had given the *Principia* a closer reading than anyone since Newton had written it and Halley had gone through it in the course of preparing the manuscript for printing and seeing the book through the press.

In the copy of these 'Notae' in Gregory's own hand, some pages have slips affixed to them which contain later addenda.[4] These are most precious since they either record remarks made by Newton or embody Newton's own emendations of the *Principia*. Sometimes, Gregory leaves no doubt of the source, as when, discussing Hooke's claims to the inverse-square law (apropos the Scholium to Prop. IV, Book I {45.5–32}), he says, 'ut Candidiss: Newtonus mihi narravit'.[5] Or when, having discussed Hypoth. III, Book III, he writes:

Hypothesis haec ab Auctore sic enunciatur.

Leges et proprietates corporum omnium, in quibus experimenta instituere licet, sunt leges et proprietates corporum universorum. Nam proprietates...

[3] Newton's statement reads (in translation): 'Quantities proportional to their differences are continually proportional.' Gregory's statement is here translated from the Latin original.

[4] In Gregory's personal copy of his 'Notae' (now in the Library of the Royal Society), these slips of paper containing Newton's alterations of the *Principia* are affixed to the appropriate pages by paste or wax; either Newton made these emendations or Gregory learned about them only after he had written his text. In the other copies the amanuensis entered these alterations into the text as he went along. After the whole of the 'Notae' had been transcribed, Gregory wrote out further such emendations of Newton's at the end of the volume.

[5] Gregory writes: 'Nay rather, if Hooke is to be believed, it was he who laid the foundations of all this philosophy and communicated it to Newton. But the truth of the matter is this, as the very straightforward

But even though Gregory thus gave the new Hypoth. III (which was to become Regula III), he did not on that account cancel his own remarks concerning the original Hypoth. III:

[Transformari posse] Hoc Cartesiani facile concedent. Peripatetici non item, qui specificam differentiam ponunt inter materiam caelestem ac terrestrem. Nec Epicureae Philosophiae Sectatores, qui atomos rerumque semina immutabilia ponunt.

This the Cartesians will easily concede. Not so the Peripatetics, who put a specific difference between celestial and terrestrial matter. Nor the Sectators of the Epicurean Philosophy, who make immutable the atoms and seeds of things.[6]

13. GREGORY'S MEMORANDA

Gregory also wrote out memoranda following his visits with Newton, and he recorded current rumours.[1] From these we may piece together his connections with the proposed new edition, and we may also date certain major alterations made by Newton. For instance, Gregory saw Newton in Cambridge on 4 May 1694, and recorded, under 38 heads, the various improvements or corrections made by Newton, including these:

2. $\frac{4\ \mathrm{SI\ cub}}{3\ \mathrm{LI}}$ is to be written in place of $\frac{8\ \mathrm{SI\ cub}}{3\ \mathrm{LI}}$ on page 209 of the second part.

5. He recognizes mistakes in the parts 'Of Water'; for instance, he recognizes that what we on his own system made *double height* in place of *height* (altitude) on page 331, ought as following from what precedes to become *quadruple height.*

8. He recognizes that 'of the Moon to the Earth' ought to be written instead of 'of the earth to the Sun'. [*Principia*, E_1, p. 439 {433.17}.]

12. He has found a theorem whereby, from a given force with which it revolves round a given centre in a given orbit in a given time, he finds the force with which it would revolve round another given centre in the same orbit: for instance, the force round C (Fig. 2) is to the force round F as AC × AF² is to AB³.

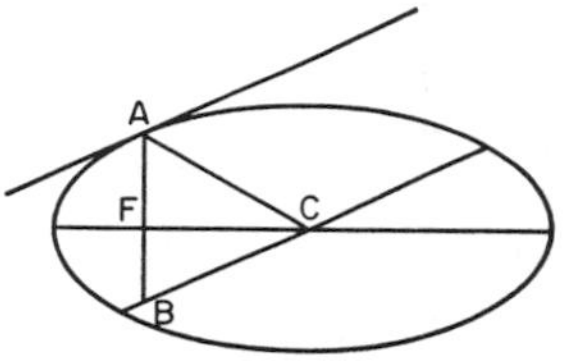

Fig. 2.

I have fully demonstrated this theorem, clearly stated or rather discovered afresh (taking only the occasion from him).

From this point he follows out with great ease the properties of conic sections so far as concerns centripetal force. For what suits both the ellipse and the parabola applies to the hyperbola as well.

Newton told me.' Then he summarizes Newton's account of the affair, how Newton had written to Hooke that a body falling from a tower on a moving Earth would not hit the ground to the west of the foot of the tower, but rather to the east, 'because, of course, of the greater impetus communicated to a stone at the top of the tower describing a greater circle'. Further, that the body would describe a curve which terminates below the Earth's surface, at the centre, 'although he did not mention that part of the figure in the letter'. Then, according to Gregory, 'Hooke wrote back that it was indeed true that it would touch the Earth toward the east, but that it would not seek the centre but passing it by and describing an elliptical curve, would again return upward to the top of the tower. And this is all that which he boasts that he first showed to Newton and that Newton's philosophy is built upon this alone.' The foregoing quotations are translated from Gregory's Latin manuscript; a more complete text is given in Supplement I.

[6] Newton, in his own manuscript notes (and, by implication, in E_2 {402.18–19}), held that this was an hypothesis of the Cartesians and the Aristotelians. See Cohen (1971), *Scientific ideas.*

[1] For instance, 28 December 1691, that Fatio was planning a new edition of the *Principia*; see §8.

16. The colour red moves more slowly, blue more quickly: this is inferred from the greater refraction of red.[2]

20. A few days ago he discovered the equation of the mean motion of the Moon, which holds when the apogee of the Moon delays in certain octants: he believes that this equation will reduce the motion of the Moon to an infinitely greater exactitude.

26. In the new edition of his *Philosophy*, he will divide it into several parts so that they can be read separately.

27. He intends to edit separately the doctrine of orbits passing through given points and with given tangents.

28. Likewise the method of quadratures by itself.

35. Throughout he has said 'subduplicated' for 'halved' and 'principal axis' for 'transverse axis'.[3]

In another memorandum (dated Cambridge 5, 6, 7 May 1694), Gregory recorded two interesting plans of Newton's:

He will spread himself in exhibiting the agreement of this philosophy with that of the Ancients and principally with that of Thales. The philosophy of Epicurus and Lucretius is true and old, but was wrongly interpreted by the ancients as atheism.

He intends to publish Sections IV and V with the Quadratures of Curves, a tract (which I have seen) of three or four sheets (if it were written on both sides) together in one small book—but after publishing the *Principia*.[4]

As to the first of these, we have seen (§11) that Newton had at one time contemplated making such an addition to his *Principia* in the form of a series of scholia. Considering how great the temptation must have been for Newton, we can only admire him the more for having recognized the importance of keeping separate from the 'mathematical principles of natural philosophy' and their applications to specific scientific problems any of his speculations on the mythical origins of science or the mystical significance of cosmic phenomena. This is all the more the case in that his contemporaries sometimes failed to make such a separation.[5] As to the second of these statements, it may be pointed out that Newton did not publish Secs. IV and V of Book I separately, and that *De quadratura* was first printed as a supplement to the *Opticks*. But it is of real interest that Newton shows himself fully

[2] The Latin original reads: 'Ruber Color tardius, Ceruleus Celerius fertur, arguitur ex Rubri majori refractione'—but, of course, the red-coloured rays are not refracted more than the blue, but less. This error may impugn the accuracy of Gregory's reports on Newton's work.

[3] Gregory's memoranda were generally written in Latin. Both the Latin original and an English translation, from which these and other extracts have been taken, are to be found in [Newton] (1959–), *Correspondence*, vol. 3; this extract occurs on pp. 311–20.

[4] *Ibid.* pp. 335–6, 338. D. T. Whiteside points out (in a private communication) that the 'tract on "Quadratures of Curves" which Gregory saw in 1694 (and of which he took a copy, now in Shirburn Castle) is not the *De Quadratura* as printed but a much abridged, considerably variant earlier version of *c.* 1692: in particular, it lacked the printed work's tables of integrals'.

[5] Gregory also recorded certain views of Newton's concerning God's role in the operation of the system of the world. Thus (Cambridge 5, 6, 7 May 1694): '[Newton says] that a continual miracle is needed to prevent the Sun and the fixed stars from rushing together through gravity: that the great eccentricity in Comets in directions both different from and contrary to the planets indicates a divine hand: and implies that the Comets are destined for a use other than that of the planets. The Satellites of Jupiter and Saturn can take the places of the Earth, Venus, Mars if they are destroyed, and be held in reserve for a new Creation.' Again (16 [?] May 1694): 'The fixed starrs not coming together is a constant Miracle'; *ibid.* pp. 334, 336, 355.

aware that Secs. IV and V of Book I (embodying contributions to the geometry of conic sections) are all but independent entities.[6] According to this memorandum, Newton planned to print *De quadratura* along with Secs. IV and V after the new edition of the *Principia* had appeared. We shall see in a moment that he also contemplated printing *De quadratura* as an appendix to the new edition of the *Principia*.

Another memorandum, perhaps of July 1694, refers to many other alterations and plans for the new edition. It is a composite of what Gregory had learned from his conversations with Newton and from a recent letter.[7] Here Gregory refers to a number of proposed alterations, many of which also appear in previous memoranda. Gregory's account of Newton's plans to change Sec. II of Book I of the *Principia* is especially fascinating since we have already (Chapter II, §7) seen the evidence from Newton's manuscripts of these same proposed alterations—for example, the reversal of the order of Prop. IV and Prop. V (see the record of this proposed change in the Apparatus Criticus {43.26 ff.}). Gregory refers to these changes as follows:

> Many corrections are made near the beginning: some corollaries are added; the order of the propositions is changed and some of them are omitted and deleted. He deduces the computation of the centripetal force of a body tending to the focus of a conic section from that of a centripetal force tending to the centre, and this again from that of a constant centripetal force tending to the centre of a circle; moreover the proofs given in propositions 7 to 13 inclusive now follow from it just like corollaries. Sections IV and V are removed and become parts of a separate treatise (of which afterwards). To proposition 10, Book II he will append another problem in which the path of a projectile is investigated according to the true system of things, that is, supposing gravity to be reciprocally as the square of the distance from a centre and the resistance to be directly as the square of the velocity—a problem which he believes to be now within his power.[8]

It will be observed in the Apparatus Criticus that, although major corrections were made in E_2 and E_3 to Prop. X, Book II, the addition mentioned by Gregory was never made; that is, gravity is considered constant and not variable according to an inverse-square law.

Gregory refers also to the notable changes in Sec. VII, Book II, almost all of which is new in E_2, and he says, 'By far the greatest changes will be made to Book III': the new Hypoth. III (of which Gregory says that Newton 'will show that the most ancient philosophy is in agreement with this hypothesis of his'), the recognition of the satellites of Saturn discovered by Cassini {391.23 ff.}, the revised discussions of the motions of the Moon and of comets. Gregory says that the new edition of the *Principia* will have two supplements, 'one about the geometry of the ancients where the errors of the moderns about the mind of the ancients are

[6] See Chapter IV, §4.

[7] Gregory refers to a letter written to him by Newton on 14 July 1694, containing among other things a statement that the 'Lem. 1 in ye third book I could not recover as tis there stated'. Newton then proceeds to give Gregory a set of 'steps' by which the solid of least resistance or the 'figure wch feels ye least resistance in ye Schol. of Prop. 35 Lib II is demonstrable'. [Newton] (1959–), *Correspondence*, vol. 3, p. 380.

[8] *Ibid.* p. 384.

detected...[and] where it will be shown that our specious algebra is fit enough for making the discoveries but quite unfit to [give them] literary [form] and to bequeath them to posterity.'[9] Now, according to Gregory, part of this work was to be 'on the discovery of orbits with a given focus or even when neither focus is given, and all this is the subject-matter of Sections IV and V'. As to the second supplement, it was to 'contain his Method of Quadratures'.

Gregory goes on, 'Most of what in early May of 1694 he had corrected or altered in his own copy has been corrected or altered at the respective places in my copy or in my notes.' Unfortunately, Gregory's annotated copy of the *Principia* is not available.[10]

14. HOPES FOR PUBLISHING THE 'NOTAE', POSSIBLY AS A RUNNING COMMENTARY TO A NEW EDITION OF THE 'PRINCIPIA'

Gregory relates that he proposed (8 May 1694) to Newton that his 'Notae in Newtoni Principia...' be published. If 'a new edition' of the *Principia* were to be 'made by the author', Gregory would omit from his own 'Notae' 'a great deal that serves to detect slips or even mistakes of Newton'; but 'otherwise the notes are to be inserted in their proper places'. Then he added, as a kind of aside, that, 'if the Author is willing', these 'Notae' might

> be interspersed everywhere on the same page or, if it seems more convenient, [be collected] at the end like Schooten's notes in the geometry of Descartes: and the book will be in folio or may be more closely printed in a quarto; or perhaps the notes will be in a separate book after the author's book has been published.

He concluded on this petulant note:

> It is only fair that I should see any new edition of the Author's beforehand or at any rate know for certain that a new one will not be produced.[1]

From these memoranda we may fairly conclude that by 1694 Gregory had not been given any assurance that he would be allowed to produce a new edition of Newton's *Principia*. Nor would it appear that Newton had even shown any willingness to include all or even a portion of Gregory's 'Notae' in any edition of the *Principia* whatever. For that matter, I have never encountered any explicit statement by Newton himself that he thought Gregory's 'Notae' worthy of publication

[9] This statement sounds just like the famous one in the *Recensio libri* (Newton's anonymous review of the *Commercium epistolicum*) in which he says that he had discovered some essential parts of the *Principia* by analysis and had then recast them in the form of 'good [Greek] geometry'. Newton (1715), 'Recensio libri', p. 206. The above-mentioned treatise on porisms survives and will be included in [Newton] (1967–), *Mathematical papers*, vol. 7.

[10] Gregory alleges that 'he [Newton] has made more changes and corrections in his treatment of comets, e.g. in Lemma 8 (and here as everywhere he has acknowledged my corrections but for one or two) and in the principal proposition 41 itself'. No such acknowledgement of Gregory's contributions occurs in E_1i or E_1a or in E_2, nor have I found any statement to this effect among Newton's papers. Perhaps Gregory meant only that Newton accepted his corrections.

[1] Printed in English translation only in [Newton] (1959–), *Correspondence*, vol. 3, p. 386.

in any form, and I have no evidence that Newton had ever seen them, much less possessed a copy. Newton never refers to Gregory's 'Notae' in any letter or other document that I know of, nor have I ever seen any direct evidence that he had even read them. There does not appear to have been a copy of these 'Notae' in his library at the time of his death; at least there is none now among his papers in the Portsmouth Collection, nor was there one among the remainder of that original collection dispersed at public auction in 1936.[2]

All of the evidence linking Newton himself to the 'Notae' derives from Gregory, as does the only statement I know in which Gregory is mentioned as possible editor of the new edition. The latter occurs in a letter written by Huygens to Leibniz (20 May 1694, following Gregory's visit to Huygens), concerning an error in Newton's 'Principes de Philosophie'; Huygens wonders whether it will be corrected 'dans la nouvelle edition de ce livre, que doit procurer David Gregorius'.[3] A copy of E_1 in the library at St Andrews contains a manuscript note reading, 'D. Gregorius...in Notis MS quas Newtoni hortatu in Principia composuit' ('Mr Gregory...in the manuscript Notes on the *Principia* which he composed on being urged by Newton), but this statement—in an unidentified hand—cannot be counted as firm evidence.

15. FURTHER MEMORANDA

On 4 March 1695/6 Gregory was still uncertain about a new edition of the *Principia* and the possible future of his own 'Notae', as shown by the following memorandum:

> I am not to revise my Notes on Mr Newton's philosophy, until there is a second edition of it by Mr Newtons self, or that we despair of having one.[1]

Finally, on 13 November 1702, Gregory recorded that Newton was 'to republish his book; & therein give us his methode of Quadratures'.[2] Perhaps Gregory had misunderstood Newton and had confused the republication of the *Principia* and the first publication of the *Opticks*. Or, possibly, Newton changed his own mind, for two days later Gregory noted that Newton promised Fatio, Halley, and a 'Mr Robarts' that he would 'publish his Quadratures, his treatise of Light, & his treatise of the Curves of the 2d Genre'.[3] Once again, on 24 March 1704, there was news of a new edition, for Newton told Gregory 'himself that he will republish his *Principia*'. He was in no hurry but (29 July 1704) 'corrects his Principia at leisure. He has come a good length in it already.'[4]

By 21 July 1706, Gregory could make the optimistic observation that 'there remains but little [for Newton] to doe, and I hope he will be persuaded to put it [the new edition] to the Press next winter. It will not be sensibly bigger than the first edition.'[5] By this time Newton had made real progress and had actually

² See [Sotheby and Co.] (1936), *Catalogue*.

³ [Huygens] (1888–), *Oeuvres*, vol. 10, p. 614.

¹ [Gregory] (1937), *Memoranda*, p. 4.

² *Ibid.* p. 13.

³ These three were in fact published together within two years, in 1704.

⁴ [Greogory] (1937), *Memoranda*, pp. 16, 19.

⁵ *Ibid.* p. 37.

shown Gregory 'a copy of his Princ. Math. Phil. Nat. interleaved, and corrected for the Press'. The job was 'intirely finished as farr as Sect. VII. Lib. II. pag. 317'. That is, the first two-thirds were completely revised to Newton's satisfaction, and so were certain other parts. For instance, the beginning of Book III had now assumed the form it took in E_2.[6] According to Gregory, 'He letts alone his Doctrine of Comets as it was.'[7] But some time later, Newton made a lengthy addition to the final discussion of the comets {519.16–526.30}. Finally, Gregory remarked that 'most of the corrections made in the book printed at Hamburg are in his copy, word for word; which he will advertise'.[8]

There is in these descriptions a resemblance to E_1i, but it is very likely that what Gregory had seen in 1706 was not E_1i. For, as we shall see in Chapter VIII, Gregory appears to have entered an erroneous 'correction' in the margin of a copy he had seen, a correction that does not occur in E_1i. A test might be possible from Gregory's remark that Newton 'takes...Sect. VII to be the hardest part of the book. He breakes the Scholium Generale at the end of this into many parts or bitts.' Actually, Newton ended up by casting out nearly all of the old Sec. VII, and replacing it in E_2 by an almost wholly new Sec. VII. The original Scholium Generale, however, is not broken up but is rather removed practically in its entirety to the end of Sec. VI. We cannot tell what alterations Newton had made in this Scholium Generale in E_1i, however, since the pages of E_1i containing the first half of this Scholium Generale were destroyed at some later date by fire.

Gregory's optimistic notes of 1705 and 1706 were succeeded by a pessimistic entry on 15 April 1707, that although there are 'only two things that Sr Isaac Newton further desires, to make a new edition of Princ. Math. Philos. Nat.', he 'does not think to goe about these things, at least the Edition, for two years yet'. According to this memorandum of April 1707, the 'two things' remaining to be done were: (1) 'the Resistance of Fluids in Lib. II. about which he [Newton] is affray'd he will need some new experiments', and (2) 'the Precession of the

[6] 'In the beginning of Lib. III. he leaves out Hyp. III, and puts another in its place as in my copy. These three he now calls *Regulae Philosophandi*. Hyp. V. &c he calls by the title of *Phaenomena*. Hyp. IV is the only one that he leaves that name to; & it comes after the Phaenomena. He has the revolutions & distances of Saturns Satellites.'

[7] *Ibid.* p. 36. 'He says his [Doctrine of Comets] (which is partly by calculation, partly by construction) comes within 10′ of the truth; Mr. Halley's (which is all by calculation) comes within 3′. He putts in Mr. Halley's Tables for Comets.' Newton had also 'recovered his first Demonstration or Investigation of Prop. XXXIX. pag. 470, which he shewed me'. See {477.19–35} and the penultimate paragraph of this section.

[8] Somewhat earlier, on 7 February 1705/6, Gregory had reported that Groening's book 'is all written to magnify Libnitz & Bernoulli. To this is subjoyned a Collection of Errors & Mistakes in Newton's Philosophy, said to be made by M. Hugens. But the greatest part of them is Sr I. Newtons own. The rest are M. Fatios, some few Hugens. They make 24 pages. They were communicated by M. Fatio to M. Hugens after whose death they fell into Volders hands, and were by him or Libnitz or Bernoulli or all three thus secretly printed' (*ibid.* p. 32). This was an amplification of an earlier note: 'In a latin book in 8vo, written by one Groninga or such a name of miscellaneous things & among others of the History of the Cycloid, printed about 1701 or 1702 at Hannover, there is printed a list of Mr. Newtons mistakes in his Princ. Philos. in about 5 or 6 pages. These were communicated by M. Fatio to M. Hugens about the year 1690, and by M. Hugens to M. Libnitz, & after M. Fatio's falling out with M. Libnitz, malitiously published by the contrivance of M. Libnitz. This is M. Fatio's account of the affair this 4 July 1705' (*ibid.* pp. 26–7).

Aequinoxes Lib. III. of which he thinks he is fully master, but has not yet written it out'.[9]

In the *Principia*, Newton had dealt with precession in Prop. XXXIX, deep in Book III, following a rather intricate and difficult presentation of the motion and orbit of the Moon, including the motion of the nodes and the inclination of the orbit to the plane of the ecliptic. Newton had presented 'these computations of the lunar motions', in the words of the Scholium to Prop. XXXV, because he 'was desirous of showing that by the theory of gravity the motions of the Moon could be calculated from their physical causes'. Then, following three propositions (XXXVI, XXXVII, and XXXVIII) on how to find 'the force of the Sun to move the sea', 'the force of the Moon to move the sea', and 'the figure of the Moon's body', Newton turns to precession.

If we may suppose with Newton that the Earth is an oblate spheroid, the plane of whose equator is inclined to the plane of the ecliptic, then the equatorial bulge acts like a ring of matter affixed to the equator of an otherwise more or less uniform sphere. The forces producing precession (that is, the attraction of the Sun and Moon tending to alter the direction of the Earth's axis or the plane of the equator) need be considered only in relation to the matter in this ring, since the matter in the uniform sphere within the ring is distributed symmetrically, with the result that the attraction of the Sun and Moon on the matter within this central sphere does not give rise to forces that produce precession.

In dealing with this topic in E_1, Newton first presents two lemmas (I and II), together with their proofs, on the rotation of the Earth. In these he supposes that the rotation of the Earth acts on all the particles composing it so as to cause them to endeavour to recede from the centre; the result is that the Earth is thereby given the shape of an oblate spheroid. Then comes a Lemma III, without proof, that if the spherical central body of the oblate Earth were taken away, so that only the ring remained, moving in the same orbit (with the same orbital motion), and rotating as the Earth does (with the axis still tilted at an angle of $23\frac{1}{2}°$ to the plane of the ecliptic), then 'the motion of the equinoctial points would be the same, whether the ring were fluid, or whether it consisted of a hard and rigid matter'. On the basis of these three lemmas and certain earlier theorems, Newton is able to show (Prop. XXXIX) that the combination of the solar and lunar precessional forces will produce a precession of 49″ 58‴ per annum, which agrees well with the figure obtained from astronomical observations of 50″. Indeed, this is an agreement to two parts in 3,000, or less than 0·1 per cent.

In order to see the changes wrought by Newton, we may turn to the Apparatus Criticus {471.15 ff.}. First of all, the original Lemma I has now split into a Lemma I and a Lemma II. The original demonstration of Lemma I, with certain alterations, including a considerable mathematical supplement, has become the demonstration of Lemma II; and a wholly new demonstration of Lemma I has been added. The

[9] *Ibid.* p. 40.

original Lemma II {475.15 ff.} has now become Lemma III, and the proof is unaltered. But the former Lemma III is no longer a lemma but has a new status: it has become Hypoth. II.[10] The text of Hypoth. II is not significantly different from that of the original Lemma III. As to Prop. XXXIX itself, the proof is altered chiefly by the introduction of new numerical data, of which some truly differ, while others are different because carried to more significant figures: as '100 ad 292369' in E_2 and E_3 for '1 ad 2932' in *M* and E_1. Furthermore, in E_2 and E_3, the final three short paragraphs {477.19–35} replace a rather more extended discussion in *M* and E_1. Newton now concludes from this theory that the total annual precession should be 50″ 00‴ 12^{iv}, 'the quantity of which motion agrees with the phenomena. For the precession of the equinoxes, by astronomical observations, is about 50″ yearly.' It is curious that Newton has altered his theoretical result in the direction of more significant figures (50″ 00‴ 12^{iv} in E_2 and E_3 for 49″ 58‴ in *M* and E_1), while at the same time he has given the astronomically determined value more approximately ('ex observationibus astronomicis est annuatim minutorum secundorum plus minus quinquaginta' for 'ex observationibus Astronomorum est 50″').

I have not been able to date these changes exactly, but there is some evidence that they may belong to the period around 1707 or thereafter. First, the Apparatus Criticus shows that the major alterations—that is, the splitting of Lemma I into Lemmas I and II, the new proof of the revised Lemma I, the altered proof of the new Lemma I, the change of the old Lemma III to Hypoth. II, the improved data in the proof of Prop. XXXIX, and the new conclusion to the proof of that Proposition—do not appear at all in E_1i or E_1a, as do most of the alterations made in the 1690s. In July 1706, furthermore, Gregory recorded that Hypoth. IV was 'the only one that he [Newton] leaves that name to'[11]—a statement that would most certainly have been amplified if Newton had already converted the old Lemma III of E_1 into an Hypothesis. But in 1706 Newton was already at work on this section, because (to quote Gregory): 'He has recovered his first Demonstration or Investigation of Prop. XXXIX. pag. 470, which he shewed me.'[12] Some of the worksheets for these changes survive (U.L.C. MS Add. 3965, ff. 332–7), including a longer unused proof of the new Lemma I, and a page containing reworked statements of the three lemmas, the revised proof of the new Lemma II, and an indication that the old Lemma III was to be Hypoth. II (with its text unaltered). The handwriting on these sheets is consistent with a dating of 1705–10.[13]

[10] Hypothesis I is the former Hypoth. IV, about the centre of the system of the world being at rest; when the original group of Hypotheses was converted into Regulae Philosophandi and Phaenomena, this one became Hypoth. I and was removed from its original location at the beginning of Book III to a place just before Prop. XI {408.1 ff.}, where it is needed. [11] [Gregory] (1937), *Memoranda*, p. 36. [12] *Ibid.* pp. 36–7.

[13] The shift from the designation of lemma to hypothesis means that Newton must have changed his mind concerning the possibility of finding a proof. I assume he wanted his readers to conclude that there could be no proof, or at least that in all likelihood this was the case; not just that Isaac Newton himself had not been able to find the proof. But at least one reader, the great Laplace, took the latter interpretation and set out to find a proof—which he did successfully. Ball (1893), *Essay*, p. 110.

16. THE NEW EDITION BEGUN

Despite the pessimistic tone of Gregory's note of April 1707, that Newton will not go about the edition for another two years, the situation was altered suddenly and dramatically within eleven months. For on 25 March 1707/8 Gregory was able to record at last that the new edition was actually under way.

[Sir Isaac Newton] has begun to reprint his Principia Philosophiae at Cambridge. He is to add his Quadratures & Curves of the second genre to the Principia.[1]

This is a reference to the edition undertaken by Richard Bentley, discussed in Chapter VIII.

One of the lasting puzzles about Gregory's relations with Newton is his continued loyalty, despite Newton's seeming rejection of his running commentary on the *Principia*. His willingness, however reluctant, to put aside his 'Notae', at least until a second edition of the *Principia* should be published, has robbed us of an introductory guide to the *Principia* that might well have enjoyed the renown of the later work of Fathers LeSeur and Jacquier.[2] These 'Notae' have never been published; by the time of the second edition of the *Principia* Gregory was dead and the 'Notae' (referring to the first edition) could no longer be published without major emendation. Newton himself took no more interest in them after Gregory's death than he had while Gregory was still alive. Horsley made some use of Gregory's comments in his presentation of the *Principia* in his edition of Newton's *Opera* in the 1770s, even quoting an occasional extract. To my knowledge, they have served no Newtonian scholar since that time in his endeavour to understand the *Principia*.[3]

[1] [Gregory] (1937), *Memoranda*, p. 41.

[2] See the bibliography, Appendix VIII to the text, pt. 2.

[3] It is planned to include selections from these 'Notae' in the commentary volume to this edition.

CHAPTER VIII

STEPS TOWARD A SECOND EDITION

1. THE DISSEMINATION OF NEWTON'S MANUSCRIPT ALTERATIONS PRIOR TO 1708: PRESSURE ON NEWTON TO PRODUCE A SECOND EDITION

During the two decades following the publication of the *Principia* in 1687, while Newton more or less continuously revised his *Principia*, rumours were frequently being circulated that a new edition might possibly appear. But not until 1708 is there any evidence that Newton was actually willing to turn over any part of the revised text to a printer, or to an editor who would see the new edition through the press. By then the old intimacy with Fatio de Duillier was only a memory, and Newton's decision not to combine forces with Gregory was apparently as firm as ever. In 1708 Newton was living in London, having moved from Cambridge in 1696 to take on his new assignment at the Mint. But the reprint or edition that David Gregory noted on 25 March 1708 (see Chapter VII, §16) was not to be printed in London under Newton's direct supervision, but in Cambridge.

In 1708 it was certainly time for a new edition (or at least a corrected and improved edition), since copies of the first one were becoming harder and harder to come by. In order to get a *Principia* (*c.* 1708) Sir William Browne, while a student, 'gave no less than two guineas for one',[1] and another student (the father of Dr James Moor of Glasgow) could not afford to pay so high a price and transcribed the whole treatise in longhand.[2] In the conclusion of the Preface to the second edition, Roger Cotes referred to the fact that Richard Bentley had urged upon Newton the need for a new edition 'since copies of the previous edition were very scarce and could only be bought at enormous prices' ('cum Exemplaria prioris Editionis rarissima admodum & immani pretio coemenda superessent').

The Cambridge 'reprint' of the *Principia* to which David Gregory referred was

[1] Rigaud, basing himself on Nichols's *Literary anecdotes*, says that two guineas 'was then thought a very cheap purchase'. He added the comment that (in 1838) copies of the first edition 'are not now of such extreme scarcity and high price, but that only proves more strongly how much they must then have been prized for their intrinsic value'; see Rigaud (1838), *Historical essay*. In 1968, a copy of the first edition was not a very rare book; a scholar could readily find one in a research library, and a sufficiently wealthy collector might easily locate one for purchase, but it is of 'high price'—of the order of $10,000 or £3,500.

[2] [Tixall Letters] (1815), vol. 2, p. 152. We shall encounter below (§5) an example of someone writing out in a copy of the first edition all the alterations introduced in the second edition, including the introduction by Cotes and the index.

begun by Richard Bentley, Master of Trinity College. Bentley, however, as we shall see (§6), had only the first eight pages composed at the University Press and printed off as a 'specimen' in twenty-five copies. Not until much later, after Bentley had enlisted the services of a young Trinity mathematician, Roger Cotes, did the printing of the new edition really begin in earnest.

In the interim, while the *Principia* was out of print, a number of scholars were able to take advantage of the revisions made by Newton. The degree to which his alterations were circulated makes it tempting to refer to them collectively as if they constituted an intermediate 'corrected' (or 'improved') 'edition' of the 1690s. We have thus far seen evidence that Newton showed such alterations to Fatio and to Gregory, and even allowed these two associates to transcribe some of them. Fatio, furthermore, has proved to have been the agent through whom these emendations were transmitted to Huygens, whence they eventually came into the hands of Leibniz, and were printed (1701) by Groening, for all to read. Gregory, for his part, entered some of Newton's improvements in his 'Notae in Isaaci Newtoni Principia Philosophiae'; I do not know how many persons had access to the four presently known manuscript copies of the 'Notae' (there may perhaps have been others), but they apparently were made to be read and used by students.

Some evidence of the distribution of Newton's manuscript emendations is provided by copies of the first edition which actually contain alterations that derive either directly or indirectly from Newton's personally annotated copies or manuscripts. Before proceeding to give an account of the final preparation and printing of the second edition under Cotes's direction, I shall describe all the copies I have found with contemporaneous manuscript alterations, whether in Newton's hand or not, and I shall also list all references I have found to yet other such copies.

This information is presented here not merely as an index of the dissemination of the revisions of the *Principia* on the eve of the second edition. The fact is that we cannot fully evaluate Cotes's actual performance as editor of the second edition without first collecting together all that we can possibly learn about Newton's own annotated (and possibly interleaved) copies of the first edition. The reason is that Cotes was given (or shown) a marked-up copy of the *Principia* when he first agreed to take on the assignment as editor, and then was sent another (possibly the same one sent for the second time) to be used by the printer; but we do not presently know the whereabouts of either of these (if they still exist). We must be content therefore to make conjectures about their contents; to do so we must first have a clear idea concerning all other marked-up copies in Newton's possession in order to see which if any may have been the one or ones sent by Newton to Cotes. Since the copies containing manuscript annotations in hands other than Newton's derive ultimately (either by a direct or an indirect line of descent) from the copies belonging to Newton, it has seemed advisable to treat all the marked-up copies together. A tabulation of annotations from these several annotated copies is given in Appendix IV to the text.

2. SOME ANNOTATED COPIES OF E_1: JOHN LOCKE'S, EDMOND HALLEY'S [?], JOHN CRAIG'S, RICHARD BENTLEY'S, AND OTHERS

An early example of a copy of the first edition of the *Principia* with annotations taken from Newton's revisions is the copy presented by Newton to John Locke some time in the early 1690s (mentioned above in Chapter II, §2, and Chapter VI, §2). This copy is now in the Trinity College Library, Cambridge (press-marked Adv. b. 1.6), and bears on the fly-leaf an inscription in an unknown hand, reading:

Liber
Johannis Locke
ex Dono
Acutissimi Authoris. Qui errores propria manu correxit.

The errors corrected by 'the most acute author' in 'his own hand' prove on examination to be only some of the misprints listed in the printed Errata. But there are a few emendations which are definitely *not* errors, and which are written in a formal scribal hand. Among them is the new Hypoth. III (printed above in Chapter II, §2), in which a careless mistake occurs in the spelling of the first word: 'Qualitas [*sic*] corporũ quae intendi et remitti nequeunt...sunt qualitates corporum universorũ.' The copyist who entered the emendations into Locke's copy was not only ignorant of the subject-matter of the *Principia*; he evidently was not even well grounded in Latin. For the above mistake (writing 'Qualitas' for 'Qualitates') is not the only one he made; he even wrote as two separate words 'Parisi' and 'ensium' {397.21} and 'sex' and 'tuplus' {1.8–9}, which in Newton's text had been divided simply because they came at the end of a line. More information concerning the corrections in Locke's copy, all of which derive from the MS Errata to E_1a, may be found in Appendix IV to the text of the present edition.

Another copy of E_1 associated with Halley is at present in Australia, in the possession of Miss Margaret Norman of Cremorne (N.S.W.), who has very kindly sent me details about its physical appearance and a microfilm of all pages containing manuscript annotations. It came from the private library of Miss Norman's 'great-grandfather, James Sprent, an Edinburgh astronomer, who migrated to Hobart in 1837, later becoming Surveyor-General of Tasmania'. This copy is in a contemporaneous calf binding, containing a tool in gold in the corners, which differs from the tool (in blind) in three copies known to have been presented by Newton and at present in the libraries of Sidney Sussex College and Trinity College (where there are two: the second being the Locke copy). The covers are separated from the spine, and pages 103/4, 503/4 have been torn out; the final unnumbered page of Errata is missing. On the title-page, immediately following Newton's name and titles, there is written in long-hand:

Μνημόσυνον carissimi Capitis Authoris longè Doctissimi Dignissimique.

This may be rendered into English as follows:

A remembrance of the cherished person of a most outstandingly learned and distinguished Author.

The handwriting seems to be Edmond Halley's.

In the margin, there is written (apparently in a different hand) the single word 'Quies.'. Could this be an abbreviation of 'Quiescat', standing for 'Quiescat anima' or 'Quiescat in pace'—'May his soul rest' or 'May he rest in peace'? Such an interpretation would imply that at that time both Halley and Newton were dead; the sentiment would be somewhat foreign to Halley, who was notorious for his lack of religion.

The remainder of the annotations, most appearing to be in Halley's hand, prove to be corrections or alterations, each of which may be found in E_1i and E_1a, but for one which I have found in no other copy of the *Principia*. It occurs in line 18 of Halley's poem; under the word 'Ulvam' ('ulva'=sedge) the word 'Algam' has been written ('alga'=seaweed). Possibly this tentative improvement in the poem may help to reinforce the conclusion that these annotations were made by Halley. Another factor supporting this conclusion is that almost every annotation is also to be found in the Texas copy, presumably also Halley's.

What is most remarkable about this copy, however, is not the fact that the alterations derive from Newton and appear to be mainly by Halley. Rather, some of them seem undisputably to have been written in by Newton himself. They include the following (E_1):

p. 10 {11.12}	quantitatum	p. 439 {433.8}	pertinentium
p. 39 {42.8}	perpetuo	p. 486 {490.18}	(per Cor. 6 Prop. XVI Lib. I.
p. 123 {121.11}	inverse	p. 487 {491.14–15}	per Cor. 7 Prop. XVI Lib. I.

With regard to these entries, it may be noted that 'perpetuo' (p. 39) is labelled by Huygens in his list as 'Manus Newt^{i}'.

Two other annotated copies prove to be especially interesting. One, in the University of Texas, has more annotations than any other that I have seen—save, of course, E_1i and E_1a. The handwriting bears some strong resemblances to Edmond Halley's,[1] although there are also some divergences. There would be a certain logic if this could be proved to be the copy that belonged to Halley, since it must have been in the possession of someone who was truly intimate with Newton; how otherwise could we account for the tremendous number of manuscript emendations deriving from Newton?

The other copy of special interest is at present in the Fisher Library of the University of Sydney, New South Wales, Australia. Purchased in 1908 by the Hon. Andrew Bruce Smith, as part of a collection largely made up of law books, the copy was discussed in the pages of *Nature* and elsewhere, the opinion being given that

[1] This was suggested to me by D. T. Whiteside, when we examined this copy together during the Newton meetings in Austin, Texas, in November 1966.

this might be Cotes's copy, or even Fatio's.[2] It is nothing of the sort. It belonged to the Scots mathematician John Craig, as proved by the absolute identity of handwriting.[3] Craig was a friend of David Gregory, and of Colin Campbell and Colin Maclaurin.[4]

On the fly-leaf of Craig's copy of the *Principia*, someone has written (in a much later hand):

> The Amendments in this book were written by Sir Isaac's own hand. See his original MSS of his Optics in Trin. Coll. Library, Cambridge.

This is a doubly erroneous statement since the 'amendments' are not in Newton's hand at all, but rather in Craig's, nor are the original manuscripts of the *Opticks* in the Trinity College Library (but in the University Library as part of the Portsmouth Collection).[5]

The preliminary pages of handwritten notes in the Sydney copy are entitled:

> ⟨A⟩dditiones & Correctiones
> ⟨Ex Author⟩is Exemplari secundae
> editioni destinato desumptae.

> Additions & Corrections
> Taken from the Author's Copy
> intended for a second edition.

There follows a list, apparently taken from the MS Errata to E_1a, begun on these preliminary pages and continued at the end of the book. The last entry occurs at the top of a page, indicating that Craig did not interrupt his list because of lack of space on which to write it: the reference is to page 271 {262.9–10}. In the text proper someone else has corrected 'ad OK' to read 'OK ad DO' on page 14 (l. 3 bott.) {16.4}, as indicated in the printed Errata. Throughout the text, there are corrections and marginal annotations here and there, of which some appear to have been made in a different hand, but others are clearly in the hand of Craig.[6] Of the latter, there is one of great interest which occurs at the end of the Scholium to Prop. XXXIV, Theor. XXVIII, Book II (=Prop. XXXV, Theor. XXVIII, Book II of E_1). The method is used to permit bodies of different 'figures' to be 'compared together as to their resistance', and to find those 'most apt [best suited] to continue their motions in resisting mediums'. Paragraph two concludes {324.19}, 'Which Proposition I conceive may be of use in the building of ships.' Alongside this sentence Craig has written in the margin of his copy:

> Hujus meditationis occasionem ipse praebui, dum Cantabridgiae de Figura navium aptissimâ invenienda, problema celeberrimo Autori proponerem.

> The occasion for this afterthought I myself provided, when in Cambridge I proposed to the celebrated Author the problem of finding the most suitable outline for ships.

[2] Smith (1908), Smith (1908*a*), Bosscha (1909).

[3] A sample of Craig's handwriting may be seen in U.L.C. MS Add. 3971, f. 79. The name of Craig as a possible owner was suggested in correspondence with D. T. Whiteside. This copy of the *Principia* was presented to the University of Sydney, Australia, on 20 December 1961, by Miss Barbara Bruce Smith, daughter of the purchaser.

[4] See [Newton] (1959–), *Correspondence*, vol. 2, p. 501; also *D.N.B.*, *s.v.* Craig, John.

[5] U.L.C. MS Add. 3970.

[6] In the literature referring to it, much is made of the fact that there are annotations in two hands. But those that were not made by Craig turn out to be in large measure the corrections listed in the printed Errata. There are a very few changes, for example, p. 107, conclusion to Lemma XXVIII, Book I, 'De ovalibus...pergentibus' {107.31–32}, possibly in yet another hand that I have not been able to identify.

So far as I know, this is the only source of our knowledge concerning any indebtedness of Newton's to Craig on this point.

In one of his mathematical books, *De calculo fluentium* (1718), Craig described a meeting with Newton at the time when the latter was writing the *Principia*:

> When I was a young man I thought out the elements of the Calculus of Fluents about the year 1685. I was then at Cambridge, and I asked the celebrated Mr Newton to read them through before I committed them to print. This he kindly did: and to corroborate some objections also, raised in my pages against D.D.T. [=Mr v. Tschirnhaus], he offered me of his own accord the quadratures of two figures.[7]

Certainly Craig had the opportunity of suggesting to Newton the thought on the best shape for ships, just as he said.

The copy of E_1 belonging to Richard Bentley is in the Trinity College Library, Cambridge. In view of the part played by Bentley in getting Newton to agree finally to the printing of a second edition (see §6), it comes as something of a surprise that his copy is not extensively annotated. Most of the emendations come from the printed Errata, while many others are mere cross-references. For example, in the proof of Prop. LXXIV, Book I, Newton writes 'per Theor. XXXI'; Bentley's marginal note reads 'p. 192'. Again he reminds himself that in Prop. LXX, Book I, the forces tend 'ad Sphaericae superficiei puncta singula', whereas in Prop. LXXIII they tend 'ad sphaerae alicujus datae puncta singula'. There are about a dozen or so stylistic revisions proposed—for instance, Bentley would have 'mihi' follow 'comparabam' {400.11}—which may be taken as an indication of the lack of depth of his study of the *Principia*. But there occur some corrections, as 'superabat' for 'superabit' {481.16}, that were later made in E_2. Chiefly, Bentley's annotations are of interest to us as proof that he actually did read the *Principia*; and the page references he added show that he did look back and that he actually did try to see the logic of certain proofs. A dash in the margin shows his attentiveness to the introduction to Sec. XI of Book I, at the point where Newton says that he has considered centripetal forces as attractions, though more probably they are impulses {160.17–20}. Bentley observes that the corollaries to Prop. LXVI, Book I, on a case of the three-body problem {173.21 ff.}, have to do with the Moon's motions, as the Preface declares ('Haec corollaria spectant ad motus Lunae. Vide praef.'). And he has drawn a vertical line to show the importance of the concluding two sentences of the first paragraph of the Preface. As we might have expected, he corrected and 'improved' Halley's poem.[8]

I have discussed earlier (Chapter VII, §9) the contents of Fatio's copy of the *Principia*, which survives. Gregory's copy has disappeared from sight and may actually have perished.[9] It is not at present in Oxford in Christ Church or the Bodleian, nor is it to be found in Edinburgh or St Andrews. I have searched for it

[7] Translated in [Newton] (1959–), *Correspondence*, vol. 3, p. 9, n. 5.
[8] The alterations of Halley's poem are presented in the Apparatus Criticus.
[9] For Gregory's reference to this copy, see Chapter VII, §13.

in vain in London (British Museum, Royal Society) and in Cambridge. Indeed, there is no hint that such a copy is among those listed in Henry Macomber's census.[10]

Two copies with manuscript emendations in hands that I have not been able to identify are to be found in the Babson Institute (Massachusetts) and the Cincinnati (Ohio) Public Library. The Cincinnati copy has many more annotations than the Babson copy.

In the University of Louisville, Louisville, Kentucky, there is a copy of the first edition with a bookplate (on the verso of the title-page), containing the coat of arms of 'The Right Honble Charles Lord Halifax 1702'. In this copy, obviously a presentation, Newton has added in his own hand a number of further emendations to the printed Errata at the end of the volume. These are as follows:

p. 47, l. 20 {51.21} reciproce ut
p. 105, l. 22. {106.9–10} areae ovalis a recta illa abscissae incrementum s.m.f. [?][11]
p. 112 {110} dele $\frac{1}{2}$D, $\frac{1}{2}$F, $\frac{1}{2}$H.
p. 118, l. 10 {115.31} pro S lege P.
p. 426, l. 8, 9 {417.21 ff., notes xxv, xxvi} minor. [This entire entry is cancelled.]
p. 485, l. 19 {490.3} verticem μ.
p. 489, l. 30 {494.20} (per Corol. Lem. X.)
p. 490, l. 4 {495.7} $\frac{4}{3}$Tτ. l. 11 {495.14} MP ad MN.
p. 488 {492.12/13 fig.} in schemate scribatur P inter N et O.

3. OTHER ANNOTATED COPIES OF E_1: THE RELATIONS OF THE ANNOTATIONS WITH ONE ANOTHER

The annotations in these various copies prove to be most interesting when grouped together so as to show either a common source or a transcription of one from another. A typical example occurs in the correction of line 16 on page 42 of E_1 (the last line of Corol. 6, Prop. IV, Book I {44.22–5}). The original version of E_1 reads:

velocitates autem in radiorum dimidiata ratione...

In E_1a, Newton has recorded two different alterations. In the MS Errata to E_1a this becomes:

velocitates autem in radiorum subduplicata ratione reciproce...

but on text-page 16 the correction takes on a slightly different form:

velocitates autem reciproce in radiorum subduplicata ratione...

The latter appears also in E_1i. The two variant forms are identical as to content; in both, 'dimidiata ratione' becomes 'subduplicata ratione reciproce'. The only difference is in the order of the words.

[10] Macomber (1953). Of course, it is possible that Gregory's copy (and perhaps others of interest) may exist in the Macclesfield Collection in Shirburn Castle, or in other private hands and in libraries.

[11] This entry is discussed in the Guide to the Apparatus Criticus; '22' altered (erroneously?) from '23'.

Groening, who combined two separate lists of corrections, records this change twice, as follows:

p. 42 l. 16. autem reciproce in
p. 42. l. 16. subduplicata ratione reciproce

in which the first derives from the text-page (but omitting the alteration of 'dimidiata' to 'subduplicata'), whereas the second comes from the MS Errata to E_1a.

In Locke's copy and the copy in the University of Texas, the correction appears as:

velocitates autem reciproce in radiorum subduplicata ratione...

which agrees with the text-page of E_1a and E_1i. But in Fatio's copy, the corrected version reads:

velocitates autem reciproce in radiorum dimidiata ratione...

Perhaps, therefore, Fatio had copied this correction after Newton had inserted 'reciproce' but before he had changed 'dimidiata' to 'subduplicata'; alternatively, Fatio may not have bothered to write out every instance of the change from 'dimidiata' to 'subduplicata', which occurs throughout E_1. In any event, Fatio's failure to include the change from 'dimidiata' to 'subduplicata' accounts for its absence in Groening's list. In an annotated copy in the Cincinnati (Ohio) Public Library, the correction is made as in Fatio's copy, with 'dimidiata' unchanged. The Sydney copy, however, has a wholly different form of correction; 'dimidiata' is unchanged, but in place of 'reciproce' the word 'inverse' is used:

velocitates autem in radiorum dimidiata ratione inverse...

This particular correction is absent from the annotated copy in the Babson Institute.

As a second example, consider the new proof of Lemma VI {31.12–14}. Groening, following Huygens's version of Fatio's notes, presents the new proof with this introduction, 'Newtoni annotatio. Deleatur Demonstratio lemm. VI, vel legatur':

> Nam si angulus ille non evanescit, continebit arcus AB cum tang. AD angulum rectilineo aequalem & propterea curvatura ad punctum A non erit continua, contra hypoth⟨esin⟩.

This demonstration appears word for word in the MS Errata to E_1a, in E_1i and E_1a, and in the Texas copy. In the Sydney and Babson copies the word 'continebit' appears before 'angulum' rather than before 'arcus'; in the Cincinnati copy, the same is true and also 'Tangenti' appears in place of 'tangente AD'. (I do not take account of abbreviations.)

A final example: in the first line of Corol. 1 to Prop. VI, Book III, 'Hinc pondera corporum non pendent ab eorum formis & texturis' ('Hence the weights of bodies do not depend on their forms and textures'). This Corol. 1 {402.8} leads to a series of corollaries referring to a vacuum; among them, in E_1 {402.27 ff.}, a Corol. 3 begins, 'Itaque Vacuum necessario datur' ('And thus a vacuum is necessarily given [that is, must be granted]'). Following the word 'corporum', in the above statement of Corol. 1, Newton has added 'sensibilium' in E_1i, E_1a, and the MS Errata to E_1a, but it is later cancelled in E_1i. This alteration also occurs in Fatio's list, and so was copied by Huygens and appears in his list too, and thus was published by Groening. It is also present in the copy presented by Newton to Locke and in the copies in the Cincinnati Public Library and the Library of the University of Texas.

This particular correction is extremely interesting from a number of different points of view. First of all, it did not last, and does not appear in E_2; Newton has considerably rewritten this corollary. Secondly, it shows us the doubts or second thoughts Newton had, once E_1 had been printed. For a very similar emendation was noted by Greves after his visit to Newton in 1702 (Chapter II, §8), saying that Newton 'designs to alter that part' of the *Principia* in which he gives 'his proofe of a vacuum'. Greves recorded that Newton 'has writ in the margin, Materia Sensibilis; perceiving his reasons do not conclude in all matter whatsoever'.[1] As I have said, a similar alteration occurs in E_1i, E_1a, the MS Errata to E_1a, and the Cincinnati and Texas copies. But in none of these copies does it appear exactly as Greves said, for in all of them Newton has added the word 'sensibilium' to modify 'corporum', whereas Greves said Newton had written 'Materia Sensibilis'. And it is not only the wording that differs from Greves's description; the location of the change in E_1a and E_1i is on the printed line of the page, whereas Greves says explicitly that Newton had put this emendation 'in the margin'. We must conclude, therefore, that—unless Greves is a poor witness—there was yet another copy of E_1 in Newton's possession in 1702, neither E_1a nor E_1i. In the next section of this chapter we shall see what further evidence exists regarding such a copy.

4. SOME NOW-MISSING ANNOTATED COPIES OF E_1

A copy of the *Principia* with manuscript annotations, mentioned in Edleston's preface to his edition of the Newton–Cotes *Correspondence*, was apparently seen last by 'the late Mr Kidd, in 1796'.[1] Since Edleston was writing in 1850, I presume he is referring to Thomas Kidd, who died in that year, a Greek scholar and schoolmaster, born in 1770, who entered Trinity College in 1789 and received his M.A. in 1797. Edleston's report states that Kidd saw a copy of the *Principia* 'in the possession of the Rev. Thomas Jones, Fellow of Trinity College... "with an astonishing quantity of additions and corrections" in Newton's hand'. Kidd's recollections, as reported by Edleston, continue: ' "Numerous loose papers of 4to form covered with diagrams and writing were placed between the leaves in different parts of the volume," which contained also "a loose copy of Halley's laudatory verses on the *Principia*, corrected throughout by the hand of Dr Bentley." ' According to Edleston, furthermore, 'Jones stated [presumably to Kidd] that this interesting volume was given to him by Mr Davies, Senior Fellow of Trinity College, who

[1] [Tixall Letters] (1815), vol. 2, pp. 149 ff.

[1] Edleston (1850), pp. xvii–xviii.

received it from Smith, and he from Newton.' This 'Smith' is Dr Robert Smith, author of important treatises on optics and harmonics, Plumian Professor of Astronomy, and Master of Trinity (1742), founder of the 'Smith's Prize' at Cambridge. Smith was Cotes's cousin, and his successor as Plumian Professor.[2]

If this were truly a copy of E_1, annotated and corrected by Newton, it must have resembled E_1a, rather than E_1i, since it is not described as interleaved throughout, although it evidently had 'loose papers of 4to form' in 'different parts of the volume' and a 'loose copy' of Halley's poem corrected by Bentley. Even today both E_1i and E_1a contain odd sheets that have been bound in at some later time, and the *Catalogue of the Portsmouth Collection* records (for example, Sec. I, IX. B:1, 4) that certain papers were removed from E_1i. Hence it would have been in keeping with Newton's practice to have put 'loose papers' in his copy.

A curious aspect of this story is that Newton should have ever made a gift of such a working copy of the *Principia* to Smith. Smith received his M.A. only in 1715; presumably the gift would have been made after 1713 or so. The problem, then, is why Newton would have given Smith a copy of the out-dated first edition, rather than a copy of the revised and current second edition. Newton kept for himself, in his own library, an annotated and an interleaved (and annotated) copy of both E_1 and E_2. These would have little value to anyone else and probably would not have been kept by Newton except for his habit of keeping all sorts of papers, drafts, and manuscripts. What value would such a copy have had for anyone except the author? I find myself tempted, therefore, to give little credence to what is contained in a report by Edleston, based on the recollections of Kidd, who was remembering what he had seen and had been told perhaps a half-century earlier by Jones, who in turn was recalling what Davies had told him of his recollections of a gift made by Smith (prior to 1768, the date of the latter's death). I do not doubt that Kidd had seen an annotated copy of the *Principia* in its first edition, of which we know that quite a number are in existence. Kidd might possibly have been shown one of these; perhaps it was even Gregory's copy, now missing, which would have been fully annotated and might very well have had in it some of the loose papers in Newton's hand that were in Gregory's possession. Alternatively, it might even have been the copy given by Rouse Ball to Trinity.[3] Such a solution requires us to dismiss Kidd's statement that the handwriting was Newton's. But it is not uncommon to attribute to Newton any seventeenth-century manuscript annotations in a copy of E_1, especially if they later appear in E_2. Both the Babson and the Sydney copies of E_1 have been described incorrectly as containing emendations in Newton's hand.

Our interest in the copy mentioned by Edleston, however, is augmented by a letter from Bentley in London to H. Sike ('at Trinity College, in Cambridge'), which was published under the date of 31 March 1706, but which Edleston ('from internal

[2] See *D.N.B.*, *s.v.* Smith, Robert. [3] See Chapter I, §6.

and external evidence, which it is not necessary to adduce here') held to be truly 31 March 1709.[4] The letter reads, in part, as follows:

Pray tell Professor Cotes, that the book in your parcel, directed to him, is presented by Sir Isaac Newton; let him read it over with care, and I will tell him further of it in a particular letter. The bundle of wood cuts were found by Sir Isaac in his study, some of which he thinks may belong to the future sheets of his book. In the printed book are folded the MS. Sheets that Sir Isaac has now finished.[5]

This letter compounds mysteries! First, there is the possibility that this copy is the very one described by Kidd, since it is said to be a 'printed book' in which are 'folded the MS. Sheets'. Edleston was 'inclined to think that it may be the identical volume', but gave no evidence therefor and was aware that, if so, 'a link must be inserted in the chain of its transmission between Newton and Smith'. Second, Edleston supposed that the 'book here alluded to was probably a copy of the *Principia*, containing Newton's MS. corrections and additions'. Bentley, however, does not say so explicitly in his letter to Sike. The only thing that might suggest this possibility is that Bentley says that Cotes should 'read it over with care, and I will tell him further of it in a particular letter'; presumably, if Bentley merely wanted Cotes to examine the *Principia*, rather than this special copy, he would have had him read the copy in the Trinity College Library, or his own copy. No doubt Cotes had already read carefully at least a good part of the *Principia*, and was familiar with the rest.

It is also to be noted that Bentley, who was a master of exact prose, directed that Sike 'tell Professor Cotes, that the book...is presented by Sir Isaac Newton', not merely being shown to him.[6] Finally, what are we to make of the statement about the 'bundle of wood cuts' which 'may belong to the future sheets of his book'?[7] I suppose Newton may have been referring to proofs of the 'wood cuts' rather than the actual blocks; that is, Newton may very well have been intending to use again in the second edition (wherever possible) the original 'wood cuts' of the first edition—thus to save time, energy, and money.

In any event, if this should have been an annotated copy of E_1 with some loose sheets folded in, ready for the printer to use in composing E_2, then—whether this is the Smith–Davies–Jones copy or not—it must have been sent back to Newton, since in May 1709 Bentley wrote to Cotes that Newton would be 'glad' to see him 'in town here [London]' and give him 'one part' of the *Principia* 'corrected

[4] Edleston (1850), p. xvi. [5] Bentley, *Correspondence* (1842), vol. 1, pp. 231–2.

[6] In a letter to Newton of 18 August 1709, written to express his impatience until he should receive the copy for printing, Cotes wrote of his 'gratitude' to Newton, being 'so much obliged to You by Y^r self & by Y^r Book'; Edleston (1850), p. 3. This could refer to a 'present' of the *Principia* and to Newton's possible aid in having secured him the Plumian Professorship; More (1934), p. 526.

[7] Does the word 'may' indicate uncertainty on Bentley's part? Or on Newton's? And does a 'bundle of wood cuts' mean the actual wood blocks to be used in printing, implying a greater degree of preparation of the new edition than even Edleston assumed? Or are these the figures, carefully drawn for the wood-cutter, or even proofs pulled from such woodcuts? We cannot tell from this letter. But Newton did have the woodcuts made in London and sent them on to Cotes in Cambridge for the second edition. See Chapter IX, §1.

for y[e] press'.[8] Actually, Newton sent Cotes the copy for the new edition in three instalments, as follows: some time between mid-August and early October 1709, on 13 September 1710, and some time before 19 July 1711. Whether the text sent to Cotes by Newton was a bound volume, broken up into three parts, or a set of unbound sheets, we cannot say. These printed pages, with manuscript corrections on them, were evidently sent directly to the printer (the University Press) and have not been seen since. But there do survive, among Cotes's papers in the Trinity College Library, some quarto sheets which evidently had once been interleaves since the additions or emendations occur at a place on the leaf that just faces the corresponding places on the printed page. Some of these were sent to the printer and bear the compositor's mark to show where the pages ended.[9] Why these sheets should have survived is as much a mystery as the disappearance of the Gregory copy, the Smith–Davies–Jones copy (unless it is one of the copies already referred to), and the annotated copy once in the possession of Cotes.

In the course of this Chapter, I have mentioned a number of references to a copy of E_1 containing manuscript annotations by Newton, and possibly interleaves or loose pages with further manuscript annotations. To see whether any or all of these could perhaps have been the same copy of E_1, or whether there must have been two or more such copies, in addition to E_1i and E_1a, I shall first list them as follows:

1. Fatio copied out a version of a proposed form of Corol. 4 and Corol. 5 to Prop. VI, Book III (in which Newton lauded a hypothesis of Fatio's) from an annotated or interleaved copy of E_1. But no trace of these can be found in either E_1a or E_1i.

2. Various copies of E_1 contain annotations that do not agree exactly with those entered by Newton in E_1i or E_1a and may presumably have come from another such altered copy of Newton's. Also Greves referred to an emendation 'Materia Sensibilis', 'in the margin' of an interleaved and annotated copy, which is not actually in the margin of E_1i or E_1a, nor does it agree exactly with the emendation as found in E_1i or E_1a.

3. Gregory saw an interleaved copy 'ready for the press'; we shall see in §5 that he presumably made a correction in pencil in the margin.

4. Bentley sent Cotes via Sike an annotated copy of E_1, apparently without interleaves but with loose sheets of manuscript corrections, emendations, and additions.

5. Newton sent Cotes a marked-up copy of E_1 in three parts, of which some interleaves survive in the volume of Newton–Cotes correspondence in the Trinity College Library. We do not know whether this was a disbound copy of E_1 with loose interleaves.

[8] Edleston (1850), p. 1.

[9] These sheets are in the volume of manuscript Newton–Cotes correspondence in the Trinity College Library. They are discussed in Edleston (1850), for example, pp. 2, 33, 50. See also McKenzie (1966), *Cambridge University Press*, vol. 1, pp. 313, 330–6.

6. Kidd saw a copy of E_1, allegedly annotated by Newton, with 'loose papers of 4to form' between the leaves; supposedly this copy had been 'received' by Davies from Smith, Cotes's cousin, who in turn had 'received' it from Newton.

From the physical descriptions, there is no reason why nos. 5 and 6 should not be the same copy, in different states. That is, no. 6 could have been the printed pages of E_1 (no. 5) marked up by Newton, and later bound up either by Newton or by Cotes after E_2 had been printed, but without the interleaves. Some of the interleaves could then have been put in loose between the pages of the bound book, while others ended up among the letters Cotes had received from Newton. It is not at all unlikely that Cotes would have wished to preserve Newton's original copy, especially since there was always the possibility that he might be blamed for errors made by Newton rather than himself. If this is so, then the only discrepancy in Kidd's recollection (that Jones said Davies had told him that Smith got it from Newton) would be that Smith would have actually got it from Cotes (along with the Newton–Cotes correspondence).[10] This is a plausible slip in a long chain of oral transmission stretching out over more than a century; understandably, the error consists in identifying the donor with the more famous author. It may also be possible, granting the foregoing reconstruction, that it was Smith himself who bound up the annotated printed sheets of E_1 which had been used by Cotes as printer's copy for E_2.

We shall see in § 5 the high probability that no. 3 may be identical to no. 5. It is very tempting to assume that no. 4 was also identical to no. 5; in this event Cotes would have sent it back to Newton for additional corrections and additions. But then no. 3 could not also have been identical to no. 5, since Gregory (no. 5) refers to an interleaved copy, whereas Bentley (no. 4) does not. It is far from definite, however, that Gregory necessarily meant that there was an interleaf bound in following each leaf of the printed book (as in $E_1 i$); perhaps he had in mind an annotated copy of E_1, with manuscript interleaves loosely inserted. The same considerations apply also to Greves's statement (no. 2) about a copy which Newton had 'interleaved'. The possibility that all of the above-mentioned copies, nos. 1–6, could be one and the same depends wholly on the interpretation of the word 'interleaved'. In the absence of further evidence, one cannot say more.

Nevertheless, it does emerge that there was at least one other copy of E_1, not $E_1 i$ or $E_1 a$, which contained annotations by Newton and which is not presently available. This was the copy used by the printer for the composition of E_2. It may or may not be identical with another such copy that was presumably seen by others before E_2 appeared.

[10] Smith proves to be an unreliable witness on one other occasion; see [Newton] (1967–), *Mathematical papers*, vol. 1, pp. 17–18.

5. THE PRINTER'S COPY FOR E_2: GREGORY'S MISTAKEN CORRECTION IN THE MARGIN

When, on 11 October 1709, Newton wrote to Cotes that he had sent him through the agency of Whiston 'the greatest part of y^e^ copy of my Principia in order to a new edition'[1] he did not describe the physical appearance of the text he had sent. No doubt it consisted for the most part of the printed pages of E_1, with alterations made right on the lines of print, between the lines, and in the margins. From the survival of interleaves among the volume of Newton–Cotes correspondence in the Trinity College Library, we may be sure that these pages were—in part or entirely—interleaved. There were also supplementary handwritten sheets that survive. (See Plate 12.) But we have no evidence as to whether Newton sent Cotes a disbound copy of E_1 (perhaps interleaved when first bound up), or a set of unbound interleaved sheets. Since the copy used for composing the second edition has disappeared from view, we must glean what information we can about it from other sources.

I believe the evidence points strongly to this printer's copy having been the interleaved copy shown by Newton to David Gregory on 21 July 1706.[2] It will be recalled that Gregory reported that the copy he had been shown was 'corrected for the Press'. Of course, Gregory may possibly have seen E_1i, which, like the copy seen by Gregory, is interleaved and which has all the appearance of being so corrected although—as may be seen in the Apparatus Criticus—it was not the actual press copy. Presumably the press copy was based on E_1i, and to some extent on E_1a, as well as on manuscript sheets such as are now preserved in the Portsmouth Collection (U.L.C. MS Add. 3965) and at the Royal Society (among the David Gregory papers).

According to Gregory, as we have seen, Newton's revision in the copy prepared for the press 'is intirely finished as farr as Sect. VII, Lib. II. pag. 317', and 'this same Sect. VII' is taken by Newton 'to be the hardest part of the book'.[3] I suppose that Gregory meant 'hardest part' to revise.

Now this Sec. VII of Book II was rather completely redone in E_2. But the first two propositions (XXXII and XXXIII) are essentially the same in E_2 (and E_3) as in E_1, save that Corols. 6, 7, and 8 to Prop. XXXIII in E_1 have been suppressed in E_2 (and E_3), with the result that the old Corol. 9 has become Corol. 6. These suppressed corollaries begin on the bottom of page 321 of E_1. Thus when Newton sent the first 320 pages to Cotes for printing in September 1708, he need not then have fully worked out the final form of the remainder of Sec. VII.[4] The difficulty

[1] Edleston (1850), p. 4.

[2] This is no. 3 on the list at the end of §4. It has been described earlier in Chapter VII, §13.

[3] [Gregory], *Memoranda* (1937), p. 36.

[4] The fate of the remainder of this Sec. VII is described in full in the Apparatus Criticus {318 ff.}. Proposition XXXIV was suppressed, Prop. XXXV (together with its scholium) became Prop. XXXIV, and Props. XXXVI–XL were dropped and replaced by a new set, Props. XXXV–XL. The concluding Scholium Generale, with some revisions, was transferred to the end of Sec. VI; see note at {307.13}.

of this Sec. VII could, in fact, account for the delay in sending on to Cotes the second instalment of the printer's copy. If these copies are the same, then Gregory's statement explains both the division at page 320 of E_1 and the subsequent delay.

But there is a different kind of evidence that points strongly to the identification of these two copies, or at least that indicates a very high probability that Gregory had seen the copy sent by Newton to Cotes.[5] The information is to be gleaned from a letter written by Cotes to Newton on 20 May 1710.[6] In it Cotes took up with Newton a proposed alteration of Corols. 5 and 6 to Prop. XXIV, Book II, of which the former reads as follows {295.14–15}:

Corol. 5. Et universaliter, quantitas materiae pendulae est ut pondus & quadratum temporis directe, & longitudo penduli inverse.	Corol. 5. And universally, the quantity of matter in the pendulous body is as the weight and the square of the time directly, and the length of the pendulum inversely.

Written as an equation, this corollary states correctly that

$$M \propto \frac{WT^2}{l}$$

or

$$T^2 \propto \frac{M}{W}\cdot l = \frac{l}{g}.$$

According to Cotes's letter, 'You have substituted in Y^r copy' the expression 'quadrato-quadratum temporis directe' in place of 'quadratum temporis directe'. This is an error, as Cotes saw on examining the new statement, since it would yield the incorrect result

$$T^4 \propto \frac{l}{g}.$$

Cotes observed to Newton that 'I find written in y^e margin of Y^r book by a different hand' the expression 'quadr. quadratum temporis (credo)'. Cotes was no doubt right in having concluded, 'This marginal note, not Y^r own judgment, was I beleive y^e occasion of Y^r making the alteration.'

In whose hand, not Newton's, was this marginal note written? We do not know for certain, since the pages of E_1 sent by Newton to Cotes, corrected for the new edition, are not available. But I believe that it was David Gregory who was responsible for the error and hence that it must have been Gregory who had entered the correction in the margin with the word 'credo' in parentheses. For in Gregory's own 'Notae', apropos this Corol. 5 (and Corol. 6), he has written:

Crediderimque perperam scriptum quadratum temporis pro quadrato-quadratum temporis in hoc et seq: Cor:	And I should think *quadratum temporis* is written incorrectly for *quadrato-quadratum temporis* in this and the following Corollary.[7]

[5] I purposely refrain from making any guesses as to whether this final press copy was the same one sent earlier by Newton to Cotes through Bentley and Sike.

[6] Edleston (1850), pp. 20–3.

[7] Royal Society copy.

I believe it would be an unnecessary and unwarranted multiplication of hypotheses to suppose that two of Newton's associates would have each independently conceived this very same error and that both would then also have used the same verb 'credo' with respect to it.

Apparently this was not the only occasion on which someone else had written in the margin of Newton's copy of E_1 a correction that Cotes found to be wrong. Writing to Newton on 16 February 1711/12,[8] Cotes called attention to another error just like the one above, and also consisting of the addition of an extra (and incorrect) 'quadratum':

The Rule delivered in this Scholium puts me in mind of a mistake in the New Edition of Your book which I did not observe till it was too late. In ye 16th Corollary of ye LXVIth Prop: of Lib: 1, or in page 166, line 9th of ye New Edition You will find *ut quadratum temporis periodici corporis P directe* &c. So You had altered it in Your Copy, but I think it should be as in ye former Edition *ut tempus periodicū*. Over against Your alteration there is written in ye margin with a black lead pencil by another hand *quadr. temporis period.* which I suppose You depended upon without considering the thing Your self.

Although this particular 'correction', unlike the former one, does not appear in Gregory's 'Notae', it is so similar (both in being written in the margin and in involving an extra 'quadratum' which is wrong) that I assume both to have come from the same author. We may, however, be astonished to find Gregory—or anyone else, for that matter—writing in Newton's copy being prepared for the press.

Cotes's reference to corrections in another hand on page 184, line 3 {180.22}, and page 304, lines 27 and 29 {295.15, 17}, proves conclusively that the copy of E_1 which Newton had sent him for printing was neither E_1a nor E_1i, since there is no such alteration in the margin of pages 184 or 304 of either of those copies. But the change from 'tempus periodicum' to 'quadratum temporis periodici' in Corol. 16 to Prop. LXVI, Book I {180.22}, appears in the text (not the margin) of E_1i, though not in E_1a.[9] And in the text of Prop. XXIV, Book II, in both E_1i and E_1a 'quadratum' is altered to 'quadrato-quadratum' in both Corol. 5 and Corol. 6 {295.15, 17}, indicating that Newton entered at least some of the textual emendations of the final copy of E_1 sent to Cotes in one or both of his own copies of E_1, which to some extent thus became his record copies.

If it was Gregory who on two pages of the *Principia* wrote in an extra 'quadratum', as seems to have been the case, he may very well have done so during his visit of 21 July 1706, when Newton showed him the corrected copy of E_1, 'intirely finished as farr as Sect. VII. Lib. II. pag. 317', for the second of these changes occurs in Sec. VI, p. 304. Of course there is the possibility that Gregory had made this change at an earlier time.

[8] Edleston (1850), p. 64.

[9] This error, caught by Cotes only after the sheet containing this corollary had been printed, appears among the printed Corrigenda to E_2 and is corrected in the text of both E_2i and E_2a.

Reference has been made here and there in this Introduction to a copy of E_1 presented to the Trinity College Library by W. W. Rouse Ball. The original owner evidently collated carefully the texts of E_1 and E_2, and added to his copy of E_1 every alteration and addition, thus producing a text generally conforming to E_2. This required interleaves or additional sheets for new material: for instance, Cotes's preface, the new Prop. X in Book II, the new part of Sec. VII of Book II. Even the index, made for E_2, has been entered, as also the revised dedication page and title-page (though with date '1714' for 'MDCCXIII'). This description at once suggests the possibility that, rather than being taken from the printed E_2 as I have indicated, it might even be the press copy, or at least a record copy made as E_1 was being printed. But there is internal evidence aplenty to the contrary: the interleaves of the copy used in printing are in the Trinity College Library, in MS R.16.38, and may be positively identified by the compositor's marks; nor could it have been a record copy, since there are clear indications—as in Prop. X, Book II—that the hand-written portions were based on the printed text of E_2. There are no alterations made to Halley's ode. The date '1714' may have arisen (consciously or unconsciously) from the copy's having been made in that year.

6. RICHARD BENTLEY AND THE 'PRINCIPIA'

The new edition of the *Principia*, to which David Gregory referred in March 1708,[1] did not get under way at once. We do not have many particulars concerning the actual first steps, and so we can only put the story together from bits and pieces. At this time Newton seems to have been in more of a mood for publishing than usual, having permitted the *Opticks* to appear in English in 1704 and in a Latin redaction by Samuel Clarke in 1706 (with some daring new Queries). Since the *Principia* had been out of print for about two decades, the pressure on Newton must have been great to produce at last the new edition (or, at least, a corrected reprint) which had been the topic of rumour and gossip for so many years.

At this opportune moment, Richard Bentley, Master of Trinity College, Cambridge, decided to take a hand. We do not know whether Bentley made the first steps, or whether Newton sought his assistance, but we may guess with some assurance that Bentley must have initiated the project. How else could we explain the choice of Bentley, since he was no scientist, nor a mathematician, but a classicist, in fact the foremost classical scholar of his age.[2] When John Conduitt, husband of Newton's niece and his successor at the Mint, asked him why Bentley had been permitted to pocket all the profits of the second edition, Newton merely replied, 'Why, he was covetous.'

In an effort to justify the choice of Bentley, his most recent biographer would have us believe that he was in some degree prepared for this assignment, since it

[1] See §1.
[2] On Bentley, see Monk (1830), and White (1965).

is 'more than probable that he attended Newton's lectures'[3] and so would have learned from Newton himself of the new physics and astronomy. But the actual evidence seems to me to make it unlikely that Bentley had gone to Newton's lectures, which may not even have been on physics and cosmology while Bentley was a student,[4] but were certainly on arithmetic and algebra.[5]

Some time in 1691, Bentley decided to study Newton's *Principia*, possibly for theological purposes. Alternatively, Bentley's interest in the *Principia* may have originated in his studies of Epicurus and Lucretius, on whom he had been working since 1689. Through William Wotton acting as an intermediary, John Craig sent to Bentley, in June 1691, a course of reading preparatory to a study of the *Principia*. Craig's directions were discouraging, and included too many titles for Bentley.[6] Bentley next turned to Newton himself, who recommended a shorter course of preparation and repeated to Bentley the instructions he had printed in the *Principia*: he should read the first three sections of Book I and then go on to Book III.[7]

In 1692 Bentley gave the inaugural Boyle Lecture,[8] consisting of six sermons 'On the Confutation of Atheism'.[9] The second set of three contained such a confutation 'from the Origin and Frame of the World'. The last two in particular were based on an interpretive exposition of Newtonian physics and cosmology. Before printing them, Bentley sent the texts to Newton for comment, and Newton favoured Bentley with a most significant set of four letters that are especially precious as a

[3] White (1965), p. 47.

[4] See Edleston (1850), pp. xciii–xciv. According to D. T. Whiteside (in a private communication), 'By statute Newton had to give 30 formal lectures a year, of which 10 were to be deposited. Since we know only Newton's deposited lectures we cannot be sure what else Newton was lecturing on when Bentley was a student.'

[5] His first biographer, James Henry Monk, argued that Bentley's cultivation of 'mathematical science with effect, may be inferred from the close and logical character of his style, as well as from his constantly recommending and patronizing such studies in others'. Aware that the 'true system of the universe, and the proper methods of philosophical investigation, had not yet become public by the writings of Newton', Monk nevertheless believed that Bentley received 'the light of the Newtonian discoveries [as it] was partially revealed to Cambridge before the rest of the world, by the lectures of the philosopher himself, delivered in the character of Lucasian Professor. These Bentley had an opportunity of attending.'

Monk was 'induced to believe' that Bentley had gone to these lectures 'by his selection of the Newtonian discoveries as a prominent subject of his Boyle's Lectures, and his familiarity with the train of reasoning by which they are established'; Monk (1830), pp. 6–7.

R. J. White similarly points to 1692, when Bentley 'came to deliver the Boyle Lectures in defence of Christianity', and 'it was to Newton that he turned for his arguments from physics and cosmology'. White believes that Bentley made an 'application of the Newtonian rigour in close and precise examination of data in advance of any indulgence in generalization', and he finds 'a profound sense in which [Bentley's] *Phalaris* may be said to represent the Newtonian revolution in classical studies, if not in literary criticism in general'. Later on, White says, 'there seems to be little doubt that he attended Newton's lectures as Lucasian Professor of Mathematics while at Cambridge', and even refers to Bentley as Newton's 'pupil'; White (1965), pp. 47, 71. But, like Monk, he cites no evidence to support his claim.

[6] Printed in Brewster (1855), vol. 1, pp. 465–9; also in [Newton] (1959–), *Correspondence*, vol. 3, pp. 150–1.

[7] [Newton] (1959–), *Correspondence*, vol. 3, pp. 155–6. Bentley certainly read the *Principia*, at least in part; see §2.

[8] These were founded by the terms of the will of Robert Boyle (d. 30 December 1691), who had provided an endowment to pay £50 for an annual Lecture (consisting of eight sermons or discourses) on the evidences of Christianity.

[9] See Miller (1958).

record of Newton's views on the necessity of God in the establishment of our universe.[10]

Bentley maintained his friendly relations with Newton,[11] which took on a certain importance after Bentley became Master of Trinity in 1700.[12] Concerned as he was in the advancement of learning in general and in the promotion of the new (Newtonian) natural philosophy, Bentley could not help but be aware of the need for a new edition. By the spring of 1708, he had evidently obtained Newton's assent, and had gone ahead to have one sheet (pp. 1–8) composed at the University Press for Newton's inspection.[13] Gregory was not the only one to hear the welcome news that the *Principia* was to become available again, for on 10 June 1708 Bentley wrote to Newton that 'the prospect' of the new edition 'has already lower'd ye price of ye former Edition above half of what it once was'. Then Bentley went on to describe the appearance of the new edition:

I have here sent you a specimen of ye first sheet, of wch I printed about a Quire; so y^t the whole will not be wrought off, before it have Your approbation. I bought this week a hundred Ream of this Paper you see; it being impossible to have got so good in a year or two (for it comes from Genua) if I had not taken this opportunity with my friend Sr Theodore Jansen, ye great Paper merchant of Britain. I hope you will like it, and ye Letter too, wch upon trials we found here to be more sutable to ye volume than a greater, & more pleasant to ye Eye. I have sent you like wise ye proof sheet, y^t you may see, what changes of pointing, putting letters Capital, &c I have made, as I hope, much to ye better. This Proofsheet was printed from your former Edition, adjusted by your own corrections and additions. The alterations afterwards are mine: which will shew & justify themselves, if you compare nicely the proof sheet with ye finished one. The old one was without a running Title upon each page, wch is deformd. Y^e Sections only made with Def. I. Def. II. which are now made full & in Capitals DEFINITIO. I &c. Pray look on Hugenius de Oscillatione, wch is a book very masterly printed, & you'l see that is done like this. Compare any period of ye Old & New; & you'l discern in ye latter by ye chang of points and Capitals a clearness and emphasis, y^t the other has not: as all y^t have seen this specimen acknowledg. Our English compositors are ignorant, & print Latin Books as they are used to do English ones; if they are not set right by one used to observe the beauties of ye best printing abroad...

Note, y^e [*sic for* y^t?] ye Print will look much better, when a book is bound & beaten.[14]

Bentley's letter does not read as if Newton had actually turned over to him the whole of the final copy of the *Principia* for the new edition. For the letter opens with an expression of Bentley's 'hope [that] you have made some progress towards

[10] The first printings of these sermons and letters are reproduced in facsimile in [Newton] (1958), *Papers and letters*, pp. 279–394. The correspondence appears in [Newton] (1959–), *Correspondence*, vol. 3, pp. 233–40, 244, 253–6. The sermons are also available in Bentley (1836–8), *Works* [Dyce].

[11] A letter of 21 October 1697 mentions a small group meeting on one or two evenings a week at Bentley's lodgings (as librarian) in St James's Palace: John Evelyn, Sir Christopher Wren, John Locke, Isaac Newton.

[12] In 1705 Queen Anne knighted Isaac Newton in the drawing-room of the Master's Lodge of Trinity.

[13] W. G. Hiscock found in the University Press Accounts for the year ending 3 November 1708 the following entries:

'From Dr. Bentley for a sheet of Sir Isaac Newton's Principia

'Compos'd one sheet of Sir Isaac Newton's Principia for a Speciment	6s	0
'Corrected ye said Sheet	1s	0
'Workt at Press 25 Sheets for Speciments	1s	0'.

See [Gregory] (1937), *Memoranda*, p. 41.

[14] MS R.4.47, Trinity College, f. 19.

finishing your great Work, wch is now expected here with great impatience'. The letter closes with Bentley's 'hope to see you in about a forthnight, & by y^t time you will have examind this Proof, and thought of what's to come next'. It is thus not clear whether Bentley had received any copy from Newton, save possibly the corrected text of pages 1–8. Since Bentley was famous as a promoter and 'operator', one may even suspect that, following a tentative assent to the new edition on the part of Newton, Bentley had rushed to get the first sheet composed and printed off in an attempt to keep pressure on Newton to go ahead with a new edition. But Newton must certainly have made some commitment or else Bentley would not have bought the paper on which the new edition was to be printed.[15]

Nevertheless, from Bentley's letter, and from the history of subsequent events, it is clear that Newton was then either not willing or not ready to turn over the whole *Principia* to the University Printer for a new edition. Newton's reluctance in this matter in June 1708 may have been due to his natural disinclination to publish his works, which may have hardened again once he had allowed the *Opticks* to appear in 1704 and 1706. Alternatively, he may have wished to wait until the whole job of revision had been completed. Furthermore, he may have had second thoughts about having Bentley in sole charge, because then he himself would have had to take on the responsibility for reading the proofs of the mathematical demonstrations. If this were the case, then we may see how wise Bentley was to have gained the services of Roger Cotes of his own College, a brilliant young mathematician who actually did edit the second edition under Newton's direction and Bentley's immediate supervision.

Bentley simply could not help editing, altering, and 'improving' every text that passed through his hands. Newton's *Principia* was no exception. Bentley wrote Newton:

> In a few places I have taken ye liberty to chang some words, either for ye sake of ye Latin, or ye thought itself. as y^t in p. 4. *motrices, acceleratrices et absolutas* I placed so, because you explain them afterwards in y^t order. But all these alterations are submitted to your better judgment; nothing being to be wrought off finally without your approbation.[16]

As may be seen in the Apparatus Criticus {5.10}, Newton accepted this correction, which appears in E_2 and E_3. Since I have been unable to find any copies of the original proof-sheet, I cannot say with certainty which other alterations in the first eight pages of E_1 were due to Bentley, but they might be any that occur in pages 1–8 of E_2 and are not to be found in E_1i or E_1a.[17]

[15] Bentley's letter says, 'This Proofsheet was printed from your former Edition, adjusted by your own corrections and additions.' It is not clear whether Bentley copied out these 'corrections and additions' or whether Newton gave him his marked-up copy of the *Principia* or a list of alterations by page and line; many such lists, at many different stages of revision of the *Principia*, are to be found in U.L.C. MS Add. 3965.

[16] MS R.4.47, Trinity College, f. 19.

[17] Examples are: {5.5}, {6.11}, {6.12}, {6.13}, {6.14}, {7.8}, {7.13}, {8.30}, {8.34}, {8.35}. But 'corpora tendunt' {3.5}, which was 'corpus tendit' in E_1, was not altered by Bentley since this change appears in both E_1i and E_1a.

Bentley's letter to Newton (10 June 1708) describing his proposed new edition was first published by David Brewster, who admitted that he had 'not been able to discover any reason why the printing of the second edition, thus fairly begun, and for which paper was purchased, should have been discontinued'.[18] A half-century or more later, Rouse Ball restated this opinion, that 'for some reason, of which we now know nothing, [Bentley] abandoned the work'. But Rouse Ball also suggested that possibly Bentley 'found the task involved a greater knowledge of mathematics than he possessed'.[19]

Possibly, Brewster and Rouse Ball may have been victims of a simple misunderstanding. Bentley sent Newton a 'specimen of the first sheet' for his 'approbation', informing him of his views about type, format, and appearance. This was in June 1708. Nine months later, in March 1709, we shall see Newton turn over to Bentley a corrected copy of the *Principia*, to be sent on to Roger Cotes in a package addressed to Professor H. Sike. Yet another eight months would pass before Newton, in October 1709, finally passed on to Cotes a part of the text for the printer. Newton was never one to rush into print in any case, and I do not see why this delay need be interpreted in any other way than as a normal lapse of time between the specimen and Newton's final willingness to have printing begin. Composing and printing an eight-page specimen is not an example of 'printing'—to use Brewster's words—'fairly begun' and then 'discontinued'. It is rather an instance of the custom, still followed by good publishers, of submitting a printed specimen to the author before going ahead with the composition of the book. There is no reason why the specimen should not be prepared well in advance of delivery of the final copy.

In the absence of evidence, we cannot know whether the decision to obtain the services of someone like Cotes was Newton's or Bentley's. Furthermore, we have no way of telling whether such a decision was made at the very start, even before the specimen was printed, or at a later time. In any event, the printing as such was not really begun until some time later, in 1709, when Newton turned over to Cotes the first of three instalments of the final revised copy. Then, since the first eight pages were already composed, the printers began with page 9.[20]

In part, Newton may have been a little reluctant to permit Bentley to be in charge of the new edition because he may have known that Bentley was famous for his freedom (perhaps licence) in altering the writings of others. Bentley ('the Cambridge Aristarchus') freely rewrote the poetry of Horace and of Milton and —to quote Conduitt—'spared neither the living nor the dead'.[21] He even revised Halley's ode at the beginning of the *Principia*, apparently without a word to the author, and rewrote the title-page in Newton's own copy (E_1a; see Plate 13 and the

[18] Brewster (1855), vol. 2, p. 250.

[19] Ball (1893), *Essay*, p. 133.

[20] This is clear from the printing records, which show that work began with signature C (pp. 9–16); hence signature B (pp. 1–8) must either have been standing in type or have been printed off. On the composition and printing of E_2, see McKenzie (1966), *Cambridge University Press*.

[21] Keynes MS 130, King's College, Cambridge.

Guide to the Apparatus Criticus, §1). Bentley was the kind of man who could be perfectly capable of ignoring Newton's objections and printing the *Principia* as he, Bentley, saw fit. Had I been in Newton's place, I would not have objected to Bentley's changing some words 'for the sake of the Latin', since Bentley was without question a master of Latin style. But I would certainly have had more than one serious *arrière pensée* about Bentley's proposal of introducing changes 'for the sake of...the thought itself'.[22]

To see Bentley in his role of editor, supervising the new edition in Cotes's care, we may turn to a minor but nevertheless revealing incident that occurred in 1709, soon after the new edition had really got under way. Almost immediately after Newton had sent Cotes the first instalment, he wrote (on 11 October 1709) that he had forgotten

to correct an error in the first sheet pag 3 lin 20, 21, & to write *plusquam duplo* for *quasi quadruplo* & *plusquam decuplo* for *quasi centuplo*.

I forgot also to add the following Note to the end of Corol. 1 pag. 55 lin 6. Nam datis umbilico et puncto contactus & positione tangentis, describi potest Sectio conica quae curvaturam datam ad punctum illud habebit. Datur autem curvatura ex data vi centripeta: et Orbes duo se mutuo tangentes eadem vi describi non possunt.[23]

At issue was a statement in Def. V {3.28, 29}, part of the discussion of Def. V {3.8–4.13} that does not appear in E_1 but is written out in full in Newton's interleaved copy (E_1i). Newton is considering the motion of a projectile fired from a mountain-top in a direction parallel to the horizon. The force of the gunpowder will give the projectile a certain velocity, so that it will move 'in a curve line to the distance of two miles before it falls to the ground'. But if the resistance of the air were taken away, and the projectile were to have double the velocity ('dupla cum velocitate'), then, says Newton, it would be carried about four times farther ('quasi quadruplo longius pergeret'), while with ten times the velocity ('decupla cum velocitate') it would be carried about one hundred times farther ('quasi centecuplo [*corrected to* centuplo] longius'). Newton's error is an easy one to make. He had assumed that the forward distance would be proportional to the square of the velocity, which would be true for uniformly accelerated linear motion, though not for uniform linear motion, in which the distance is proportional to the velocity. In the analysis of the path of a projectile, it is only the component of descent that is accelerated, and the linear forward component of the motion of projection is (save for air resistance) uniform. Newton had not yet found this error when he gave Bentley the text on which the specimen of eight pages had been composed. Evidently, he looked at these proofs after sending the copy on to Cotes and realized that he had mistakenly assumed that the forward component of the projectile's motion would be accelerated rather than uniform. In his own

[22] See Bentley's letter to Newton (10 June 1708), printed above, in which he wrote that he was enclosing the first sheet.

[23] Edleston (1850), pp. 4–5. This is the very same letter in which Newton announced to Cotes that he had sent on 'the greatest part of y^e copy of my Principia in order to a new edition'.

copy of E_1i he corrected the numbers, writing in 'plusquam duplo' and 'plusquam decuplo' (as in the letter to Cotes) for 'quasi quadruplo' and 'quasi centuplo [*changed from* centecuplo]'. Compare the previous alteration on a badly burned page of manuscript in Newton's hand (U.L.C. MS Add. 3965, f. 733), containing a long list of corrections. There he cancelled 'quasi duplo' and 'quasi decuplo', replacing them by 'quasi quadruplo' and 'quasi centecuplo [*sic*]'. I assume that Newton was taking into consideration the slight increase in distance arising from the descent to the curved surface of the Earth (considered at rest), rather than to a plane.

In his letter to Cotes, Newton did not refer to his manuscript text, or to the first edition (E_1), but rather to 'the first sheet pag 3 lin 20, 21'—which accords with the printed pages of the second edition (E_2). Hence there can be no doubt that Newton had before him one of those specimen sheets (pp. 1–8) composed a year earlier by Bentley. Newton assumed that, since the new copy for printing had barely arrived, there would be plenty of time in which to make the new correction. But he reckoned without Bentley, who rushed ahead with all possible expedition, perhaps fearing that Newton might change his mind about the new edition.

The response to Newton's letter of 11 October 1709 is dated 20 October. Although Newton had addressed himself to Cotes, it was Bentley who replied, explaining:

> Mr Cotes, who had been in ye Country for about a month returnd hither ye very day, y^t^ Dr Clark brought your letter, In which, I perceive, you think we have not yet begun your book; but I must acquaint you y^t^ five sheets are finely printed off already...
>
> Your new Corollary, which you would have inserted, came just in time: for we had printed to the 50th page of your former Edition; & y^t^ very place, where the Insertion is to be, was in the Compositors hand.

This letter not only shows how rapidly the printing was moving ahead, but implies that the printing of the first forty pages was being done in part during Cotes's absence. As with the first edition, sheets were printed off as they were approved, since otherwise there would be required vast quantities of type and facilities for storing standing type for 500 pages.

Newton's error had already been caught, presumably by Cotes, so that the printed page read correctly 'quasi duplo' and 'quasi decuplo', as in the second and third editions. Bentley did not know that this had actually been Newton's first revision, before he had decided to change 'quasi' to 'plusquam'. Bentley now had to get Newton to accept his own first correction, or else reprint a two-page leaf in order to make a cancel. Obviously, this would be troublesome, and there would be an extra cost. Bentley therefore wrote to Newton as follows:

> The correction in the first sheet, wch you would have *plusquam duplo*, & *plusquam decuplo*, was provided for before: for we printed it *quasi duplo* and *quasi decuplo* which you know amounts to ye same thing; For *Quasi* tendres either the Excess or ye Defect; & in my opinion, since in y^t^ place you add no reason why it will be plusquam, tis neater to put it *quasi* undetermind, and leave ye reader to find it out.[24]

[24] MS R.4.47, Trinity College, f. 20.

Newton hardly needed such a schoolmasterish explanation of the meaning of 'quasi' and 'plusquam'. Bentley, however, was quite correct that Newton had not in that place given any 'reason why it will be plusquam'. Newton apparently accepted Bentley's ruling. In any event, he did not insist on having the sheet reprinted and he did not include this 'quasi' among the printed Errata. Furthermore, he did not even alter the text of either E_2i or E_2a from 'quasi duplo' and 'quasi decuplo' to 'plusquam duplo' and 'plusquam decuplo', nor did he revise these phrases in E_3 {3.28, 29}. An amusing sequel is that Andrew Motte, in his English translation (1729), left out this adverb altogether, writing: 'the same [projectile], if the resistance of the Air was took away, with a double or decuple velocity, would fly twice or ten times as far'.

PART FOUR

THE SECOND AND THIRD EDITIONS OF THE 'PRINCIPIA'

CHAPTER IX

THE SECOND EDITION OF THE 'PRINCIPIA'

1. ROGER COTES AS EDITOR OF THE SECOND EDITION (1709–1713)

NEWTON was extremely fortunate in having Roger Cotes as editor of the new edition. Cotes was not merely an able mathematician, fully capable of coping with the difficulties of the *Principia*; he was sufficiently gifted to make important suggestions to Newton which improved the second edition immeasurably. Perhaps of even greater significance for the job at hand was Cotes's gentle and modest personality. He was able to criticize Newton and correct his errors without ever antagonizing him—no mean feat! In 1709 Cotes (who had been elected a Fellow of Trinity College on receiving his M.A. in 1705) had held for two years the inaugural appointment as Plumian Professor of Astronomy in Cambridge. He died young (aged 34) in 1716, and of him Newton said, 'If Mr. Cotes had lived we might have known something.'[1]

We do not know anything at all concerning the new edition between June 1708, when Bentley had one sheet of the *Principia* composed, and the spring of 1709, when Newton gave Bentley a copy of the *Principia* to be sent on to Roger Cotes through the agency of Professor Henry Sike. I have discussed in the preceding chapter the impossibility of determining precisely which copy of E_1 this may have been. But whatever copy it was, it did contain 'folded' in it 'the MS sheets that Sir Isaac has now finished'. The latter, and perhaps the book too (especially if it was the copy corrected for printing), must have been returned to Newton, since six months later he sent Cotes the first portion of the final version so that printing could be got under way.[2]

Two months after Bentley had sent Cotes a copy of the *Principia* through Sike, he wrote directly to Cotes (on 21 May 1709) that he had 'waited to day on Sr Isaac

[1] Brewster (1855), vol. 1, p. 461; Edleston (1850), p. lxxvii.

[2] There is an exemplary edition by Edleston, to which reference has been made frequently in these pages. The letters that passed between Cotes and Newton will appear in vol. 5 of the Royal Society's edition of Newton's *Correspondence*. Edleston had available to him only the volume of Cotes–Newton manuscripts in the Trinity College Library. Hence his edition is based on drafts of Cotes's letters to Newton, of which the originals are in the Portsmouth Collection, now in the University Library, Cambridge, but then still at Hurstbourne Castle. Some few letters in the Portsmouth Collection do not exist in draft in the volume in the Trinity College Library and so are missing from Edleston's edition. In the transcripts given of these letters in this Chapter, and elsewhere in the present edition, I have adjusted the text to conform to the final version whenever the letter in question may be found in U.L.C. MS Add. 3983.

Newton, who will be glad to see you in town here, and then put into your hands one part of his Book corrected for ye press'.[3] Evidently Newton was not quite so willing as Bentley indicated to set the new edition in train, for three months later (18 August 1709) we find Cotes writing to Newton of his 'earnest desire' to 'see a new Edition of Y^r Princip.' and his consequent impatience to 'receive Y^r Copy of it which You was pleased to promise me'. He assured Newton that he was 'so much obliged to You by Y^r self & by Y^r Book' that he would 'take all the Care I possibly can that it shall be correct'.[4] He then told Newton he had been examining Corol. 2 to Prop. XCI, Book I, and had 'found it to be true' by using the table of integrals printed by Newton in the treatise *De quadratura* accompanying the *Opticks*, and said that he had checked over a portion of Newton's tract and believed that he had found two errors. In reply (11 October 1709) Newton told Cotes that he had sent him a great part of the copy for the new edition.[5] He acknowledged the correction of the two errors which Cotes had found, and then stated his policy about editorial revisions in a rather astonishing but characteristic way:

> I would not have you be at the trouble of examining all the Demonstrations in the Principia. Its impossible to print the book w^th^out some faults & if you print by the copy sent you, correcting only such faults as occurr in reading over the sheets to correct them as they are printed off, you will have labour more then it's fit to give you.[6]

As we shall see, although Newton repeated similar instructions to Cotes more than once, he reckoned without Cotes's extreme conscientiousness. Cotes simply could not permit so much as a single paragraph to be printed unless he had first checked every step of the demonstration or calculation, or had read it through carefully, line by line, for sense.[7]

Newton's letter to Cotes ended with a reference to a Mr Livebody, 'who made the wooden cutts', and who 'thinks that he can sett the cutts better for printing off then other composers can, and offers to come down to Cambridge & assist in composing'.[8] From this letter we may gather not only that Newton was having the cuts made in London, but also that he had sent some cuts to Cotes along with the

[3] Edleston (1850), p. 1.

[4] *Ibid.* p. 3.

[5] See Chapter VIII, §6.

[6] Edleston (1850), p. 5. Newton's attitude may be contrasted with Bentley's. Bentley replied to Newton's statement about Cotes not 'examining all the Demonstrations' as follows: 'You need not be so shy of giving Mr Cotes too much trouble: he has more esteem for you, & obligations to you, than to think y^t trouble too grievous: but however he does it at my Orders, to whom he owes more than y^t. And so pray you be easy as to y^t; we will take care y^t no little slip in a Calculation shall pass this fine Edition.'

[7] The only occasion on which Cotes decided not to check Newton's text in this way was at the very end of the job, when Newton sent him the new conclusion to the discussion of comets, completing Book III; see Cotes's letter to Newton of 23 October 1712, printed in Edleston (1850), p. 143, reading in part, 'I have not observ'd anything of moment to be altered in the Theory of Comets.'

[8] Bentley replied for Cotes ('who had been in ye Country for about a month') on 20 October 1709, concerning this Livebody or Lightbody: 'I proposed to our master Printer to have Lightbody come down & compose, which at first he agreed to; but the next day he had a character of his being a mere Sot, & having plaid such pranks y^t no body will take him into any Printhouse in London or Oxford; & so he fears he'll debauch all his Men. So we must let Him alone; and I dare say we shall adjust the Cuts very well without him.' MS R.16.38, Trinity College.

printer's copy.[9] For he concluded the letter: 'When you have printed off one or two sheets, if you please to send me a copy of them I will send you a further supply of wooden cutts.' This sentence also implies that Newton did not necessarily expect to receive proofs, but he did hope that Cotes would send him at least the first, and presumably later, sheets as they were 'printed off'.[10]

Within six months Cotes was able to report to Newton (on 15 April 1710) that the University Press had 'printed so much of y^e Copy You sent us y^t I must now beg of You to think of finishing the remaining part, assoon as You can with convenience'.[11] That is, 224 pages had been printed off, corresponding to the first 251 pages of E_1. Since E_2, when completed, was to contain 484 pages of text, the job was nearly half done. Evidently Cotes had not continued to send Newton each set of sheets as they were printed off, for he informed Newton that the 'whole y^t is finished shall be sent You by y^e first oportunity'.

The first 224 pages of E_2 contain the whole of Book I, the first seven propositions of Book II, and the first page (of three) of Lemma II on Newton's method of 'moments', of finding derivatives of terms raised to negative or positive integral and fractional powers, with and without coefficients, and the sums and products of such terms. Immediately to come was the famous 'Leibniz Scholium'.

Cotes recognized that in this scholium, and in the succeeding parts of Book II, more care was required than in Book I. Apparently he had been taking some liberties in preparing the final copy for printing, and, as a result, many of the alterations of E_1 found in E_2 must have originated with Cotes the editor, rather than with Newton the author.[12] We must keep in mind that in those days publishers, printers, and editors made changes in an author's text of a sort that today we would hold to be beyond the bounds of propriety. Cotes explained his editorial procedure to Newton in his letter of 15 April 1710, as follows:

> I have ventured to make some little alterations my self whilst I was correcting ye Press, such as I thought either Elegancy or Perspicuity or Truth sometimes required. I hope I shall have Y^r pardon if I be found to have trusted perhaps too much to my own Judgment, it not being possible for me without great inconvenience to the Work and uneasiness to Y^r self to have y^r approbation in every particular.[13]

The next part, however, Cotes found to be 'somewhat more than usually intricate', and so, he told Newton, 'I have been looking over them before-hand'.

The remainder of this letter gives us a measure of both the intensity of Cotes's devotion to his task and the depth of his critical acumen. A few samples will show

[9] Apparently all the cuts were made in London. Presumably these are the very cuts referred to in Bentley's letter to Sike of 31 March 1706, printed in Chapter VIII, §4, in which Bentley referred to a 'bundle of wood cuts'.

[10] Apparently, 'printed' usually meant only 'set up in type', whereas 'printed off' meant that the sheet in question had gone through printing to the number of copies (750) required for the edition. This usage is still current at the Cambridge University Press.

[11] Edleston (1850), p. 8.

[12] This is the reason why the annotations from E_1i and E_1a in the Apparatus Criticus are so precious, since they enable us to identify positively some of Newton's own alterations.

[13] Edleston (1850), pp. 8–9.

with what attention to every detail he approached this assignment. First he proceeded to take up with Newton the Rules following Prop. X, Book II, for the motion of projectiles in a resisting medium such as air {261.7 ff.}. Cotes began:

Pag: 270 Reg. 1 I think should begin thus——Si servetur tum Medii densitas in A tum velocitas quacum corpus projicitur, & mutetur——I must confess I cannot be certain y^{t} I understand ye design of Reg: 4 & ye last part of Reg: 7 and therefore dare not venture to make any alteration without acquainting You wth it. I take it thus, y^{t} in ye 4th Rule You are shewing how to find a Mean among all ye Densitys through which the Projectile passes, not an Arithmeticall Mean between ye two extream Densitys ye greatest & least, but a Mean of all ye Densitys considered togather, which will be somewhat greater y^{n} that Arithmeticall Mean ye Number of Densitys which are greater y^{n} it being somewhat more y^{n} ye Number of Densitys which are lesser y^{n} ye same. If this be y^{r} design I would thus alter the 4th Rule wth Y^{r} consent. Quoniam densitas Medii prope verticem Hyperbolae major est quam in loco A, ut habeatur densitas mediocris, debet ratio minimae tangentium GT ad tangentem AH inveniri, & densitas in A augeri in ratione paulo majore quam semisummae harum tangentium ad minimam tangentium GT...

About y^{e} end of the 8th Rule are these words which I would either leave out or print thus, quorum minor eligendus est quorum minor potius eligendus est. Pag: 274. lin: 2 should be $\frac{2TGq}{nn-n \times GV}$. There are others like this which I will not trouble You wth.

Cotes then discussed some possible alterations in Prop. XIV and its corollary, and finally turned to Corol. 3 to Prop. XV, Book II {277.14 ff.}, as follows:

Pag: 286: lin. 5 must be thus corrected——Rr & TQ seu ut $\frac{\frac{1}{2}VQ \times PQ}{SQ}$ & $\frac{\frac{1}{2}PQq}{SP}$ quas simul generant, hoc est, ut VQ & PQ, seu OS & OP. This Corollary being thus corrected, the following must begin thus. Corol: 4. Corpus itaque gyrari nequit in hac Spirali, nisi ubi vis resistentiae minor est quam vis centripeta. Fiat resistentia aequalis vi centripetae & Spiralis conveniet cum linea recta PS, inque hac recta—&c. Tis evident (by Corol. 1.) y^{t} ye descent along ye line PS cannot be made wth an uniform Velocity. Tis as evident I think y^{t} it must be wth an uniform Velocity because ye Resistance & force of Gravity, being equall, mutually destroy each others effect, & consequently no Acceleration or Retardation of Motion can be produced. I cannot at present see how to account for this difficulty & I choose rather to own my ignorance to You y^{n} to run the hazard of leaving a Blemish in a Book I so much esteem.

Then, following a critique of Corol. 8, and a suggestion for altering Prop. XVI (with a proposed corollary), Cotes concluded with an expression of his hope that Newton would 'pardon my freedom in this Letter'. Before Newton had a chance to reply, Cotes wrote again about further difficulties and added:

The difficulty which I proposed to You concerning the 4th Corollary of Prop. XV I have since removed. Upon examination of that Proposition I think I have observed another mistake in ye 3^{d} Corollary which ballances that I before mentioned to You in the same Corollary.[14]

Newton replied to the first letter on 1 May 1710, thanking Cotes for his 'remarks upon the papers now in the Press' and especially expressing gratitude for Cotes's 'corrections' and his 'care of the edition'.[15] Almost all of Cotes's suggested 'corrections may stand', he wrote, and he then went on to give an altered Prop. XVI, and also two new corollaries as Cotes had proposed. In a postscript he replied to

[14] *Ibid.* p. 12. [15] *Ibid.* pp. 14–15.

Cotes's second letter, proposing to him 'that yor difficulty will be removed by the words & motus corporis cessabit'. Cotes did not agree, however, and he responded on 7 May 1710, 'I am not satisfied y^{t} Y^{r} words [& motus corporis cessabit] will remove the difficulty proposed. They cannot in my opinion be reconciled with Cor: 1.'[16] Cotes acknowledged 'Y^{r} objection to be just against those words of mine', but he remained fixed in his opinion that he had been right in his objection to what Newton had written. 'I am yet of opinion', he said, '...that the resistance is to the Centripetal force as $\frac{1}{2}$Rr to TQ. Y^{r} own objection does, I think, if carefully considered, prove it to be so. To avoid further misunderstanding I will put down my Demonstration more at large, thus...' Newton capitulated, beginning his next letter in these words: 'I have reconsidered the 15th Proposition with its Corollaries & they may stand as you have put them in yor Letters.'[17]

The tone and level established in this first exchange continued to the very end of the correspondence, until the edition had been completed. Cotes, in these letters, shows himself always respectful but nevertheless firm when sure that he is right. Newton, in his replies, does not use the weight of his authority to try to crush the younger man, but responds by reason and proof. Prodded by Cotes, Newton enriched the *Principia* to a degree that would never have been achieved but for Cotes's intervention. It is clear, I believe, from a reading of the Newton–Cotes correspondence that Newton had originally intended a far less drastic revision of Books II and III than he eventually produced. The credit is Cotes's.

I do not propose to give here a detailed analysis of Cotes's actual contribution to E_2. That belongs more properly in a volume of commentary on the *Prinicpia*, and depends in the first instance on the availability of the correspondence between Newton and Cotes. In Appendix V, printed at the end of the text of the present edition, all of the emendations to the *Principia* that originated in letters either from Cotes to Newton or from Newton to Cotes are tabulated in sequence, with a reference to the particular letter containing each suggestion. In what follows, I shall attempt no more than to give the reader the atmosphere in which the revisions were made during the period from 1710 to 1713, when much of Book II and all of Book III were being edited and printed.

The correspondence between Newton and Cotes is available in the Portsmouth Collection (U.L.C. MS Add. 3983), and the volume of Newton–Cotes letters in the Trinity College Library, comprising drafts of Cotes's letters, Newton's replies, and various textual material sent by Newton to Cotes for the printer. But there is a major break in the exchanges between Cotes and Newton prior to Letter IV, in which Cotes informed Newton (15 April 1710) that all the copy in hand had been composed and 'printed off', i.e. everything up to the end of page 251 of E_1 or page 224 'of y^{e} new Edition'.[18] Edleston observed a gap between this letter and the previous one extant, dated 11 October 1709, and made conjectures as to

16 *Ibid.* p. 16. 17 *Ibid.* p. 19.

18 Corresponding to {243/244} of E_3. See Edleston (1850), p. 8.

what must have passed between Cotes and Newton. We have some evidence on this subject from the draft of a much later letter (1719) to Johann Bernoulli, reading (in the original Latin and in translation) as follows:

In editione secunda Libri mei Principiorum postulabat D. Cotes ut Corol. 1 Prop. XIII Lib. 1 demonstratione munirem, et ea occasione Corollarium illud verbis nonnullis auxi: Sed hoc factum est antequam hae lites coeperunt. Nam schedae primae viginti octo illius editionis, (id est usque ad pag. 224 inclusive) impressae fuerunt ante 13 Apr., 1710, et schedae primae triginta septem (id est usque ad pag 296 inclusive) impressae fuerunt ante 30 Junii 1710, et prelum subinde quievit usque ad mensem Junum anni proxime sequentis, ut ex Literis D^{ni} Cotes eo tempore ad me missis et adhuc asservatis intelligo. Scheda igitur septima in qua Corollarium illud extat impressa fuit anno 1709. Et hoc annoto ut intelligas me animo candido Corollarium illud auxisse et hactenus nullas tecum lites agitasse. Litibus autem componendis quas cum amicis meis habuisti, quantum in me est operam dabo.[19]

In the second edition of my Book of Principles Mr Cotes asked me to strengthen with a demonstration the first corollary to Prop. XIII Book I, and on that occasion I augmented that corollary by a few words. But this was done before these quarrels began. For the first 28 sheets of that edition (that is up to page 224 inclusively) were printed before 13 April 1710, and the first 37 sheets (that is up to page 296 inclusively) were printed before 30 June 1710, and the press then was at rest until June of the following year, as I understand from Mr Cotes's letters sent to me at that time and still preserved. Therefore the seventh sheet in which that corollary stands was printed in 1709. And I make this remark so that you may understand that I augmented that corollary with every good intention and hitherto have had no quarrels with you. And I shall try in so far as in me lies to settle the quarrels that you have had with my friends.

In this letter we may see Newton reconstructing the printing time-table of the second edition from the information contained in Cotes's letters. But we do not have, in manuscript, any direct communication from Cotes about Prop. XIII, Book I.

2. DELAYS IN GETTING ON WITH THE JOB

It must not be thought that Cotes always yielded to Newton—whether on matters of detail or of major content. For instance, in a letter of 15 June 1710, Newton told Cotes: 'In pag. 348 lin 7, 14, 15, 16 for A & C put other letters suppose F & G, writing, Designet jam $FV+GV^2$ resistentiam Globi &c because $AV+CV^2$ was used before for the differentia arcuum.'[1] Cotes, however, as Edleston has pointed out, 'altered this part of the Scholium in conformity with his [own] remarks at the close of the preceding Letter'.[2] But Cotes hastened to send Newton (on 30 June 1710) the sheets Oo and Pp (corresponding to {280.15–295.21} in E_3) which

are not yet printed off, but will stay for Your corrections if You shall think fit to make any. I could wish You would be pleased to look 'em over, for I fear I may possibly have injured You.

[19] From U.L.C. MS Add. 3968, f. 614.

[1] Edleston (1850), p. 30.

[2] *Ibid.* Cotes had written, on 11 June 1710: 'Page 348. l: 7 &c. You seem to confound y^e *Differentia arcuum* with the *Resistentia Globi*; the former is represented by $AV+CV^2$ & y^e latter ought I think to be represented by $\frac{7}{11}AV+\frac{3}{4}CV^2$.' For the printed version, see {314.3–5, 7, 8, 10}.

Newton seems once again to be concerned by the excessive trouble Cotes was taking and he warned Cotes not to be too much at pains to verify every calculation:

> You need not give your self the trouble of examining all the calculations of the Scholium. Such errors as do not depend upon wrong reasoning can be of no great consequence & may be corrected by the Reader.[3]

In the letter of 30 June Cotes told Newton that the 'Press being now at a stand I will take this opportunity to visit my Relations in Lincoln-shire & Leicester-shire'. He hoped to return 'to College' in about five or six weeks and would then write to Newton 'to desire the remaining part of Y^r^ Copy'.[4] It was not, however, until 4 September 1710 that Cotes wrote to Newton of his imminent return, and asked Newton 'to send the remaining part of Y^r^ Copy assoon as You can'. Nine days later, on 13 September 1710, Newton sent on to Cotes 'the next part' of the *Principia*.[5]

For one reason or another, the edition tended to drag on rather than to move along swiftly to its conclusion. On 24 March 1710/11, Newton wrote to Cotes:

> I send you at length the Paper for wch I have made you stay this half year. I beg your pardon for so long a delay. I hope you will find the difficulty cleared, but I know not whether I have been able to express my self clearly enough upon this difficult subject, & leave it to you to mend any thing either in the expression or in the sense of what I send you. And if you meet wth any thing wch appears to you either erroneus or dubious, if you please to give me notice of it I will reconsider it. The emendations of Corol. 2 Prop 38 & Prop 40 are your own. You sent them to me in yours of Sept. 21, 1710, & I thank you for them.[6]

---

[3] *Ibid.* p. 31.

[4] Newton wrote in reply (on '31 June', that is, 1 July, 1710) that the 'remaining part of the copy will be ready against your return from the visit you are going to make to your friends'. He ended the letter with 'many thanks to yor self for your trouble in correcting this edition'. *Ibid.* p. 32.

[5] 'Beginning at p. 321, with part of Cor. 2, Prop. XXXIII. Lib. 2, and ending at p. 432 with Prop. XXIV. Lib. 3'; that is, {321.20}–{427.14}. *Ibid.* p. 33.

[6] *Ibid.* p. 38. In Corol. 2, Prop. 38 {342.24–343.3}, the final sentence {342. 28–343.3} was changed by Cotes from Newton's original 'Nam Globus tempore casus sui cum velocitate cadendo acquisita partes octo tertias diametri suae describet per demonstrata Galilei, id est (per hanc Propositionem) vis ponderis qua Globus hanc velocitatem acquisivit aequalis est vi resistentiae ejus, & propterea Globum accelerare non potest' to 'Nam Globus tempore casus sui, cum velocitate cadendo acquisita, describet spatium quod erit ad octo tertias diametri suae, ut densitas Globi ad densitatem Fluidi; & vis ponderis motum hunc generans, erit ad vim quae motum eundem generare possit quo tempore Globus octo tertias diametri suae eadem velocitate describit, ut densitas Fluidi ad densitatem Globi: ideoque per hanc Propositionem, vis ponderis aequalis erit vi Resistentiae, & propterea Globum accelerare non potest' (punctuation and capitalization given as in E_2). This was suggested by Cotes in a letter of 21 September 1710, where the emendation is introduced by the words, 'With Y^r^ leave I would alter the latter part of Cor. 2. Prop. 38 thus'. On the sheet sent by Newton, Cotes has crossed out Newton's version after *acquisita* and written *Corrige* in the margin. On another sheet the alteration occurs written out in Cotes's hand.

In the same letter of 21 September, Cotes says, 'I propose the following alteration in Prop. 40'; there follow the third and fourth paragraphs of Prop. 40 {344.18–345.13} essentially as they appear in E_2 and E_3, with these exceptions: 1,3862943611 F of E_2 is 1,3862943613 F in Cotes's letter {344.23, 25}; 'Libri secundi Propositionem nonam' of E_2 is 'Prop. IX Lib II' in Cotes {344.26}; and the final paragraph in Cotes's version lacks the last sentence in E_2. This in turn is followed by Cotes's statement, 'I think this alteration or some other to the same effect (which You may be pleased to send me) will make the theory more easy. I have not as yet had time to go over the Calculation of the Scholium which is annexed to this Proposition.'

In the Newton–Cotes correspondence (page 75), marked with a dagger, the two paragraphs are written out, as in Cotes's letter, in Newton's hand with *secundis* {344.20} replacing an original *primis*, and with *aequalia*

Things began to move ahead once again, albeit slowly. A few letters (and months) later, on 23 June 1711, Cotes announced that he had 'received Y^r^ Letter & delivered ye Papers to ye Printer. I hope we shall now go on without any further intermission.'[7] But by 19 July Cotes was again getting desperate at not having heard from Newton, and began a letter to him in these words:

I wrote to You about a Month ago concerning the 48th Proposition of Yr second Book, & the last week I desired the Printer to send You all the sheets which were printed off. If You have received those sheets, You will perceive by them that ye Press is now at a stand. But having no Letter from You I fear the sheets have miscarried. The Compositor dunn's me every day, & I am forc'd to write to You again to beg Yr Resolution.[8]

Newton's reply, on 28 July, was frank enough:

I received your Letters & the papers sent me by the Printer But ever since I received yours of June 23 I have been so taken up with other affairs that I have had no time to think of Mathematicks. But now being obliged to keep my chamber upon some indisposition wch I hope will be over in a day or two I have taken your letter into consideration.[9]

On 30 July Cotes returned to the question yet once more, in a letter beginning:

I have read Yr Letter & find my self obliged to trouble You once more. I must beg leave to tell You I am not as yet satisfied as to the Inconsistency which I mentioned in my former Letter.[10]

But not until six months later, on 2 February 1711/12, did he finally get word from Newton about the forty-eighth proposition of Book II. At the same time Newton informed Cotes:

I have at length got some leasure to remove the difficulties wch have stopt the press for some time, & I hope it will stop no more. ffor I think I shall now have time to remove the rest of your doubts concerning the third book if you please to send them...

You stuck at a difficulty in the third Proposition of the third Book. I have revised it & the next Proposition & sent you them inclosed as I think they may stand. What further Observations you

deleted after *spatia* {345.7}. Also *Demittatur* {344.18} does not begin a new paragraph. Cotes has added in his own hand the sentence that has been mentioned above as missing {345.11–13} and has also added and crossed out an additional 'vel etiam spatia in fluido descripta, si modo a fluido cadentis pondus diminui non autem alia ulla resistentia oriri intelligatur'.

The only surviving part of these paragraphs from Newton's original version sent to Cotes apparently occurs on page 76 in the volume of Newton–Cotes correspondence in the Trinity College Library (MS R.16.38), containing (crossed out) the final bit {345.7 beginning with *velocitate* to 345.11 ending with *descripta*}, again without the extra sentence. It has the following variants from Cotes's suggested version: '1,3862944–4,6051702 L' {345.9} is '1|3862943'; and 'in tertia columna' {345.9} is 'in posteriore parte tertiae columnae quamproxime'. In MS Add. 3965, however, ff. 253–4, a version of the paragraphs occurs that is probably very close, if not identical, to the one sent to Cotes. The most significant variant between these papers and the text of E_2 probably reveals Cotes's contribution. The eight lines at {344.20–27}, 'Inveniatur...ex hypothesi quod', were in Newton's version simply

$$\text{'Et erit } \frac{2PF}{Gg} - 1|3862943 \text{ F altitudo cadendo descripta quamproxime, si modo'.}$$

Frequently, as in this example, one can identify Cotes's actual contributions with precision through Newton's long lists of corrections to E_1 in MS Add. 3965. On f. 211 of MS Add. 3965 there exists a draft of Newton's letter of 30 May 1710, apparently written after the version in MS Add. 3984, f. 5^v^, but before the version on f. 5^r^.

[7] Edleston (1850), p. 44. [8] *Ibid.* p. 50. [9] *Ibid.* p. 51. [10] *Ibid.* p. 52.

have made upon the third Book or so many of them as you think fit if you please to send in yor next Letters, I will dispatch them out of hand. I shall be glad to have them all because I would have ⟨the⟩ third Book correct.[11]

3. THE END IN SIGHT

Cotes's tenacity may be seen in relation to Prop. XXX, Book II. On 20 May 1710 Cotes sent Newton an alteration which in his opinion was 'necessary to make the Demonstration accurate'.[1] Newton in reply expressed his 'fear least that wch relates to Prop. XXX may render the Demonstration thereof too obscure'. He proposed, therefore, 'that the Proposition with its Demonstration may stand, & in the end of it, after the words et sic eidem aequabitur quam proxime, may be added these two sentences. Quinimo eidem aequabitur accurate, ideoque conclusiones praedictae sunt accuratae. Nam si ad alteras partes...'[2] Cotes returned to this subject in his next letter (1 June 1710): 'I beg leave of You to express my sense freely, I fear it will be look'd upon as a Blemish in Y^{r} book, first to demonstrate y^{t} ye Proposition is true quamproximè & afterwards to assert it to be true accuratè.'[3] Though he still thought that 'ye alteration which I proposed... does make ye Demonstration compleat to an intelligent Reader', he now proposed another way in which 'it may be put down more at large'. But, if Newton were to 'think the demonstration will even this way be too obscure, a new Scheme may be cut with the addition of ye lines here drawn & the demonstration may end thus ...' He concluded, 'I think ye first of these two ways sufficiently clear; but will wait for Y^{r} resolution.' Newton capitulated at last in a letter of 8 June, beginning, 'I have reconsidered your emendation of the XXXth Proposition wth the Demonstration & approve it after the manner you propose in the first of the two ways set down in your Letter of June 1st.'[4]

Two of the topics in which Cotes became deeply involved were the theory of the tides and the theory of the Moon's motion: the former made complex by the joint action on the Earth's waters of the Sun and the Moon, the latter by the joint action of the Sun and the Earth on the Moon. Principally the lunar theory occurs in Props. XXV–XXXIII and XXXIV–XXXV of Book III and in a lengthy scholium following Prop. XXXV beginning, 'By these computations of the lunar motions I was desirous of showing that by the theory of gravity the motions of the moon could be calculated from their physical causes.' The tides and the lunar theory are discussed in a series of some two dozen letters that were exchanged from 7 February 1711/12 to 15 September 1712, in which many other topics are also introduced.[5]

Cotes's critical acumen in dealing with problems of mathematical physics is perhaps best demonstrated in his examination of the 'New Scholium to Prop. XXXV'

[11] *Ibid.* pp. 56–7.
[1] Edleston (1850), p. 21.
[2] *Ibid.* p. 24.
[3] *Ibid.* p. 26.
[4] *Ibid.* p. 27.
[5] *Ibid.* pp. 57–141.

{459.32 ff.}, dealing with 'the origin and quantity of various Lunar Inequalities', and occupying 'three sides of a sheet of foolscap...which seems to have been doubled up and placed loosely between the pages of Newton's interleaved copy of the 1st Ed.'[6] This scholium, later referred to by Newton as a 'first draught of the Moon's theory', is analysed in some detail by Edleston.[7] Newton's final version profited enormously from Cotes's intervention. What is perhaps of most interest, however, is that the series of exchanges between Newton and Cotes concerning this scholium was the last instance of a meeting of these two minds on a technical problem, although there were to be some further interactions with regard to philosophical or methodological questions.

The fact of the matter is that by this time Newton was reaching the end of the job, and possibly the end of his patience. On 23 September 1712 he wrote to Cotes about 'the last Paragraph of the Scholium': 'The description of the Variatio secunda is derived only from phaenomena' and 'wants to be made more accurate by them that have leasure & plenty of exact observations.' And then he added the comment, 'The public must take it as it is.'[8]

On 14 October, however, he was able to write in a different vein: 'I send you the conclusion of the Theory of the Comets to be added at y^e end of the book after the words [Dato autem Latere transverso datur etiam tempus periodicum Cometae Q. E. I.].'[9] That is, he sent Cotes the new material in E_2 beginning 'Caeterum Cometarum revolventium' {519.16} and ending 'primus omnium quod sciam deprehendit',[10] namely, the conclusion of the final proposition of Book III, just before the Scholium Generale. Then he announced that there was 'an error in the tenth Proposition of the second Book', which would have to be corrected, even though that part of the *Principia* had already been printed and the type distributed. Newton advised Cotes that there would be required 'the reprinting of about a sheet & an half'. Newton was in the process of 'correcting' this proposition and would, of course, 'pay the charge of reprinting it, & send it to you as soon as I can make it ready'.

4. PROPOSITION X OF BOOK II

On 1 November 1712, and again on 23 November, Cotes wrote to Newton, asking him to send on the alterations to be made in Prop. X.[1] Cotes was anxious to finish his assignment, and (23 November) had been able to report that the whole treatise had been printed off, save for 'about 20 lines' (actually 19) on the first page of the

[6] *Ibid.* p. 109. Edleston's description may be readily confirmed by examining this sheet (nos. 169–71) in the volume in the Trinity College Library. As Edleston observed, there was just not room enough on the quarto interleaf, also to be found in the above-mentioned volume (no. 190), headed 'Scholium', and containing the opening words, followed by '&c'. This was Newton's method of referring to a supplementary sheet. The presence of such loose folded sheets recalls the description of the volume sent to Cotes by Bentley in a parcel addressed to Sike; see Chapter VIII, §4.

[7] *Ibid.* pp. 110–12. [8] *Ibid.* p. 141. [9] *Ibid.* pp. 141–2.

[10] The second half of the final paragraph, 'Hujus generis...paulatim migrare' {525.31–526.30}, was altered in the third edition.

[1] Edleston, pp. 144–5.

final sheet (Qqq), which was also to contain whatever conclusion Newton would choose to compose. But it was not until 6 January 1712/13 that Newton sent Cotes 'the tenth Proposition of the Second book corrected'.[2] He reckoned that the alteration would 'require the reprinting of a sheet & a quarter from pag 230 to pag. 240', and would need a new 'wooden cut' which he intended to send to Cotes 'by the next Carrier'. It was Newton's hope that 'this Proposition as it is now done will take up much the same space as before. If not, the space about the cuts may be made a little wider or a little narrower, or the number of lines in a page may be increased or diminished by a line.' In E_2, this rewritten Prop. X occupies pages 232–9, and is followed by a scholium (p. 240) completing Sec. III of Book II. Page 232 is the verso side of the fourth leaf of signature 'Gg', and pages 233–9 occupy the first three leaves and the recto side of the fourth leaf of signature 'Hh'. There is no marked difference between the lengths of the printed pages for this new material and the rest of the book (save that page 235 is a line shorter than other pages), and I do not find the type any more crowded around the figure (p. 232) than is the case for other similar figures (for example, p. 222).

A glance at the Apparatus Criticus {252.7 ff.} shows that such a change was actually made. In E_2, there is seen to be a wholly new proof of Prop. X and a wholly new Corol. 1. Corollary 2 of E_2 is merely a revision of the former Corol. 3 of *M* and E_1. The scholium is the same in E_1 and E_2, save for two new additional introductory paragraphs {259.17–260.7}.

This Prop. X displays 'the Uses of the Terms of a converging Series for solving of Problemes' and 'the Method of second Differences', but not fluxions (and not dotted or 'pricked' letters[3]), and not even 'moments', a topic exhibited in E_2 just prior to Prop. X, in the same Sec. II of Book II, just before the famous 'Leibniz Scholium'. Proposition X is thus notable because its proof is purely analytical and not geometric (or synthetic), as most proofs in the *Principia* are. That there must be an error in Prop. X was brought to Newton's attention by Nicolaus Bernoulli,[4] who had been visiting England during September and October 1712.[5] Since the

[2] *Ibid.* p. 145.

[3] Many readers are apt to think of fluxions wholly in terms of dotted or 'pricked' letters, as $\dot{x}$ or x', or $\ddot{z}$ or z''. The dotted letters represent but one of a great many systems of notation invented by Newton. In Prop. X, Book II, a wholly different system is used. But we must not, because of the great success of the Leibnizian algorithm, confuse notation and concept. Newton himself stressed this last point in the many drafts and the printed version of the 'Recensio libri' (1715), p. 204. In a letter of 20 May 1710, Cotes solved Corol. 1 to Prop. XXVII, Book II, by the method of 'fluxions', using dotted letters. See Edleston (1850), pp. 21–3.

[4] The result in question was due to Johann Bernoulli and was communicated to Leibniz in a letter of August 1710, and to the Académie Royale des Sciences, Paris, in January 1711 (printed in their *Mémoires* for 1711, published in 1714). Edleston (1850), p. 142.

[5] There is often a lack of precision in describing this episode. D. T. Whiteside (in a private communication) writes: 'Bernoulli had not found the error in Prop. X of the first edition. What he had done was to derive the result for a semicircular trajectory by an alternative approach and he found that his result was different from Newton's (in this particular case) by a factor of $\frac{3}{2}$. But he *couldn't* fault Newton's 1687 proof. Indeed, neither could Newton himself for many days, but he at last saw why *and was the first to do so*: thereafter, it was child's play for him to correct his howler.' The truly remarkable aspect of this feat becomes evident when we recall that Newton was then within a month of his seventieth birthday.

whole sheet of signature 'Hh' was reprinted, the only clue to the last-minute rewriting of the proof of Prop. X is the fact that the leaf containing pages 233–4 is a cancel pasted on the stub of the original leaf.[6]

5. WHAT TO DO WITH THE LAST SHEET: A POSSIBLE MATHEMATICAL APPENDIX

On 1 November 1712 Cotes had sent Newton 'the Sheets as far as they are Printed off, that Your self or some freind may revise them, in order to see what Errata may be put in a Table'.[1] As yet there had been no firm decision as to what disposition should be made of the last sheet, of which part of one side of the first leaf contained the conclusion of the theory of comets. Perhaps, Cotes wrote (23 November 1712), he could 'fill up the...Sheet' by adding 'a Table of the Contents of each Section, if You think fit'. Bentley had proposed that there be 'subjoyned an Index to the whole, but particularly to the Third Book. If You approve of it, such an Index may soon be made.'[2] In the event, Cotes made an 'Index Capitum Totius Operis', printed as part of the front-matter, just prior to the Definitions, and also an 'Index Rerum Alphabeticus', printed following the concluding Scholium Generale. (The table of contents and index are reprinted in E_3, the only changes being in the page numbers.) By any reasonable modern standards, neither of these indices is very satisfactory.[3]

Evidently Newton had at last given up any idea he had held about having a mathematical appendix to the new edition. In a letter of the preceding July (1712), Cotes reported his enthusiasm for Newton's plans, as told him by Bentley on the latter's return from London,[4] of adding 'a small Treatise concerning the Methods

[6] On 13 January 1713 Cotes reported that he had considered the 'alteration of Prop. X...and am well satisfied with it'. He had received the new cut sent by Newton and had altered 'some things in Your Paper', but these were hardly 'worth Your notice, being only faults in transcribing'. Edleston, however, observed (p. 147) that Cotes, 'besides making the alterations alluded to here, has (perhaps from want of room) omitted a paragraph at the beginning of the Scholium of the Prop. (p. 269, Ed. 1, p. 240, Ed. 2.) in which Newton points out another mode of viewing the problem which is the subject of the Proposition'. This paragraph, which would have gone just before {259.17}, is printed by Edleston as follows: 'Fingere liceret projectilia pergere in arcuum *GH, HI, IK* chordis & in solis punctis *G, H, I, K* per vim gravitatis & vim resistentiae agitari, perinde ut in Propositione prima Libri primi corpus per vim centripetam intermittentem agitabatur, deinde chordas in infinitum diminui ut vires reddantur continuae. Et solutio Problematis hac ratione facillima evaderet.'

[1] Edleston (1850), p. 144. A half-page of 'Corrigenda' in small type appears on the verso of the last leaf of the Index Rerum.

[2] *Ibid.* p. 145.

[3] Newton told Cotes on 6 January 1712/13, 'As for making a Table to the book I leave it to you to do what you think. I beleive a short one will be sufficient.' On 5 March Newton asked for the exact page where the diagram of the Comet's path was to go, 'that it may be graved upon the Plate for directing the Bookbinder where to insert it'. Bentley added a note to the letter reminding Cotes by 'Sr Isaac's Leave' of what Cotes and Bentley 'were talking of, An alphabetical Index, & a Preface in your own Name'. Cotes at once (8 March 1712/13) sent Newton the page number for the cut of the Comet, promised 'in a day or two to set about the Alphabetical Index', and said he would 'write to Dr Bentley concerning the Preface by ye next Post'. Edleston (1850), pp. 146, 148.

[4] *Ibid.* p. 118.

of Infinite Series & Fluxions'.[5] Cotes certainly hoped Newton would 'go on with Your design: it were better that the publication of Your Book should be deferr'd a little, than to have it depriv'd of those additions'.[6] There is no further reference in the Newton–Cotes correspondence to the possibility of printing this mathematical tract along with the *Principia.*

6. A PREFACE BY COTES

Cotes had agreed to write a Preface to the new edition, but he was not at all sure what he ought to say in it. Hence, he wrote to Bentley (10 March 1712/13) to ask him 'with what view' Sir Isaac 'thinks proper to have it written'.[1] Understandably, he wanted the Preface to accord with Newton's desires and he did not plan to interject into Newton's book any topics of which the author might not fully approve. Cotes himself thought a main thrust of the Preface should be an attack on Leibniz, and he especially singled out the latter's *Tentamen de motuum coelestium causis*[2] as one of those 'peices of his...which deserve a censure'. Cotes said he would be glad if, 'whilst I am making the Index', Sir Isaac 'would be pleasd to consider of it [the Preface] & put down a few notes of what he thinks most material to be insisted on'. He said this on the 'supposition that I write the Preface my self'. But in actual fact he held it much more advisable that either Bentley or Newton or both 'should write it' while Bentley was 'in Town'. He assured Bentley:

> You may depend upon it that I will own it, & defend it as well as I can, if hereafter there be occasion.

In the twentieth century we would think it odd, to say the least, that Cotes should propose that Bentley or Newton—or Newton and Bentley jointly—should write a Preface for Cotes to sign as if it had been his own, and to promise that thereafter he would pretend he had written it and as such defend it.

Bentley's reaction, following a discussion with Newton, is to be found in a letter to Cotes dated at London, 'At Sr Isaac Newton's', 12 March 1712/13.[3] As to Cotes's proposal of 'a Preface to be drawn up here, and to be fatherd by you', as Bentley wrote, 'we will impute it to your Modesty; but You must not press it further, but go about it your self'. Yet he did give Cotes a few suggestions, such as

[5] See Supplement VIII to this Introduction for Newton's plans to have a mathematical supplement to the *Principia*. In a draft of this letter, Cotes also proposed a more ambitious programme of publishing Newton's mathematical writings, including 'Your Algebra', 'Your Treatise of ye Cubick Curves', '& what others You have by You of the like nature'. Edleston (1850), p. 119.

[6] Cotes expressed a somewhat similar sentiment in a letter to William Jones (30 September 1711): 'I am very desirous to have the Edition of Sr Isaac's Principia finish'd, but I never think the time lost when we stay for his further corrections & improvements of so very valuable a book, especially when this seems to be the last time he will concern himself with it. I am sensible his other Business allows him but little time for these things & therefore I ought not to hasten him so much as I might otherwise do, I am very well satisfied to wait till he has leasure.' *Ibid.* pp. 209–10. Newton was then approaching 70 years of age: who would ever have expected that a dozen years later he would undertake yet another revision of his *Principia*?

[1] Edleston (1850), p. 149.

[2] For Newton's views concerning Leibniz's *Tentamen*, see Chapter VI, §5. [3] Edleston (1850), p. 150.

to 'give an account...of ye work it self' and 'of ye improvements of ye New Edition'. Sir Isaac gave his 'consent' for Cotes to add whatever he might 'think proper' about the controversy over the 'first Invention' of the calculus. Cotes certainly knew this subject well enough to write it up without any 'hints', and both Newton and Bentley would be glad to read whatever Cotes wrote, 'to suggest any thing y^t^ may improve it'. Newton and Bentley were in agreement, however, that Cotes was 'to spare y^e^ *Name* of M. Leibnitz'. Newton evidently had second thoughts on the whole question, and in a letter of 31 March,[4] enclosing a short 'Account of this new Edition', which is printed in E_2 as a second Preface of the author's, Newton announced to Cotes that he would definitely not read Cotes's Preface before it was printed, 'for I find that I shall be examined about it'.

It has been suggested that Newton may very well have had in mind the proposed discussion by Cotes of the priority of Newton and Leibniz in the matter of the inventions of fluxions. As it turned out, Newton on at least one occasion indicated that he had not seen this Preface before publication. We have a manuscript draft in which he writes:

> The Preface of the Editor praefixed to the second Edition of my Book of Principles ⌊at Cambridge⌋ I did not see till the Book came abroad but [it seems to be in answer to M^r^ *del. and* I find it *del.*] what he saith there in relation to Mr Leibnitz is in answer to his accusing ⌊me⌋ of introducing occult qualities & miracles & therefor Mr Leibnitz is still the aggressor.[5]

Newton was referring here to an attack upon his principle of gravitational attraction, made by Leibniz in a letter to Hartsoeker, published in England in 1712.

7. NEWTON'S CONCLUDING SCHOLIUM GENERALE

When Newton transmitted to Cotes the new proof of Prop. X, Book II (6 January 1712/13), he said, 'I shall send you in a few days a Scholiu⟨m⟩ of about a quarter of a Sheet to be added to the ⟨end⟩ of the book.'[1] He mentioned that 'some are perswading me to add an Appendix concerning the attraction of the small particles of bodies'. This Appendix, he said, 'will take up about three quarters of a Sheet, but I am not yet resolved about it'. If, indeed, the Appendix was to have occupied six printed pages, it must have been an essay of some 2,000 words. But it was never sent on to Cotes, nor does it appear in either E_2 or E_3. Furthermore, I have not found among Newton's papers any such essay which may positively be identified as this proposed Appendix.[2] When Newton finally did send Cotes the concluding Scholium Generale, on 2 March 1712/13, he explained that the Appendix was not going to be included in the *Principia* after all:

> I intended to have said much more about the attraction of the small particles of bodies, but upon second thoughts I have chose rather to add but one short Paragraph about that part of Philosophy. This Scholium finishes the book.[3]

[4] *Ibid.* pp. 156–7. [5] U.L.C. MS Add. 3968, f. 506. [1] Edleston (1850), p. 146.

[2] There are many drafts of Newton's on this topic, which is discussed in his essay *De natura acidorum* and in the final Queries of the *Opticks*. See Hall and Hall (1960), and Boas (1958). [3] Edleston (1850), p. 147.

Cotes read the Scholium, and discussed one part of it in a letter of 18 March 1712/13, in which he also outlined what he planned to put into the Preface. He was concerned, however, how he might reconcile Newton's doctrine about hypotheses in the Scholium Generale with his actual procedure in Corol. 1 to Prop. V, Book III, and notably the words 'Et cum attractio omnis mutua sit...' {399.17–18}.

> I shall be glad to have Your resolution of the difficulty, for such I take it to be. If it appeares so to You also; I think it should be obviated in the last sheet of Your Book, which is not yet printed off, or by an Addendum to be printed with ye Errata Table. For till this Objection be cleared I would not undertake to answer any one who should assert that You do *Hypothesim fingere*, I think You seem tacitly to make this Supposition that the Attractive force resides in the Central Body.[4]

Newton's reply explained that

> as in Geometry the word Hypothesis is not taken in so large a sense as to include the Axiomes & Postulates, so in Experimental Philosophy it is not to be taken in so large a sense as to include the first Principles or Axiomes wch I call the laws of motion. These Principles are deduced from Phaenomena & made general by Induction: wch is the highest evidence that a Proposition can have in this philosophy. And the word Hypothesis is here used by me to signify only such a Proposition as is not a Phaenomenon nor deduced from any Phaenomena but assumed or supposed wthout any experimental proof.[5]

Newton did not approve of Cotes's suggestion of an addendum to be printed among the Errata. He countered by sending Cotes an emendation of the Scholium Generale.

The concluding Scholium Generale, possibly the most famous of all of Newton's writings, exists in a number of manuscript drafts.[6] In the version that Newton had sent to Cotes on 2 March 1712/13,[7] the fifth paragraph ended with these three sentences (see Plate 14):

> Rationem vero harum gravitatis proprietatum ex phaenomenis nondum potui deducere, & hypotheses non fingo. Quicquid enim ex phaenomenis non deducitur, hypothesis vocanda est; & *Hypotheses* seu Metaphysicas seu Physicas, seu qualitatum occultarum seu Mechanicas non sequor.[8] Satis est quod gravitas revera existat,

> Indeed, I have not yet been able to deduce the reason of these properties of gravity from phenomena, & I do not feign[9] hypotheses. For whatever is not deduced from phenomena is to be called an hypothesis; & I do not follow *Hypotheses*, whether Metaphysical or Physical, whether of occult qualities or Mechanical. It is

[4] *Ibid.* p. 153. This letter is actually dated 'Febr. 18th $171\frac{2}{3}$'.

[5] *Ibid.* pp. 154–5.

[6] Hall and Hall (1962), *Unpublished papers*, pp. 348–64.

[7] Newton's transcript of the Scholium Generale (in his own hand) is at present in the volume of Newton–Cotes correspondence (Trinity College Library, R.16.38); see Plate 13.

[8] In the first version of what eventually became Book III—the *De mundi systemate* (*lib.* 2us), for which see Chapter IV, §6, and Supplement VI—Newton used the deponent verb 'sequor' in a non-pejorative sense in relation to the 'hypothesis' of Copernicus, revised by Kepler, writing: '& propterea Hypothesis, quam Flamstedius sequitur, nempe Keplero-Copernicaea...' ('...and therefore, the Hypothesis which Flamsteed follows, namely the Keplero-Copernican'. The sense of 'Hypotheses non sequor' is probably 'I am not a follower of hypotheses', or even possibly 'I am not one of those who pursue hypotheses'.

[9] We are indebted to Alexandre Koyré (1957) for having reminded us that 'feign' is Newton's own usage, at least in the *Opticks*. Andrew Motte rendered this slogan by 'I frame no hypotheses'. I have shown elsewhere that at that time 'frame' often had a pejorative sense. In certain manuscripts, Newton used 'frame' and 'feign' in exhibiting his disdain of hypotheses. Cohen (1966), 'Hypotheses in Newton's philosophy', and Cohen (1971), *Scientific ideas*. The word 'occult' is used here in the seventeenth-century (and not the twentieth-century) sense. The phrase 'reason of' implies a 'cause of'.

et agat secundum leges a nobis expositas, et ad corporum coelestium et maris nostri motus omnes sufficiat.

enough that gravity should really exist, & act according to the laws expounded by us, & should suffice for all the motions of the celestial bodies & of our sea.

Following the correspondence with Cotes about hypotheses, in the course of which Cotes had raised the possibility that Newton might have 'feigned' an hypothesis, Newton told Cotes to change the second sentence. It was no longer to read '& *Hypotheses* seu Metaphysicas seu Physicas, seu qualitatum occultarum seu Mechanicas non sequor'.[10] Rather, Newton would now say,

et ejusmodi Hypotheses seu Metaphysicae seu Physicae seu Qualitatum occultarum seu Mechanicae in Philosophia experimentali locum non habent.

and Hypotheses of this kind, whether Metaphysical or Physical or of Occult Qualities or Mechanical, have no place in experimental Philosophy.

Then he went on with three new sentences:

In hac Philosophia Propositiones deducuntur ex phaenomenis & redduntur generales per Inductionem. Sic impenetrabilitas mobilitas & impetus corporum & leges motuum & gravitatis innotuere. Et satis est quod Gravitas corporū revera existat & agat secundum leges a nobis expositas & ad corporum coelestium et maris nostri motus omnes sufficiat.[11]

In this Philosophy Propositions are deduced from phenomena and rendered general by Induction. Thus the impenetrability, mobility, and impetus of bodies and the laws of motions and of gravity became known. And it is enough[12] that the gravity of bodies should really exist and act according to the laws expounded by us, and should suffice to explain all the motions of the heavenly bodies and of our sea.

In this sequence we may see Newton's expression of disdain for hypothetical explanations becoming stronger and stronger. In the first versions he had never made a statement of general policy for science as extreme as *Hypotheses non fingo*. We may examine, on this point, the five surviving manuscript drafts. In one of these (*D*: U.L.C. MS Add. 3965, f. 363^{v})[13] we find Newton concluding:

Causam vero harum gravitatis proprietatum ex phaenomenis nondum potui deducere: et hypotheses seu mechanicas seu qualitatum occultarum non sequor [*originally* Nam hypotheses seu mechanicas seu qualitatum occultarum fugio]. Satis est quod gravitas revera detur et agat secundum leges a nobis expositas, & ad corporum coelestium et maris nostri motus omnes sufficiat.

Indeed I have not yet been able to deduce the cause of these properties of gravity from the phenomena; and I do not follow hypotheses whether mechanical or of occult qualities [*originally* For I flee from hypotheses whether mechanical or of occult qualities]. It is enough that gravity should really be granted and act according to the laws which we have explained, and should suffice to account for all the motions of the celestial bodies and of our sea.

Another version (*E*: U.L.C. MS Add. 3965, f. 365^{v}) has the final form of the above passage copied out word for word, differing only in such minor matters as punctua-

[10] Edleston (1850), p. 155.

[11] *Ibid.*

[12] Motte, and following him Cajori, would have this read 'And to us it is enough...', as if Newton were talking for himself editorially, rather than in general ('It is enough...' or 'It suffices...').

[13] The assignment of letters follows the suggestion made by A. R. Hall and M. B. Hall (1962), *Unpublished papers*, pp. 348 ff., who have published the text of versions *A* and *C* with English translations; I have used their translations below.

tion, the use of '&' for 'et', and the underlining of the words 'causam', 'hypotheses', 'mechanicas', and 'qualitatum occultarum'. Hence it is a later redaction. Afterwards, Newton has cancelled the word 'detur' in the final sentence and has replaced it by 'existat'.

But in a third version (*A*: U.L.C. MS Add. 3965, f. 357ᵛ), we observe some changes being made. Originally Newton had written:

Nam hypotheses seu mechanicas seu metaphysicas seu qualitatum occultarum [fugio ut praejudicia *del.*] fugio.	For [I flee as harmful *cancelled*] I flee from hypotheses, whether mechanical or metaphysical or of occult qualities.

Then he altered the sentence, so as to make it read:

Nam hypotheses seu metaphysicas seu physicas seu mechanicas seu qualitatum occultarum fugio.	For I flee from hypotheses, whether metaphysical or physical or mechanical or of occult qualities.

This led to the concluding sentence:

Praejudicia sunt et scientiam non pariunt.	They are harmful [prejudices] and do not engender science.

Here we may see Newton increasing the number of types of hypothesis he will shun from the original pair (mechanical, and of occult qualities), first to three (by adding those that are metaphysical), and then to four (including physical ones too).[14]

There is another draft (*C*: U.L.C. MS Add. 3965, f. 361ᵛ), which differs very considerably from all the others in its form and content; for instance, it includes a series of 'propositions' about the 'attraction of particles', 'attraction of the electric kind' and various actions of 'the electric spirit' (even including 'nutrition'). In this document, printed by A. R. and M. B. Hall, we find:

Causam vero harum proprietatum ejus ex phaenomenis nondum potui invenire. Nam hypotheses seu mechanicas seu qualitatum occultorum [*read* occultarum] fugio. Praejudicia sunt et scientiam non pariunt. Satis est quod gravitas revera detur, & agat secundum leges a nobis expositas & ad corporum coelestium et maris nostri motus omnes sufficiat.	Indeed, I could not discover the cause of these properties of it [gravity] from phenomena. For I flee from hypotheses whether mechanical or of occult qualities. They are harmful and do not engender science. And it is enough that gravity should really be granted and act according to the laws which we have explained, and should suffice to account for all the motions of the celestial bodies and of our sea.

[14] Observe that Newton had at first limited himself to the statement that he was not a follower of mechanical hypotheses or of hypotheses of occult qualities, and only later added physical and metaphysical qualities to the list. The reason may be that Newton had read the exchange of letters between Hartsoeker and Leibniz, of which a portion had been published in England in spring 1712, and in which he had been accused of having failed to give a mechanical explanation of gravity and of having introduced attraction, which—as presented—could be nothing other than an occult quality. In the very same letter in which Cotes took up the question of 'hypothesin fingere', he called Newton's attention to these letters, and in particular to Leibniz's implication that the *Principia* 'deserts Mechanical causes, is built upon Miracles, & recurrs to Occult qualitys'. See Edleston (1850), p. 153, and Koyré (1965), *Newtonian studies*, III, B. A draft of a letter to the editor, dated May 1712, exists in manuscript (U.L.C. MS Add. 3968, f. 257) and shows that Newton had actually read Leibniz's criticism before composing the General Scholium.

This particular group of sentences resembles version *D*, quoted above, save that in place of the word 'invenire' at the end of the opening sentence Newton has 'deducere'; furthermore, the second sentence reads like the first, rather than the second, state of the same sentence in version *D*. While the final sentence is the same in both versions, this version *C* has an additional sentence, 'Praejudicia sunt et scientiam non pariunt', which does not occur in version *D*, but is found in version *A*. As mentioned, versions *D* and *E* are practically identical.

In a fifth version (*B*: U.L.C. MS Add. 3965, f. 360), the paragraph concludes in three sentences which are word for word the same as the first text sent by Newton to Cotes. This is undoubtedly a draft of the Scholium Generale immediately prior to the one sent to Cotes. In this version *B*, however, the opening word 'Rationem', in the antepenultimate sentence, was at first 'Causam'. Newton has at last added the famous phrase *Hypotheses non fingo*. In a sense, the presence of *Hypotheses non fingo* (a strong and absolutely pejorative expression) made it unnecessary to say also *Hypotheses non sequor* (a weaker and more personal statement), and so we can understand why Newton later wrote to Cotes to take it out and replace it by the dictum: *Hypotheses...in Philosophia experimentali locum non habent*. But the latter was more general than either the phrase it replaced or the one that remained; Newton was now saying that he was not talking about his own personal credo (in the sense of what he himself would not 'feign' or would not 'follow'), but about the new science as a whole.

Newton also sent Cotes an additional sentence to be added at 'the end of the last Paragraph but two':

Et haec de Deo: de quo utique ex phaenomenis disserere, ad Philosophiam experimentalem pertinet.	And so much concerning God: to discourse of whom from phenomena[15] surely belongs to experimental philosophy.

Thus Newton's statements about not 'feigning' hypotheses and about discussions of God in experimental philosophy were both additions to the text of the Scholium Generale, and both were tacked on at the same time! It would seem that Newton, once having decided (in answer to Cotes's objection) to append a strong statement that hypotheses of whatever kind have no place in experimental philosophy, felt he must also add a sentence to show that God was not to be considered an hypothesis in experimental philosophy—because one could discuss him 'ex phaenomenis'.[16]

Later on, after mature reflection, Newton decided that he had been careless and so, as we may see from the Apparatus Criticus {529.36–37}, in E_3 he toned down his statement about God to read 'ad Philosophiam naturalem pertinet' rather than 'ad Philosophiam experimentalem pertinet'. In the end, then, as Gerald

[15] Motte and Cajori here translate 'phaenomena' by 'the appearances of things', although elsewhere they translate it by 'phenomena'.

[16] Often it is said that God appears in the *Principia* only in E_2, in the concluding Scholium Generale. But in E_1 {405.35} there is a reference to God and his Divine Providence, removed in E_2. See Cohen (1969), 'Newton's *Principia* and divine providence'; also Chapter VI, §4.

Holton has remarked, Newton thus states that phenomenologically based discussions of God do have a place in *natural* philosophy, while hypotheses have no place in *experimental* philosophy.[17]

The letter to Cotes containing the emendations of the Scholium Generale, written from London, Saturday, 28 March 1713, ended with a statement that Newton did not have 'time to finish this Letter'[18] and intended to write again on Tuesday. And indeed, on the following Tuesday, 31 March 1713,[19] Newton wrote again, and repeated in summary form the statements he had made:

> Experimental Philosophy proceeds only upon Phenomena & deduces general Propositions from them only by Induction.[20] And such is the proof of mutual attraction. And the arguments for ye impenetrability, mobility & force of all bodies & for the laws of motion are no better. And he that in experimental Philosophy would except against any of these must draw his objection from some experiment or phaenomenon & not from a mear Hypothesis, if the Induction be of any force.[21]

This letter is the last one in the series, as found in the bound volume of Newton–Cotes correspondence in the Trinity College Library.[22]

8. THE JOB COMPLETED

Cotes finished the index in April 1713, and sent a copy of it to William Jones in London, who returned him 'many thanks for the very Instructive Index, that you have taken the pains to add'.[1] Cotes in reply, on 3 May, expressed his pleasure that Jones 'can approve of the Index to the Principia', but, he explained, 'It was not design'd to be of any use to such Readers as Your self, but to those of ordinary capacity.'[2] He affirmed his hope that 'the whole Book may be finished in a fortnight or three Weeks. I have lately been out of Order, or it might have been done by this time.'

The last job to be done was to gain approval of the preface. Cotes, as we have seen, had been told not to send the preface to Newton for approval, but he was able to show it to Samuel Clarke, who read it and returned it with comments. In a letter of 25 June 1713, Cotes thanked Clarke for 'Your corrections of the Preface', and 'particularly for Your advice in relation to that place where I seem'd to assert

17 Holton (1965).

18 Edleston (1850), p. 156.

19 *Ibid.* pp. 156–7.

20 This sentence is of great interest since so much has been written concerning what Newton may have meant by *deducing* 'general Propositions' from phenomena 'by Induction'. Edleston (1850), p. 156.

21 The final sentence is all but an anticipation of the Regula IV {389.6–13} which Newton added in E_3.

22 There are two other letters written a couple of years later, in 1715, in one of which (April 29) Cotes sent Newton some observations of 'the late Eclipse' and spoke out 'plainly' for Newton to 'let that excellent Clock be now sent down to us which You order'd to be made for the use of our Observatory'. Shortly thereafter Cotes wrote again, chiefly to thank Newton for having given orders that 'the Clock may be sent to Cambridge', and also to give Newton 'an account of what was observ'd by Us during the time of the sun's total obscuration in the late Eclipse'. Edleston (1850), pp. 179–84.

1 Edleston (1850), pp. 223–4.

2 *Ibid.* p. 224.

Gravity to be Essential to Bodies'.[3] Happily, he informed Clarke, 'The impression of the whole Book was finished about a week ago.' Hence, the date of completion of the printing of the second edition was about 18 June 1713.[4]

9. THE DISTRIBUTION OF THE SECOND EDITION

In a letter apparently without date (except for 'Tuesday'), but postmarked 1 July 1713, Bentley wrote to Newton: 'At last your book is happily brought forth; and I thank you anew y[t] you did me the honour to be its conveyer to the world.'[1] Although he was sending Newton only six copies, he hoped that Newton would 'be so free as to command what more you shall want'. In Cambridge, he wrote, there were 'no Binders... y[t] either work well or quick; so you must accept of them in Quires'. Bentley 'gave Roger a dozen', of which two were presents for 'D[r] Clark and Whiston'. Bentley himself had given copies to the Treasurer, Lord Trevor, and the Bishop of Ely; he 'thought it was properest for you to present D[r] Halley: so you will not forget him'. Then he gave some details of the sale arrangements:

> I have sent (though at great abatement) 200 already to France & Holland. The Edition in England to y[e] last buyer is 15[s] in quires: & we shall take care to keep it up so, for y[e] honour of y[e] Book.

It would seem that bound copies were selling for a guinea (or £1. 1*s*.).

The number of copies of the second edition may be determined from a letter of Cotes to Newton (1 November 1712) concerning the plate facing page 465, showing the path of the comet of 1680 from 4 November 1680 to 9 March 1680/81. In both E_1 and E_2, this large diagram was printed from a copper plate on a pull-out sheet. But in E_3 the size of the diagram was reduced so that it could be included on a regular printed page {506}. In so small a diagram, the dates could no longer be included at the appropriate places in the representation of the comet's orbit; these positions were indicated in the diagram by letters, which were then linked with dates in the text. Since the second edition was being printed in Cambridge and 'the Copper-plate of the Comet' was being engraved and printed in London under Newton's supervision, Cotes was asked by Newton for the number of copies needed. On 1 November 1712, Cotes wrote to Newton, 'The Printer tells me there will be 750 requisite.'[2]

I have not attempted to locate presentation copies of the second edition. Newton did work on a list of prospective recipients, which is so informative as an index of his plan of distributing his book that I print it in its entirety.[3] Today's reader will

[3] *Ibid.* p. 158. Clarke's comments were evidently transmitted orally to Cotes, since in this same letter Cotes wrote, apropos the offending passage, 'I am fully of Your mind that it would have furnish'd matter for Cavilling, & therefore I struck it out immediately upon D[r] Cannon's mentioning Your Objection to me, & so it never was printed.'

[4] Jones thanked Cotes for a presentation copy on 11 July, and on 27 June Newton waited on the Queen to present a copy of the new edition to her. *Ibid.* p. 225.

[1] From MS R.16.38, Trinity College.

[2] Edleston (1850), p. 144.

[3] From U.L.C. MS Add. 3965, f. 358.

be most interested in the fact that the first entry is that six copies were to be sent to Muscovy for the Czar and the libraries of the realm.[4] In transcribing the list I have added within square brackets the number of copies in each group. It will be seen that in all there are some seventy possible recipients of copies listed by Newton.

The Czar 6 for himself & ye principal Libraries in Muscovy [6]. The Abby Bignon eight for himself, young Monsr Cassini, De la Hire, Maraldi Varignon, & the Libraries of the Academy & the Observatory & the King [8]. The University Library in Cãbridge & the libraries of Trinity St Johns Kings & Queens [5]. The University Library in Oxford, & the Libraries of Christ Church &c. [2+]. The four Professors of Math. [4]. The libraries of the Duke of Tuscany, Venice, the Duke of Savoy the Kings of Denmark & Prussia, the Elector of Hanover, the Universities of the Low countries vizt Leiden, Utrecht, Franeker, Groeningen [10].

The Universities of France at Paris besides the Princes friends at home [1+6?], Toulouse, Burdeaux, Poictiers, Orleans, Montpellier, Rhemes, Douay, Avignon, Lyons, Aix [10]. The kings Library & the library of the Academy [*already mentioned*].

In Germany at Vienna, Liege Leipsic Prague Mentz Collogn Triers [7].

In Switzerland the public Library of Zuric, Bern, & Basil [3].

In Millain the publick library [1]. In Turin the Duke of Savoys library [*already mentioned*]. The publick library at Venice [*possibly already mentioned*]. The University of Upsal in Sueden [1].

Mr Taylor, Mr Machin, The two Professors at Cambridge the two at Oxford two at Edinburgh & St Andrews [4, *the Professors at Cambridge and Oxford having already been mentioned*]. Abby Bignon, young Monsr Cassini, De la Hire, Maraldi, & Varignion [*all have been mentioned*] Feuillée at Paris [1]. Mr Bernoulli at Bazil, Mr Leibnitz, Count Herberstein at the Emperors Court [3].

10. ERRATA

The relations between Cotes and Newton seem to have ended a little rudely and abruptly. Some six months after the publication of the second edition of the *Principia*, Cotes wrote to Newton a letter expressing considerable chagrin. Well he might have! For there is no record that Newton had ever so much as written him a letter of thanks for the enormous care he had spent on the revision of the text of the *Principia* and the job of seeing the work through the press.[1] In a draft of one of Newton's early letters to Cotes (30 May 1710; U.L.C. MS Add. 3984, f. 5^v) Newton wrote at the bottom: 'ut me admonuit D. Cotes acutissimus Astronomiae Professor apud Cantabrigienses'—that is, 'as Mr Cotes, the very acute Professor of Astronomy in Cambridge, informed me'. But this acknowledgement, apparently intended for the text of the *Principia*, does not appear in the letter actually sent to Cotes, and hence is not to be found in the *Principia* itself; Cotes apparently never saw this particular compliment.

Cotes, we must remember, received no remuneration for the job, all the royalties going to Bentley. In place of a letter of thanks written by Newton, Cotes received

[4] There appears to be some evidence that the Czar (Peter the Great) may have met Newton in London at a meeting of the Royal Society. See Grigorian (1964).

[1] If Newton ever wrote such a letter to Cotes, expressing his gratitude, there is no trace of it today, for among Newton's drafts of his letters to Cotes there is none of this sort, nor is there such a message in the collection of letters received by Cotes from Newton and preserved; nor does there exist a reply from Cotes (or a draft of a reply) that would indicate that he had ever received a final letter of thanks from Newton.

through the University Printer, Cornelius Crownfield, and not directly from the author, a lengthy paper of corrections and additions written out by Newton. Presumably Newton intended that these should be printed and bound up with copies of the book.[2] When Cotes received the list from Crownfield, he could hardly have helped feeling that he was being given a public chastising for the mistakes he had overlooked. Certainly he must have asked himself why Newton could not have written to him personally. Even Edleston, who usually cannot support even a suspicion of criticism of Newton,[3] wrote of 'so formidable a list', and he admits that 'Cotes does not seem to have been altogether pleased at the receipt' of it.[4]

Cotes's response lays bare the state of his feelings. In contrast to the warm and friendly letters which had been written during the preparation of the edition, this one is cold and formal. He has received the 'paper of *Errata, Corrigenda* & *Addenda*', he writes, and 'I take leave to send You some Observations upon them' as follows: 'By comparing Y^r Catalogue with my Table of *Corrigenda*, I find You have omitted that of Pag. 3. lin. 14...You have also omitted that of Pag. 47. lin. penult...'[5] The list goes on and on:

Your addition in pag. 47 lin. 4, should I think be omitted. For if that addition be made, the 8 preceding lines are to no purpose & ought to be omitted. Tis very evident by lin. antepenult. pag. 46 that PV is equal to $\frac{2DCq}{PC}$.

In Pag. 109 You direct to put H in the Figure instead of O. You mean instead of the lower O, which bisects the transverse diameter of the Hyperbola...

In Pag. 148 lin. 7 I think the alteration should not be made. There are three different *distantiae* & three different *termini* & one common angular motion.

In Pag. 151. You change *prima* the Feminine into *primum* the Neutre. Tis my Opionin that this alteration is not necessary...

Pag. 191. lin. 7 I think wants no correction. I cannot understand by what reasoning You make one; You will be pleas'd to reconsider it. If Your correction be true, it will be very necessary to explain it more fully.

In the conclusion of his letter, Cotes really gave vent to his feelings:

I observe You have put down about 20 Errata besides those in my Table. I am glad to find they are not of any moment, such I mean as can give the Reader any trouble. I had my self observ'd

[2] Edleston (1850), pp. 160–5.

[3] For instance, we have seen that Cotes raised a real difficulty in accepting Newton's statement in 'the first Corollary of the 5th' Proposition of Book III: 'it lyes in these words *Et cum Attractio omnis mutua sit* I am persuaded they are then true when the Attraction may properly be so call'd, otherwise they may be false'. Edleston refused to take such criticism seriously, commenting in a footnote (p. 152) that 'the difficulty raised by Cotes here affords an instance of the temporary haze which may occasionally obscure the brightest intellects'. Happily, Alexandre Koyré has restudied the question, to test the validity of Cotes's objection; Koyré (1961), reprinted in Koyré (1965), *Newtonian studies.*

[4] Edleston (1850), p. 166.

[5] *Ibid.* As to the last one {54.19}, Cotes said that it 'is requisite to determine Your meaning. Whilst that Sheet was printing I remember I did not understand what it was that You there asserted, & not having then time to examin the thing to the bottom, I was forc'd to let it go. Soon after I consider'd it, & found in what sense Your words could be true & accordingly made the alteration. Since Y^r Book has been published I have been ask'd the meaning of that place by one who told me, he knew not what sense to put upon Y^r words; I referr'd him to the Table of Corrigenda & then I perceiv'd he understood You.' This correction is included among the printed corrigenda.

several of them, but I confess to You I was asham'd to put them in the Table, lest I should appear to be too diligent in trifles. Such Errata the Reader expects to meet with, & they cannot well be avoided. After You have now Your self examined the Book & found these 20, I beleive You will not be surpris'd if I tell You I can send you 20 more, as considerable, which I have casually observ'd, & which seem to have escap'd You. And I am far from thinking these forty are all that may be found out, notwithstanding that I think the Edition to be very correct. I am sure it is much more so than the former, which was carefully enough printed; for besides Your own corrections & those I acquainted You with whilst the Book was printing, I may venture to say I made some Hundreds with which I never acquainted You.

Certainly the reader may share Cotes's dismay at his treatment.

The list of 'Corrigenda et Addenda' that Newton sent to the University Printer in 1713 appears never to have been printed up for inclusion in the second edition of the *Principia*.[6] For the most part, the items on the list were entered into both E_2a and E_2i; but there are some instances where the entry was made in one rather than both of these two copies of E_2, or made in a different form. See also Appendix V to the text.

We have seen, above, a number of examples of Newton's having told Cotes specifically not to be over-diligent 'in trifles';[7] now the roles were reversed. For Cotes's letter to Newton makes it evident that he had never held it his aim to correct every trifling error; especially, he was not concerned with those which would give no trouble to an intelligent reader. Newton appears not to have replied to Cotes's letter, and the only further known correspondence between them is a pair of letters written to Newton by Cotes in 1715 in regard to a total eclipse of the Sun.[8]

11. CHANGES IN THE SCHOLIUM GENERALE

When Newton saw the Scholium Generale in print, he changed his mind about some parts of it. He evidently at once sent to Cotes a pair of corrections which he wished to have made and which are preserved on a small sheet of paper along with the manuscript of the Scholium Generale sent to the printer (Trinity College Library, R.16.38, f. 271). The page in question must already have been printed off since the leaf containing pages 483–4, the second and final leaf of the final half-sheet of the text, is a cancel.[1] To see what changes Newton made I print below the final versions sent to Cotes, together with the variants (indicated by *S*) of the

[6] Edleston never found a copy of E_2 with such sheets, nor have I.

[7] Yet another instance occurred in 1710, after Cotes had received the corrections for the Scholium Generale at the end of the sixth section of Book II. He wrote Newton that he had not had leisure 'to examine all the Calculations of y[t] Scholium'. Newton replied, 'You need not give your self the trouble of examining all the calculations of the Scholium.' But he reckoned without the measure of his correspondent. And before two weeks had passed, Cotes wrote that he had 'examined the whole Calculation' and had 'done it anew where I thought it necessary'. Edleston (1850), p. 31.

[8] *Ibid.* pp. 179–84.

[1] In E_2, the text ends with a signature (Qqq) which consists of a half-sheet containing pp. 481/482, 483/484. Then comes a set of four unnumbered leaves, with a tiny signature mark, barely discernible on the recto of the third leaf (Rrr); the Index Rerum Alphabeticus occupies the first seven pages, while the eighth contains the corrigenda.

original Scholium Generale previously sent to Cotes. These changes would hardly appear to be of so great a significance as to warrant a cancel. We must remember, however, that the Scholium Generale was an obvious target for any critic and that Newton would be somewhat apprehensive about its general reception.

Pag. 483. l. 1, 2, 3. *Aeternus* est et *Infinitus*, *Omnipotens* & *Omnisciens* [Omnipotens & Omnisciens *om. S*] id est, durat ab aeterno in aeternum [ad aeternum *S*] et adest ab infinito in infinitum, omnia regit & omnia cognoscit quae fiunt aut sciri possunt [*altered from* omnia cognoscit quae fiunt aut sciri possunt & omnia regit quae sunt]. Non est aeternitas vel infinitas sed aeternus & infinitus, non est [*altered from* Non locus est] duratio vel spatium, sed durat & adest. Durat semper & adest ubique, & existendo semper et ubique durationem et spatium, aeternitatem et infinitatem constituit [omnia regit. . .infinitatem constituit: Non est locus, non spatium, sed est in loco et in spatio idque semper et ubique *S*]. Cum unaquaeque spatii particula sit *semper*, et unumquodque durationis indivisibile momentum *ubique*; certe rerum omnium Fabricator ac Dominus non erit *nunquam nusquam*. Omnipraesens est non per *virtutem* solam, sed etiam per *substantiam*: nam virtus sine substantia subsistere non potest. In ipso* continentur & moventur universa, sed absque mutua *passione*. Deus nihil patitur ex corporum motibus: illa nullam sentiunt resistentiam ex omnipraesentia Dei. Deum summum necessario existere in confesso est. Et eadem necessitate *semper* est et *ubique*. Unde etiam totus est sui similis, totus oculus, totus auris, totus cerebrum, totus brachium, totus vis sentiendi, intelligendi et agendi; &c

Pag. 484 lin. 17. Adjicere jam liceret nonnulla de spiritu quodam subtilissimo [subtilissimo *om. S*] corpora crassa [crassa *om. S*] pervadente, & in iisdem [iisdem: corporibus *S*] latente; cujus vi et actionibus particulae corporum ad [parvas *del.*] minimas [minimas: parvas *S*] distantias se mutuo attrahunt, & contiguae factae cohaerent; & corpora Electrica [attrahunt. . .& corpora Electrica: attrahunt, et corpora electrica *S*] agunt ad distantias majores, tam repellendo &c.

* Ita sentiebant veteres, *Aratus* in Phaenom. sub initio. *Paulus* in Act. c. 7, v. 27, 28. *Moses* Deut. 4.39. & 10.14. David Psal. 139.7, 8. Solomon Reg. 8.27. Job c 22. v. 12. Jeremias Propheta c. 23. v. 23, 24 [*Aratus*. . . v. 23, 24: Act. 17. 27, 28. Deut. 4. 39. & 10. 14. 1 Reg. 8. 27. Job. 22. 12. Psal. 139. 7. Jer. 23. 23, 24. *S*].

By the time Cotes received these final revisions from Newton, the corrigenda, which are to be found on the last page of the final signature (the rest of which contains the index), must also have been printed off. Therefore Cotes, who had one further correction to be made, wrote it out in his own hand underneath Newton's two corrections, and, as a result, it is printed at the bottom of the last page of the text (p. 484) rather than among the other corrigenda. It reads:

Pag. 478, lin. 25, *lege*: duodecim minuta secunda.

(See the Apparatus Criticus {522.25}.)

Within six months, Newton had decided to introduce yet further changes in the Scholium Generale. In the list of Corrigenda et Addenda sent by Newton to the University Printer we find:

P. 482 l. 2, *post* spatiis *adde* ob defectum aeris [{527.15} *alteration not made in* E_3].

Ib. lin 18 *lege* ut se mutuo quam minime trahant [{527.30–1} *alteration made in* E_3].

Ib. l. 29 *lege* non in corpus proprium (uti sentiunt quibus Deus est anima mundi,) sed in servos [{528.5–6} *alteration made in* E_3 *but with commas in place of parentheses*].

P. 483 l 36, *post* Fatum et Natura. *adde*, A necessitate Metaphysica, quae utique eadem est semper et ubique, nulla oritur rerum variatio. Omnis illa quae in mundo conspicitur pro locis ac temporibus diversitas a voluntate sola Entis necessario existentis oriri potuit. Dicitur autem Deus per Allegoriam videre, audire, loqui, ridere, amare, odio habere, cupere, dare, accipere, gaudere, irasci, pugnare, fabricare, condere, construere, & intelligentes (vitam infundendo)* generare. Nam sermo omnis de Deo a rebus humanis per similitudinem aliquam desumi solet. Et haec de Deo; de quo utique ex phaenomenis disserere ad Philosophiam experimentalem pertinet [{529.27 ff.} *alteration made in* E_3, *but for some small differences*: *the first sentence begins* A caeca necessitate metaphysica..., *the second sentence reads* Tota rerum conditarum pro locis...diversitas, ab ideis & voluntate entis...existens solummodo oriri..., *the next sentence* Dicitur...*ends with*...construere, *so that the postil is not present, the next sentence concludes*...aliquam desumitur, non perfectam quidem, sed aliqualem tamen, *and the final sentence has* ad philosophiam naturalem *for* ad Philosophiam experimentalem].

* Job. 38.7. Luc. 3.38.

When Fatio de Duillier saw E_2, he was quick to note that the leaf containing the final part of the new concluding Scholium Generale was printed on a cancel. What had the original leaf contained?—he must have asked himself. His reaction may be ascertained from an isolated fragment among his manuscripts in the Library of the University of Geneva (Papiers Fatio N° 3 (Inv. 526), f. 66), in which he describes how he had read 'an Abridgement of my French Treatise about the Cause of Gravity' at the Royal Society in February 1689/90. Fatio mentions the approbation his ideas had won from Newton, and refers to 'some Additions written by himself at the End of his own printed Copy of the first Edition of his Principles'. This portion of Fatio's note has been quoted earlier (Chapter VII, §10). As printed by Gagnebin, this extract concludes:

> And he gave me leave to transcribe that Testimony. There he did not scruple to say *That there is but one possible Mechanical cause of Gravity, to wit that which I had found out*: Thô he would often seem to incline to think that Gravity had its Foundation only in the arbitrary Will of God.[2]

The original, however, goes on:

> As I was absent from England in 1713, when the second Edition of those Principles came forth, I cannot tell why that Passage was suppress in it. But I have some reason to suspect it had been printed off with the rest of the Book, inasmuch as I see the last leaf, wherein it was to have been placed in that Edition, is torn off to make room for another leaf.

Little did Fatio realize that Newton had apparently abandoned any intention of mentioning Fatio's hypothesis at least a decade and a half before the second edition appeared.

[2] Fatio (1949), *La cause*, p. 117.

CHAPTER X

THE RECEPTION OF THE SECOND EDITION: THE TWO REPRINTS

1. REVIEWS IN FRANCE

THE second edition of the *Principia* was published in June 1713 and news of its appearance spread rapidly among the learned. This time there was no account of the book in the *Philosophical Transactions*, and the *Bibliothèque Universelle* was no longer being published. But both the *Journal des Sçavans* and the *Acta Eruditorum* carried a discussion of the second edition, just as they had done when the first edition had appeared in 1687. A third review was published in a new journal, the *Mémoires pour l'Histoire des Sciences et des Beaux Arts* (de Trevoux), which had been started in 1701. Both the *Journal des Sçavans* and the *Mémoires* (de Trevoux) reviewed the Amsterdam reprint of 1714, showing that the book itself had aroused sufficient interest to demand a Continental reprint even before any accounts of the new edition had been published. But the review in the *Acta Eruditorum* was based on the original second edition (Cambridge, 1713), no doubt because Newton had sent a copy directly to the editor, as well as to Leibniz. All three reviews were anonymous.

The review in the *Journal des Sçavans*[1] is extremely interesting, in that it is so different in tone from the sharp comments with which the first edition had been greeted in this famous periodical.[2] Thus, the reviewer fully understood that in the first two books Newton had treated the 'mouvement des corps avec la précision des Geometres, en sorte neanmoins qu'on applique les principes établis à plusieurs problêmes de Physique', whereas in the third book he had demonstrated 'par les mêmes principes le systême du monde'. Following a description of the Definitions and Axioms, the reviewer gave a précis of Newton's views on absolute and relative time, space, place, and motion, to conclude: 'Ce qu'il explique d'une maniere qui ne laisse rien à desirer.' A lengthy paragraph was given to Section I, but the reviewer then hastened on to give only a few highlights of the rest, expressing his regrets as follows: 'Nous voudrions pouvoir suivre M. Newton dans toutes les parties de son Ouvrage, & faire voir plus en particulier l'élevation de son génie.'

[1] *Journal des Sçavans* (March 1715), pp. 157–60.

[2] See Chapter VI, §6.

The review, save for such occasional outbursts of praise, is matter-of-fact and descriptive. Nowhere is there a derogatory remark, or even a pejorative adjective. Even when the reviewer finally stated that Newton came out directly in opposition to vortices, he merely gave the reason for Newton's position,[3] and said, 'c'est la raison pour laquelle notre Auteur rejette absolument cette hypothese [des tourbillons]'. He then closed, without so much as a reference to the new concluding Scholium Generale, by observing that 'c'est ici une nouvelle Edition de ce Traité', and listing some of the principal differences between it and the first edition.[4]

Quite different in tone was the account in the *Mémoires* (de Trevoux).[5] This began with a contrast between the reputation of the *Principia* among geometers ('qui admirent la force & la profondeur du génie de l'Auteur') and among 'les Physiciens', by whom 'la reputation de cet ouvrage...est contestée' and who for the most part 'ne sçauroient s'accomoder d'une attention naturelle, qu'il prétend être entre tous les Corps'. Since the reviewer held that 'M. Gregory a travaillé son Astronomie Physique & Geometrique sur les principes de M. Newton', and since Gregory's book had been discussed at length in two previous issues (February and March 1710),[6] there was no point in repeating at the present time what had been said of (or, more truly, against) Newtonian science. As to the improvements of the new edition, the reviewer merely said that 'cette nouvelle édition n'est augmentée que de deux ou de trois éclaircissemens'.

At last the reviewer comes to the 'proof' of the inverse-square law of attraction, namely, the Moon test in the Scholium to Prop. IV, Book III: 'La preuve semblera aller jusqu'à la conviction, si l'éxperience peut montrer qu'en effet la terre & la lune sont, pour ainsi parler, en commerce d'attraction.' The exposition concludes with these remarks:

> Toutefois ce raisonnement; qui d'abord paroist plausible, ne prouve pas; parceque cette conformité entre le mouvement de la lune & celui des corps pesants ne se rencontrant que dans l'arc proposé LC, c'est d'un cas particulier tirer une conclusion generale.[7]

There follow 'quelques calculs assez curieux de M. Newton, fondez en partie sur les Observations Astronomiques & en partie sur ses propres principes'. These have led Newton to claim that 'les degrez des meridiens sont plus petits vers l'équateur,

[3] 'Nous observons que la force par laquelle l'Auteur explique tous les phenomenes dont il traite, est la pesanteur[, une force qui,] à mesure qu'elle s'en éloigne...diminuë exactement dans la raison doublée des distances...Et...cette proportion, qui se verifie par les observations, ne se trouve point dans l'hypothese des tourbillons.' Furthermore, Newton had shown that the motions of comets are 'trés-reguliers, qu'elles suivent les mêmes loix que les Planetes, & qu'on ne peut pas expliquer ces mouvemens par le moyen des tourbillons'.

[4] 'Enfin...la principale difference...c'est qu'on a retouché dans [la nouvelle Edition] divers endroits, surtout la section seconde du premier Livre, la section septiéme du second Livre, & dans le troisiéme Livre la theorie de la Lune, la precession des Equinoxes, & la theorie des Cometes.' This, of course, is a more or less direct translation of the conclusion to Newton's Preface to E_2.

[5] *Mémoires pour l'Histoire des Sciences et des Beaux Arts* (de Trevoux) (February 1718), pp. 466–75.

[6] *Ibid.* (February and March 1710), pp. 252–74, 415–34.

[7] The reason given shows a rather complete lack of understanding of the Newtonian method: 'Il faudroit que les sinus verses des arcs ou plus petits, ou plus grands fussent entre eux en raison doublée des arcs; ce qui ne sçauroit être dans les arcs sensibles, comme le sont ceux aux environ de 33″ par rapport à l'orbite de la lune.'

& vont en croissant vers les poles'. To which the reviewer counters with the remark that, following Cassini, there is 'un sentiment tout contraire touchant la figure de la terre'. Thereupon, the review terminates with the observation that 'On s'est dispensé de suivre M. Newton dans les autres partis de son systeme', since the whole system has already been expounded 'tout entier & assez exactement' in the previously published account of Gregory's *Astronomy*.

2. THE REVIEW IN THE 'ACTA ERUDITORUM'

Of a wholly different kind is the review in the *Acta Eruditorum*.[1] As mentioned in Chapter I, §2, the anonymous reviewer took it upon himself to make a careful collation of the two editions, and to describe in detail the differences between them. Thus he calls attention to changes in terminology[2] and to the expansion of Def. I {1. 8–9}. He refers to the new corollaries to Props. I and II, Book I, and observes how they simplify the proof of Prop. IV. He mentions the two new corollaries to Prop. IV, giving the contents of Corol. 7 {44.26–8}. He paraphrases the new Prop. VI, and shows that one of its several new corollaries was the old Prop. VI of E_1. The following extract (presented here in translation) gives the tone of the review:

> The author renders the seventh proposition more general; for while in the former edition he had determined the centripetal force of a body revolving in the circumference of a circle tending toward some point in the circumference, now he seeks the centripetal force tending toward any point. In the ninth proposition there is added a new, and that very brief, demonstration concerning the centripetal force tending toward the centre of a spiral; and the same thing occurs [*that is*, there is a new brief demonstration] concerning the centripetal force tending toward the centre and focus of an ellipse in props. 10 and 11; also to the scholium of the former certain words are added, and to prop. 17 a general scholium is added, concerning the centrifugal force which is directed toward any point situated within a conic section.[3]

When the reviewer comes to the 'Leibniz Scholium' (Lemma II, Book II {246.10–26 ff.}) he quotes rather than paraphrases, and says editorially (p. 133) that 'when in lemma 2 book 2 pages 224 and following he had set forth the simpler rudiments of the differential calculus by his own method, even in this new edition he does not deny that the illustrious Leibnitz had communicated its foundations to him, although he [Newton] zealously concealed a certain technique of his own'.

It is of special interest to observe the reviewer's remarks concerning Prop. X, Book II {252. 1 ff.}, since we have seen that Newton corrected this only after the

[1] *Acta Eruditorum* (March 1714), pp. 131–42.

[2] For example, 'subduplicate' for 'dimidiate', and the change in form from 'bodies are resisted' to 'it is resisted to bodies' ('nunc legitur: *corporibus resistitur*, ubi antea extabat, *corpora resistuntur*').

[3] It is amusing to find that although Newton uses the expression 'centripetal force', which he had invented and named in honour of Huygens (see Supplement I), the reviewer in his paraphrase (p. 132) writes of 'centrifugal force'. Thus unconsciously he shows his allegiance to the Continental physics of Descartes, Huygens, and Leibniz and his essential rejection of the Newtonian point of view. R. S. Westfall quite correctly points out (in a private communication) that it is meaningless to write of such a 'centrifugal force' directed 'toward' the point; the reviewer should have written 'away from' it.

edition had been all but completed, which therefore required the reprinting of a whole sheet plus a leaf of a second sheet.[4] The reviewer writes:

> The renowned Bernoulli, in the *Acta* for the year 1713, page 121, noted that Newton, opportunely advised by himself (through his nephew Nicolaus Bernoulli), before the completed printing of the new edition, had corrected what he had erroneously stated in the former edition concerning the ratio of the resistance to gravity, and had inserted them into his book on a separate sheet of paper. That this was done, a comparison of the new edition with the previous one and with Bernoulli's remarks on it shows; and the sheets cut out and the new ones substituted in their place certainly declare that the errors of the first edition already had crept also into the second.[5]

As mentioned earlier (Chapter IX, §4), anyone who looks closely at a copy of E_2 will see that the stub of the former pages 231–2 (the last leaf of the signature Gg) is plainly visible.

Naturally, during the course of reporting the alterations in the *Principia*, the reviewer calls attention to the change from Hypotheses to Rules and Phenomena in Book III; and he stresses the significance of the new Rule III. Considerable space is devoted to a presentation of other emendations, omissions, and additions to be found in Book III. Finally, the reviewer calls attention to the new General Scholium as the 'concluding flourish upon a work of profound erudition'. Here, he says (p. 141), there 'are expounded the difficulties with which the hypothesis of vortices seems to the most celebrated author to be pressed; some thoughts are adduced concerning God; certain suggestions concerning the cause of gravity are indicated and mention is interjected of a certain new hypothesis concerning a certain most subtle spirit pervading gross bodies (the same perhaps as the hylarchic principle of Henry More)'.[6] Then, following a summary of the General Scholium, the reviewer turns to Newton's admission that 'the force of gravity arises from some cause', and his subsequent denial 'that it is mechanical, since it does not act in proportion to the quantity of surfaces, but of solid matter'. The reviewer repeats Newton's statement that 'he cannot deduce the cause of the properties of gravity

[4] On the rôle of Johann Bernoulli, see Chapter IX, §4, note 4.

[5] The reviewer notes: 'Although, moreover, Newton has introduced Bernoulli's correction, nevertheless he will not deny that his own solution of the problem is still far from the simplicity of Bernoulli's solution published in last year's *Acta*. Indeed it will seem strange to some that he still commends on page 236 that very rule through which, as Bernoulli observes in last year's *Acta*, p. 95, he falls into errors, inasmuch as [the rule is] opposed to the true laws of differential calculus established by the illustrious Leibniz.' It is of great interest that in E_3 the contents of that p. 236 of E_2 here criticized {255.19–256.28} were reprinted without change. Newton evidently (and rightly) did not agree that he had fallen into error.

The review then goes on (p. 134): 'To proposition 15 he has not given that fullness in the new edition which the renowned Bernoulli observed should be given to it (*loc. cit.* p. 125); however, he has emended the demonstration of the immediately following proposition 16, which Bernoulli recommended him to reconsider, in the way that the renowned Hermann taught that it should be corrected, in the Venetian Journal vol. VII, published in the year 1711, page 217. He adds also three corollaries which do not appear in the former edition, and emends a lemma, which is premised for the sake of those [i.e. that sort of] propositions; whence necessarily in the demonstration to prop. 75 [a mistake for 15] some changes had to be made. But he did not attend to the things which Bernoulli advised concerning the corollaries of the same proposition, since they were not seen by him before publication.'

[6] I have mentioned earlier that one is tempted by this reference to Henry More to guess that the author of the review may in fact have been Leibniz himself. See Koyré and Cohen (1962), 'Newton, Leibniz and Clarke', p. 68, note 16.

from the phenomena' and 'professes that he does not feign hypotheses' ('hypotheses...non fingere profitetur'), 'to which he does not concede a place in experimental philosophy'. And then the reviewer concludes with the following sentence, which particularly infuriated Newton:

> Nevertheless, it is to be feared that most people may assign a greater value to hypotheses than to the author's most subtle spirit...pervading gross bodies and lying hidden in them; by whose force and actions the particles of bodies attract one another at the least distances; and, made contiguous, cohere; electric bodies act at greater distances as much by repelling as by attracting neighbouring small bodies; light is emitted, reflected, refracted, inflected, heats bodies; all sensation is excited; the limbs of animals are moved at will...unless you might say that it is the same as the ether or subtle matter of the Cartesians.

Newton replied in the 'Recensio libri', the anonymous book review that he wrote in English of the *Commercium epistolicum*.[7] He concluded with a rebuttal of the review of the second edition of the *Principia* in the *Acta*. Mention has already been made (Chapter II, §3) of the fact that in E_2i Newton altered the final paragraph of the Scholium Generale by adding {530.34} the words 'electrici et elastici' to modify 'spiritus'. Furthermore, A. R. Hall and M. B. Hall have demonstrated that Newton was indeed led to conceive of this 'spirit' by thinking about electric 'effluvia' and the phenomena of electrical repulsion and attraction.[8] May not Newton have contemplated adding the two qualifying adjectives in order to make it clear that *his* 'spiritus' was neither the hylarchic spirit of Henry More nor the ether (or subtle material) of Descartes?

3. THE AMSTERDAM REPRINTS (1714, 1723)

The second edition of the *Principia* in 1713 did not exceed 750 copies.[1] Rigaud observed in 1838 that copies did not then 'occur in the market much more frequently than [copies of] the first [edition]', a condition equally true in our own day.[2] It was apparently—from the very beginning—a difficult book to obtain on the Continent; witness a letter from C. R. Reyneau to William Jones mentioning the problems of finding a copy in Paris in November 1714: 'La seconde édition [des Principes de la Philosophie Naturelle]...est encore si rare ici, que je n'ai pu la voir que des instants par le moyen de ceux, qui ont enlevé ce qui en étoit venu d'abord.'[3] As early as the summer of 1713, the *Journal Littéraire* announced an Amsterdam reprint by 'une compagnie des libraires' of Newton's *Principia*, 'sur la seconde édition qui vient de paroître en Angleterre. Deux presses roulent continuellement pour avancer cet ouvrage.'[4] This reprint appeared in 1714, and

[7] De Morgan (1848), 'The second edition of the *Commercium epistolicum*'; De Morgan (1852), 'The account of the *Commercium epistolicum* in the *Phil. Trans.*'; see Jourdain's discussion of these papers in De Morgan (1914), *Essays on Newton*, editor's preface, pp. vii–viii.

[8] Hall and Hall (1959), 'Newton's electric spirit'; cf. Koyré and Cohen (1960), 'Newton's "electric and elastic spirit" '.

[1] See Chapter IX, §9. [2] Rigaud (1838), *Essay*, p. 106. [3] *Ibid.* [4] *Ibid.*

evidently it too did not satisfy the needs of Continental scholars. Once again a company of booksellers combined to issue another reprint of the *Principia* (Amsterdam, 1723), together with a separately paginated supplement, containing four of Newton's mathematical tracts and a preface by W. Jones.[5]

It would be interesting to find out whether the first Amsterdam reprint (1714) was undertaken so soon after the English edition because it was more profitable to reprint the volume than to import copies, or because copies of the English edition were difficult to obtain—either because of the conditions of international trade or because the whole edition went quickly out of print. The notice in the *Journal Littéraire* appeared in the issue for July–August 1713, shortly after the completion of the printing of the second edition in England in June 1713. So far as I know, this Amsterdam reprint was wholly unauthorized. In any event, the estimate of the size of the prospective market was continually too small, since the original issue (1713) plus the first Amsterdam reprint (1714) did not satisfy the demand; apparently even the second Amsterdam reprint (1723) was not issued in a large enough number of copies.

These Amsterdam reprints were a major undertaking, requiring the cutting of new wood-blocks for the figures and a new setting of type. The second reprint (1723) contains not only four tracts by Newton and W. Jones's 'Praefatio Editoris', but also extracts from four letters of Newton's.[6] These tracts are: *De analysi per aequationes infinitas* (first published by Jones in 1711), *De quadratura curvarum* and *Enumeratio linearum tertii ordinis* (published with the *Opticks* in 1704 and the *Optice* in 1706, but eliminated from the second English edition of the *Opticks* in 1717/18), and the *Methodus differentialis*. But this whole collection—the four tracts, the extracts from Newton's letters, and Jones's Praefatio—was merely a reprint, without alteration, of a collection that was first published as a small book in London in 1711.[7] It was reissued separately in Amsterdam in 1723 as well as being included as a supplement to the reprint of the *Principia*.[8] How curious indeed that Newton's long-cherished plan of publishing *De quadratura* together with the *Principia*[9] should have been realized only in this presumably unauthorized Amsterdam reprint of 1723!

I have not attempted to collate this reprint with either the original second edition or the earlier reprint of 1714, but it appears to be an unrevised new printing, differing only in the text of the title-page, which now refers to the mathematical

[5] See the bibliographical account in Appendix VIII to the text, part two.

[6] 'Fragmenta Epistolarum ad D. Oldenburgium, 13 Jun. & 24 Oct. 1676, ad D. Wallisium, 1692, ad D. Collinsium, 8 Nov. 1676.'

[7] See Gray (1907), *Newton bibliography*.

[8] Some authors (for instance, Rigaud) write as if this edition of the *Principia* contained as supplement only Newton's *De analysi*. The mathematical part has a separate title-page (which lists all the contents in full) and its own pagination, but the title-page of the *Principia* concludes: 'EDITIO ULTIMA *Cui accedit* ANALYSIS *per Quantitatum* SERIES, FLUXIONES *ac* DIFFERENTIAS *cum enumeratione* LINEARUM TERTII ORDINIS.' Hence this title-page, taken by itself, would not tell that the *De quadratura* was also present.

[9] See Chapter VII, §13; Chapter IX, §5; and Supplement VIII.

works printed as a supplement. Nor have I collated the first Amsterdam reprint of 1714 with the original second edition. But it is worthy of our notice that a third printing of E_2 was required—with the cutting of diagrams and setting up of the type—since the very need for this printing gives some indication of the progress the Newtonian natural philosophy had been making on the Continent.

4. PREPARATION FOR A THIRD EDITION

In the same year, 1723, that saw the publication of the second Amsterdam reprint of the *Principia*, Newton decided to undertake a new revised edition of his great work. At the time he was 81 years of age, and was just recovering from an 'incontinence of urine'. At first diagnosed simply as 'the stone' and thought to be incurable, Newton's problem was later found to have been 'owing merely to a weakness in the sphincter of the bladder'.[1] By avoiding travel by carriage, and by following a strict diet prescribed by his physician and friend, Dr Richard Mead, Newton regained his health.

No doubt Newton was stimulated to begin a new edition by the availability of a competent editor who was willing to do the hard work, Dr Henry Pemberton. But another stimulus may have been the fact of the Amsterdam reprints, made without any corrections or improvements of the author's.

During the years following the appearance of the second edition of the *Principia*, Newton had continued to revise and improve both of his physical treatises: the *Principia* and the *Opticks*. In 1717 he published a third edition of the *Opticks* (also issued in 1718) with a set of interesting new Queries.[2] I have not been able to discover exactly when Newton actually committed himself to the production of a third edition of the *Principia* (published in March 1726), which had got under way by November 1723. Possibly, during the decade following the printing of the second edition in 1713, he may merely have been collecting errors and misprints for his own interest or for record purposes. We simply do not know. But there is no question of the fact that he entered some proposed alterations in both an annotated and an interleaved (and annotated) copy of the second edition of the *Principia*, which are thus the analogues of the earlier annotated and the interleaved (and annotated) copies of the first edition.[3] We have seen that he wrote out in these copies almost all of the errors he had included in the list he sent to the publisher soon after the second edition had been printed.[4] And, of course, he continued to

[1] Brewster (1855), vol. 2, p. 377.

[2] The *Opticks* was published in English in 1704. The Latin version, published in 1706, is a new edition because it is revised; for example, it has seven new Queries. These and eight further new Queries were added to the second English edition, or the 'third edition' in all, in 1717 (1718). The new Queries 17–23 of the Latin edition became, with some revisions, Queries 25–31 of the second English edition; and the eight new Queries of that edition were thus numbered 17–24. See Koyré (1960), 'Les Queries'. Professor Henry Guerlac has announced that he is preparing a 'variorum' edition of the *Opticks*.

[3] For an account of these copies and their use in the Apparatus Criticus of the present edition, see Chapter II.

[4] See Chapter IX, §9, and Appendix V.

essay improvements and revisions on loose pieces of paper, of which a great number have been preserved in the Portsmouth Collection. No student of the *Principia* can help being overwhelmed by the sheer bulk of such material that Newton preserved. Whoever examines E_1i and E_1a, E_2i and E_2a, and Newton's manuscript remains in the Portsmouth Collection and elsewhere will continue to be impressed by this physical testimony to Newton's apparently unceasing endeavour throughout his maturity to make every line he wrote as exact an expression of his thoughts as possible. The manuscript alterations in E_2i and E_2a (like those in E_1i and E_1a) are given in the Apparatus Criticus of the text of the present edition. Those which Newton wrote out on separate pieces of paper will be discussed in the Commentary (see Chapter II above). Certain aspects of these loose manuscripts are of general interest, particularly in so far as they show Newton's later working habits and general point of view, and they may properly be discussed here.

5. SOME EMENDATIONS OF THE SECOND EDITION

In the last decades of his life Newton seems to have wished to clarify his philosophical position, as may be seen in a number of different examples. One such occurs in the Regulae Philosophandi. In the second edition of the *Principia* there are three of these Regulae, of which we have seen that the first two had already been present in the first edition as Hypotheses I and II, while the third was a replacement for the old Hypothesis III.[1] It will be recalled that toward the end of the job of revising E_1 and printing E_2, Cotes had written to Newton concerning a difficulty he had found in Corol. 1 to Prop. V (Book III), in which Newton says, 'Et cum attractio omnis mutua sit...' ('And since every attraction is mutual...'). To explain his meaning, Cotes gave an example of two globes A and B placed on a table; A is at rest and 'B is moved towards it by an invisible Hand'. According to Cotes,

> A by-stander who observes this motion but not the cause of it, will say that B does certainly tend to the centre of A, & thereupon he may call the force of the invisible Hand the Centripetal force of B, or the Attraction of A since ye effect appeares the same as if it did truly proceed from a proper & real Attraction of A. But then I think he cannot by virtue of this Axiom [Attractio omnis mutua est] conclude contrary to his Sense & Observation, that the Globe A does also move towards the Globe B & will meet it at the common centre of Gravity of both Bodies. This is what stops me in the train of reasoning by which as I said I would make out in a popular way the 7th Prop. Lib. III.[2]

Newton's resolution of the problem had been to explain to Cotes that, 'as in Geometry the word Hypothesis is not taken in so large a sense as to include the Axiomes & Postulates, so in experimental Philosophy it is not to be taken in so large a sense as to include the first Principles or Axiomes wch I call the laws of

[1] See Chapter II, §2. This topic is explored in Cohen (1971), *Scientific ideas*, chap. 2, 'From Hypotheses to Rules'.

[2] Edleston (1850), pp. 152–3; Prop. VII is: *That there is a power of gravity pertaining to all bodies, proportional to the several quantities of matter which they contain.*

motion',[3] and he proceeded to direct Cotes to alter the concluding Scholium Generale 'for preventing exceptions against the use of the word Hypothesis'.[4]

Three days later Newton wrote again to Cotes on this topic (and others), concluding that 'he that in experimental Philosophy would except against any of these must draw his objection from some experiment or phaenomenon & not from a mere Hypothesis, if the Induction be of any force'.[5] It was then too late to incorporate such a statement in the *Principia*, since the pages at the beginning of Book III had long since been printed off, and we have seen that Newton did not take kindly to Cotes's suggestion of 'an Addendum to be printed with y^e^ Errata Table'. But the need to have such a statement in the *Principia* remained, and apparently Newton wrote out an additional Regula Philosophandi (IV) for this purpose. Too late for inclusion in E_2, the new Regula found a natural place in E_3. This Regula is a formal and more complete statement in Latin of what Newton had written to Cotes:

RULE IV.

In experimental philosophy we are to look upon propositions collected by general induction from phenomena as accurately or very nearly true, notwithstanding any contrary hypotheses that may be imagined, till such time as other phenomena occur, by which they may either be made more accurate, or liable to exceptions.

This rule we must follow, that the argument of induction may not be evaded by hypotheses.[6]

This fourth Regula appears with some minor variants on an interleaf of E_2i, as may be seen in the Apparatus Criticus.

Among Newton's manuscripts, there are a number of sheets containing drafts or revises of this Regula. One of them even goes on to the proposed Regula Philosophandi V, which has been printed above in Chapter II, §2. It is astonishing to see how Newton worked over and reworked the text of this Regula Philosophandi; but he never entered it in either E_2a or E_2i, nor did he print it in E_3. The essential contents of this Regula Philosophandi V appear in the manuscripts in yet another guise, among a series of Definitions of such fundamental concepts as Phaenomenon, Hypothesis, Regula, Corpus, and Vacuum. On one manuscript sheet (U.L.C., MS Add. 3965, f. 420), Newton wrote out a definition of 'Phaenomenon' that is all but the same in content as this proposed and abandoned Regula V.[7]

[3] *Ibid.* pp. 154–5.

[4] See Chapter IX, §7.

[5] Edleston (1850), p. 156. See also Koyré (1956), 'L'Hypothèse et l'expérience chez Newton', translated in Koyré (1965), *Newtonian studies*.

[6] For the Latin text, see {389.6–13}.

[7] On another sheet (f. 422^r, 422^v) Newton has written out in full a section of Definitions, three in number, of Corpus, Vacuum, and Phaenomenon, with rather full explanations of each. On yet another sheet (f. 428), largely covered with computations, he has placed above a version of Reg. IV a short statement of what he means by 'bodies'. A torn fragment (f. 437^v), following some sheets of data and other proposed revisions, contains an early version of Defs. I and II, Corpus and Vacuum, together with explanations. I have transcribed, translated, and discussed these definitions in the Wiles Lectures for 1966; Cohen (1971), *Scientific ideas*. J. E. McGuire has been making an independent study of these same texts.

Beginning on f. 491, there is a set of sheets containing the 'Corrigenda et addenda in Lib. I...in Lib. II...in Lib. III', of which Newton sent a copy to the printer on the termination of the printing of E_2.[8] (In his own copy, however, he has entered some later additions to the list.) This is followed (in MS Add. 3965, ff. 493–4^v) by several additional pages of 'Errata, Corrigenda & addenda', similarly tabulated for each of the three books of the *Principia* by page and line. The final entry contains a revision of the paragraph of the Scholium Generale concerning God. Newton has begun to write the final sentence not as it had been printed in E_2 but as it was to appear in E_3 (except for punctuation and capitalization), namely, 'Et haec de Deo, de quo utique ex phaenomenis disserere ad Philosophiam Naturalem pertinet', rather than 'ad Philosophiam experimentalem pertinet' {529.35–37} ('And thus much concerning God, to discourse of whom from phenomena belongs to Natural Philosophy', or, in the version of E_2, 'belongs to experimental Philosophy'). But he has then crossed out 'Naturalem', has then written it again 'Natura...' and crossed it out once more, and has finally written out 'experimentalem' as in E_2. On the following folio (f. 495) there is another set of these same alterations, much of it all but identical to the one just described, but less complete and possibly an earlier draft. This is followed (ff. 497–8) by yet another set of 'Corrigenda et addenda' (at first labelled 'Errata, corrigenda et addenda') which include many of the entries in the previous lists and many more besides. For instance, Newton has put into this list the new paragraph, first published in E_3, explaining how the Second Law of Motion may be applied (and, in context, presumably had been applied by Galileo) to discover the law of freely falling bodies and the motion of projectiles {21.22 ff.}, and a somewhat similar paragraph intended to follow the presentation of the Second Law itself, of which there are earlier drafts on a loose scrap of paper (f. 367). These two alterations, and two others, also of paragraph length, are neatly copied out on a single sheet of paper (f. 500).

There follow further such lists, in which the 'Corrigenda et addenda' include paragraphs of considerable interest. Thus (f. 504; see Plates 15 and 16) we find Newton's intention to have two definitions precede the Regulae Philosophandi:

DEFINITIO I.	DEFINITION I.
Corpus voco rem omnem quae moveri et tangi potest et qua tangentibus resistitur...	I call body everything which can be moved and touched and which resists things that touch [it]...
DEF. II.	DEF. II.
Vacuum voco spatium omne per quod corpus sine resistentia movetur...	I call vacuum all space through which a body moves without resistance...

These definitions, and their explanations, are cancelled.

[8] See Chapter IX, §9.

There is also a revision of Regula II, which in E_2 reads just as in E_1 (save that in E_1 it was denoted as Hypothesis II):[9]

Ideoque Effectuum naturalium ejusdem generis eaedem sunt Causae. [E_1, E_2]	Therefore for natural effects of the same kind the causes are the same.

so as to become:

Ideoque...eaedem assignandae sunt Causae quatenus fieri potest.	Therefore for natural effects of the same kind the same causes are to be assigned, as far as possible.

This is the form in which it is printed in E_3.

Here one finds also the two new paragraphs, introduced in E_3 at the end of Phaenomenon I, in which Newton presents the recent determinations of the elongations and diameters of the satellites of Jupiter made by James Pound. This material appears in E_3 {390.21–391.22} substantially as on these sheets and is referred to by Newton in the preface to E_3. There follow other new astronomical data, also incorporated in E_3.

Another alteration proposed by Newton is that in Corol. 2, Prop. VI (Book III) {402.16} 'Pro Aether scribe Aer', but in E_3 this change is not made, although it appears in E_2i. A few pages later (ff. 521ᵛ–522) there are the revised Corols. 7 and 8 and the new Corols. 9 and 10 to Prop. XXXVII (Book III) substantially as printed in E_3 {469.9–470.21}.

On ff. 507 and 509, Newton directs that Prop. XXXIII be followed by a scholium and other material indicated as follows:

Schol.

Hanc Propositionem D. Machin aliter investigavit ut sequitur.

Lemma

Motus Solis medius a Nodo est ad motum medium Solis (caeteris paribus) in subduplicata ratione motus horarii mediocris Solis a Nodo in quadraturis existente, ad motum horarium Solis.

In recta quavis - - - - in Octantibus erit 1° 29′ 57″.

The foregoing mention of 'D. Machin' (= Mr Machin; see note 5 to Chapter XI, §7) and the statement of the lemma are references to two propositions and a short scholium written by John Machin and presented by Newton in E_3 within a set of continuous quotation marks on pages 451–4. It will be observed, however, that, although the final words in Newton's manuscript correspond to the final words of the scholium following Machin's Prop. II, the lemma quoted by Newton in these manuscript sheets resembles, but differs somewhat from, Machin's Prop. I in E_3; the opening words of the proof are wholly different.

Some light may be cast on the cause of this discrepancy by referring to another group of manuscripts elsewhere in the Portsmouth Collection (U.L.C. MS Add. 3966, §11). Here (ff. 99–101) one finds a beautifully written-out discussion, 'De Motu Nodorum Lunae', obviously in the hand of a professional amanuensis, containing a 'Propositio 1' and a 'Prop. 2' plus a supplementary section. Proposi-

[9] For the change from Hypotheses to Regulae see the Apparatus Criticus, {387}.

tio 2 corresponds to the Prop. II in E_3 (pp. 453–4), and the supplement similarly corresponds to the scholium (p. 454). But Prop. 1 of this manuscript differs from Prop. I as printed, although it is stated almost exactly as in the lemma in the above-mentioned manuscript list of addenda of Newton's. The proof (in MS Add. 3966) begins, as in Newton's manuscript list of addenda, with the words 'In recta quavis'. For some reason this whole document exists in a second version (MS Add. 3966, f. 97), written out in full in Newton's hand. (In this one the first proposition is stated exactly as in the manuscript list of addenda.) Since Newton was a great copier of all sorts of things, including his own writings as well as those of others, there is probably no special significance to his having copied out Machin's two propositions. Although Newton called the first of these propositions 'Lemma' in his manuscript list of addenda, this expression does not appear (in MS Add. 3966) in either Newton's copy of 'De Motu Nodorum Lunae' or the amanuensis's copy. Since the statement of Prop. I and its proof differ in E_3 from the version to be found in both of these copies and in Newton's manuscript list, we may conclude that at some time later on, and before E_3 was printed, Machin forwarded to Newton a substitute Prop. I and a new proof. The figure remained unaltered.

We shall see in Chapter XI, §7, that Henry Pemberton submitted an independent proof of the 'motion of the moon's nodes', to which Newton refers in E_3 in the scholium {451.12–17} introducing Machin's two propositions. There is a manuscript, apparently in Pemberton's hand (MS Add. 3966.92–4), to be found together with the manuscripts just described, containing what I presume to be the proof in question.

In MS Add. 3965 there is yet another copy of the 'Corrigenda et addenda in Lib. 1 Princip.' (f. 515), including both the new paragraph following Lex II (on the application of this Lex to continuous forces, as gravity producing free fall) and the new paragraph (on how Galileo has used the first two Laws of Motion and the first two corollaries to find the laws of falling bodies and of projectiles) intended for the scholium following the Leges Motus. The first of these additions was later discarded; the second appears in E_3 {21.22 ff.}. In this set of alterations Newton has copied out a revision of all but the first paragraph of Exper. 13 on the speed of fall of different globes in air {352.4 ff.}, and has tacked on an account of some more recent experiments he had made with J. T. Desaguliers and others in 1719 and which later was to become a separate Exper. 14.[10] In the materials in this list corresponding to Book III, we find the new Regula IV. But now the definitions of body and vacuum are wholly absent. The new astronomical data are even more plentiful, the corollaries to Prop. XXXVII are present too, and the revised material on the comet of 1680 {495.20 ff.} is written out much as printed in E_3, together with other and revised cometary material. The revisions of the Scholium Generale are fuller than in the previous version, including the new references to

[10] Desaguliers (1719). These new experiments are discussed in Chapter XI, §4.

authorities as printed in the middle of page 528 in E_3. The final sentence of this long paragraph has 'Philosophiam naturalem' as in E_3 {529.36–7}, but Newton had at first written 'ad Philosophiam naturalem spectat' and then crossed out 'spectat' and gone back to 'pertinet.'

There follow a number of pages (ff. 527 ff.) of revised cometary material, including further drafts of the subject as presented in E_3, and a draft of the revised paragraph on God for the Scholium Generale. The latter (f. 539) is evidently a predraft of the version mentioned in the preceding paragraph, since it contains some tentative phrases. It still ends 'ad Philosophiam experimentalem pertinet'. On a small bit of paper (f. 544) Newton has been trying out some alterations of this Scholium, and has written out the sentence in question at the top of the sheet in the form 'ad philosophiam naturalem pertinet', but has then crossed out 'naturalem' and has written 'experimentalem' above the latter.[11]

These lists of 'Errata, Corrigenda, et Addenda' do not have the rough aspect of work-sheets, but rather appear to be carefully written copies, of the kind that could have been given to a printer to serve as a basis for a new revised printing or a new edition. It may be that much of the text given to Pemberton for the third edition consisted of just such sheets, perhaps as a supplement to an annotated or an interleaved copy of the second edition. We simply have no way of telling. If in part Newton gave Pemberton a set of sheets such as I have been describing, then perhaps they contained a version based on a synthesis of all these different sets. But in such a case, there must have been supplements as well, because sometimes —as in the case of the two propositions of Machin's—Newton indicated that there was such a supplement by a series of dashes. Of course Newton must have given Pemberton a copy of the second edition for the printer to use in composing the new edition. It would have been in accordance with Newton's custom to have entered on the text pages themselves those corrections or alterations that could be most easily indicated in this fashion.[12]

[11] Actually, Newton had begun by writing: 'Et haec de Deo: de quo utique ex Phaenomenis disserere non', then crossed out 'non', and proceeded to 'ad philosophiam naturalem pertinet', and only then changed 'naturalem' to 'experimentalem'.

[12] For further information on the materials sent to Pemberton, see Appendix VI to the text.

CHAPTER XI

THE THIRD EDITION OF THE 'PRINCIPIA'

1. HENRY PEMBERTON AND THE THIRD EDITION

I HAVE not been able to ascertain with any precision the date when Newton either decided to undertake the preparation of the third edition or engaged Henry Pemberton to superintend such an edition. The difficulty in attempting to work out an accurate chronology stems primarily from the fact that we do not know the whereabouts of Pemberton's papers, if indeed they still survive.[1] We do have available in the Portsmouth Collection (U.L.C. MS Add. 3986) a set of twenty-three letters written by Pemberton to Newton during the preparation of E_3, together with seven sets of queries; and yet another letter of Pemberton's is to be found among the Jeffery Ekins papers in the Bodleian. Unfortunately, there are no surviving drafts or copies (at least there are none in the Portsmouth Collection) of Newton's replies to these letters, if he ever wrote any.[2] It is likely, as we shall see shortly, that Newton's responses may have taken the form of comments on the proof-sheets and so there may not have been any letters from Newton to Pemberton.

Of the surviving communications from Pemberton, only seven bear dates. The earliest of these was written on 11 February 1723/4, the latest on 9 February 1725/6. The letter of 11 February 1723/4 discusses page 63 of E_3 and is apparently the sixth in the sequence. Since the remainder of the book (some 480 pages) took twenty-four months to compose, revise, and print, or progressed at an average rate of some twenty pages per month, we may safely assume that the printer began work on the third edition of the *Principia* in November 1723, and perhaps—since the first essay at composing any book is apt to be more difficult than the later stages—in October.[3] Such a conjectural reconstruction accords with the fact that the printer, William Bowyer, received his first consignment of paper for E_3 in November 1723;[4] further supplies of paper came in as needed during the next nineteen months.

[1] On Pemberton, see Wilson (1771). The most recent study of Pemberton and the *Principia* is Cohen (1963), 'Pemberton's translation'.

[2] For a sheet in MS Add. 3965 (f. 499) directly related to Pemberton's fourth letter, see Appendix VI to the text.

[3] See Brewster (1855), vol. 2, pp. 380–1.

[4] See Davis (1951), p. 85.

Henry Pemberton, who died without issue in 1771, left his 'Books to my friend Dr James Wilson', and—but for £20 to his servant—'all the Rest and Residue of my Estate' to a 'Nephew in Law', Henry Mills, a timber merchant in Rotherhithe, on the London waterfront. James Wilson was Pemberton's biographer and literary executor. In his own will he asked 'that all the papers of my late friend Doctor Pemberton that shall be in my Custody be delivered to Mr Henry Mills of Rotherift [*sic*] who married the Doctor's niece'.[5] Three appeals have been published,[6] without result, in hopes of finding these manuscripts and letters of Pemberton's, which would cast light on Newton's ultimate revision of his *magnum opus*; even more interesting might be the manuscripts of Pemberton's English translation of the *Principia* and his commentary thereon.

Known today primarily for his association with Isaac Newton, Pemberton (*b.* 1694) was a physician who studied medicine at Leyden, after which he went to Paris 'to perfect himself in the practice of anatomy'.[7] While in Paris, he attended the auction of the library of the Abbé Gallois and purchased a 'good store' of mathematical books 'both ancient and modern'. Returning to England, he made the acquaintance of some members of Newton's circle, notably John Keill, Professor of Astronomy at Oxford, and Richard Mead, Newton's physician. Although he returned to Leyden (1719) 'to take his doctor's degree' under Hermann Boerhaave, he 'rarely' practised medicine, 'owing to the fickle state of his health'.

While still resident in Leyden, Pemberton had studied the methods and subject-matter of Newton's *Principia* and had apparently developed considerable skill as a mathematician. A paper that he wrote confuting 'Mr. Leibnitz's notion of the force of moving bodies' was shown to Newton by Dr Mead. According to Wilson, Newton 'was so well pleased with it, that, as great a man as he was, he condescended to visit the doctor at his lodgings, bringing along with him a confutation of his own, grounded on other principles'. The two papers were published together in the *Philosophical Transactions*[8] of the Royal Society in 1722:

> Hence followed a free intercourse between these two persons, whose conversation turned upon mathematical and philosophical subjects. Sir Isaac Newton was one of the modestest men in the world, so that he even solicited Dr. Mead to prevail on Dr. Pemberton to assist him in making a new edition of the *Principia*.[9]

Although Newton referred to Pemberton in the Preface as 'Vir harum rerum peritissimus', he did so only as an afterthought (see Appendix VII) and he did not have Pemberton write an introduction as he had previously allowed Roger Cotes to do in the second edition (1713).

[5] Cohen (1963), p. 339.
[6] Rigaud (1836), De Morgan (1854), Cohen (1963).
[7] This and later quotations are all taken from Wilson (1771).
[8] Pemberton (1771).
[9] Wilson (1771), p. xiii.

Pemberton published an interpretation of Newton's ideas, entitled *A view of Sir Isaac Newton's philosophy*,[10] which he first began to write in about 1721 or 1722. Wilson relates that Pemberton 'desired nothing more than to be acquainted with that great man [Newton]'; it was 'to bring this about' that 'he was composing... a treatise giving a familiar account of Sir Isaac's discoveries in philosophy'.[11] Newton died in March 1727 and Pemberton's book was published in the following year. This book presents the general principles of the Newtonian philosophy, followed by an exposition of the main parts of Newton's *Principia* and *Opticks* that established Pemberton's reputation as interpreter of Newtonian science. But Pemberton's major work was probably the preparation of the 'London Pharmacopoeia' for the Royal College of Physicians, published in 1746 under the title: '*The Dispensatory of the Royal College of Physicians*, translated into English with remarks... by Henry Pemberton'. A French version in two volumes appeared in Paris in 1761–71.

At the conclusion of the preface to his *View of Sir Isaac Newton's philosophy*, Pemberton discusses two of his own works in progress. The first is a commentary on the *Principia*. He writes:

> As many alterations were made in the late edition of his [Newton's] Principia, so there would have been more if there had been a sufficient time. But whatever of this kind may be thought wanting, I shall endeavour to supply in my comment on that book. I had reason to believe he expected such a thing from me, and I intended to have published it in his life time, after I had printed the following discourse, and a mathematical treatise Sir Isaac Newton had written a long while ago... As to my comment on the Principia, I intend there to demonstrate whatever Sir Isaac Newton has set down without express proof, and to explain all such expressions in his book, as I shall judge necessary.[12]

Pemberton concludes this account by announcing his translation of the *Principia*: 'This comment I shall forthwith put to the press, joined to an english translation of his Principia, which I have had some time by me.'

In the above-mentioned preface to the *View of... Newton's philosophy*, Pemberton says that a 'more particular account of my whole design has already been published in the new memoirs of literature for the month of March 1727'. This is a reference to the journal published in London by William and John Innys, under the title: *New memoirs of literature, containing an account of new books printed both at home and abroad, with dissertations upon several subjects, miscellaneous observations, &c.* In the issue for March 1727 (volume 5), the month in which Newton died, there is an announcement, written by Pemberton:

> LONDON
>
> We have lost the glorious Founder of the best Philosophical School, that ever appeared in the World. The Celebrated Sir Isaac Newton died on Monday the 20th of this month, about one of the clock in the morning, in the 85th year of his age...

[10] Pemberton (1728), *View*.
[11] Wilson (1771), p. xii.
[12] Pemberton (1728), *View*, pp. a2–a3.

Then Pemberton says that he doubts 'not but the following Advertisement will be very acceptable to the Readers of that great Man's Works'. This advertisment reads as follows:

ADVERTISEMENT

Speedily will be published,

SIR ISAAC NEWTON'*s Mathematical Principles of Natural Philosophy*. Translated from the last Edition, with a Comment, by H. PEMBERTON, M.D. F.R.S.

Notwithstanding the Disappointments, which have unexpectedly delayed the Account of Sir *Isaac Newton*'s Philosophical Discoveries, which I drew up for the use of such as are unacquainted with Mathematical Learning; I shall now be able in a little time to deliver that Book to the Subscribers. And I did intend, immediately after the Publication of that Treatise, to put to the Press a Translation, which I have now by me, of this great Philosopher's Mathematical Principles of Natural Philosophy, with a Comment for the Use of Mathematical Readers. And whereas it is probable, that many will now give their Explanations of this great Man's Writings; I having had a very particular opportunity of being fully informed of his real Mind from his own Mouth, do intend to proceed in my Design with all Expedition: wherein I shall present the Publick with such a Translation of Sir *Isaac Newton*'s words, as shall comprehend, in the fullest manner I am able, his true Sense. And, besides many other occasional Remarks, I shall illustrate at large the Meaning of the difficult Passages, by explanatory Notes; and shall demonstrate in form those numerous Corollaries and Scholiums, which he, for brevity, has set down without proof.

H. PEMBERTON

March 27.
1727.

Few will fail to agree that Pemberton indeed had had a unique opportunity 'of being fully informed of [Newton's] real Mind from his own Mouth'. As a translator, Pemberton would certainly have been in a position to present Newton's 'true Sense'. And how valuable indeed would be Pemberton's 'occasional Remarks', his 'explanatory Notes' on 'the Meaning of the difficult Passages', and his demonstrations of the 'numerous' corollaries and scholia which Newton 'set down without proof'. We do not know why this important translation and commentary was never published, but we may surmise that the rapid publication of Andrew Motte's translation (1729) caused Pemberton to abandon his efforts to complete his own English version and commentary.[13]

2. THE TEXT OF THE THIRD EDITION DELIVERED TO THE PRINTER

I have mentioned earlier that Newton must have given Pemberton either a corrected set of sheets of E_2 or a bound corrected copy of E_2, together with loose papers containing either new material that would not fit in the margin of the printed sheets or material that Newton had written out on such sheets and did not wish to recopy. Most likely Pemberton was given a set of unbound sheets, since it was in sheets that he gave the corrected copy to the printer; if he had been given a bound copy, he must have broken it up subsequently.

[13] See Cohen (1963).

It does not appear that Newton gave the revised text to Pemberton in instalments, as he had done while preparing the second edition (see Chapter VIII, §4). Then Cotes had had to wait on Newton's pleasure and convenience, and the printing was interrupted for lack of copy from the author. But in the case of the third edition, any interruption of the printing was due to a quite different cause. In one of his letters to Newton (U.L.C. MS Add. 3986, Letter 10) Pemberton discusses such a delay, as follows:

I just now called in at Mr Innis's[1] and was very much surprized to hear that the printer complained to you yesterday; that he had no Copy, whereas he hath had four sheets a full week. I suppose you remember that I waited on you on tuesday sevennight with the draughts of thirty figures, I had just then received from Mr Senex. The next day I spent in examining them, and corrected several faults; and on thursday I delivered them with four sheets of the Copy; and the printer has now in his hands all those cuts, except only one I have here inclosed (being the 22d) to have some directions from you concerning it. As soon as I receive this from you with the sheet belonging to it, I shall deliver these likewise to the printer, and the other sheets that come between the sheet I here speak of and the others here inclosed.

A rather long letter (Letter 14), without date, shows us another kind of problem that was apt to hold up printing. In this letter, Pemberton refers to a conversation he had had with Newton, concerning Corols. 7, 8, and 9 to Prop. XXXVI, Book II, in relation to Prop. XXXVII.[2] These propositions were part of that same Sec. VII of Book II that Newton had completely rewritten in E_2. Proposition XXXVI deals with the motion of water flowing out of a small circular hole in the bottom of a cylindrical container. As to Prop. XXXVII, Pemberton said that he had worked out two propositions of his own, which he was enclosing, and which might serve, 'if I have reasoned truly', to bring out Prop. XXXVII,

that when a cylinder moves in a fluid contained in a vessel of so wide extent as to be in effect infinite in respect of the cylinder; then the resistance of the cylinder moving in the direction of its axis will be neither greater nor less but just equal to the weight of a cylinder of the fluid, whose base shall be equal to the base of the cylinder resisted and altitude half the height whence a body must fall to acquire the velocity of the cylinder. But then you'l see that the case of the second corollary of your proposition comes out by my reasoning a little different...

Pemberton then went on to suggest that perhaps Newton might wish to resolve the question by having an experiment made. Since Pemberton also wanted to revive Newton's old argument in E_1 about the motion of water running out of a vessel, but with some revisions, a set of related experiments seemed called for.[3]

[1] The third edition was printed for William & John Innys, printers to the Royal Society.

[2] This letter begins: 'According to your order I have here made bold to set down a few thoughts upon the subject I had the honour last to discourse with you upon.

'Whereas in corol. 7, 8, 9. of prop. 36 the force of the water upon the circle PQ is set down with some latitude, it being only collected to be greater than one third and less than two thirds of the weight of the incumbent column; so that the following proposition, as you were pleased to observe, is left as much at large; I here have presumed to enclose two propositions, which I deduced from considering the method of reasoning on the running of water out of a vessel, which you gave in your first edition.'

[3] Pemberton wrote, 'If, Sir. you should chance not to discover any paralogism in my propositions, might it not be worth while to consider of making some experiments in globes so large as might make the forementioned difference manifest; provided that you also think the considerations, I am now going to propose, to be of any weight.'

Pemberton was especially sensitive on the subject of this proposition, possibly since it was so prominently displayed in the review of E_2 in the *Acta Eruditorum* (see Chapter X, §2). Hence we may understand the basis for his concern, expressed as follows:

> I have one difficulty further about this 36th proposition, that the method of exhibiting the velocity of the water gives the velocity so much the greater, as the hole bears a greater proportion to the bottom of the vessel; even so that by widening the hole, the velocity here exhibited may be increased to any magnitude whatever short of infinite.
>
> Would it be amiss to examine this proposition by some experiments, where the hole might bear a considerable proportion to the breath of the vessel? especially since several foreigners have thought fit to controvert this proposition, and are at present disputing about it; so that in all probability such experiments will be made by others.

The letter concluded with a statement that Pemberton, while waiting for Newton's reply, would 'proceed to prepare for the press some of the following sheets'. There is no evidence concerning Newton's reactions to Pemberton's proposals; Pemberton's two propositions do not appear in E_3, nor does Newton seem to have had anyone make the suggested experiments.

3. PEMBERTON AS EDITOR

Pemberton's preparation of the press copy seems to have involved various types of jobs. First, he examined carefully the mathematical and scientific (that is, astronomical and physical) content to see if he might discern any faults that should be corrected. In particular, he appears to have paid rather close attention to the supplementary material Newton had given him on separate sheets of paper. Thus, in the letter discussed above (at the end of §2) Pemberton refers to 'the paper you have given me to be added in pag. 305'. Second, Pemberton acted as a good editor should, and sought to give the *Principia* consistency of spelling. Thus, he wrote to Newton (U.L.C. MS Add. 3986, Letter 12) that in the sheet Pp, which he was sending him in proof, 'the word funependulum is spelt, as I have here written it, with an e between the n and p', but in previous sheets 'it is spelt with an i, thus funipendulum'.[1] Both, of course, are permissible. Pemberton wrote that he had 'put this mark qr' alongside the occurrences of the word in the sheet in question (pp. 294, 295), 'and if you chuse to have the word spelt everywhere, in the same manner, be pleased to correct it accordingly'. Newton evidently could not have cared less, and in E_3 we find both 'funipendulum' and 'funependulum'.[2]

As editor, Pemberton also paid close attention to style. One feature of the third

[1] In this sheet, the word occurs only on p. 294, lines 13, 15, and p. 295, line 29.

[2] Alterations in mere spelling forms, as (corpus) 'funipendulum' vs. (corpus) 'funependulum', are not included among the 'variae lectiones' composing our Apparatus Criticus. Instances of the spelling with an 'e' ('funependulorum') occur in all editions, for example, at {294.13, 15} and {295.29}, while the spelling with an 'i' ('funipendulorum') occurs in all editions in the title of Sec. X {142.13–14} and again ('funipendula') at {153.1}.

edition is the general absence of capital letters to begin words that are not proper names. In the concluding Scholium Generale, the words 'Deus' and 'Dominus' are neither capitalized nor italicized in E_3 as they were in E_2.[3] Even 'sacris literis', capitalized in E_2 in the scholium on space and time, following the Definitions, is given without any capitals in E_3 {11.16–17}. This change in style may be seen by comparing a few lines {528.3–14} as printed in E_2 and in E_3:

E_2	E_3
Nam *Deus* est vox relativa & ad servos refertur: & *Deitas* est dominatio Dei non in corpus proprium, sed in servos. *Deus summus* est Ens aeternum, infinitum, absolute perfectum; sed Ens utcunque perfectum sine dominio, non est *Dominus Deus*. Dicimus enim *Deus meus*, *Deus vester*, *Deus Israelis:* sed non dicimus *Aeternus meus*, *Aeternus vester*, *Aeternus Israelis*; non dicimus *Infinitus meus*, *Infinitus vester*, *Infinitus Israelis*; non dicimus *Perfectus meus*, *Perfectus vester*, *Perfectus Israelis*. Hae appellationes relationem non habent ad servos. Vox *Deus* passim significat *Dominum*, sed omnis Dominus non est Deus.	Nam deus est vox relativa & ad servos refertur: & deitas est dominatio dei, non in corpus proprium...sed in servos. Deus summus est ens aeternum, infinitum, absolute perfectum: sed ens utcunque perfectum sine dominio non est dominus deus. Dicimus enim deus meus, deus vester, deus *Israelis*...sed non dicimus aeternus meus, aeternus vester, aeternus *Israelis*, aeternus deorum; non dicimus infinitus meus, vel perfectus meus. Hae appellationes relationem non habent ad servos. Vox deus passim significat dominum: sed omnis dominus non est deus.

We have no documentary evidence enabling us to know for certain that these changes in capitalization and italicization were in fact due to Pemberton. They may have originated—at least in some measure—with the printer; and the same is true for changes in punctuation, abbreviation, and so on. But it is notable that Pemberton's letters to Newton are carefully punctuated; furthermore, Pemberton himself uses no underlining (corresponding to italics), and he capitalizes nouns (other than proper names) very rarely.

From the letters sent by Pemberton to Newton, we cannot be certain that he sent every sheet in proof to Newton. It is possible, of course, that the printers sent proofs directly to Newton, and that Pemberton personally sent Newton only such sheets as required his attention, either to answer a query or to approve a correction made by Pemberton. These letters, as may be seen from the entries in Appendix VI to the text,[4] refer to both proof-sheets of E_3 and pages that had actually been printed off (and of which the type had been distributed); in some cases they raise questions about text not yet delivered to the printers for composition, and hence are indicated by the page numbers of E_2 rather than of E_3.

As to Pemberton's sending Newton proof-sheets, in one letter (Appendix VI, entry no. 4A) Pemberton writes that he is enclosing 'a proof of your third sheet, which I send you before the printer has corrected it, to ask your opinion...'. In

[3] In the Apparatus Criticus such alterations (capitalization, italicization) are not included among the 'variae lectiones'.

[4] This Appendix contains extracts from Pemberton's letters, arranged in sequence according to their contents, to make precise his contribution to the third edition of the *Principia*.

other letters, Pemberton refers to 'this revise' (entry no. 4B), or even a page of a sheet, as (entry no. 16): 'Be pleased to take particular notice of the two last lines in page 63; because I have made a small alteration in them.' Again, he writes (18 February 1723/4, entry no. 21) of 'a small alteration' he has directed 'to be made in page 71'. If Newton does not approve, he is to strike out 'the correction in the margent and at the bottom of the page', and then the 'printers will leave it as it is'. Or (entry no. 23), Newton was to show his approval of a proposed correction by inserting it 'in the margent'. Pemberton both 'in the margent directed' that a certain change be made (entry no. 24) and called Newton's attention (entry no. 26) to the fact that he had made some change 'in this sheet'. Sometimes, Pemberton would suggest an alternative expression 'in the margent' of the sheet (entries nos. 49, 57); and on at least one occasion (entry no. 70) Pemberton sent Newton 'six sheets of your Book, with a few questions upon them'.

In the absence of any replies by Newton to Pemberton, we cannot tell what Newton's response to these questions may have been. Presumably, he replied to many of Pemberton's queries and suggestions on the actual margins of the proof-sheets, and these he may have returned directly to the printer. We may see from Appendix VI that many of Pemberton's suggestions were accepted by Newton, although a considerable number were not. In one case, Pemberton altered the figures 'according to the opinion I propose', not with ink 'but with black lead only, so that wherever you do not approve, they are easily restored to what they were before'.

In a letter dated 9 February 1725/6 (Letter 23), Pemberton recalled that he had 'formerly observed' to Newton that the demonstrations of Props. XXIII and XXIV, Book I, were faulty since 'the line XY will not always cut the section'; this 'occasioned your adding a short paragraph at the end of prop. 24, in pag. 82'. Pemberton now proposed 'the cancelling the leaf, and instead of the present paragraph to insert something like this which follows'.

In hac propositione et casu secundo [propositionis *del.*] superioris constructiones eaedem sunt sive recta XY trajectoriam secet sive non. Si vero non secet demonstrationes perfici possunt sumendo in tangente AI punctum tam prope puncto contactus ut duae rectae [*replacing* lineae] duci possint trajectoriae occurrentes; quarum altera in hac propositione parallela sit rectae DC, altera tangenti PL; et in casu secundo propositionis praecedentis harum rectarum altera parallela sit rectae BD, et altera rectae PC.

In this proposition and the second case of the preceding [proposition] the constructions are the same whether the right line XY shall cut the trajectory [that is, the conic] or not. If indeed it should not cut it the demonstrations can be accomplished by taking a point on the tangent AI sufficiently near to the point of contact that two right lines may be drawn meeting the trajectory; of these, in this proposition, let one be parallel to the right line DC, the other to the tangent PL; and in the second case of the preceding proposition let one of these right lines be parallel to the right line BD, and the other to the right line PC.

Pemberton assured Newton that this 'paragraph will stand nearly in the same room as the present, and will perhaps answer your purpose somewhat more fully'.

Newton, however, preferred to let matters remain as they were, with the result that the paragraph, newly written for E_3 {87.1–6}, reads as follows:[5]

In hac propositione, & casu secundo propositionis superioris constructiones eaedem sunt, sive recta XY trajectoriam secet in X & Y, sive non secet; eaeque non pendent ab hac sectione. Sed demonstratis constructionibus ubi recta illa trajectoriam secat, innotescunt constructiones, ubi non secat; iisque ultra demonstrandis brevitatis gratia non immoror.

In this proposition, and case 2. of the foregoing, the constructions are the same, whether the right line XY cuts the trajectory [*that is*, the conic] in X and Y, or not; neither do they depend upon that section. But the constructions being demonstrated where that right line does cut the trajectory, the constructions, where it does not [cut it], are also known; and therefore, for brevity's sake, I omit any farther demonstration of them.

In his letter Pemberton gave a proof that the 'paragraph here proposed is true'. Newton's refusal of Pemberton's paragraph is only in part a critical rejection of the proposed proof; for it may also be an instance of an aversion to cancels if there were no gross error.[6]

Newton's apparent unwillingness to cancel the leaf containing the foregoing paragraph {87.1–6} might have been stiffened by Pemberton's suggestion that yet another leaf be cancelled, in this case to permit the introduction of a new form of Corol. 3 to Prop. VI of Book I. This corollary, as may be seen in the Apparatus Criticus {47.28–33}, was absent in *M* and E_1, but was written into E_1i and printed in E_2. In E_2i and E_2a, Newton inserted the words 'circulum concentrice tangit, id est' {47.28–29} so that the final version in E_2i and E_2a begins as follows (I have enclosed within square brackets the words added in E_2i and E_2a to the text of E_2):

Corol. 3. Si orbis vel circulus est, vel [circulum concentrice tangit, id est *add.* E_2i E_2a] angulum contactus cum circulo quam minimum continet, eandem habens curvaturam eundemque radium curvaturae ad punctum contactus P; & si PV chorda sit...

Corol. 3. If the orbit either is a circle, or [is tangent to a circle concentrically, that is] contains a least possible angle of contact with a circle having the same curvature and the same radius of curvature at the point of contact P; & if PV be the chord...

In the last paragraph of the above-mentioned letter suggesting that the leaf containing page 87 be cancelled, Pemberton proposes that the leaf containing this Corol. 3, and two other leaves, be cancelled as well:

> In the last paragraph of page 47 there occurr these words circulum concentrice tangit, aut concentrice secat; and this double expression is here made use of, because the circle here spoke of does more frequently cut the curve than touch it. Notwithstanding this afterwards the word contactus twice occurrs without any other expression joined with it to include the more frequent case, when the circle and curve cut. This impropriety may be avoided by cancelling this leaf and writing this 3d corollary after some such manner as this...
>
> If you approve of these corrections; these two leaves together with two others which are likewise to be cancelled, will make one compleat sheet to be reprinted.

[5] In Motte's translation.

[6] He may very well have remembered with some bitterness the review of E_2 in the *Acta Eruditorum* (see Chapter IX, §4, and Chapter X, §2), in which special emphasis was given to a cancel as evidence that Newton had allowed an error to go undetected until it had been called to his attention by Nicolaus Bernoulli.

I assume that the version sent by Newton for E_3 must have read much as the one quoted above from E_2i and E_2a does; at least, this version does go on to refer twice to a single point of contact, as if the curve 'touches' the circle, without introducing the alternative expression for 'the more frequent case' when the curve cuts the circle. Pemberton suggested to Newton that he change the present indicative ('circulus est' and 'concentrice tangit, aut concentrice secat') to present subjunctive ('circulus sit' and 'concentrice tangat aut concentrice secet'), and that he drop out the words 'angulum contactus cum circulo quam minimum continet'; also that the participial phrase 'eandem habens curvaturam...ad punctum contactus P' be rewritten, somewhat more fussily, as 'eandem habeat curvaturam eundemque radium curvaturae ad punctum P cum circulo per istud punctum descripto' ('if it should have the same curvature and the same radius of curvature at the point P as a circle described by that point').

Newton did not, however, accept Pemberton's solution, but apparently made a simpler alteration, which took care of Pemberton's objections to a reference to 'contact' only. Thus, 'angulum contactus' became 'angulum contactus aut sectionis' and 'punctum contactus P' became 'punctum P', so as to produce the reading {47.28–31}:

Corol. 3. Si orbis vel circulus est, vel circulum concentrice tangit, aut concentrice secat, id est, angulum contactus aut sectionis cum circulo quam minimum continet, eandem habens curvaturam eundemque radium curvaturae ad punctum P...

Corol. 3. If the orbit either is a circle, or is tangent to or cuts a circle concentrically, that is, contains [*or* makes] with a circle the least angle of contact or section, having the same curvature and the same radius of curvature at the point P...[7]

That a cancel was made may be seen in the fact that there is a small figure 8 on the bottom of page 48 (the 8th page of signature G, and the verso of page 47 containing Corol. 3) which appears on the 8th page of no other gathering or signature in E_3. Furthermore, in almost every copy I have examined, the stub of the old leaf (G4) is visible.

Of course, we do not know whether Newton had made the change forthwith or waited till later on. In any event, his correction did not answer all of the objections implied in Pemberton's proposed revision; notably it ignored Pemberton's transformation of the phrase 'a circle having the same curvature and radius of curvature at the point P'. Pemberton dealt with this subject in another letter (Letter 7), as follows:

> In this sheet occurrs the expression concentrice tangit, which I have once already presumed to make a query upon. But by what you were pleased to answer, I am in doubt, whether I sufficiently explained my meaning. What I scrupled was this: That in the case where the circle does not fall on the same side of the curve on each side of the point in the curve, through which it is described, but does really intersect the curve, although the angle of intersection be less than what any two

[7] More freely: 'If the orbit either is a circle, or touches or cuts a circle concentrically, that is, touches or cuts the circle at the least possible angle, having the same curvature or the same radius of curvature at P...' Motte's translation is quoted above.

circles can make, when they touch each other; and though every circle either greater or less than this would actually touch the curve in this point, and intersect it in another in such manner, that in this concentrical circle these points of contact and of intersection may properly be considered as coalescing: yet when this circle crosses the curve in this point, which in all other circles is a point of contact; should not this point be considered as really converted from a point of contact into a point of intersection by the coalescence of the forementioned point of intersection with it? and if so, should not this concentrical circle receive a name strictly consistent with the present condition of this point; rather than such a denomination, that has reference to that condition, which here this point has, as it were, laid down; so that in the exactest propriety of speech it does not really [*changed from* in reality] belong to it. Methinks if Circulus concentricus, Circulus concentrice descriptus, Circulus qui concentrice describitur, or some expression of like sort, would be sufficient to represent fully, what is here intended, there would be less room for any cavil of this nature.

Newton, however, appears to have remained silent, perhaps not wishing to have to be harsh in judging Pemberton.

In his previous letter to Newton about this question Pemberton referred to two other leaves 'which are likewise to be cancelled', the four making 'one compleat sheet to be reprinted'. We have just seen that Newton agreed to reprint page 47, which is a cancel in all copies of E_3 that I have seen, and that he apparently rejected the suggestion that page 87 be cancelled. We shall see below that Pemberton was also unsuccessful in attempting to get Newton to alter Exper. 13 on page 351 and introduce a cancel for that page. In at least one copy of E_3 pages 65–6 appear as a cancelled leaf, or a replacement pasted on to the original stub. Pemberton did not discuss this leaf in his letters to Newton.

4. PEMBERTON'S CONTRIBUTIONS TO THE 'PRINCIPIA' (THE EXPERIMENTS ON DROPPING OBJECTS)

The list of Pemberton's contributions to the *Principia* leaves no doubt that he was a conscientious editor, with a keen eye for details and the niceties of Latin style. But of course he was neither imaginative enough nor a sufficiently skilled mathematician to do any editing in a real creative sense; he was certainly a very poor third to Halley and Cotes in the league of editors of Newton's *Principia.* Unlike Cotes, Pemberton appears to have consulted Newton about even the most minor alterations. Thus (U.L.C. MS Add. 3986, Letter 1) he asked permission to shift the word 'vas' in

at postquam, vi in aquam paulatim impressa, effecit vas, ut haec quoque sensibiliter revolvi incipiat

so that it would follow 'postquam' {10.16}.

In the Scholium following the Leges Motus, Newton had originally (M and E_1) written of Wren, Wallis, and Huygens as 'hujus aetatis Geometrarum facile principes' {22.15}. Reading the proofs, Pemberton recognized that it would be no longer proper in 1726, as it had been in 1687, to refer to Wren (*d.* 1723), Wallis (*d.* 1703), and Huygens (*d.* 1695) as 'easily the foremost geometers of this age', and suggested that Newton make a change. Accordingly, Newton proposed 'aetatis novissimae' (or 'of the most recent age'), but Pemberton then informed Newton that he had 'taken the liberty to change aetatis novissimae into aetatis superioris, the latter phrase being directly an expression in Cicero', and this is the way it stands in E_3 {23.32–3}.

In another instance (Letter 2), Pemberton (having first sought Newton's permission) was responsible for altering 'corpus B' to 'corpus B quiescens' on the grounds that Newton later says 'Si corpora obviam ibant'. In yet another {58.2}, Newton accepted Pemberton's correction of 'opposita' for 'conjugata', since, as Pemberton explained (Letter 4), the 'two hyperbola's belonging to the same transverse axis are never called conjugate hyperbola's by Apollonius, or by any other conic writer of authority...but are always called opposite hyperbolas, or rather opposite sections'. Newton also agreed with Pemberton that 'conisectionis' should become two distinct words, 'coni sectionis' {63.28}, and that 'Sit istud L' {63.34} should become 'Sit L coni sectionis latus rectum' so as to remove the ambiguity as to 'whether this line L be the latus rectum of the conic section, or of the other orbit' (Letters 4 and 5).

In reading over the Scholium following Prop. XL (Book II), presenting investigations of 'the resistances of fluids from experiments', Pemberton found a faulty expression in the report on the experiments made in July 1719 by J. T. Desaguliers. With respect to 'the words demittendo ab altitudine pedum 272',[1] Pemberton wrote, suppose 'some such expression as this were used, demittendo ab altiori, quam prius, templi loco, ut caderent spatium 272 pedum'. That is, instead of saying 'dropping down from a height of 272 feet', Pemberton would have Newton say something like 'dropping down from a higher place of the church than before, so as to fall through a space of 272 feet'. The problem was that in Exper. 13 Newton had said that in June 1710 two glass globes had been let fall from 'the top of St. Paul's church in London' and 'in their fall described a height of 220 London [English] feet'. But since, as Pemberton explained to Newton, 'it is said that in the preceding experiment the globes were let fall from the top of the church (a culmine) should not some expression here be made use of, that might let the reader understand, that there was in the church a place still higher, that was chose for this experiment?' How else could the globes now (Exper. 14) fall through 272 feet, 52 feet more than from 'the top'? Newton, however, did not accept Pemberton's emendation word for word, but wrote out on the reverse side of Pemberton's letter a new expression, which was printed in E_3 {353.13–15}, '[demittendo] ab altiore loco in templi ejusdem turri rotunda fornicata, nempe ab

[1] Since this whole experiment was performed and written up after E_2, and was not entered in E_2i or E_2a, we do not have Newton's original version as given to Pemberton. But the quotation in Pemberton's letter corresponds exactly with the versions of this Exper. 14 in the manuscripts in the Portsmouth Collection (U.L.C. MS Add. 3965, ff. 398, 399, 402^v, 410^v, 411, and 518), all of which contain the reading 'demittendo ab altitudine pedum 272'.

altitudine pedum [272]', that is, 'dropping down from a higher place in the round vaulted tower of the same church, namely, from a height of 272 feet'.[2]

The correction was made and new proofs were sent to Newton. But now Pemberton introduced another problem, discussed as follows in a letter addressed 'To S^r^ Is. Newton to be sent along with the proof of the sheet Zz' (Letter 16):

> Mr Blackborn the correcter of the press having observed to me that he met with some difficulty in the line against which he has put this mark * * in the first page of the sheet whose signature is Zz; I have directed the ⟨om⟩ission of the word ambientis, and a comma ⟨only⟩ to be put in its stead. I desire the ⟨fa⟩vour of you to cast your eye upon this alteration, and if you approve not of it; to adjust the sentence to your mind. The word ambientis seems to me to burden the sentence, and I think the sense of it is very fully expressed by what follows. The word fornica in the second line following I suppose should be fornicata, as Mr Blackbourn has proposed.

Of course Mr Blackbourn (or Blackborn) was right: 'fornicata' was intended.[3] As to 'ambientis' and the comma, we have four early manuscript drafts (U.L.C. MS Add. 3965, ff. 398, 399, 402^{v}, 412) in which the word 'concavae' appears without 'ambientis': Newton is referring to the experiments of Desaguliers, who constructed special falling objects to be dropped from the top of St Paul's,

formando vesicas porcorum in orbem sphaericum ope sphaerae ligneae concavae quam madefactae implere cogebantur inflando aerem.	forming hogs' bladders into spherical orbs with the help of a hollow wooden sphere, which having been moistened[4] they were forced to fill by being inflated with air.

Then, in another copy (MS Add. 3965, f. 51), the word 'ambientis' is inserted following 'concavae' and in two further copies (3965, ff. 410^{v}, 411) Newton wrote out 'concavae ambientis'. Thus he wrote that these hogs' bladders[5] were given a spherical shape by means of a 'sphaerae ligneae concavae ambientis [a surrounding

[2] Motte renders this 'let fall from the lantern on the top of the cupola of the same church; namely, from a height of 272 feet'; Newton (1729), *Principia*, vol. 2, p. 158. Motte obviously did not find this in Newton; it comes from J. T. Desaguliers's account of these experiments in the *Phil. Trans.* in 1719. The 'vaulted tower' is the 'dome' and 'St. Paul's church' the Cathedral.

[3] We do not know, of course, whether 'fornica' was a proof error, or whether it actually appeared in the text Newton sent to the printer. Newton appears to have returned the corrected proofs directly to the printer and not to Pemberton for delivery to the printer. Hence, the error must have been Newton's, not seen earlier by Pemberton.

[4] Motte translates 'madefactae' by 'being wetted well first'. Newton (1729), *Principia*, vol. 2, p. 157. In the account of these experiments published in the *Phil. Trans.* in 1719, J. T. Desaguliers described how he 'contrived a way to make dryed Hogs Bladders perfectly round, by blowing them (when moist) within a strong Spherical Box of *Lignum Vitae*, and letting them dry in the said Box before I took them out: which I did by opening the Box that screw'd in the middle, and had a hole in the Pole of one of its Hemispheres to let the Bladder pass thro', in order to tye it after blowing; and some few holes all over the Box, that in blowing no Air might be confin'd between the inside of the Box and the Bladder, so as to hinder it from taking a Spherical Figure. Besides, I took off the ends of the Ureters, the Fat and a great deal of the upper Coats of the Bladders, before I blowed them in the Box, to render them still lighter.

'The Bladders I used were some of the thinnest I cou'd find ready blown at a *Druggists*, which I moistened in water, taking care to leave none in the inside. I chose those rather than Green ones, which in drying wou'd have stuck so fast to the inside of the Box, that it wou'd scarce have been possible to have got them out without tearing.' Desaguliers (1719), pp. 1075–6.

[5] Although Desaguliers referred specifically to hogs' bladders (see preceding note) as Newton did too in E_3 {353.11}, 'vesicas porcorum', the early versions of this Exper. 14 (MS Add. 3965, ff. 398, 399, 402^{v}) describe these spheres as made of bladders of calves ('vitulorum').

hollow wooden sphere] quam madefactae implere cogebantur inflando aerem'. I do not see why Mr Blackbourn was disturbed, since 'ambientis' clearly modifies 'sphaerae', being in the genitive singular as 'ligneae' and 'concavae' are. Perhaps he was troubled by the presence of three modifiers in a row. Pemberton was logically right that 'ambientis' is unnecessary and does 'burden the sentence', since the 'hollow wooden sphere' had to be 'surrounding' the bladder. A comma before 'quam' would, as he said, clearly be helpful in showing where the modifiers end. In E_3, however, 'ambientis' is kept and the comma is placed between 'concavae' and 'ambientis,' thus separating 'ambientis' from the other two modifiers, whereas it should have been put between 'ambientis' and 'quam'.

Wholly apart from questions of sense, however, this last example is particularly notable as an instance of the true care with which some proof-readers at that time read their copy in an intelligent manner. Mr Blackbourn not only verified the accuracy with which the compositors had translated Newton's handwritten expressions into type, but even queried the sense as affected by punctuation—and he evidently read Latin. Of course, in the sentence under discussion, ordinary prose was being used to describe how bladders were soaked, put into moulds, and inflated, and then let fall.

In the letter addressed to Newton on the question of the new Exper. 14, referring to a height of 272 feet when previously the top of the same church has been said to be only 220 feet high, Pemberton took up two further problems. His first question was

in relation to the words horologium cum elatere ad singula minuta secunda quater oscillante; which rendred verbatim is a clock with a spring that oscillated four times in a second. Now since the spring of the clock is not the part which oscillates, but the pendulum suppose elatere were changed into pendulo.

Pemberton obviously missed the point. The 'elater' was not a spring driving the clock, but rather a vibrating or oscillating spring-regulator.[6] Newton surely did not need instruction as to which part of a clock oscillates. In any event, perhaps as a sop to save Pemberton's pride, he made a different change from the one suggested, and the text {353.22–3} contains 'vibrante' in place of the original 'oscillante'.

Curiously, Motte translated this sentence as follows: 'And one of those that stood upon the ground had a machine vibrating four times in one second; and another had another machine made with a pendulum vibrating four times in a second also.' Thus Motte referred to both the 'machine' and the pendulum as 'vibrating four times in one second'.[7] I leave to the reader the choice whether it is worse to conceive the whole 'machine' vibrating or merely its spring! Motte's translation ignores the fact that Newton says expressly that one of those who stood

[6] In the manuscripts (U.L.C. MS Add. 3965, ff. 398, 399^v, 402^v, 410^v, 518) we do, in fact, find: 'horologium cum elatere ad singula minuta secunda quater oscillante'.

[7] Newton (1729), *Principia* [Motte], vol. 2, p. 158.

on the ground 'habebat horologium cum elatere ad singula minuta secunda quater vibrante' while another had a 'machinam aliam affabre constructam cum pendulo etiam ad singula minuta secunda quater vibrante'. Newton thus indicates a clear distinction which he expected his reader to make between a clock regulated by a spring ('horologium cum elatere') and 'another machine skilfully made with a pendulum' ('machinam aliam affabre constructam cum pendulo'). Motte not only suppressed the distinction (and eliminated 'affabre') but, by translating 'horologium cum elatere' simply as 'machine', left it to the reader to guess that the device was a timer—and he omitted the reference to the spring.[8]

In the letter we have been discussing, Pemberton expressed his regret that he had not thought earlier of his query concerning the heights in St Paul's. For if, in Exper. 14, the height is 272 feet, and in Exper. 13 it is 220 feet, then not only did Newton have to alter the statement in Exper. 14 (as we have just seen him do); he would also have to correct the previous statement in Exper. 13, which 'cannot now otherwise be rectified than by reprinting the last leaf of the preceding sheet, and by changing the expression A culmine in the beginning of Exper. 13. into A superiori parte'. Newton evidently demurred. Page 351 is not a cancel in any copy of E_3 that I have seen. Furthermore, Newton did not even consider this change from 'a culmine' (from the top) to 'a superiori parte' (from a higher place) to be of sufficient importance to warrant inclusion in the printed errata.

5. THE COMPLETION OF THE THIRD EDITION

The foregoing account gives the reader some impression of the give and take between Pemberton and Newton during the years of preparation of the third and final edition of the *Principia*. Although Newton clearly was grateful to Pemberton for his many suggestions, and his care with the proofs, he did not go along with each and every proposed alteration. Appendix VI lists all the changes mentioned in Pemberton's letters to Newton, from which the reader may at once see the extent of Pemberton's contribution and the degree to which Newton accepted his suggestions. Generally speaking, Pemberton's role was limited to questions of style (including both consistency and accuracy of expression) and of clarity of logical reasoning and explanation; only occasionally did he take up with Newton questions of mathematics or even physics. Pemberton may therefore be compared more

[8] It does not seem the case that Newton was indulging in 'elegant variation' in not repeating 'horologium' in the second part of the sentence and using 'machinam' instead. For he goes on, in the next sentence, to refer to a 'similem machinam', and, in the sentence after that one, he states that all of these 'instrumenta' were so contrived as to be readily stopped and started. In the prior drafts (MS Add. 3965, ff. 398, 399^{v}, 402^{v}, 410^{v}, 411, 518) Newton had merely written 'machinam affabre constructam' and not 'machinam aliam affabre constructam' as in E_3 {353.23–24}. Pemberton asked Newton whether in 'the 12th line of the paper annext to page 326 [the page of E_2 where this new material is inserted], after machinam might not the word aliam be properly inserted, as I have done with black lead? because a clock is also a machine, and has been mentioned just before.' This query implies that the second 'machine' was not a clock at all. Possibly it was some sort of timer without a dial.

with Halley than with Cotes, who had assumed a more active participation, challenging Newton on the most profound plane of mathematical exactness. In a real sense, Pemberton was not the man to be such a critical editor as Cotes had been, but who else ever was? Furthermore, in 1709–13 Newton was but 66–70 years old. It is not very remarkable that he was then still capable of the mental exercise that Cotes demanded from him, but the situation was very different in the autumn of 1724. Newton was then past eighty;[1] by the time the edition was completed in the spring of 1726 he had already passed the eighty-third anniversary of his birth.

The production of the third edition was not very rapid, even by comparison with the times taken to compose and to print the earlier two editions. We have seen that the printer began work on the third edition in October or November 1723, and that in February 1723/4 Pemberton was discussing with Newton the proof-sheets that included pages 63 and 71. By May 1725 proofs had advanced to page 451. In mid-July of 1725 Newton saw the proofs of sheet Ttt (containing pages 506 and 509), and only three more remained to be done: Uuu, Xxx, and Yyy (containing the Index Rerum), plus the front-matter. Since the preface is dated January 1725/6, presumably the printing must have been completed by some time in February 1726. The whole job thus took approximately two and a quarter years from the time that composition began until the completion of the last part of the printing. Although this is somewhat less than the time taken by the second edition, we must remember that the latter had been delayed by Newton's inability to complete in short order the final revised copy for delivery to Cotes. As to the first edition, the manuscript of Book I was presented to the Royal Society on 28 April 1686[2] and Halley reported to Newton on 5 July 1687 that he had 'at length brought your Book to an end', a shade over fifteen months later.

6. ASPECTS OF THE THIRD EDITION

There are no bold and exciting innovations in E_3 in the sense that there were in E_2, in which Sec. VII of Book II was all but entirely new, as were also the concluding Scholium Generale at the end of Book III and the final five pages on comets, and in which the Hypotheses of Book III were radically altered to achieve a presentation chiefly as Regulae Philosophandi and Phaenomena. Of course the two propositions by Machin, on the motion of the nodes of the Moon's orbit, were new in the third edition, as was the final set of experiments (Exper. 14) in the scholium at the end of Sec. VII (although the latter had previously been published in the *Phil. Trans.* in 1719), and here and there Newton had introduced new and better astronomical data. But none of this was so exciting as to provoke much discussion.

[1] He had been born on Christmas Day 1642; 25 December 1642 O.S. = 4 January 1643 N.S.

[2] Halley sent Newton a proof of the first sheet on 7 June 1686; the last sheet was printed thirteen months later.

Indeed, what has most attracted attention in the third edition as such has been the elimination of the original 'Leibniz Scholium' in Book II, following Lemma II {246.11–26}, and the substitution of a wholly new one. I give the two scholia below in English translation:[1]

[E_1]

In letters which passed between me and that most skilful geometer G. G. Leibnitz ten years ago, when I signified that I had a method of determining maxima and minima, of drawing tangents to curves, and the like, which would apply equally to irrational as to rational quantities and concealed it under transposed letters which would form the following sentence—'Data aequatione quotcunque fluentes quantitates involvente, fluxiones invenire, et vice versa' that eminent man wrote back that he had fallen upon a method of the same kind, and communicated his method, which hardly differed from mine in anything except language and symbols.

The foundation of both is contained in the preceding Lemma.

[E_3]

In a letter of mine to Mr. *J. Collins*, dated *December* 10 1672 having described a method of Tangents, which I suspected to be the same with *Slusius*'s method, which at that time was not made publick; I subjoined these words: *This is one particular, or rather a corollary, of a general method, which extends itself, without any troublesome calculation, not only to the drawing of Tangents to any Curve lines, whether Geometrical or Mechanical, or any how respecting right lines or other Curves, but also to the resolving other abstruser kinds of Problems about the crookedness, areas, lengths, centres of gravity of Curves, &c. nor is it (as* Hudden's *method* de Maximis et Minimis) *limited to equations which are free from surd quantities. This method I have interwoven with that other of working in equations, by reducing them to infinite series.* So far that letter. And these last words relate to a Treatise I composed on that subject in the year 1671. The foundation of that general method is contained in the preceding Lemma.

In the second edition of the *Principia*, as may be seen in the Apparatus Criticus, Newton altered the penultimate sentence, by adding at the end '& Idea generationis quantitatum', so as to make it read 'which hardly differed from mine, except in the forms of words and symbols, and the concept of the generation of quantities'. Florian Cajori quite properly pointed out that this addition gives 'greater precision to the statement'. So we may be all the more surprised to see, by the Apparatus Criticus, that among the revisions entered by Newton in his own copies of E_2 he crossed out this addition (in E_2i, but not in E_2a). Certain loose sheets relating to this scholium and bound up in E_2i are printed in Appendix III to the text.

Among the manuscripts in the Portsmouth Collection relating to the dispute concerning the invention of the calculus, there are a number of drafts of this scholium and of alterations to it (U.L.C. MS Add. 3968 §5). From these manuscript pages it is clear that Newton at times contemplated a mere revision of the scholium, leaving it more or less as it stands in the second edition, and that at other times he thought of adding a page or more of evidence to prove his case

[1] De Morgan (1914), *Essays on Newton*, pp. 148–9, for the translation from E_1; the translation from E_3 is Motte's.

against Leibniz.[2] There is also (f. 33) a manuscript version in Newton's hand of the new scholium for E_3.[3]

I mention the existence of this manuscript version primarily because Pemberton has been accused of having been responsible for the new scholium. In one of the manuscripts (U.L.C. MS Add. 3968, f. 25), Newton wrote,

> In the second Lemma of the second Book of Principles I demonstrated the Elements of the Method of fluxions synthetically. And because Mr Leibnitz had published those Elements in another form two years before, without acknowledging the correspondence wch had been between us, eight years before, I added a Scholium not to give away that Lemma but to put Mr Leibnitz in mind of making a publick acknowledgm^t of that correspondence.

Of course he failed to get Leibniz to make any admission that the calculus was not his own independent discovery, and—whatever Newton's intention—the scholium was (as Newton wrote) interpreted by the Leibnizians as if 'in my book of *Principles*, pp. 253, 254, I allowed him the invention of the *Calculus Differentialis* independently of my own; and that to attribute this invention to myself, is contrary to my knowledge'.[4] In that paragraph, Newton then said, 'I do not find one word to this purpose'. The reader must judge for himself to what degree this final statement accords with the text of the scholium in the first and second editions.

7. NEWTON AND PEMBERTON: THE REFERENCE (IN BOOK III) TO MACHIN

Pemberton has described Newton's attitude toward suggested corrections:

> Neither his extreme great age, nor his universal reputation had rendred him stiff in opinion, or in any degree elated. Of this I had occasion to have almost daily experience. The Remarks I continually sent him by letters on his Principia were received with the utmost goodness. These were so far from being any ways displeasing to him, that on the contrary it occasioned him to speak many kind things of me to my friends, and to honour me with a publick testimony of his good opinion...As many alterations were made in the late edition of his Principia, so there would have been many more if there had been a sufficient time.[1]

[2] For instance, f. 30 is headed in Newton's hand: 'In the end of the Scholium in Princip. Philos. pag. 227 after the words Utriusque fundamentum continetur in hoc Lemmate, add...', following which there is a lengthy addition in Latin. The page number (227) makes this an unambiguous reference to E_2 rather than to E_1; the quotation 'Utriusque...Lemmate' corresponds to the final sentence of the Lemma in E_2 (and in E_1).

[3] It contains, among other worked-over parts, the following concluding matter which was then crossed out: 'per quod utique resolvitur pars prior hujus problematis; *Data aequatione fluentes quotcunque quantitates involvente, invenire fluxiones*; *& vice versa*.' The final sentence, 'Given any equation involving fluent quantities, to find the fluxions; and vice versa', was stated by Newton as an anagram:

6a 2c d ae 13e 2f 7i 3l 9n 4o 4q 2r 4s t 12v x.

Here, 'the information given' is that 'whoever can form a certain sentence properly out of six *a*'s, two *c*'s, a *d*, and so on, will see as much as one sentence can show about Newton's mode of p^roceeding'; De Morgan (1914), *Essays on Newton*, p. 93. Surely no one can believe that Leibniz, or anyone else, could have derived any real information concerning Newton's discovery from such an anagram, and we may well agree with De Morgan that Joseph Raphson was certainly 'unscrupulous' when he 'declared that Leibniz had first deciphered the anagram, and then detected the meaning of the word fluxion, after which he forged a resemblance'.

[4] See Chapter X, §2, for a statement of this kind in the review of E_2 in the *Acta Eruditorum*.

[1] Pemberton (1728), *View*, p. [a2^v].

Pemberton himself is mentioned by name in the text of the *Principia* in Book III, in the beginning of a scholium following Prop. XXXIII, on the motion of the nodes of the moon's orbit, just before Newton introduces the pair of propositions and the scholium thereto, 'De motu nodorum lunae' by John Machin.[2] Originally, to judge by Newton's manuscript versions (U.L.C. MS Add. 3965, ff. 507, 509, 521^v), these were to have been introduced simply by this single short sentence:

Hanc Propositionem [i.e., Prop. XXXIII. Invenire motum verum lunae.] D. Machin aliter investigavit ut sequitur.	This Proposition Mr Machin investigated in another fashion as follows.

We have seen above that what we know in E_3 as Machin's Prop. I (followed by a Prop. II and a scholium) was at first denoted by 'Lemma' (MS Add. 3965, ff. 507, 509); only in a later draft (MS Add. 3965, f. 521^v) did it become 'Prop. I'. Newton must have given Pemberton yet another version of the above-mentioned introduction, because in discussing it with Newton Pemberton wrote (MS Add. 3986, Letter 19):

In the second line, after the word invenerunt, I have added a short sentence alluding in the slightest manner I could to my having published something of this method in the epistle upon Mr Cotes's book: for as it does not seem to me improper to hint at that particular; so to the best of my remembrance you took some Notice of it in the Scholium, which you once drew up; and which, as far as I can recollect it, I have made the pattern of this here written. But before you return the sheet, you'l please to adjust the whole of this Scholium as shall be most agreeable to your mind.

Presumably, since Pemberton's addition was a short sentence following 'invenerunt', it must have been in substance at least the second sentence of the scholium as printed, although perhaps it referred more exactly to Pemberton's publication.[3] As printed {451.13–17}, the first sentence, with a further slight alteration introduced by Pemberton, reads:

Alia ratione motum nodorum *J. Machin Astron. Prof. Gresham.* & *Hen. Pemberton M.D.* seorsum invenerunt.	*J. Machin*, Prof. of Astron. at Gresham, and *Hen. Pemberton* M.D. separately found the motion of the nodes by a different method.[4]

[2] See Chapter XI, §5.

[3] This publication is entitled: *Epistola ad amicum J. W.* [Jacobum (=James) Wilson] *de Rogeri Cotesii inventis, curvarum ratione, quae cum circulo et hyperbola compara ionem admittunt, cum appendice* (London: 1722).

[4] Pemberton (U.L.C. MS Add. 3986, Letter 19) wrote to Newton: 'After Mr Machin's name I have put his title of Professor of Astronomy at Gresham College; and have put out the letter D before his name, judging it useless.' The subsequent history of this sentence is like a great joke. In the first edition of Motte's translation (1729), we find: 'Mr. *Machin* Astron. Prof. Gresh. and Dr. *Henry Pemberton* separately found out the motion of the nodes by a different method.' The form was altered slightly in the edition of 1803 by the introduction of some commas, to become: 'Mr. *Machin*, Astron., Prof. Gresh., and Dr. *Henry Pemberton*, separately...'. Whoever made this change did not realize that Machin was not 'Astronomer' and 'Professor at Gresham', but 'Astronomical Professor at Gresham' or 'Professor of Astronomy at Gresham'. Motte's title-page (volume 1) thus referred to him correctly as 'JOHN MACHIN *Astron. Prof. Gresh.* and *Secr. R. Soc.*', but the title-page (volumes 1 and 2) in 1803 followed the new punctuation of the text: 'JOHN MACHIN *Astron.*, Prof. at Gresh., and Sec. to the Roy. Soc.'. In the edition of 1819 the text sentence in question was printed exactly as it had been in 1803, but the title-page now referred correctly to Machin as 'Astronomical Professor at Gresham College, and Secretary to the Royal Society'. The Cajori text of 1934 continued the earlier practice of italicizing all proper names. But the word 'Astron.' was dropped altogether, carelessly, and the abbreviations 'Prof. Gresh.' spelled out as 'Professor Gresham'. Then, evidently to make this new character's name conform to

The remainder reads, in translation:

> Mention has been made of this method in another place. Their several papers, both of which I have seen, contained two propositions, and exactly agreed with each other in both of them. Mr. *Machin*'s paper coming first to my hands, I shall here insert it.

There is one other reference to Pemberton in E_3, in Newton's Preface to the new edition, beginning:

In editione hacce tertia, quam Henricus Pemberton M.D. *vir harum rerum peritissimus curavit...*	In this third edition, the care of which was undertaken by *Henry Pemberton M.D.*, a man of the greatest learning in these subjects...[5]

Pemberton's friend James Wilson has recorded that Pemberton valued this compliment given him by Newton in the preface far more than the liberal present Newton gave him of 200 guineas.[6] As may be seen in Appendix VII, which consists of the successive versions of Newton's preface to E_3, this compliment is not to be found in the early versions of the preface. Even in the final version, it was added only as an afterthought.

8. THE THIRD EDITION

The third and final edition of the *Principia* was printed in quarto in Caslon English in an edition of 1,250 copies: 50 on superfine Royal, 200 on General Royal, and 1,000 on Demy.[1] Following the half-title, there is a statement that the edition was licensed and granted the royal privilege 'for the Term of fourteen Years' on 25 March 1726. The frontispiece is a three-quarter-length portrait of 'Isaacus Newton Eq. Aur. Aet. 83.' painted in 1725 by John Vanderbank and engraved in 1726 by George Vertue.

Newton sent six copies to the Académie Royale des Sciences in Paris, with a covering letter to the Secrétaire Perpétuel,[2] Fontenelle (who was soon to produce the first published biography of Newton[3]), and he sent off various presentation copies. Within a year of the publication of the ultimate approved edition of his great masterpiece, Newton died (20 March 1726/7); he was buried in Westminster Abbey.

the usage adopted in the book, Gresham's 'name' was italicized, so that the sentence became: 'Mr. *Machin*, Professor *Gresham*, and Dr. *Henry Pemberton*, separately...'. One wonders whether Thomas Gresham would have considered himself demoted or promoted, thus to become a mere professor at his college rather than to have remained the founder! See Cohen (1963), 'Pemberton's translation'.

[5] This is taken from Robert Thorp's translation; [Newton] (1777), *Principia* [Thorp], p. [d 3^r]; it is omitted from Motte's. This omission is striking, and may have arisen from Motte's annoyance at Pemberton's announcement of a rival translation of the *Principia* to his own. See Cohen (1963), 'Pemberton's translation', and the introduction to the reprint (1968) of Motte's original English version of 1729. This reprint, and Thorp's translation, may be found listed in Appendix VIII (part 2) to the text.

[6] Wilson (1771).

[1] Macomber (1953), 'Census', p. 25; Davis (1951), p. 86.

[2] See [Sotheby and Co.] (1936), *Newton papers*, numbers 135, 136.

[3] This was the official *éloge* prepared for the Académie Royale des Sciences, based to some extent on biographical information sent to Fontenelle by Conduitt. See Gillispie (1958), in [Newton] (1958), *Papers and letters*, pp. 427–43.

In 1729, Andrew Motte's English translation appeared,[4] reprinted in 1803 and again in 1819. A volume of annotated selections in Latin was printed in England in 1765, but after Newton's death the Latin text was printed only once in Newton's own land during the eighteenth century: in Samuel Horsley's edition of Newton's *Opera* (vols. 2 and 3, 1779 and 1782). In 1822, 1833, and 1871 there were Latin reprints in Glasgow, and in 1954 a facsimile of the first edition was published in London. But the chief printings in the Age of Newton, or in the Age of the Enlightenment, were not in Britain but on the Continent. These include the Geneva editions of 1739–42 and of 1760, with the commentary of Fathers LeSeur and Jacquier, reprinted again in Prague (1780–5), and the French translation made by the Marquise du Chastellet (Paris, 1756–9).[5]

Faithful to its tradition, the *Acta Eruditorum* published a review of E_3, just as it had done of E_1 and E_2.[6] Most of the review consists of a description or summary of the new matter in E_3, according to the stated philosophy that, 'Since every single thing that comes from Newton deserves the attention of learned men of the first order, we thought it our duty to preserve in these *Acta* the record of the third edition also, although enough has been said about the first in the *Acta* of 1688, pages 304 ff., and about the second in the *Acta* of 1714, pages 131 ff.'[7] Much space is devoted by the reviewer to the new material on the comets of 1680 and of 1723, and to Desaguliers's experiments of 1719 on the fall of spherical hog bladders. Of course the reviewers noted the alteration of the 'Leibniz Scholium', and discoursed on it as follows:

> That Scholium which concerns Leibniz and which is mentioned in the *Acta* of 1714, page 133, is now omitted and in its place another appears, in which the author exhibits an excerpt from a letter written to Collins on 10 December 1672, in which he pointed out to Collins that he had a general method which extends itself without any bothersome calculation not only to drawing tangents to any curves whether geometrical or mechanical or in any way respecting right lines or other curves but also to resolving other more abstruse problems about curvatures, areas, lengths, centres of gravity of curves, etc., and is not restricted to those equations only which are free from surd quantities, and that he had interwoven this method with that other which sets out the solution of equations by reducing them to infinite series.

Having recorded the fact itself, the reviewer made no comment.

Then Newton's new Regula IV is presented, followed by an account of the new data on the sizes of the planets and the elongations of their satellites (chiefly based on the micrometric studies of Pound, mentioned in the preface to E_3). Reference is made to Machin's propositions, and the new numerical data for the theory of the Moon's motion. Halley's corrections of Flamsteed's data for the comet of 1680 and 1681 lead to the remark, 'We pass over what we have already mentioned above in Newton's own words.'

[4] For the abortive attempts of Pemberton to complete and publish an independent translation of his own, see §1.

[5] For an account of all these editions see Gray (1907), Babson (1950), and Appendix VIII to the text.

[6] *Acta Eruditorum*, February 1726, pp. 73–6.

[7] This extract and those that follow have been translated from the Latin printed version.

We may fittingly terminate this Introduction to a new edition of Newton's *Principia* with a selection from the *Acta*. It is the conclusion of the review of the final edition, produced at the end of Newton's life, the main text of the present edition. According to the reviewer:

This edition carries off the prize from the former ones, not only because of the splendour of the type and of the figures, but also because, besides the new additions, both the text and the figures have undergone revision throughout. Furthermore, there is prefixed to the edition an elegant portrait of the grand old man himself, venerable and illustrious for his outstanding merits in the republic of letters.

Finis coronat opus!

SUPPLEMENTS

SUPPLEMENT I

SOME AUTOBIOGRAPHICAL STATEMENTS BY NEWTON ABOUT THE COMPOSITION OF THE 'PRINCIPIA'

AMONG Newton's manuscripts there are many autobiographical statements, almost all of which seem to have been stimulated by one or another aspect of the controversy with Leibniz concerning priority and independence in the matter of inventing or discovering the method of fluxions. Some of these are drafts either for the *Commercium epistolicum*, or for the *Recensio libri* which Newton wrote about the *Commercium* for the *Philosophical Transactions*, both of which appeared without any indication of his authorship. Others are drafts of letters, as to Des Maizeaux, who was printing a collection of Newton–Leibniz letters, and other documents. Yet others are drafts or parts of drafts of proposed prefaces to either the *Principia* or *De quadratura*, explaining how Newton had used *De quadratura* in discovering some major propositions of the *Principia*. These documents are often mutually contradictory on matters of detail, yet they convey a firm impression of Newton's preferred version of his course of discovery. They deal with Hooke and Halley, express Newton's dependence on Kepler and Galileo, assert his admiration for Huygens, and attempt to explain the delay in publishing the method of fluxions. In presenting these extracts, I have pared them down to the bone, for the most part omitting any statements that do not relate directly either to the *Principia* or to dynamics, such as dates of discovery of the direct and inverse methods of fluxions. I have generally put aside the direct references to *De quadratura*, which are included in Supplement VIII; but some of the following fragments were doubtless intended to introduce one or another version of that tract.

The extracts that are given below in English only were written in English (with the exception of one paragraph quoted from the published translation of the *Epistola posterior*); those that are given in both Latin and English were written in Latin and have been translated for this edition. It should be kept in mind that I have made a selection from the many such autobiographical documents in the Portsmouth Collection (almost all in U.L.C. MS Add. 3968). As elsewhere in the present work, the symbols ⌞ ⌟ enclose words or passages that are later insertions.

1

MS Add. 3968, §41, f. 85

The following statement has often been printed, first in the *Catalogue of the Portsmouth Collection* (p. xviii), again by Rouse Ball (p. 7), and elsewhere. To the best of my knowledge, however, the true character of this document has never been fully shown by printing the text *in extenso*. It should be pointed out that both paragraphs have been cancelled. Above all it must be kept in mind that both sides of the page on which this statement occurs are covered with tentative versions that were never approved by Newton for publication.

In particular, the two paragraphs printed below (from f. 85^r) are not consecutive. To give them their proper value, they must be seen as part of the matrix of manuscripts in which they are embedded. They are preceded by a paragraph beginning in the middle of a sentence, 'was with difficulty that I...'. This paragraph concludes: 'I do not think [it *del.*] ⌊the business⌋ of such consequence that I should [meddle with this controversy any further *del.*] part with a quiet life for the sake of it, & therefore I intend to meddle with this controversy no further./ But whatever is done, I do not think this business of such consequence that I should meddle wth it any further.' Following the two paragraphs printed below, there are two others, each ending in the middle of a sentence, of which the first begins: 'And Mr Collins in a Letter to Mr Bertet of Feb 21 $16\frac{70}{71}$ said...' and ends abruptly: 'And in his Letter to Mr Strode'. The second reads: 'And this is demonstrated by my having in those days series for squaring of figures wch in some cases breake off & become finite. And in a letter to Mr David Gregory the brother of Mr James Gregory he said'. This David was the father of the David Gregory who was associated with Newton and wrote the 'Notae'.

Finally, to see the character of this page as a repository of notes and tentative versions, attention may be called to the fact that, either before or after writing the foregoing paragraphs, Newton wrote some comments at what is now the bottom of the page. If we rotate the page through 180° we can read these proposed revisions, both of which refer to the second edition of the *Principia*: 'Pag. 219. lin. 22. dele [Obtinet haec ratio quamproxime ubi corpora in Mediis rigore aliquo praeditis tardissime moventur.] Et scribe [In Mediis quae rigore omni vacant, & quorum vis resistendi oritur a sola vi inertiae partium, resistentiae corporum sunt in duplicata ratione velocitatum &c.]' and 'Pag. 252. lin. 14. Adde Scholium sequens.' (Newton often used square brackets to indicate quoted material.) The first of these refers to the second sentence of the scholium following Prop. IV, Book II, in E_2. In E_3 (see text {239.3–5}) the sentence was in fact deleted, but the replacement differs from the one quoted above in that it lacks the clause '& quorum vis resistendi oritur a sola vi inertiae partium'. The second reference is also to Book II, the end of Sec. III, following Prop. XIV; the scholium, present in E_3 {274.2 ff.}, does not appear in the earlier editions.

Everything on the page written in English (that is, all except the five lines in Latin referring to revisions of E_2) has been completely cancelled, apparently at successive times. Possibly the two paragraphs printed below were cancelled when Newton made a summary statement to the same effect in a letter to Des Maizeaux (written between 1718 and 1720), a relevant part of which is printed below as document 5. It is fairly certain from the cancellations that the present document contains versions that those in document 5 were intended to replace.

The editors of the Portsmouth *Catalogue* have pointed out that in the sentence beginning 'At length in the winter between the years 1676 & 1677 ..' the dates should probably have been 1679 and 1680, and that in the sentence 'And in the winter between the years 1683 & 1684 this Proposition...' the years should probably have been 1684 and 1685. In the remaining selections I have not indicated what errors Newton may have made in assigning dates, nor the contradictions between one document and another, my intention being less to give the reader a Newtonian chronology than a Newtonian view of the process and stages of composition of the *Principia*.

If it be asked why I did not publish this book [De quadratura] sooner, it was for ye same reason that I did not publish the theory of colours sooner, & I gave the reason in my Letter of 24 Octob. 1676.[1]

And the testimony of these two ancient, knowing & credible witnesses[2] may suffice to excuse me for saying in the Introduction to the book of Principles [*sic*: *read* Quadratures] that I found the Method by degrees in the years 1665 & 1666. In the beginning of the year 1665 I found the Method of approximating series & the Rule for reducing any dignity of any Binomial into such a series. The same year in May I found the method of Tangents of ⌞Gregory &⌟ Slusius, & in November had the direct method of fluxions & the next year in January had the Theory of colours & in May following I had entrance into ye inverse method of fluxions. And the same year I began to think of gravity extending to ye orb of the Moon, & [having deduced *del.*] ⌞having found out how to estimate the force with w^ch^ ⟨a⟩ globe revolving within a sphere presses the surface of the sphere: from Keplers Rule⌟ from Keplers rule [*should have been del.*] of the periodical times of the Planets being in a sesquialterate proportion of their distances from the centers of their Orbs, [I having *del.*] ⌞I⌟ deduced that the [⌞centripeta⌟ *del.*] forces wch keep the Planets in their Orbs [tend to *del.* are as *del.* about the centers of *del.* are *del.*] ⌞must⌟ ⟨be⟩ reciprocally as the squares of their distances from the centers about wch they revolve: [I also *del.*] ⌞& thereby⌟ compared the force requisite to keep the Moon in her Orb with the force of gravity at the surface of the earth, and found them answer pretty nearly. ⌞All this was in the two plague years of 1665 & 1666.⌟ ffor in those days I was in the prime of my age for invention & minded [Philoş *del.*] ⌞Mathematicks & Philosophy⌟ more then at any time since. ⌞What Mr Hugens has published since about centrifugal [*changed from* centripetal] forces I suppose he had before me.⌟ At length [*changed from* And some year⟨s later⟩] in the winter between the years 1676 [*changed from* 1666] & 1677 I found the [*changed from* out] [Keplers *del.*] Proposition that by a centrifugal force [the *del.*] reciprocally as the square of the distance a Planet

[1] This paragraph may be compared with the following statement in a draft of a letter to Des Maizeaux (MS Add. 3968, ff. 389^v^, 390^v^): 'And as the notation used in this Book [of Quadratures] is the oldest, so it is the [best *del.*] shortest ⌞& most expedite,⌟ but was not known to the Marquess de l'Hospital when he recommended the differential Notation. If it be asked why I did not publish this book sooner, it was for the same reason that I did not publish the Theory of colours sooner, & I gave the reason in my Letter of 24 Octob. 1676. The first Proposition of the Book of Quadratures is certainly the foundation of the method of fluxions.'

[2] The two 'witnesses' are Barrow and Collins; see Whiteside (1966), p. 38, n. 3.

must revolve in an Ellipsis [& so *del.*] about the center of the force placed in the lower umbilicus of the Ellipsis & with a radius drawn to that center describe [equal *del.*] areas proportional to the times. And in the winter between the years 1683 & 1684 this Proposition wth the Demonstration was entered in the Register book of the R. Society. And this is the first instance upon record of any Proposition in the higher Geometry [solved *del.*] ⌊found out⌋ by the Method in dispute. In the year 1689 M[r] Leibnitz ⌊endeavouring to rival me⌋ published a Demonstration of the same Proposition upon another supposition but his Demonstration [was *del.*] ⌊proved⌋ erroneous, [& thereby it was discovered that he was *del.*] for want of skill in the Method.[3]

The letter of 24 October 1676, the 'Epistola posterior', is an account of Newton's mathematical work, written to Oldenburg but in fact a reply to 'M[r] Leibnitz's ingenious letter'. Since the Latin text is readily available, in this case—as an exception to the general rule stated earlier—I publish the relevant passage in translation only, from [Newton] (1959–), *Correspondence*, vol. 2, p. 133.

And five years ago when, urged by my friends, I had planned to publish a treatise on the refraction of light and on colours, which I then had in readiness, I began again to think about these series and I compiled a treatise on them too, with a view to publishing both at the same time. But on the occasion of the Reflecting Telescope, when I had sent you a letter in which I briefly explained my ideas of the nature of light, something unexpected caused me to feel that it was my business to write to you in haste about the printing of that letter. Then frequent interruptions that immediately arose out of the letters of various persons (full of objections and of other matters) quite deterred me from the design and caused me to accuse myself of imprudence, because, in hunting for a shadow hitherto, I had sacrificed my peace, a matter of real substance.

In another passage, MS Add. 3968, f. 122[v], Newton is more explicit about why he did not publish his *De quadratura* and other writings.

Anno 1666 incidi in Theoriam colorum, et anno 1671 parabam Tractatum de hac re, aliumque de methodo serierum & fluxionum ut in lucem ederentur. Sed subortae mox disputationes aliquae me a consilio deterruerunt usque ad annum 1704.	In 1666 I came upon [*or* chanced upon] the theory of colours, and in 1671 I was preparing a tract concerning this subject, and a second [tract] about the method of series & of fluxions intending to publish them. But certain disputes which arose soon after discouraged me from this design till the year 1704.

2

MS Add. 3968, f. 101

The following memorandum has two parts, of which the first is wholly cancelled. In the manuscript, Newton wrote down in a single column the sequence from 1 to 11, evidently intending at some later time to identify the actual propositions in question; such a list is given in document 4 below. At a later date, he did not so identify these propositions, but instead he added a further paragraph in the blank

[3] This paragraph may be compared with the versions in the drafts of the letter to Des Maizeaux, for example (f. 390[v]): 'And the testimony of these three ancient knowing & credible witnesses may suffice to excuse me for saying in the Introduction to the Book of [Principles *del.*] ⌊Quadratures⌋ that I found the method by degrees in the years 1665 & 1666...I was then in the prime of my age for invention & most intent upon mathematicks & philosophy & found out in those two years the methods of series & fluxions.' The three witnesses (f. 390[r]) are Barrow, Collins, and Wallis.

space to the right of these numbers: 'In the end of the year 1679...to the east.' In this memorandum Newton says that he 'set upon writing the Book of Principles' following the request of the Royal Society that his earlier communication (*De Motu*) 'might be printed'. Note that Halley's visit is here dated 'Spring 1684'. It is evident from the text that this document, printed here in its entirety, was intended as a preface to an edition or printing of the *Principia*, which was to contain *De quadratura*. See Supplement VIII.

At the Request of Dr Halley I sent to him in October or November 1684 the following Propositions demonstrated, & soon after gave him leave to to [*sic*] communicate them to the R. Society

1
2 In the end of the year 1679 I communicated to Dr Hook then Secretary of the R.
3 Society, that whereas it had been objected ⌞represented⌟ that if the Earth had a diurnal
4 motion from west to east, it would leave falling bodies behind & cause them to fall to the
5 west; the contrary was rather true. ⌞In falling⌟ They would ⌞keep⌟ the motion wch they
6 had from west to east before they began to fall, & this motion compounded with the
7 motion of descent arising from gravity would carry them to the east.
8
9
10
11

[And by consulting the Minute Books ⌞Journal Books⌟ of the R. Society I find that they were communicated then to the R. Society ⌞read before the R. Society⌟ *del.*] ⌞And⌟ find that they were ⌞entred in the Register Book of the Society Decem 10 1684⌟ Decem. 10 following. [And thereupon at *changed to* At *del.*] ⌞And at⌟ the request of the R. Society that they might be printed I soon after set upon writing the Book of Principles & sent it to the Society ⌞Apr ⌟ & they ⌞ordered it to be printed May 19⌟ 1686 as I find by [another *del.*] a minute in their Jornal book. And their President [ordered it to be printed *del.*] ⌞wrote an Imprimatur⌟ July 5^{t} following. And the book came abroad the next year in as I find by the Account given of it in the Philosophical Transactions

In the end of the year 1679 [I com *del.*] in answer to a Letter from Dr Hook [I wrote that *del.*] then Secretary of the R. S. I wrote that whereas it had been objected against the diurnal motion of the earth that it would cause bodies to fall to the west, the contrary was true. ffor bodies in falling would keep the [old *del.*] motion [from west to east *del.*] which they had from west to east before they began to fall, & this motion being added to their motion of falling would carry them to the east. Dr Hook replied [that *del.*] soon after that they would do so under the [line & that *del.*] ⌞Equator⌟ but in our Latitude they would fall not exactly to the east but decline from the east a little to the south. And that he had made some experiments thereof & found that they did fall in that manner. And he added that [they *del.*] they would not fall down to the center of the earth but rise up again & describe an Oval as the Planets do in their orbs. Whereupon I computed what would be the Orb described by the Planets. ffor I had found before [that *del.* the *del.*] by the sesquialterate proportion of the tempora periodica of the Planets ⌞with respect⌟ to their distances from the Sun, that the forces wch kept them in their Orbs [were *del.*] about the Sun were as the squares of their mean distances from the Sun reciprocally: & I found now that whatsoever was the law of the forces wch kept the Planets in their Orbs, the areas described by a Radius drawn from them to the Sun would be proportional to the times in wch they were described. And [upon *del.*] ⌞by the help of⌟ these two Propositions I found that their Orbs would be such Ellipses as Kepler had described.

In Spring 1684 Dr Halley coming to Cambridge & asking me if I knew what figure the Planets described in their Orbs about the Sun was very desirous to have my Demonstration & in autumn following I sent to him the following Propositions demonstrated & soon after gave him leave to communicate them to ye R. S. And they were entred in the Register Book of the R. Society Decem 10. 1684. Dr Hook complained that I had the hint from him, but he producing no demonstration the R. Society desired that the Paper containing thos Propositions might be printed. And thereupon I soon after set upon writing the Book of Principles, & sent it to the R. Society & they made an order May 19 1686 that it should be printed, as I find by a minute in their journal book; & their President wrote an Imprimatur July 5^t following.

By reason of the short time in wch I wrote it, & of its being copied by an Emanuensis who understood not Mathematicks there were some faults besides those of the Press. By measuring the quãtity of water wch ran out of a vessel through a round hole in the bottom of a given magnitude I found that the velocity of the water in the hole was that wch a body would acquire in falling half the height of the [vessel *del.*] water in the vessel. And by other Experiments I found afterwards that the water accelerated after it was out of the vessel untill it arrived at a distance from the vessel equal to the diameter of the hole, & by accelerating acquired a velocity [equal to the *del.*] equal to that wch a body would acquire in falling [⌊almost⌋ through *del.*] the whole height of the water in the vessel. or thereabouts. [In t *del.*] The Demonstration of the first Corollary of the 11th 12th & 13th Propositions [I ⟨??⟩ *del.*] being very obvious, I omitted it in the first edition & [instead thereof inserted *del.*] ⌊contented my self with⌋ adding [*originally* added] the 17th Proposition whereby it is proved that a body [in all cases *del.*] ⌊in⌋ going from any place with any velocity will in all cases describe a conic Section: wch is that ⌊very⌋ Corollary. [But at the desire of Mr Cotes I *del.*] In the 10th Proposition of the second Book the tangent of the Arch GH was drawn from the wrong end of the arch, [but is now put right *del.*] wch made some think that there was a error in second [differences *del.*] fluxions.

The following is not cancelled.

The Ancients had two Methods in Mathematics wch they called Synthesis & Analysis, or Composition & Resolution. By the method of Analysis they found their inventions & by the method of Synthesis they [published them *del.*] composed them for the publick. The Mathematicians of the last age have very much improved [Analysis & *del.*] Analysis [& laid aside the Method of synthesis *del.*] but stop there [in so much as *del.*] & think they ⌊have⌋ solved a Problem when they have only resolved it, & by this means the method of Synthesis is almost laid aside. The Propositions in the following book were invented by Analysis. But considering that [they were *del.*] the Ancients (so far as I can find) admitted nothing into Geometry [but wha *del.*] before it was demonstrated by Composition I composed what I invented by Analysis to make it [more *del.*] ⌊Geometrically authentic &⌋ fit for the publick. And this is the reason why this Book was written ⌊in words at length⌋ after the manner of the Ancients without Analytical calculations. But if any man who undestands [*sic*] Analysis will [resolve de *del.*] reduce the Demonstrations of the Propositions [back *del.*] from their composition back into Analysis (wch is ⌊very⌋ easy to be done,) he will [esily *del.*] see by ⌊what⌋ method of Analysis they were invented. ⌊And ⌋By this means the Marquess de l'Hospital was able to affirm that this this [*sic*] Book was [presque tout de ce Calcul.] almost wholy of the infinitesimal [calculus *del.*] Analysis.

The expression 'presque tout de ce Calcul.' was enclosed within square brackets (indicating a quotation) by Newton himself. The manuscript goes on to discuss aspects of the controversy with Leibniz. The reference is to the preface to L'Hospital's *Analyse des infiniment petits*... (Paris, 1696).

3

Newton published a somewhat similar statement, anonymously, in the *Recensio libri* or: 'An Account of the Book entituled *Commercium Epistolicum Collinii & aliorum, De Analysi promota*, published by order of the Royal-Society, in relation to the Dispute between Mr. *Leibnits* and Dr. *Keill*, about the Right of Invention of the new Geometry of Fluxions, otherwise call'd *the Differential Method*'. The following paragraph comes from *Phil. Trans.* vol. 29 (1714, 1715, 1716), p. 206.

By the help of the new *Analysis* Mr. *Newton* found out most of the Propositions in his *Principia Philosophiae*: but because the Ancients for making things certain admitted nothing into Geometry before it was demonstrated synthetically, he demonstrated the Propositions synthetically, that the Systeme of the Heavens might be founded upon good Geometry. And this makes it now difficult for unskilful Men to see the Analysis by which those Propositions were found out.

4
MS Add. 3968, f. 106

The following note has been printed in Chapter III, §6.[4]

In the tenth Proposition of the second Book there was a mistake in the first edition by drawing the Tangent of the Arch GH from the wrong end of the Arch ⌞which caused an error in the conclusion⌟; but in the second Edition I rectified the mistake. And there may have been some other mistakes occasioned by the shortness of the time in which the book was written & by its being copied by an Emanuensis who understood not what he copied; besides the press-faults. ffor I wrote it in 17 or 18 months, beginning in the end of December 1684 & sending it to ye R. Society in May 1686: excepting that about ten or twelve of the Propositions were composed before, vizt the 1st & 11th in December 1679, the 6th 7th 8th 9th 10th 12th, 13th 17th Lib. I & [Prop. *del.*] ⌞the⌟ 1, 2, 3 & 4 Lib. II, ⌞in [1684 *del.*] June & July 1684⌟ [*originally* composed in the summer time of the year 1684].

5
MS Add. 3968, f. 402

The following extract is taken from a draft of a letter to Des Maizeaux, written in 1720 (or a little earlier). This part of the letter was copied by Newton almost verbatim from a set of numbered comments on a letter of Leibniz's. In the draft of the letter these paragraphs have been cancelled, but the comments are not similarly cancelled. It should be observed that this text is a revision of document 1, printed above.

By the inverse Method of fluxions I found in the year 1677 the demonstration of Keplers Astronomical Proposition, vizt that the Planets move in Ellipses, wch is the eleventh Proposition of the first book of Principles; & in the year 1683 at the importunity of Dr Halley I resumed the considera-

[4] It is printed here again, so as to make the record complete. A variant form of the same statement may be found, *infra*, in Supplement VIII, §7. Note that Newton, although referring to an error in M (which appears in E_1), uses the letters 'GH' from the figure in E_2.

tion thereof, & added some other Propositions about the motions of the heavenly bodies, which were by him communicated to the R. Society & entred in their Letter-book the winter following; & upon their request that those things might be published, I wrote the Book of Principles in the years 1684, 1685 & 1686, & in writing it made much use of the method of fluxions direct & inverse, but did not set down the calculations in the Book it self because the book was written by the method of composition, as all Geometry ought to be. And this Book was the first specimen made public of the use of this method in the difficulter Problems. [*This final sentence is a later insert. In a prior draft, f. 405, Newton had concluded this paragraph thus:* 'And ever since I wrote that Book I have been forgetting the Methods by which I wrote it.']

[When Mr Leibnitz was *del.*]

If it be asked why I did not publish the Book of Quadratures before the year 1704, I answer: ffor the same reason that I did not publish the Theory of colours before that year. I found that Theory in the beginning of the year 1666, & in the year 1671 was upon a designe of publishing it together with the methods of series & fluxions, but the next year for a reason given in my Letter of 24 Octob. 1676, I first suspended & then laid aside [*in the prior draft* left off] my designe of publishing them till the year 1704. And for the same reason I am very averse from medling with these matters any further [*altered in the prior draft from* I intend to meddle...no further].

The error of 1683 for 1684 occurs elsewhere; for example, on f. 80^{v} of MS Add. 3968, §41:

In [the win *del.*] Autum 1683 Mr Newton sent the Principal Propositions of his Principia Mathematica to London where they were communicated to the R. Society & in ye year 1686 he sent up the rest of the book, & two years after an Epitome thereof was printed in the Acta Eruditorum, & ye year following Mr Leibnitz published three papers relating to that book.

In a supplementary draft, f. 403, Newton wrote of his discovery as follows:

About three years after ⌊the writing of these Letters⌋ [*that is three years after 1676*] by the help of this method of Quadratures I found the Demonstration of Keplers Propositions that the Planets revolve in Ellipses describing with a Radius drawn to the sun in the lower focus of the Ellipsis, areas proportional to the times. And in the year 1686 I set down the Elements of the Method of ffluxions ⌊& moments⌋, & demonstra⟨ted⟩ them synthetically ⌊in the second Lemma of the second Book of Principles⌋ in order to make use of the Lemma in demonstrating some following Propositions.

6

MS Add. 3968, f. 415^{v}

The following passage contains Newton's views on his contribution to dynamics, in relation to Galileo, Huygens, and Leibniz.

Galileo began to consider the effect of Gravity upon Projectiles. Mr Newton in his Principia Philosophiae improved that consideration into a large science. Mr Leibnitz christened the child by ⟨a⟩ new name as if it had been his own ⟨,⟩ calling it *Dynamica.* Mr Hygens gave the name of vis centrifuga to the force by w^{ch} [bodies *del.*] revoling [*sic*] bodies recede from the centre of their motion. Mr Newton ⌊in honour of that author⌋ retained the name & called the contrary force vis centripeta. Mr Leibnitz to explode this name calls it sollicitatio Paracentrica, a name much more improper then that of Mr Newton. ⌊But his mark must be set upon all new inventions⌋ And if one may judge by the multitude of new names & characters invented by him, he would go for a great inventor.

The following (MS Add. 3968, f. 412^v) is more typical of Newton's statements about Leibniz.

[Leibniz] changed the name of vis centripeta used by Newton into that of sollicitatio paracentrica, not because it is a fitter name, but to avoid being thought to build upon M[r] Newton's foundation... he has set his mark upon this whole science of forces calling it Dynamick, as if he had invented it himself, & is frequently setting his mark upon things by new names & new Notations...

7

Extract from John Conduitt's 'Memorandum relating to S[r] Isaac Newton given me by M[r] Abraham Demoivre in Nov[r] 1727'

As mentioned earlier (Chapter III, §1, note 9), a small extract was printed by L. T. More (the three sentences given below, 'S[r] Isaac in order to make good... calculations agree together') and a large part of the text exists in a nineteenth-century transcript by H. R. Luard (U.L.C. MS Add. 4007, ff. 706–7). The version below has been made available through the kindness of the owner of the manuscript, Mr Joseph Halle Schaffner of New York City. I have printed only an extract.

The word *sic* has been introduced twice, where the text is confusing and may therefore seem erroneous, and where the Luard transcript differs from this one. In one case, 'Kepler's notion' was—understandably—misread as 'Kepler's motion'; in the other the word 'coursely' became 'cursorily'—which may have been what Newton intended, unless he meant 'coarsely'.

Since Conduitt tended to end a sentence with a comma and did not always begin the next sentence with a capitalized first letter, his text is difficult to comprehend, especially on first encounter. I have printed this document with extra spaces between sentences, to make it easier to read.

In 1673 D[r] Hook writt to him [S[r] Isaac] to send him something new for the transactions, Whereupon he sent him a little dissertation to confute that com̄on objection which is, that if it were true that the Earth moved from West to East, all falling bodies would be left to the West, & maintained that on the contrary they would fall a little Eastward, & having described a curve with his hand to represent the motion of a falling body he drew a negligent stroke with his pen, from whence D[r] Hook took occasion to imagine that he meant the Curve would be a Spiral, Whereupon the D[r] writt to him that the Curve would be an Ellipsis & that the body would move according to Kepler's notion [*sic*], w[ch] gave him an occasion to examine the thing thoroughly & for the foundation of the Calculus he intended laid down this proposition that the areas described in equal times were equal, which tho[u] assumed by Kepler was not by him demonstrated, of which demonstration the first glory is due to S[r] Isaac. In 1684 D[r] Halley came to visit him at Cambridge, after they had been some time together,[5] the D[r] asked him what he thought the Curve would be that would be described by the Planets supposing the force of attraction towards the Sun to be reciprocal to the square of their distance from it. S[r] Isaac replied immediately that it would be an Ellipsis, the Doctor struck with joy & amazement asked him how he knew it, Why saith he I have calculated it, Whereupon D[r] Halley asked him for his calculation without any further delay, S[r] Isaac looked among his papers but could not find it, but he promised him to renew it, & then to send it him,

[5] Conduitt has here made a later insertion, as follows: 'In ⌞May-Quaere⌟ 1684 D[r] Halley made S[r] I. a visit at Cambridge & there in a conversation.'

S[r] Isaac in order to make good his promise fell to work again, but he could not come to that conclusion w[ch] he thought he had before examined with care, however he attempted a new way which tho[u] longer than the first, brought him again to his former conclusion, then he examined carefully what might be the reason why the calculation he had undertaken before did not prove right, & he found that having drawn an Ellipsis coursely [*sic*] with his own hand, he had drawn the two Axes of the Curve, instead of drawing two Diameters somewhat inclined to one another, whereby he might have fixed his imagination to any two conjugate diameters, which was requisite he should do, that being perceived, he made both his calculations agree together.

After this D[r] Halley was (I think) sent down to Cambridge by the Royal Society to prevail with S[r] Isaac to print his discoveries w[ch] gave rise to the Principia.[6]

8

DAVID GREGORY'S COMMENT ON HOOKE'S CONTRIBUTION TO THE NEWTONIAN PHILOSOPHY

The following paragraph occurs in Gregory's 'Notae' as a comment on the Scholium to Prop. IV, Book I, of the *Principia*. In the Royal Society copy, it is inserted on a small sheet. I have included this statement because, as Gregory says, it is a report of what Newton told him ('ut Candidiss. Newtonus mihi narravit'). The translation is intentionally as literal as can be, so as to preserve the precise sense of the original. On Gregory's 'Notae', see Chapter VII, §12.

Immo si Hookio fidendum, ille omnis hujus philosophiae fundamenta jecit et ⌊cum⌋ Newtono communicavit. Verum Res revera ita se habet ut Candidiss. Newtonus mihi narravit. Cum forte inter alia Newtonus Hookio scriberet tantum abesse ut Grave e summitate turris decidens, ex motu diurno terrae ad occidentem a pede turris ⌊erecti⌋ caderet ut e contra ad orientem a pede turris terram subjectam attingeret (scil. ob maiorem impetum lapidi ad summitatem turris ⌊circulum maiorem describentis⌋ communicatum) et Casu Curvam manu duceret a lapide inter cadendum descriptam, hancque Curvam infra terrae superficiem ad morem Spiralis in Centro terminatam describeret, cum tamen istam figurae partem in epistola minime attingeret. Rescripsit Hookius verum ⌊quidem⌋ esse quod ad orientem terram attingeret sed quod non Centrum peteret sed illud praeterlapsum et Ellipticam curvam describens rursus in altum reverteretur ad summitatem Turris. Atque hoc Omne illud est quod se Newtono primum ostendisse jactat, et Newtoni philosophiam huic soli esse superstructam.

Indeed if Hooke is to be trusted it was he who laid the foundations of all this philosophy and communicated it to Newton. But the situation is really thus, as the very frank Newton told me: when by chance—among other things—Newton was writing to Hooke that, far from falling to the west of an upright tower's foot because of the Earth's diurnal motion, a body falling down from the top of a tower would on the contrary reach the ground beneath it eastward from the tower's foot (namely, because of the greater impetus communicated to the stone at the top of the tower as it describes its greater circle), he negligently drew freehand the curve described by the stone while falling and depicted this curve below the earth's surface as a spiral terminating at the centre (although he did not mention that part of the figure at all in the letter); Hooke wrote back that it was true indeed that it [*that is*, the stone, the body] would reach the Earth toward the east, but that it would not seek the centre but passing that by, and describing an elliptical curve, would again return upwards to the top of the tower. And this is all that he boasts that he first showed to Newton and that Newton's philosophy is built upon this alone.

[6] This final paragraph is a later addition made by Conduitt.

SUPPLEMENT II

HUMPHREY NEWTON AND THE 'PRINCIPIA'

HUMPHREY NEWTON—who copied out *LL*, *De mundi systemate* (*liber secundus*), and *M* (see Plates 5 and 6)—served as Newton's amanuensis for some five years: 1683–9 according to David Brewster, 1685–90 according to L. T. More. Humphrey himself said he came to Cambridge from Grantham in the 'last year of K. Charles 2^{d}', who died on 2 February 1685/6. Since the earliest document in his hand appears to be the tract *De Motu*, presumably he began his services in the late summer or autumn of 1685. Brewster calls him 'Dr. Newton'. He was no doctor in the sense of formal degrees but was a 'manmidwife', as we learn from the following extracts. Humphrey had come from Grantham; according to Stukely, he was a 'relation' of Newton's.

Humphrey's recollections of serving his illustrious namesake were written, soon after Newton's death, at the request of John Conduitt, who was gathering information for an intended biography. I print below the opening portion of one of Humphrey's letters to Conduitt, in which he discusses the *Principia* directly. The text is taken from the manuscript version in the Keynes Collection, King's College Library, Cambridge, dated 'Jan. 17.–2$\frac{7}{8}$' from Grantham.

> Receiving Y^{rs}, return as perfect & as faithful Account of my deceased ffriend's Transactions, as possibly does at this time occur to My Memory. Had I had y^{e} least Thought of gratifying after this Manner S^{r} Is'.s ffriend, I should have taken a much stricker view of his Life & Actions.
>
> In ye last year of K. Charles 2^{d} Sr Isaac was pleas'd through ye mediation of Dr Walker, (then Schoolmaster at Grantham) to send for me up to Cambridge, of whom I had the opportunity as well Hon^{r} to wait of [*sic*], for about 5 years, in wch Time he wrote his Principia Mathematica, wch stupendous work, by his Order, I copied out, before it went to ye Press. After ye Printing S^{r} Is. was pleas'd to send me wth several of Them, as Presents, to some of ye Heads of Colledges, & others of his Acquaintance, some of w^{ch} (particularly Dr Babington of Trinity) ⌞said⌟ That They might study seven years, before They understood any thing of it...

William Stukeley, M.D., F.R.S., an antiquary of some importance in the history of British archaeology, and a writer on earthquakes, wrote in 1752 a book of *Memoirs of Sir Isaac Newton's life*, which he intended for publication, but which was not printed until 1936. Stukeley wrote:

> What I have to say on his life is divided into three parts. I. What I knew of him personally, whilst I resided in London, in the flourishing part of my life. II. What I gathered of his family and education at Grantham, after I went to live there. III. Of his character.

The following extracts contain information on Stukeley's relations with Newton and the discovery of the principle of universal gravitation, and give further information about Humphrey Newton.

[Pp. 9–10.] In April 1705 Sir Isaac Newton came to Cambridg, to offer himself a candidate to represent the University in parliament. In 16th of that month Queen Ann was pleasd to visit the University of Cambridg, from Newmarket, whither a deputation of the heads of the Colleges had been, to invite her. I was then in what we call there junior sophs or 3d year after admission, residing in C.C. [Corpus Christi] college. The whole University lined both sides of the way from Emanuel college, where the Queen enter'd the Town, to the public Schools. Her Majesty dined at Trinity college where she knighted Sir Isaac, and afterward, went to Evening Service at King's college chapel...

We had then in our college, under the instruction of Dr Robt Dannye (who dy'd in the month of March 1730, rector of Spofforth in Yorkshire) gone thro' an excellent course of lectures in mathematics, and philosophy, particularly the Newtonian. And I own, upon this Royal visit, my curiosity was mostly excited, and delighted, in the thought of beholding Sir Isaac; who remain'd some time with us, and no joy could equal that which I took in seeing the great man, of whom we had imbibed so high an idea, from being conversant in his works. We always took care on Sundays to place ourselves before him, as he sat with heads of the colleges; we gaz'd on him, never enough satisfy'd, as on somewhat divine. The University was well sensible of the proposed honor, and readily chose him thir representative, who was thir greatest boast and ornament.

[Pp. 12–13.] On the 20 March 1717–8 whilst I practised physick in London, I was admitted a fellow [of the Royal Society] by Sir Isaac, at the recommendation of Dr Mead. Being Sir Isaac's countryman of Lincolnshire and pretty constant in attendance at the weekly meetings, from that time I was well receiv'd by him, and enjoy'd a good deal of his familiarity and friendship. I often visited him, sometime with Dr Mead, Dr Halley, or Dr Brook Taylor, Mr W. Jones or Mr Folkes and others; sometime alone; and we discours'd upon divers curious matters, as well as on country news; I being acquainted with many of his friends and relations there, and my brother being at that time apprentice to his old intimate friend and schoolfellow, Mr Chrichloe, of Grantham. Being generally of the Council of the Royal Society, and upon the casual absence of a Secretary, I was sometime order'd by him to take his seat, for that sitting. Several times I was propos'd by him, and elected an auditor of the yearly accounts of the Society, when we dined with him at his house by Leicester fields.

In the year 1720 Sir Isaacs picture was painted by Sir Godfry Kneller to be sent to Abbè Bignon in France, who sent his picture to Sir Isaac. Both Sir Isaac and Sir Godfry desir'd me to be present all the times of sitting, and it was no little entertainment to me, to hear all the discourse that passed between these two great men. Tho' it was Sir Isaac's temper to say little, yet it was one of Sir Godfrys arts to keep up a perpetual discourse, to preserve the lines and spirit of a face. I was delighted to observe Sir Godfry, who was not famous for sentiments of religion, sifting Sir Isaac to find out his notions on that head, who answered him with his usual modesty and caution.

[Pp. 19–20.] On 15 April 1726 I paid a visit to Sir Isaac at his lodgings in Orbels buildings in Kensington, dined with him and spent the whole day with him, alone. Among other discourse he mentioned to me, that he was born on Christmas day 1642. Some have observed, this time was particularly fruitful of great genius's. I acquainted him with my intentions of retiring into the country and had pitched on Grantham...

After dinner, the weather being warm, we went into the garden and drank thea, under the shade of some appletrees, only he and myself. Amidst other discourse, he told me, he was just in the same situation, as when formerly, the notion of gravitation came into his mind. It was occasion'd by the fall of an apple, as he sat in a contemplative mood. Why should that apple always descend perpendicularly to the ground, thought he to him self. Why should it not go sideways or upwards, but constantly to the earths centre? Assuredly, the reason is, that the earth draws it. There must be a drawing power in matter: and the sum of the drawing power in the matter of the earth must be in the earths center, not in any side of the earth. Therefore dos this apple fall perpendicularly, or towards the center. If matter thus draws matter, it must be in proportion of its quantity. Therefore the apple draws the earth, as well as the earth draws the apple. That there is a power, like that we here call gravity, which extends its self thro' the universe.

And thus by degrees he began to apply this property of gravitation to the motion of the earth and of the heavenly bodys, to consider their distances, their magnitudes and thir periodical revolutions; to find out, that this property conjointly with a progressive motion impressed on them at the beginning, perfectly solv'd their circular courses; kept the planets from falling upon one another, or dropping all together into one center; and thus he unfolded the Universe. This was the birth of those amazing discoverys, whereby he built philosophy on a solid foundation, to the astonishment of all Europe.

[P. 22.] Humphry [*sic*] Newton, a physician and manmidwife, a relation of Sir Isaacs, now lived at Grantham. In the beginning of the year 1728 Mr Conduit was sponsor to a son of his, who was baptized Isaac. He was named Isaac in honor to the memory of Sir Isaac, [he is] now a physician there. I was deputy at his baptism.

[Pp. 56–7.] D^r [Humphrey] Newton of Grantham, aforemention'd, was sizer to Sir Isaac; lived under his tuition 5 years; was assistant to him, particularly in his chymical operations, which he pursu'd many years. He often admir'd Sir Isaacs patience in his experiments, how scrupulously nice he was in weighing his materials, and that his fires were almost perpetual. In the year 1705–6 when I was at C.C. College, I went a course of chymical lectures with Seignior Vigani, in Sir Isaac's room, where he made his chymical operations, being backward, towards the Master's lodg. Dr Newton says that all the time he was with him, he scarce ever observ'd him to laugh, but once. He remembers it was on this occasion: he ask'd a friend to whom he had lent an Euclid to read, what progress he had made in that author, and how he liked it? He answerd by desiring to know what use and benefit in life that kind of study would be to him. Upon which Sir Isaac was very merry. According to my own observation, tho' Sir Isaac was of a very serious and compos'd frame of mind, yet I have often seen him laugh, and that upon moderate occasions. He had in his disposition a natural pleasantness of temper, and much good nature, very distant from moroseness, attended neither with gayety nor levity. He used a good many sayings, bordering on joke and wit. In company he behavd very agreably; courteous, affable, he was easily made to smile, if not to laugh.

[Pp. 60–1.] D^r [Humphrey] Newton tells me that several sheets of his [Isaac Newton's] Optics were burnt, by a candle left in his room. But these I suppose he was able, by a little pains, to recover again; or if there be any imperfection in that work, we may reasonably suspect it was owing to this accident. He says Sir Isaac constantly went to church on Sundays, to St. Mary's, though not always to the College chapel. In mornings he was up at study. He seldom went to the hall to dinner, but had his victuals brought to his chamber; and then very often, so deeply intent was he, that he never thought of it till supper time. When he was busy in study he never minded his meal times. And when he took a turn in the Fellows Gardens, if some new gravel happen'd to be laid on the walks, it was sure to be drawn over and over with a bit of stick, in Sir Isaac's diagrams; which the Fellows would cautiously spare by walking beside them, and there they would sometime remain for a good while.

SUPPLEMENT III

NEWTON'S PROFESSORIAL LECTURES

One of the conditions of appointment of the Lucasian Professor was that he give at least one lecture per week in each term of the academic year. He was especially ordered to revise and polish the texts of the ten (or more) lectures for each term and he was then required to deposit the final version in the University Library.[1] For the period of Newton's tenure of the Lucasian Professorship, there are, in the University Library, manuscripts of lectures that start in October, except for the first series, which began in January 1669/70, and the last, which began in September 1687. All were thus given during the Michaelmas Term, save for the inaugural series in the Lent Term of 1670.

These lectures were given on optics (1670-2), on arithmetic and algebra (1673–83), on a part of Book I of the *Principia* (1684–5), and on the System of the World (1687). There is no record of what lectures, if any, Newton gave in 1686, or during the period from 1688 until he removed to London in the summer of 1696. The absence of any texts of lectures deposited for these years may serve as a starting-point for our queries concerning the reliability of these manuscripts as texts of lectures actually given and then placed on deposit. Shall we merely assume that in 1686, and from 1688 to 1695, Newton did not comply with the requirements of giving lectures, or that he only failed to adhere to the other requirement of subsequently depositing a text? We know nothing whatever as to whether he really did give any lectures during these years 1686, 1688–95, and we are equally ignorant as to what, if any, were the subjects on which he lectured during the other two terms of the seventeen years 1670–85, 1687.

An examination of the numbers assigned by Newton to individual lectures in these deposited texts, as listed conveniently by Edleston,[2] causes increased dubiety. These are as follows. On optics: January 1669/70, Lect. 1–8; October 1670, Lect. 9–15, Lect. 1–3; October 1671, Lect. 4–13; October 1672, Lect. 14–16; on arithmetic and algebra: October 1673, Lect. 1–7; October 1674, Lect. 1–10; October 1675, Lect. 1–10; October 1676, Lect. 1–10; October 1677, Lect. 1–10; October 1678, Lect. 1–6; October 1679, Lect. 1–6; October 1680, Lect. 1–8; October 1681, Lect. 1–10; October 1682, Lect. 1–10; October 1683, Lect. 1–10; *De motu corporum*: October 1684, Lect. 1–9; October 1685, Lect. 1–10; *De motu corporum, lib. 2us* [De mundi systemate]: 29 September 1687, Lect. 1–5.

[1] See the regulations concerning the Lucasian Professorship, printed by D. T. Whiteside in [Newton] (1967–), *Mathematical papers*, vol. 3, pp. xx–xxvii.

[2] See Edleston (1850), pp. xci–xcviii.

These lists indicate that the set of lectures beginning in October 1670 consisted of Lectures 9–15 and Lectures 1–3, the latter making up the beginning of 'Opticae pars 2da...' in the manuscript, while in the set beginning in October 1672 there were only three lectures given. This manuscript was 'put into the hands of the Vice-Chancellor and delivered by him to Robert Peachy to be placed in the University Library, Octob. 21, 1674'.[3] We may assume, therefore, that the text deposited must roughly coincide with the state of Newton's knowledge while delivering the lectures on optics in the Lent Term of 1669–70 and the Michaelmas Terms of 1670–1, 1671–2, and 1672–3. But we have no grounds for believing that Newton read the text marked Lect. 1–8 in the Michaelmas Term of 1669–70, or that he necessarily gave only three lectures in 1672–3 (and those during October 1672),[4] particularly in view of the requirement that he was to deposit the lectures not as read, but only after they had been duly 'revised and polished'. And the same considerations apply equally to the other sets of lectures.

It is not difficult to find cause to doubt that these manuscripts could ever have been the actual texts of the lectures as given. Consider first the lectures on algebra and arithmetic. Flamsteed owned a page of manuscript marked 'Mr Newton's paper given at one of his lectures, Midsummer, 1674'.[5] This 'paper' is printed by Edleston, who notes that 'Flamsteed was at Cambridge, from the end of May until July 13'.[6] But, if we had limited ourselves to the texts of these lectures deposited in the University Library, we would not even have known that Newton was lecturing at that time!

From the contents of the 'paper' given to Flamsteed, unmistakably written out in Newton's hand, we may identify the lectures on arithmetic and algebra in which —supposedly—these topics were treated. The subject-matter corresponds to Lectures 6 and 7 of the 1674 series, which, according to MS Dd.9.68 (the text deposited by Newton in the University Library), began in October and ended in December. Yet Flamsteed attended the lecture in the Easter Term, which began in late April and ended early in July of 1674. Hence, if we accept the dates given in the deposited text, we must conclude either that Newton repeated the lecture series from the Easter Term, or that he never gave these lectures at all in the Michaelmas Term (or at least not as indicated in MS Dd.9.68).[7]

If the deposited lectures on arithmetic and algebra are thus open to suspicion

[3] *Ibid.* p. xcii.

[4] Humphrey wrote two letters to John Conduitt, after Newton's death, in which information is given about Newton's lectures, as follows: 17 January 1727/8, '[Newton] seldom left his Chamber, unless at Term Time, when he read in ye Schools, as being Lucasianus Professor, where so few went to hear Him, & fewer y^t^ understood him, y^t^ oftimes he did in a Manner, for want of Hearers, read to ye Walls.' 14 February 1727/8, 'When he read in ye Schools, he usually staid about half an hour, when he had no Audit^rs^ he comonly return'd in a 4th Part of that time or less.' More (1934), *Newton*, pp. 247, 249.

[5] I have examined this paper, still preserved among the Flamsteed papers in the archives of the Royal Greenwich Observatory, Herstmonceux Castle, vol. 42.

[6] Edleston (1850), pp. 252–3.

[7] Like other deposited 'lectures', these are paralleled by a manuscript version in the Portsmouth Collection (U.L.C. MS Add. 3993).

as a complete and reliable index to the lectures given by Newton as Lucasian Professor, what about the lectures on optics? Are they more reliable, or less? Let the record speak for itself. There are two manuscript copies, not just one, of Newton's *Lectiones opticae*. One of these was deposited in the University Library, but the other remained in Newton's possession and is now in the Portsmouth Collection (U.L.C. MS Add. 4002). The latter is in Newton's hand, the former is mostly in the hand of a copyist, almost certainly John Wickins. These versions differ not only in their actual texts, but even in their alleged division into 'lectures'.

A comparison of the two shows that MS Add. 4002 is an earlier version than MS Dd.9.67. While the latter is divided into two sets of lectures, Lect. 1–15, Lect. 1–16, the former is divided into one continuous sequence, Lect. 1–18 (but the material corresponding to 'Lect. 14 & 15' and to 'Lect 16 et 17' is not divided into two 'lectures').

The manuscript deposited in the University Library (MS Dd.9.67) is not a mere revision; it is more extensive in scope and presents the subject-matter in a partial rearrangement; it is furthermore divided into two 'parts', of which the first occupies pages 1–177, and the second (entitled 'Opticae Pars 2da De Colorum Origine') has its own pagination from 1 to 101 and its own sequence of figures from 1 to 62. One paragraph of the early version (MS Add. 4002) is not to be found in this revised text at all. Some sections (§§24–9 and portions of §§30–96) were revised and occur as Parts 1 and 2 of 'Opticae Pars 2da' in MS Dd.9.67. Thus the presentation of the origin of colours, in 'Pars 2da' of MS Dd.9.67, was originally the first portion of the manuscript and hence appears at the beginning of MS Add. 4002.[8] The contents of MS Add. 4002 (paginated continuously from 1 to 129, with the figures drawn in where they belong on the text pages) correspond primarily to the first three of the four sections of 'Pars 1^{a}' of MS Dd.9.67,[9] or the first 64 out of a total of 77 pages; but in MS Dd.9.67 (unlike MS Add. 4002), the figures are not drawn on the text pages but gathered together at the end of each of the two parts of the manuscript. Although MS Dd.9.67 is divided into sections with titles, MS Add. 4002 is not. In both manuscripts there are divisions into small 'articles' indicated in the margin by numbered postils. As is to be expected from the foregoing account, these divisions (and the numbered postils) are not the same in the two manuscripts.

Now it may appear to the reader that, since MS Dd.9.67 was actually deposited, it must represent the 'authoritative' text, and hence correspond more closely to the lectures as given than MS Add. 4002 does. Possibly, then, the division of the earlier draft, MS Add. 4002, into eighteen lectures, corresponding to Lect. 1–12 in MS Dd.9.67, should not be at all disturbing to us, since it may imply only that when Newton gave the lectures he found he did not need eighteen sessions, and delivered the whole text in twelve. This may be so, but I seriously doubt it. The reason

[8] See Westfall (1963), 'Newton's reply', pp. 83–4.

[9] The remainder, as mentioned, is in good part to be found in 'Pars 2da' of MS Dd.9.67.

is that Lect. 1 in MS Add. 4002 is dated January 1669 (O.S.; 1670 N.S.), just as in MS Dd.9.67, but in MS Add. 4002 Lect. 9 has a date written above the heading, July 1670, which may be presumed to be when this lecture was given, and not just a record of Newton's intentions. But according to MS Dd.9.67, the revised version, Newton gave Lect. 9 on a very different subject! Obviously both cannot be correct.

I think the question may be easily resolved in a very different manner: far from attempting to decide whether to credit MS Dd.9.67 or MS Add. 4002, I suggest that neither one nor the other is fully trustworthy as to subjects lectured on at given times. The manuscripts deposited by Newton as 'lectures' on optics and on the system of the world are finished treatises, written out in final form, even to carefully drawn figures; they appear to have been made ready for the press without needing any further intermediate editing. They are revised, rewritten, and polished versions, prepared by Newton according to the strict terms of his professorship—presumably *after* they had been delivered—and accordingly deposited. That Newton chose merely to leave them on deposit rather than submit them to a printer is merely idiosyncratic, but there can be no doubt that these texts themselves are written as treatises: they do not even bear the word 'lectures' in the title. There is no title whatsoever to MS Add. 4002, while MS Dd.9.67 begins 'Opticae pars 1ª, De radiorum Lucis Refractionibus'. We know the work as *Optical lectures read in the publick schools of the University of Cambridge, Anno Domini, 1669* or *Lectiones opticae, annis MDCLXIX, MDCLXX & MDCLXXI. In Scholis publicis habitae: et nunc primum ex MSS. in lucem editae*,[10] but these titles were added by the editors or publishers and do not appear in either manuscript version. The very first page, however, does mention (in both manuscripts) 'the *Dissertations*, which you have not long since heard from this Place', and which contained 'so great a Variety of optical Matters'[11] ('Dissertationes, quas hic non ita pridem audivistis, tanta rerum opticarum varietate...fuerint compositae'). This reference to Isaac Barrow's 'Lectiones opticae' is all the more interesting in that Barrow mentioned Newton in the preface, thanking him for having read at least a portion of the text.[12] It is generally believed that Barrow had resigned the Lucasian chair so that Newton might have it; hence, it is pleasing to find that the opening page (second sentence) of Newton's inaugural set of lectures should thus refer graciously to his predecessor. In addition to this sentence, there are a number of other traces (either in direct statement or in style)[13] in the texts deposited

[10] The texts of the Latin and English versions are notably different; here I am only concerned with the fact that both the Latin and the English versions were printed after Newton's death with the word 'Lectures' or 'Lectiones' added to the title by the editor, publisher, or printer. [11] Newton (1728), *Optical lectures*, p. 1.

[12] Other references to Barrow occur in this work, for example, in MS Add. 4002, ff. 49, 95, 99.

[13] For instance: 'Verùm ne videar officii limites excessisse...' (p. 23); '...nè demonstratio haec, quae longiuscula futura est, vos itaque taedio afficiat...' (p. 94); '...his itaque paucis...in gratiam sequentium obiter notatis...pergo actutum disserere...' (p. 95). The word 'Lectiones' does not appear as such in the title of *Arithmetica universalis; sive de compositione et resolutione arithmetica liber*. But this work is described as 'In usum juventutis academicae', and a statement in the preface informs the reader that the book originated in Professorial Lectures.

by Newton which indicate that they were treatises written out to be given as lectures.[14]

How different all this is from Whiston's lectures, given when he replaced Newton, first as his deputy, then as Lucasian Professor. These were entitled respectively *Astronomical lectures, read in the publick schools in Cambridge* and *Sir Isaac Newton's mathematick philosophy more easily demonstrated:...Being forty lectures read in the publick schools at Cambridge.* Not only are these two works actually divided into 'Lectures', but in the printed versions each one is dated; and there is other information appropriate to a lecture series, as that 'Lecture I. *was only an Oration to the University of* Cambridge; *and is here omitted*'. Again, Lect. XVI is prefaced by the announcement:

Chosen Professor, Jan. 8. 1701–2.
Admitted Professor, May 21. 1702.[15]

Other lectures of that time and even later were also printed in a fashion that showed their origin as lectures, being divided into 'lectures' rather than 'chapters'.[16] Conversely, some works printed as 'lectures' were never actually read publicly as lectures; a good example is Barrow's *Lectiones geometricae.*

I turn now to the lectures for 1687. The manuscript deposited, entitled 'De motu Corporum Liber secundus' (MS Dd.4.18), was copied out by Humphrey

[14] I should like to repeat that I do not doubt that these deposited manuscripts correspond *grosso modo* to the subjects on which Newton may have been lecturing, as this sentence shows. But what I question is that Newton gave the contents of a portion marked off as Lecture 'so and so' on a particular date!

With respect to Barrow and Newton, there is some further information to be gleaned from the manuscripts. As to Barrow's having resigned his chair in favour of Newton, there exists some contemporaneous opinion on this subject; for example, a letter from John Collins to James Gregory (25 November 1669): 'Mr Barrow hath resigned his Lecturer's place to one Mr Newton of Cambridge, whome he mentioneth in his Optick Praeface as a very ingenious person' ([Newton], *Correspondence* (1959–), vol. 1, p. 15). Among Newton's manuscripts I have found a fragment (MS Add. 3968, § 41, f. 117) reading: 'Upon account of my progress in these matters he procured for me a fellowship in Trinity College in the year 1667 & the Mathematick Professorship two years later.' No name is mentioned, but the circumstances would agree with the accepted story. The fragment is undated, but it is immediately preceded by a mathematical text which may be dated January or February 1716.

As to Newton's contribution to Barrow's 'Lectiones Opticae', here is what Barrow wrote in the preface: 'Isaac Newton, a Fellow of our College (a man of exceptional ability and remarkable skill) has revised the copy, warning me of many things to be corrected, and adding some things from his own work, which you will see annexed with praise here and there.' (Quoted from Barrow (1916), *Geometrical lectures*, translated by J. M. Child.) By 'revised the copy' ('exemplar revisit'), Barrow may very likely have meant 'looked over a [press] copy', that is, that Newton read through proof-sheets, rather than that he served as an editor, 'revising the copy' in the present-day sense of this expression.

Newton's copy of Barrow's 'Lectiones opticae' (the actual title reads: *Lectiones XVIII, Cantabrigiae in scholis publicis habitae; in quibus opticorum phaenomenωn genuinae rationes investigantur, ac exponuntur*) is at present in the Trinity College Library (press-marked NQ.16.181). On the fly-leaf it bears the inscription: 'Isaaco Newtono Reverendus Author hunc dono dedit July 7th 1670.'

[15] Whiston (1728), *Astronomical lectures*, Lec. XVI, p. 192.

[16] For example, Barrow's *Lectiones opticae, Lectiones mathematicae*, or Keill's *Introductio ad veram physicam, seu lectiones physicae habitae in schola naturalis philosophiae Acad. Oxoniensis* (divided into 'Lectiones' numbered 1–16), translated as *An introduction to natural philosophy: or philosophical lectures read in the University of Oxford*, and *Introductio ad veram astronomiam, seu lectiones astronomicae habitae in schola astronomiae Acad. Oxoniensis* (divided into 'Lectiones' 1–30), translated as *An introduction to the true astronomy: or, Astronomical lectures read in the astronomical school of the University of Oxford.*

from Newton's own complete version, and corresponds to the first 27 sections only, something less than half of the manuscript retained by Newton and presently in the Portsmouth Collection (MS Add. 3990). The latter is not only complete; it is the working copy, extensively revised and emended by Newton. The text itself of MS Add. 3990 is partly in Humphrey's hand and partly in Newton's.[17] In MS Add. 3990, Newton has written in the margin in his own hand the postils giving a title to each section, or a summary of its contents. The first 27 of these are numbered in roman numerals,[18] but the remaining sections are not numbered at all. These first 27 sections correspond to MS Dd. 4. 18, in which Humphrey has copied out the postils as well as the text and has included the numbers.

In MS Add. 3990, at the end of section 27, Newton has entered a marginal note, 'ffinis anni 1687', altered to 'ffinis Lectionum anni 1687'. This alteration no doubt means that Newton saw that his original note might imply that he had only in 1687 finished writing the first part of *De mundi systemate* (*liber secundus*)—as we have called it in Chapter IV, §6—and so he inserted the word 'Lectionum', making it unambiguous that he was referring to the end of the lectures. But MS Add. 3990 is not divided up into lectures, as MS Dd. 4. 18 is. The latter has five divisions, of which the first is entitled 'Sept. 29 Praelect. 1', followed by 'Lect. 2', 'Lect. 3', 'Lect. 4', 'Lect. 5'. Unlike the other four sets of deposited 'lectures', MS Dd. 4. 18 has no indication of a year, although presumably we may assign the year with some assurance because of the note in MS Add. 3990. This manuscript also differs from the others in the use of 'Praelect.' for 'Lect.' in 'Praelect. 1', and in having the day as well as the month. The lecture would have been given on 29 September 1687 (=9 October N.S.), which was a Thursday.

We have no way of telling whether Newton's note ('ffinis Lectionum anni 1687') in MS Add. 3990 was a way of marking a point in the manuscript that corresponded to the end of the lectures given in 1687. At once, however, we are faced with a whole series of puzzles. Why did Humphrey not copy this note into the last page of MS Dd. 4. 18? Shall we assume that Newton added the note at some later time, that it was not there when Humphrey made the copy for deposit? Or, was this entry intended merely to mark for purposes of record how much of MS Add. 3990 Newton had asked Humphrey to copy out? It is worthy of notice, however, that the first 27 sections of MS Add. 3990, though numbered, are not divided into lectures. Was the division then made arbitrarily in the copy after the lectures had been given? Indeed, why did Newton have Humphrey make a copy at all? Why not have handed in the whole manuscript, as he had done with the optical and mathematical lectures? Had this copy been undertaken when Newton was still thinking of using this version in the *Principia*? And had Humphrey stopped making this copy when Newton had decided to compose a different version of *De mundi systemate*, the one that was published as Book III? Very likely this was the case,

[17] See Chapter IV, §6, and Supplement VI.

[18] Owing to a mistake in numbering, these run from I to XXVIII.

since MS Dd. 4. 18 is still entitled *De motu corporum, liber secundus*. If this is so, then Newton may merely have deposited a convenient text.

Whether this is so or not, we cannot tell without further evidence. But I rather suspect that Newton marked off a place arbitrarily, noted that he had done so in the margin, and then assigned numbers to the sections in this part. Afterwards—in the new copy, and only in that copy—he would have divided the text into five lectures. Surely, if he had been carefully planning to select a complete portion of *De mundi systemate*, he would have concluded it more sensibly with the next section, XXVIII (27 in the printed edition), that 'All the planets revolve around the Sun', thus giving a fuller and more exact understanding of 'the true disposition of the whole system', than with XXVII, on 'The intensities of the forces and the resulting motions in individual cases'.

While many of the questions about the texts of these lectures necessarily must involve conjectures, I believe the evidence is overwhelming that the deposited treatises on optics, mathematics, and the system of the world are in no case to be regarded as containing the very texts of actual lectures given on certain indicated dates. Above all, it seems to me that anyone who reads these manuscripts carefully cannot escape the general impression that the division of these treatises into 'lectures' was in each case superimposed on a completed manuscript (revised according to the statutes governing the professorship) and consists of a rather arbitrary division of the text and no doubt an equally arbitrary assignment of dates in the margins. Surely we are not to believe that even Newton would have given a first 'lecture' in October 1671 that began with the proof of Prop. III, the text of which had been stated in the previous academic year! Again, why were there only three lectures in 1672–3? Then there is a contradiction, as we have seen, between the date and subject of a true lecture which Flamsteed attended and the schedule given by Newton in the manuscript deposited in the University Library. We have also seen that an actually dated lecture (July 1670) in MS Add. 4002 cannot be reconciled with the division into lectures presumed in MS Dd. 9. 67. And so on.

These things being so, there is no need for us to give too much credence to the two sets of lectures 'De motu corporum' deposited for 1684–5. No doubt Newton may have been lecturing then on these very topics, or on very similar ones, but there is no obligation for us to take this manuscript, which we have here designated *LL*, as representing the actual state of the *Principia* as of the dates Newton has written in the margin.[19] It would have been consistent with his habits—if our foregoing critique is valid—for him to have put dates and divisions into 'lectures' on a later manuscript before depositing it. The only difference in actual form and content between *LL* and the manuscripts that Newton deposited on optics, and on

[19] Furthermore, if the dates in LL_α/LL_β are not false, we must assume that Newton was lecturing on the identical topics (*Principia*, Book I, pp. 32–5) in his 'Lectio 9' of 1684 and again in his 'Lectio 5' of October 1685.

the system of the world, is that *LL* is in some part rough and unpolished, and obviously incomplete. Futhermore, in the single case of dynamics, Newton actually did contemporaneously publish a treatise on the subject of the lectures (*Principia*, Book I, and possibly part of Book II). Hence he could not deposit the final complete manuscript (since it had gone to the printers and presumably was no longer available to him). Instead of making a complete manuscript copy of the printed book, or of parts thereof, he merely deposited the jumble preserved as MS Dd.9.46. The state of this manuscript—incomplete, containing two overlapping versions of one part, and ending in the middle of a sentence—is certainly proof in itself that we should not regard *LL* as lectures given in specified seasons so much as a working draft of the beginning of the *Principia*.

SUPPLEMENT IV

NEWTON'S LUCASIAN LECTURES: 'DE MOTU CORPORUM, LIBER PRIMUS'

THE manuscript that I have described in Chapter IV, §2, as *LL*, deposited by Newton in the University Library as his professorial lectures for 1684 and 1685, is still preserved (MS Dd.9.46). For convenience I refer to the deposited texts as Newton's Lucasian Lectures. The Lucasian Lectures that are specifically related to the text of the *Principia* exist in two sets: one (*LL*) corresponds to parts of Book I of the *Principia*, while the other is an early and partly rejected form of Book III (see Supplements III and VI).

The gatherings of sheets composing the volume (*LL*) seem to have been arranged and bound in a somewhat random fashion, and then numbered sequentially by a modern librarian. Today one may rearrange these pages at will by the use of photocopies, playing a kind of game of solitaire, so as to put them in a meaningful order. It then turns out that *LL* contains two major sequences, one of which begins with the Definitions, proceeds to the Axioms or Laws of Motion, and goes on to Book I, ending in the middle of a sentence in Prop. XXIV (=Prop. XXXV), corresponding to page 118 of E_1, and {116.18} of E_3. The second sequence in part overlaps the first, beginning with Corol. 6 in Prop. XVI and ending in the middle of a sentence in Prop. LIV, corresponding to page 159 of E_1 and {156.28} of E_3. Each of these two sequences is written out in the hand of Humphrey Newton on the recto side of each leaf, with corrections and additions made by Newton himself, sometimes on the verso side.[1] The two sequences derive from two drafts or states, one earlier than the other. I have designated these LL_α and LL_β. That part of LL_α now in *LL* shows a revision by Newton in three stages. He discarded the first eight leaves and replaced them by a set of twenty leaves; he then revised leaves 9–24 of LL_α; and finally, he added a new supplementary sequence. Leaves 25–32 of LL_α remain today as part of *LL*; but they should have been discarded, since they are replaced by parts of LL_β. Further leaves of LL_α are to be found here and there throughout Newton's manuscripts, and I have been able to identify a number of them and to draw inferences concerning the contents of the remainder. Hence, for

[1] Although both LL_α and LL_β are written out in Humphrey Newton's hand, certain parts of the manuscript have a different appearance from others. Thus pp. 20, 21, and 23 (in the modern librarian's enumeration) are more crowded than either the preceding or the following pages, as if they were a replacement and the text had to be squeezed into a given space. The slope of the writing differs slightly in LL_α from that in LL_β, and the pages of LL_β( ) are a little closer or more squeezed together than the rest. I do not know what significance, if any, this may have.

ease of description, we may divide LL_α into four parts, as follows (I have indicated by the symbol [*LL*] those portions which are presently contained in the bound volume: MS Dd.9.46):

$\langle LL_\alpha^{1-8}\rangle$, *the original first eight leaves* of LL_α, now missing, presumably written out by Humphrey and numbered by Newton in the upper right-hand corner of each recto page;

[*LL*]: LL_α, *original leaves* 9–24, written out by Humphrey and numbered by Newton in the upper right-hand corner of each recto page;

[*LL*]: LL_α, *original leaves* 25–32, written out and numbered as above;

LL_α, *further pages* that are not presently part of *LL*.

The contents of LL_β may be similarly itemized:

[*LL*]: LL_β (*1*), *twenty leaves*, replacing $\langle LL_\alpha^{1-8}\rangle$, unnumbered, with the exception of the ninth leaf, bearing a '9' in Newton's hand;

[*LL*]: LL_α, *original leaves* 9–24, heavily revised by Newton on being transformed from LL_α to LL_β; the number of the first leaf of each set of 8 has been changed by Newton: '9' to '21', '17' to '29';

[*LL*]: LL_β, *the concluding* 56 *leaves* of *LL*; the first leaf of each set of 8 is numbered by Newton: '37', '55' (!), '63', '71', '79', '87', '95'.

LL_β, *further pages* that are not presently part of *LL*.

Because, generally speaking, the text of LL_α and LL_β is written out by Humphrey on the recto sides only, the page numbers are numbers of the leaves, or folio numbers. Occasionally, a correction or expansion, or an addition, is written out (usually by Newton) on a facing verso page, and the last leaf of all of *LL* (f. [100] in a proper sequence) is written on both the verso and recto sides. (For two pages of *LL* see Plates 7 and 8.) *M* is apparently a fair copy of LL_β with some amendments.

If the sets of leaves bound up in MS Dd.9.46 were placed in their proper order and numbered in sequence, starting with 1, the relation of such numbers to the numbers written on the actual pages is given in the accompanying table, which also contains the numbers assigned by a librarian (in the order in which the leaves are bound up, starting with 4) and the numbers written on some of the pages by Newton himself. I have also given Newton's own indications of the division into lectures.[2]

From the table it is seen how, in the part of LL_α now in MS Dd.9.46, which originally contained thirty-two leaves, the first eight pages were discarded and replaced. Since the new expanded beginning comprises twenty pages unnumbered by Newton, the old page '9' was no longer the ninth but the twenty-first page; Newton accordingly altered the number '9' to '21'. But when the original first set of eight leaves was replaced by twenty new leaves, Newton did not renumber every succeeding leaf; only the number on the first page of each successive group of eight was altered, page 9 becoming 21, page 17 becoming 29, page 25 becoming 37, and so on. Because the first twenty leaves of *LL* represent a different state of

[2] Edleston (1850), pp. xcv–xcvii. Edleston has made a somewhat similar outline, referring the contents of *LL* to E_1.

composition from the original LL_α, apparently belonging to the stage of writing and revising that characterize LL_β, I have called them $LL_\beta(1)$, thus distinguishing them from both LL_α and the remainder of LL_β. As we may see from the table, the fragment of LL_β preserved in *LL* was numbered in eights; that is, each eighth leaf is numbered by Newton, but the intervening seven do not bear any numbers other than those of the modern librarian.

Comparative table of page numbers in LL

LL_β (*1*) and LL_α			LL_β		
Proper order of pages	Modern librarian's numbers	Newton's numbers	Proper order of pages	Modern librarian's numbers	Newton's numbers
1	4		45	48	37
2	5		⋮	⋮	
3	6		52	55	
⋮	⋮		53	40	55
8	11		⋮	⋮	
9	12	9	60	47	
10	13		61	88	63
⋮	⋮		⋮	⋮	
20	23		68	95	
21	24	21 [changed from 9]	69	96	71
			⋮	⋮	
22	25	10	76	103	
23	26	11	77	32	79
24	27	12	⋮	⋮	
⋮	⋮	⋮	84	39	
28	31	16	85	80	87
29	64	{17	⋮	⋮	
		{29	92	87	
30	65	18	93	72	95
31	66	19	⋮	⋮	
⋮	⋮	⋮	100	79	
36	71	24	[100^v]		
37	56	{25			
		{37			
38	57	26			
39	58	27			
⋮	⋮	⋮			
44	63	32			

Only the numbers in the two right-hand columns appear on the actual manuscript pages. A set of suspension points indicates that the intervening numbers are on the appropriate pages.

The first page of LL_β is numbered 37 because this page was intended to replace the page in LL_α with the very same number 37, which was given to it by Newton when he renumbered every eighth page of LL_α after replacing those first eight pages by a set of twenty new unnumbered pages. The first page of LL_β begins with Corol. 6 to Prop. XVI just as the renumbered page 37 of LL_α does. As Edleston points out, the next number in the sequence of eights should have been 45, rather than

Newton's 55, which is 'apparently...a clerical error, which is propagated through the remainder of the MS.'.

The relation of LL_α and LL_β may most easily be seen in a table listing the equivalent parts of $LL_\beta(1)$, LL_α, and LL_β. First, *LL* contains, in the order found in *M*, E_1, and the later printed editions,

Definitions	Props. I–IX
Scholium	Lemma 12
Axioms or Laws of Motion	Props. X–XII
Corollaries	Lemmas 13–14
Scholium	Props. XIII–XV
Lemmas 1–11	Prop. XVI (plus Corols. 1–5).

This material is composed of $LL_\beta(1)$ and the revised part of LL_α, and should have been followed by LL_β.[3] Newton, however, included (perhaps carelessly) another eight leaves of LL_α which correspond to LL_β as follows:

LL_α	LL_β
Corols. 6–9 to Prop. XVI	Corols. 6–9 to Prop. XVI
Prop. XVI[4]	Prop. XVII
	Lemma 15
Props. XVII–XVIII	Props. XVIII–XIX
	Prop. XX
	Lemma 16
Prop. XIX	Prop. XXI
	Lemmas 17–21
	Props. XXII–XXIV
	Lemma 22
	Props. XXV–XXVI
	Lemmas 23–25
	Prop. XXVII
	Lemma 26
	Prop. XXVIII
	Lemma 27
	Prop. XXIX
Prop. XX	Prop. XXX
	Lemma 28
	Prop. XXXI
Props. XXI–XXIV	Props. XXXII–XXXV
	Props. XXXVI–LIV

Although LL_β, after Newton's manuscript emendations, is so close to *M* that the latter can be considered a slightly improved copy of LL_β, this is not quite true of LL_α.[5] The structure is the same (the numbering and order of the parts: Definitions,[6]

[3] In other words, a continuous sequence exists in *LL*, made up of $LL_\beta(1)$, the revised part of LL_α (but not the last eight leaves), LL_β, as far as it goes.

[4] On the reason for two Props. XVI, see Chapter IV, §2; and p. 000 below.

[5] An exception is the opening set of twenty pages which, in their ultimate revision, do not differ notably from *M* and E_1.

[6] Originally there were two extra Definitions ('Axis materiae' and 'Centrum materiae'). These are printed in Herivel (1965), *Background*, pp. 321–2. Herivel does not point out that the above-mentioned Defs. II and III have been cancelled.

Laws, Lemmas, Propositions, Scholia), but more notable alterations, expansions, and additions were made between LL_α and M than between LL_β and M. I shall begin this analysis by first looking closely at $LL_\beta(1)$ and LL_α, then seeing what relation LL_α bears to LL_β. Finally I shall discuss the remaining parts of LL_α and LL_β that are either missing or to be found elsewhere among Newton's papers.

The first page of *LL* bears the general title *De motu Corporum, Liber primus*, which might give the impression that at this stage both the Definitions and the Laws of Motion were regarded as belonging to Book I, in distinction to M and E_1, where Book I begins *after* the Definitions and the Laws of Motion. Nevertheless, following the Scholium to the Laws of Motion, the title appears again, *De Motu Corporum, Liber primus*, as in M and E_1, followed by 'Artic. I. De Methodo [*altered from* continens Methodum] Rationum primarum et ultimarum cujus ope sequentia demonstrantur', as in M, and as in E_1 with 'Artic.' changed to 'Sectio'.[7]

The first twenty leaves of *LL* (those numbered from 4 to 23 by a modern librarian), or $LL_\beta(1)$, include everything in E_1 from page 1 to the end of the proof of Lemma VI {31.14}. These are the twenty pages that replace an earlier set of eight pages, which would have been numbered by Newton from 1 to 8.[8] There can be no doubt about these pages having been pages 1–8, for next comes Lemma VII beginning on a page originally given the number 9, which Newton subsequently altered to 21. But the succeeding page is still numbered 10 in Newton's hand, followed by 11, 12, up to 32.

Generally speaking, LL_α corresponds to M up to the middle of page 25/[37], where Prop. XVII, Prob. VIII, is mistakenly numbered as Prop. XVI, Prob. VIII, even though this number had just been used for the previous proposition on page 27/[36]. The remainder consists of the propositions in the following list, a variant of the previous list in Chapter IV, §2, in which I give both the numbers assigned in LL_α and those in M and E_1:

Prop. XVII, Prob. IX	= Prop. XVIII, Prob. X: M and E_1
Prop. XVIII, Prob. X	= Prop. XIX, Prob. XI
Lemma XV	= Lemma XVI
Prop. XIX, Prob. XI	= Prop. XXI, Prob. XIII
Prop. XX, Prob. XII	= Prop. XXX, Prob. XXII
Prop. XXI, Prob. XIII	= Prop. XXXII, Prob. XXIV
Prop. XXII, Theor. IX	= Prop. XXXIII, Theor. IX
Prop. XXIII, Theor. X	= Prop. XXXIV, Theor. X
Prop. XXIV, Theor. XI	= Prop. XXXV, Theor. XI

Manuscript LL_α ends in the middle of the first sentence of the proof of this Proposition.

Of course LL_α did not originally break off in the middle of a sentence. I have identified another eight pages numbered by Newton 41–8, which apparently

[7] See Chapter V and Newton's letter to Halley of 2 June 1686, 'The Articles are wth y[e] largest to be called by that name. If you please you may change y[e] word to *Sections*, thô it be not material.' [Newton], *Correspondence* (1959–), vol. 2, p. 437.

[8] For another source of information concerning the original pages '1' to '8' now missing, see Supplement VII.

belonged to LL_α, which I have designated in Chapter IV, §2, as MS_x; and I have shown what the ten missing propositions must have been in $\langle LL_\alpha^{33-40}\rangle$.[9]

Following my reconstruction of the missing eight pages (33–40) and the identification of the succeeding set of eight pages (41–8),[10] we may list, as follows, the contents of the sixteen pages that at one time came directly after the present manuscript LL_α. First there was:

$\langle LL_\alpha^{33-40}\rangle$: Props. XXV, XXVI, XXVII, also designated Probs. XIV, XV and Theor. XII.

We have seen (Chapter IV, §2) that these may be identified as follows:

$\langle LL_\alpha{}^{3-40}\rangle$: Prop. XXV, Prob. XIV = Prop. XXXVI, Prob. XXV: *M* and E_1
Prop. XXVI, Prob. XV = Prop. XXXVII, Prob. XXVI
Prop. XXVII, Theor. XII = Prop. XXXVIII, Theor. XII

the reason being that Prop. XXVII of LL_α must be Theorem XII, and hence Prop. XXXVIII of *M* and E_1. Under these circumstances, there follows an unambiguous identification of what must have been Props. XXV, XXVI (Probs. XIV, XV) of LL_α. Then, there must have followed

$\langle LL_\alpha^{33-40}\rangle$: Prop. XXVIII, Theor. XIII = Prop. LVIII, Theor. XXI: *M* and E_1
Prop. XXIX, Theor. XIV = Prop. LIX, Theor. XXII
Prop. XXX, Theor. XV = Prop. LX, Theor. XXIII
Prop. XXXI, Theor. XVI = Prop. LXI, Theor. XXIV
Prop. XXXII, Prob. XVI = Prop. LXII, Prob. XXXVIII
Prop. XXXIII, Prob. XVII = Prop. LXIII, Prob. XXXIX
Prop. XXXIV, Prob. XVIII = Prop. LXIV, Prob. XL

after which comes another portion of LL_α (that is, LL_α continued, which I have found and identified in U.L.C. MS Add. 3965, ff. 7–14). These pages are numbered by Newton in the original sequence 41–48, and made up as follows:

$\langle LL_\alpha$ cont.$\rangle$: Prop. XXXIV, Prob. XVIII = Prop. LXIV, Prob. XL: *M* and E_1
Prop. XXXV, Theor. XVII = Prop. LXVI, Theor. XXVI
Prop. XXXVI, Theor. XVIII = Prop. LXVII, Theor. XXVII
Prop. XXXVII, Theor. XIX = Prop. LXVIII, Theor. XXVIII
Prop. XXXVIII, Theor. XX
Prop. XXXIX, Theor. XXI = Prop. LXX, Theor. XXX
Prop. XL, Theor. XXII = Prop. LXXI, Theor. XXXI
Prop. XLI, Theor. XXIII = Prop. LXXII, Theor. XXXII
Prop. XLII, Theor. XXIV = Prop. LXXXIII, Theor. XXXIII
Prop. XLIII, Theor. XXV = Prop. LXXIV, Theor. XXXIV

The proof of Prop. XXXIV, Prob. XVIII, began on the last page of the previous set of eight pages $\langle LL_\alpha^{33-40}\rangle$ but almost all of it is on page 41 (f. 7) of the present set. In Prop. XXXV (Prop. LXVI of E_1) there is present only the first of the twenty-two corollaries of E_1.[11] The Corollary to Prop. XXXVII (Prop. LXVIII) appears on these pages as a scholium. Proposition XXXVIII, Theor. XX of this manuscript

[9] It may be observed that the missing part is a set of eight pages; we have seen that these sets of eight pages are characteristic of *LL* throughout, save for the initial twenty pages of LL_α, which, however, replaced an earlier set of pp. 1–8.

[10] For details concerning MS_x, see Chapter IV, §2.

[11] But there is an intermediate version in U.L.C. MS Add. 3990, where Prop. XXXV of LL_α (= Prop. LXVI of E_1) has at least sixteen corollaries.

is replaced in M and E_1 by a wholly different proposition and three corollaries, but the ensuing scholium is largely the same; in these sheets it is rather completely rewritten by Newton between the lines of Humphrey's copy. This portion of manuscript breaks off just before the whole of the proposition has been stated.

There are several ways of guessing how far LL_α must have gone. The first is to examine the references in 'De motu corporum, liber secundus', the early draft of the 'System of the world' which eventually became Book III of the *Principia* (and for which see Supplements III and VI). In the *Principia* (M, E_1, E_2, and E_3), Book II deals with the motion of bodies in fluids, or resisting mediums.

At the stage of LL_α, Book II had not yet become the Book II on motion in fluids that we know today, but was then devoted to the application of Book I to the problems of the 'System of the world'. A small portion of Book I (LL_α) evidently contained some propositions on motion in fluids. Newton took this material from the end of Book I and enlarged it into a wholly new Book II, and then recast the old Book II as a Book III. Presumably this was done at the LL_β level, since we have evidence that would indicate that at one time Newton intended to have LL_β conclude with some material on motion in fluids and then decided to have Book I end more or less as in the *Principia.*

The references to LL_α in *De mundi systemate* (*lib.* 2us) are tabulated below, and they both supplement and corroborate the information presented in this Supplement and in Chapter IV, §2. In the list printed on page 314 the final entry was

Prop. XXIV (LL_α) = Prop. XXXV (LL_β).

Hence, we should expect that Prop. XXV of LL_α would be Prop. XXXVI of LL_β; since the propositions of LL_β are the equivalents of propositions in M and E_1, this should also be Prop. XXXVI of the *Principia.* That this is indeed the case may be seen in the following table, listing all references to propositions given in *De mundi systemate* (*lib.* 2us), and giving their equivalents in LL_β and the *Principia*, starting from Prop. XII/Prop. XIII.

LL_α	LL_β and *Principia* (Book I)
Prop. XII	= Prop. XIII
Prop. XIV	= Prop. XV
Prop. XV	= Prop. XVI
Prop. XXV	= Prop. XXXVI
Prop. XXVII	= Prop. LVII
Prop. XXX	= Prop. LX
Prop. XXXV	= Prop. LXV
Corol. 3, Prop. XXXV	= Corol. 3, Prop. LXV
Prop. XXXVI	= Prop. LXVI
Corol. Prop. XXXVIII	= Corol. Prop. LXVIII
Prop. XXXIX	= Prop. LXIX
Prop. XLIII	= Prop. LXXIII
Prop. XLIV	= Prop. LXXIV
Prop. XLVI	= Prop. LXXVI

There are also two references to 'Schol. Prop. XII'. These may be identified as Corol. Prop. XIII and Corollaries, Prop. XLV. Furthermore, there is a reference to a 'Prop. LXXII' which turns out to be identifiable as Prop. XXII, Book II. Since Props. LXXIII, LXXIV, and LXXVI of the *Principia* (and LL_β) occur in Sec. XII of Book I, LL_α would have included (with omissions) material corresponding to this section. But we have no way of telling whether there would also have been any parts of the remaining two sections (XIII and XIV).

At the stage of LL_α, there was no separate book on motion in resisting mediums, the Book II that we know in the *Principia*. Hence any discussion of this topic would have had to appear as a final section (or 'article') in Book I or LL_α. Since in the prior *De Motu* the concluding Problems 6 and 7 are of this sort, we may guess that Newton would have contemplated including them (at least) in his 'Liber primus', and possibly even more of what was to become Sec. I of Book II of the *Principia*. Apparently, as mentioned above, there was also in LL_α a Prop. LXXII, equivalent to Prop. XXII of Book II of the *Principia*. Between Prop. XLVI and this Prop. LXXII, there would have been Props. XLVII–LXXI, but we have no sure and simple way of telling which of the propositions at the end of Book I or the beginning of Book II of the *Principia* would correspond to these twenty-five propositions of LL_α.

I have said earlier that the manuscript which I have called LL_β is a later and revised version of LL_α, no doubt contemporaneous with the first twenty pages of the present *LL*. This part of *LL*, called LL_β, resembles LL_α in form, in that it too consists of a finished manuscript, written on the recto side of the page only, in Humphrey Newton's hand, with corrections made by the author on these pages and occasionally on the verso pages, which are otherwise blank. These pages are numbered in eights, as we have seen was the case for some of LL_α and for Newton's renumbering; the numbers are in Newton's hand.

LL_β begins at the top of a page numbered 37 by Newton, and containing Corols. 6, 7, 8, 9 to Prop. XVI, Theor. VIII. I believe on two grounds that there were never any earlier pages of LL_β—save, of course, the above-mentioned first twenty pages, now called $LL_\beta(1)$. First, the text of $LL_\beta(1)+LL_\alpha$ from the Definitions, Laws of Motion, and Book I through the end of Corol. 5 to Prop. XVI is in order of contents just like M and E_1; hence LL_β begins where the final version (M and E_1) begins to diverge from LL_α. That is, Newton apparently replaced the original first eight pages of LL_α by the present first twenty pages, revised the remaining material through Corol. 5 to Prop. XVI, and then rewrote the remainder. Second, I believe it is quite clear that Newton intended LL_β to begin right after Corol. 5 to Prop. XVI, since, as mentioned above, he gave the page beginning with Corol. 6, which is the first page of LL_β, the number 37. The manuscript breaks off in the middle of the third sentence of the proof of Prop. LIV, Prob. XXXVI; obviously, there must have been more to LL_β at some time.

Clearly, had Newton not included in *LL* the last eight pages of LL_α (that is, those leaves numbered consecutively from 25 to 32 in his first enumeration, of which leaf 25 has also the number 37), then $LL_\beta(1)+LL_\alpha+LL_\beta$ would make a continuous version of the Definitions, Laws, and most of Book I, that is all but the same as *M*, a version that with a few improvements was ready to be copied out in a fair copy for the printer. The appearance of a set of *disjecta membra* was perhaps not due to Newton at all, but may very well have been the fault of a careless librarian who sent Newton's manuscript to the binder in an arbitrary and illogical grouping of eights, or who never checked the bound volume to make certain the pages were in proper order. There is even some evidence that possibly the extra eight pages of LL_α were never intended by Newton to be a part of *LL* at all, for the first page (25/[37]) is dirty, as if it had been exposed, as might have been the case if these eight pages had once been separate from the rest, so that this front page would not have been protected in the manner that it would have been if it had been an inside page of the whole sequence.

I remind the reader (see Supplement III) that we must not assume that *LL* is a reliable index to the actual text of lectures and so a source of dated knowledge of Newton's progress at given times. Newton's carelessness shows itself in the inclusion of these extra pages, in the abrupt termination of the last page of LL_β in the middle of a proof (nay, in the middle of a sentence!). Furthermore, these eight extra pages of LL_α contain postils allegedly indicating that they comprise the end of Lect. 7, the whole of Lect. 8, and the beginning of Lect. 9. The last item in LL_α is the start of a 'Prop. XXII'. Newton indicated in LL_β that 'Lect. 1' in the next series, starting 'Octob. 1685', began with 'Prop. XXII'.[12] But these are very different propositions, despite their having the same numbers. Proposition XX (Prob. XII) of LL_α is Prop. XXX (Prob. XXII) of LL_β (and of *M* and E_1) and was allegedly given in Lect. 8 of the first series and repeated in Lect. 3 of the second. Propositions XXI, XXII, and XXIII of LL_α are Props. XXXII, XXXIII, and XXXIV of LL_β (and of *M* and E_1) and were allegedly a part of both Lect. 9 of the first series and Lect. 5 of the second.[13]

This analysis confirms the general impression that these sheets of *LL* do not

[12] Herivel has noted, quite correctly, that 'the second set [of Lecture numbers] was evidently numbered 10, 11, 12, ... originally in place of 1, 2, ...'.

[13] The 'lectures' 'De motu corporum' for 1685 (LL_β) were numbered at first in the right-hand margin, in arabic numerals with no word 'Lect.' preceding each number. This set of numerals runs from 10 to 19, making a continuous sequence with the nine 'Lectures' (1–9) for 1684 (LL_α). When Newton altered these designations '10'–'19' to 'Lect. 1'–'Lect. 10' he did not keep the same divisions of LL_β into 'Lectures'. Thus (using the modern librarian's folio numbers), on f. 43 there appears to have been originally a small '10' changed to read 'Octob. 1685 Lect. 1'; on f. 86 the '2' in 'Lect. 2' is written over an original '11'; on f. 65 a small uncancelled '12' is still found in the margin, and on the following leaf (93 according to the librarian) a small '12' is cancelled and replaced by 'Lect. 3'; on f. 96 there is still a small uncancelled '13', while on f. 98 no small number appears at 'Lect. 4'; on f. 101 a '14' is cancelled and replaced by 'Lect. 5'; on f. 36 (librarian's, be it remembered) the '6' in 'Lect. 6' is written over a small '15'; on f. 39 the '7' in 'Lect. 7' replaces a small '16'; on f. 83 an '8' similarly replaces '17'; on f. 72 a '9' replaces '18'; and on f. 76 a '10' replaces '19'.

contain the texts of lectures given as indicated. It is clear, I believe, that—as required—Newton rewrote his lectures, actually composing a treatise 'De motu Corporum', and merely handed over the penultimate draft, arbitrarily and even carelessly divided into alleged lectures. That this was so seems further indicated by the fact noted by Herivel, that such phrases as 'unde caveat Lector' and 'Hunc ...in finem Tractatum sequentem composui' are those of a treatise and not a series of lectures.[14]

One or two examples[15] from the overlapping parts of LL_α and LL_β will show that LL_β is not only an expanded but a revised (and improved) version. Corollary 6 to Prop. XVI in LL_α contains three corrections in Newton's hand: (1) following the words 'dimidiata ratione distantiae' Newton inserted 'corporis ab umbilico figurae' {62.21}; (2) Newton crossed out the final four words 'quàm in hac ratione' following 'Hyperbola major' {62.22}; (3) in the parenthetical reference {62.22–3} Newton altered '(per Lemma)' to '(per Cor. 2, Lem XIV)'. In LL_β the text incorporates these alterations in the copy. In LL_α, following the second Prop. XVI (that is, Prob. VIII), which is Prop. XVII in LL_β, *M*, and E_1, Newton added Corols. 3 and 4 on the facing verso page. These two corollaries appear within the text page of LL_β. Again in the statement of Prop. XVII in LL_α (Prop. XVIII in LL_β, *M*, and E_1), Newton added the words 'transversis et umbilico' following 'Datis axibus'. In LL_β, this is therefore copied by Humphrey as 'Datis axibus transversis et umbilico...'. But then Newton himself altered this to read {66.20} as in *M* and E_1, 'Datis umbilico et axibus transversis...'.

The relation between LL_α and LL_β may be seen in yet another example: Prop. XVI, Prob. VIII of LL_α, an erroneous number for Prop. XVII, Prob. VIII, which is the number assigned in LL_β, *M*, and the printed editions. In LL_α, this proposition reads as follows:

Posito quod vis centripeta sit recipro ceproportionalis quadrato distantiae a centro et cognita vis illius quantitate; requiritur linea quam corpus describet, de loco dato cum data velocitate secundum datam rectam emissum.

Supposing that the centripetal force is reciprocally [that is, inversely] proportional to the square of the distance from the centre, and that the quantity of that force is known; it is required [to find] the line [that is, curve] which a body will describe, [when] started off from a given place with a given velocity along a given right line.

14 Herivel, the only author since Edleston who has had anything new to say about *LL*, has treated this material somewhat differently. He divides what I have called *LL* into 'four sections α, β_1, β_2, γ'. Although he does not define the limits of these four parts, I believe that his 'α' corresponds to my $LL_\beta(1) + LL_\alpha$ up to the last eight pages of LL_α, beginning with p. 25/[37]. His 'β_1' is apparently these eight pages, and his 'β_2' is the beginning of what I have called LL_β, his 'γ' being the remainder. After stating that either 'β_1 or β_2 could have been the original continuation of α', he gives some reasons for favouring 'β_1', but not including the evidence of the original system of numbering on pp. 25/[37], 26, 27, 28, 29, 30, 31, 32, which puts the question so out of doubt that to separate LL_α thus into 'α' and 'β_1' is wholly misleading; it is equally misleading to divide LL_β into 'β_2' and 'γ', since it is a continuous manuscript.

15 Others have been given in Chapter IV, §2.

The translation of 'emissum' by 'started off' (in the sense of being 'projected' or 'sent off') is in keeping with the physical sense of the proposition; the word 'emitted' is not very satisfactory, however literal, and I do not like Motte's 'a body... that is let go'. Newton himself was not sure that 'emissum' was the best choice of word, for over it (in LL_α) is written in his own hand the alternative word 'migrans', even though he did not cross out 'emissum'. In LL_β, Humphrey copied out this version exactly, keeping the original word 'emissum'. In LL_β Newton then introduced some alterations in his own hand. First, 'et cognita vis illius quantitate' was altered so as to read 'et quod vis illius quantitas sit cognita'; then, the final 'emissum' became 'egrediens'. This is the version that Humphrey next copied in *M*. But after *M* had been written out, Newton added in his own hand the word 'absoluta' after 'vis illius quantitas' (as may be seen in our Apparatus Criticus {63.12–13}), and also added this in LL_β so as to make LL_β a conforming copy, or copy of record.

This example shows how *M* is related to LL_β, being generally no more than a fair copy. The relation of LL_β to LL_α may be shown by an alteration on the same respective pages 37 of LL_α and LL_β. Corollary 7 to Prop. XVI (Theor. VIII) originally began with the following sentence in Humphrey's hand:

In Parabola velocitas ubique est ad velocitatē corporis gyrantis in circulo ad eandem distantiam, in dimidiata ratione numeri binarii ad unitatem, in Ellipsi minor est, in Hyperbola major.	In a parabola, the velocity is everywhere to the velocity of a body gyrating in a circle at the same distance, in the dimidiate ratio [square root] of the number two to unity, in an ellipse it is less, in an hyperbola greater.

Then Newton introduced two alterations. First he changed 'gyrantis' to 'revolventis' and then added a final phrase, 'quam in hac ratione' ('than according to this ratio'). This is the form in which Humphrey copied out the sentence in LL_β and as such it appears in *M*. Curiously enough, Newton later decided to omit the added phrase 'quam in hac ratione', but he crossed it out in LL_α rather than in LL_β so that it appeared in *M* and accordingly was printed in E_1 and then remained in all later editions of the *Principia*.[16]

I have indicated earlier in this Supplement the basis for conjecturing how far into Book I (and Book II) of the *Principia* LL_α may have extended. We have better evidence concerning the limits of LL_β. Since *M* was copied from $LL_\beta(1)$, pp. 21 ff. of LL_α, and LL_β, we may guess that LL_β most likely would have continued at least as far as the end of Book I as we find it in *M* and E_1. Evidence to support such a conjecture is to be found in some odd pages of LL_β (U.L.C. MS Add. 3970, ff. 428*b*, 615–17), which correspond to the concluding portion (end of Sec. XIV) of Book I of the *Principia*. (These pages have been discussed in Chapter IV, §10.)

At one time, Newton evidently still contemplated introducing into Book I of LL_β a portion (however small) of what we know today as Book II of the *Principia*.

[16] There are a number of such examples in the overlapping parts of LL_α and LL_β; but I have given only these few instances to show how these two stages of Book I are related to each other and to *M*.

This important inference concerning LL_β may be drawn from the final sentence at the bottom of a rejected page (U.L.C. MS Add. 3970, §9, f. 617[r]), 'Hactenus exposui motus corporum in spatiis liberis'. This cancelled sentence, opening a new paragraph, could have two possible implications. One is that Newton, having dealt in Book I with motion in free spaces, will now go on to motion in resisting spaces in a separate Book II; alternatively, he may have intended to proceed in Book I to motion in spaces where there may be resistance. And, indeed, these are the same words that are antecedent to the Scholium to Prob. 5 of *De Motu* which (MS *D*: U.L.C. MS Add. 3965, §7, f. 50[r]) serves to introduce the two problems (6, 7) which are there devoted to this subject. We may thence conclude that most probably at this stage these two propositions (or possibly three: *De Motu* Prob. 6 = Prop. II, Book II, *M*, E_1; *De Motu* Prob. 7/Scholium = Props. III/IV, Book II, *M*, E_1) were intended to be part of LL_β, possibly together with other propositions of Book II as we now know it, such as Prop. XXII, Book II. These two propositions, according to the numerical sequence on ff. 615–17, would have been numbered XCIX and C. (For further information see Chapter IV, §10.) Newton did not for long continue with his plans to include these propositions in Book I (LL_β). At the time of preparing *M*, they were assigned to a separate and new Book II. Thus LL_β included at least all fourteen sections of Book I of *M* and E_1, and—at least for a moment—also a small portion of Book II.

SUPPLEMENT V

THE RESISTANCE OF SPHERICAL BODIES

On f. 128^v of MS Add. 3965, Newton (see Chapter IV, §5, and Plate 9) has worked out the beginning of the general topic defined in the following opening statement:

Corporum Sphaericorum in Mediis quibuscunque fluidissimis resistentiam ex inertia [*written* iniertia] materiae fluidi oriundam determinare.

To determine the resistance of spherical bodies in any very fluid mediums arising from the inertia of matter of the fluid.

But only the first short paragraph deals with so general a topic, beginning:

Si globus uniformi cum motu in directum movetur quantitas fluidi quod versus posticas globi partes [movetur] ad spatium a globo relictum implendum movetur: aequalis est quantitati spatii cylindrici quod globus circulo suo maximo ad lineam motus sui perpendiculari describit.

If a globe moves straight forward with a uniform motion the quantity of fluid which moves toward the rear portion of the globe to fill the space left by the globe is equal to the quantity of cylindrical space which the globe describes by its great circle perpendicular to the line of its motion.

At once, however, Newton limits the discussion to spindle-shaped[1] or fusiform bodies, which he defines as follows:

Fusum vel corpus fusiforme voco quod planum arcu et chorda comprehensum rotando circa chordam illam generat.

I call a spindle or a spindle-shaped [fusiform] body one which a plane surface comprehended by an arc and [its] chord generates by rotating about that chord.[2]

The property of the motion of such bodies in a resisting medium is given in the following statement:

Corporis fusiformis [*written above* circularis, *which is thus cancelled*] secundum longitudinem [*written* longitudiem] axis sui in medio fluidissimo progredientis resistentia aequalis est vi qua totus corporis motus generari posset quo tempore maximus corporis circulus longitudinem cylindri corpori aequalis describere posset.[3]

The resistance of a spindle-shaped [circular] body proceeding along the length [i.e. in the direction] of its axis in a highly fluid medium is equal to the force whereby the whole motion of the body could be generated in the time that the great circle of the body could describe the length of a cylinder equal to the body.

There follow a series of statements, without proof, which are heavily worked over. Newton has inserted 'Lem. 1' before the first of these, but has not similarly

[1] In a letter to Collins (20 August 1672), Newton writes in English about such a body as a 'spindle'. See [Newton] (1959–), *Correspondence*, vol. 1, p. 230.

[2] Earlier on the same page we find: 'ffusum voco solidum quodvis quod rotatione arcus cujuscunque circa chordam suam ceu axem generatur.' That is, 'I call a "fusus" any solid which is generated by the rotation of any arc about its chord as an axis.'

[3] I have omitted some cancelled parts of this passage.

added numbers to the others. Following Lemma 1, there is a construction in which Newton defines the 'diaphragm' of a spindle-shaped body, but it is not entirely clear what he intends. At first reading, the impression is confusing since Newton first states an equality between the motion of the spindle-shaped body in one direction and of the fluid in a contrary direction; then he declares that the total motion of the fluid is greater than the motion of the spindle-shaped body. Attention must be paid here to the word 'total'. I print these in full, in their ultimate form (that is, without indicating all cancellations and insertions), so that they may be compared with a more general version of the same statements—thus showing the progress made by Newton from one state to another. For instance, in the first version Newton does not introduce densities, as he does in the second.

Lem. 1. Motus corporis fusiformis secundum longitudinem axis sui AB in fluido ejusdem secum densitatis progredientis, & motus fluidi in partes contrarias, aequantur inter se.

Lem. 1. The motion of a spindle-shaped body proceeding lengthways along its axis AB in a fluid of the same density as itself, and the motion of the fluid in the opposite direction, are equal among themselves.

Sit ADBE ffusus quem planum ACBD arcu ADB et chorda ACB rotando circa chordam ACB describit. sitque DE diameter circuli maximi illa rotatione geniti et C centrum hujus circuli. Progrediatur fusus ab A versus B secundum longitudinem chordae AB quae Axis est ffusi et tempus dividi intelligatur in partes aequales innumeras & quamminimas: Sit Cc longitudo quam centrum circuli maximi unica temporis parte describit, et per punctum c agatur recta ecd circuli maximi diametro ECD parallela et ffusi superficiebus AEB et ADB hinc inde in e ac d occurrens. Et ffusi pars DEedD quae circulo maximo DE, et circulo de centro c intervallo cd vel ce descripto, continetur, dicatur ffusi Diaphragma.

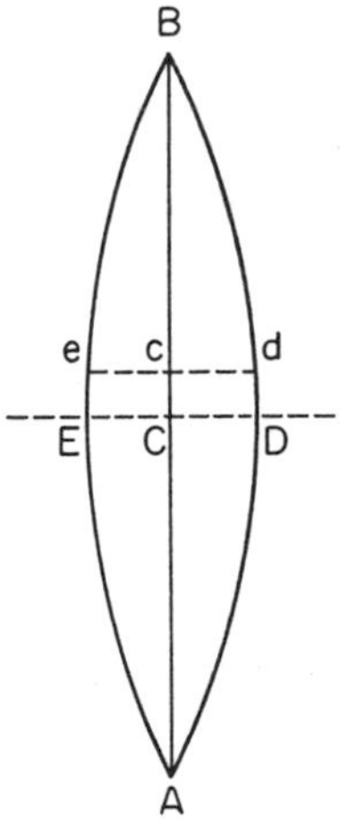

Let ADBE be a spindle which the plane ACBD[4] by the arc ADB and the chord ACB describes by rotating about the chord ACB. And let DE be the diameter of the great circle produced by that rotation and C the centre of this circle. Let the spindle proceed from A toward B along the length of the chord AB which is the axis of the spindle and let the time be understood to be divided into innumerable least possible equal parts. Let Cc be the length which the centre of the great circle describes in a single part of time, and through the point c let the right line ecd be drawn parallel to the diameter ECD of the great circle and meeting the spindle's surfaces AEB and ADB on the one side in e and on the other in d. And let the part DEedD of the spindle which is contained between the great circle DE, and the [small] circle de described about the centre c with the radius cd or ce, be called the Diaphragm of the spindle.

Motus totus ffluidi major [*changed from* paulo major] est quam motus ffusi sed non triplo [*changed from* duplo] major.

The total motion of the fluid is greater than the motion of the spindle-shaped body, but not thrice greater.

[4] The sense is clearer if we assume that 'comprehensum' (or 'comprehended') is to follow 'chorda ACB' (or to precede 'by the arc ADB').

Motus ffluidi oritur ex majori pressione ante ffusum quam post, et maximus est in plano circuli maximi infinite producto.

The motion of the fluid arises from the pressure in front of the spindle-shaped body being greater than that behind it, and is greatest in the infinitely extended plane of the great circle.

Resistentia qua motus totus ffluidi & ffusi singulis temporis partibus diminuitur aequalis est vi quae motum Diaphragmatis eadem temporis parte vel generari vel tolli posset [*this should have been either* qua motus *or* generare vel tollere].

The resistance whereby the total motion of the fluid and the spindle-shaped body is in individual parts [instants] of time diminished is equal to the force whereby the motion of the Diaphragm could be either generated or destroyed in the same part of time.

Motus corporis singulis temporis partibus amissus est ad motum Diaphragmatis ut motus ffusi solius ad motum totum ffusi et fluidi, id est minor quam 1 ad 2 major quam 1 ad 3.

The motion of the body lost in individual parts of time is to the motion of the Diaphragm as the motion of the spindle-shaped body alone to the total motion of the spindle-shaped body and the fluid, that is less than 1 to 2, greater than 1 to 3.

The facing page of the other leaf of this same sheet (f. 129^{r}) has a more formal presentation, in which each statement is numbered, either as a lemma or as a corollary. This group of lemmas and corollaries is more general, dealing with any solid ('Si solidum quodvis...') and not merely spindle-shaped bodies, although the original version of Lemma 1 had 'fusus' or 'fusus solidus' throughout for 'solidum'. Furthermore, the actual statements are more precise. Yet it is to be noted that there are no proofs or explanations, merely declarations that are far from obvious, possibly a program to be developed. The whole text reads as follows:[5]

Lemma. 1. Si solidum quodvis in ffluido quocunque progrediatur, motus solidi est ad motum fluidi in partes contrarias factum, ut densitas solidi ad densitatem fluidi.

Lemma 1. If any solid proceed in any fluid, the motion of the solid is to the motion of the fluid effected in the opposite direction, as the density of the solid to the density of the fluid.

Lem. 2. Si solidum quodvis in ffluido quocunque progrediatur motus ffluidi in partes contrarias oritur ab excessu pressionis partium fluidi ad partes anteriores solidi supra pressionem fluidi ad partes posteriores ejusdem solidi.

Lem. 2. If any solid proceed in any fluid the motion of the fluid in the opposite direction arises from the excess of pressure of the fluid's parts [*or* matter] on the front portion of the solid over the pressure of the fluid on the rear portion of the same solid.

Corol. 1. Igitur pressio partium fluidi ad anteriores partes solidi progredientis semper major est pressione partium fluidi ad posteriores partes solidi.

Corol. 1. Therefore the pressure of the fluid's parts on the front portion of the solid as it advances is always greater than the pressure of the fluid's parts on the rear portion of the solid.

Corol. 2. Et excessus pressionis tantus est quantus sufficit ad motum partium fluidi in partes contrarias generandum.

Corol. 2. And the excess of the pressure is just sufficiently great to generate the motion of the fluid's parts in the opposite direction.

Lem. 3. Si solidum quodvis in ffluido quocunque progrediatur, resistentia qua solidum

Lem. 3. If any solid should advance in any fluid, the resistance whereby the solid is re-

[5] Again, just the final version is given, without indication of insertions, alterations, or cancellations. I have supplied some missing periods.

retardatur oritur ab excessu pressionis partium fluidi ad anteriores solidi partes supra pressionem partium fluidi ad posteriores partes solidi, & huic excessui proportionalis est.

tarded arises from the excess of the pressure of the fluid's parts on the front portion of the solid over the pressure of the fluid's parts on the rear portion of the solid, and is proportional to this excess.

Lem. 4. Resistentia solidi progredientis aequalis est vi qua motus fluidi in contrarias partes generatur. Nam vis illa oritur ex reactione corporis cui resistitur.

Lem. 4. The resistance of the advancing solid is equal to the force whereby the motion of the fluid in the opposite direction is generated. For that force arises from the reaction of the body which is resisted.

Lem. 6 [*changed from* 5]. Motus solidi singulis temporis momentis ex resistentia amissus est ad motum motum [*sic*] fluidi iisdem momentis genitum, ut motus totus solidi ad motum totum solidi et fluidi conjunctim. Nam solidum retardari non potest nisi fluidum simul retardatur.

Lem. 6. The motion lost by a solid in each single moment of time from resistance is to the motion of the fluid produced in the same moment as the total motion of the solid to the total motion of the solid & fluid together. For the solid cannot be retarded unless the fluid is retarded simultaneously.

Lem. 5 [*changed from* 6]. Motus fluidi major est quam motus solidi sed non triplo major.

Lem. 5. The motion of the fluid is greater than the motion of the solid but not three times greater.

Prop.

Prop.

In attempting to identify and hence to date these fragments, we must keep in mind that nowhere in Book II does Newton discuss such general problems. Sections VIII and IX deal respectively with motions (pressures or waves) 'propagated through fluids' and the 'circular motion of fluids'; only Secs. I–VII are devoted to resistance to motion in fluids. But each of Secs. I, II, and III is devoted to a separate and special case of resistance, which may be proportional to the velocity (Sec. I) of the moving body, to the square of the velocity (Sec. II), or partly to the velocity and partly to the square of the velocity (Sec. III). Next, Sec. IV presents the 'circular motion of bodies in resisting mediums', while Sec. V introduces the subject of 'hydrostatics', in considerations of the 'density and compression of fluids'. Section VI, on the 'motion and resistance of pendulous bodies', is followed by the only general treatment of 'motion of fluids, and the resistance made to projected bodies' (Sec. VII). The contents of the propositions and lemmas printed in the present Supplement V indicate that they were intended as a revision either of the original Sec. VII of M and E_1 or of the new Sec. VII of E_2 and E_3, or possibly were part of an independent exercise on the very same subject. The greater degree of generality rules out the possibility that these could have been preliminary to M or E_1.

It is, however, extremely difficult to find any way of telling whether Newton produced these statements just before E_2 or just after E_2, when (in 1715–16) he was hard at work on a revision, still carried forward by the momentum of the job of preparing E_2. One argument in favour of the later date is the presence of the discussion of spindle-shaped or fusiform bodies, something new and not part of the known preparation of E_2. Also favouring the later date is the greater degree of generalization. The sequence of lemmas listed above deals with bodies in general, whereas in Sec. VII in E_2 Newton deals primarily with the resistance of fluids to the motions of globes and cylinders, with a few notable exceptions: the scholium following Prop. XXXIV on the solid of least resistance, Lemma V (which introduces a spheroid 'of equal breadth' with a given cylinder and sphere), and the scholium following Lemma VII (applying to 'all convex and round bodies' of a given size). Hence these pages (ff. 128–9) may very well represent a final attempt, after the completion of E_2, to generalize the results that he had given in E_2, notably in the corollaries to Prop. XXXV, Book II.

These texts were never completed—at least to Newton's satisfaction—and have never been published until now. On purely internal grounds, therefore, these attempted lemmas seem to us like an essay at generalization following the completion of the new text of Sec. VII but before (rather than after) the printing of E_2. In any event, the sequence of presentations shows dramatically the stages of Newton's accelerating maturity in dealing with the motion of bodies in resisting mediums.

I do not mean to imply that Newton was ever as much the master of the mechanics of fluids, or of motion in fluids, as he was of non-energetic particle dynamics. Professor Clifford Truesdell has very properly called our attention to Newton's limitations and failures in dealing with this subject. From any point of view—the physical nature of the assumptions, the scope and usefulness of concepts, the applicability of refined mathematical methods, or the harmony with actual experimental data—these lemmas and corollaries show us by how far Newton had surpassed his predecessors and contemporaries. In the degree of generalization for which he was striving, Newton gives us a gauge of the progress he had made in his treatment of fluids between 1684 and 1715.

SUPPLEMENT VI

NEWTON'S 'SYSTEM OF THE WORLD'

IN Chapter IV, §6, it was stated that some fourteen sections of *De mundi systemate* (*lib. 2us*) were taken over, practically *verbatim et litteratim*, into the third book of the *Principia*, namely, *De mundi systemate* (*lib. 3^{ius}*). The Latin text, printed from Newton's manuscript (U.L.C. MS Add. 3990), was first published in London in 1728 (Impensis J. Tonson, J. Osborn & T. Longman) and again in 1731 (Impensis J. Tonson, J. Osborn & T. Longman, T. Ward & E. Wickstead, & F. Gyles), under the title '*De mundi systemate liber* Isaaci Newtoni'. The 1731 edition was reprinted in 1744 in volume 2 of the collection of Newton's *Opuscula* edited by Johannes Castillioneus (Giovanni Francesco Salvemini). The fourth printing of this tract in Latin was in 1782 in volume 3 of Newton's *Opera*, edited by Samuel Horsley, of which the whole five-volume set was reprinted in facsimile in 1964. In Horsley's version, for the first and only time, the sections were numbered throughout, each paragraph having a postil and being assigned a number in sequence from 1 to 78, followed by an unnumbered paragraph stating a problem in cometary motion[1] and its solution in Lemmas I–V and Probs. I, II. These same numbers are used in the version printed by Florian Cajori as an appendix to his revision of Motte's translation of the *Principia*. In my discussion below, I shall use these numbers and the pages of the first edition, so that the reader may conveniently identify each passage; the printer indicated each eighth page of the first edition and the signature letter (for example, 'G. Prim./41') on the margin of Newton's manuscript and marked the beginning of each page by a symbol ⊏ right in the line of text, so that the portion of MS Add. 3990 corresponding to any given page of the first edition may be readily found. The existence of MS Add. 3990, partly in Newton's hand, may silence forever Augustus De Morgan's allegation, singled out for approval by David Eugene Smith: 'I greatly doubt that Newton wrote this book.'[2]

I shall not discuss here the many interesting aspects of Newton's manuscript, since this would take us far afield from the *Principia* as such. (I have undertaken a critical edition of the Latin and English versions of this work, with an analytical commentary.) But there are one or two features that must be noted in relation to the *Principia*. Newton's manuscript contains references to propositions and corollaries; for instance, in §38 (p. 42), the manuscript reads 'per Coroll. 19 Prop.

[1] 'The relation between the velocity of a comet and its distance from the sun's centre being given, the comet's orbit is required.'

[2] De Morgan (1954), *Budget of paradoxes*, vol. 1, pp. 139–40.

XXXVI'. In the printed version this has been altered to 'per Coroll. 19 & 20. Prop. lxvi'. In the previous paragraph the manuscript reference, 'in Coroll. 19, Prop. XXXVI', has been altered to 'in Coroll. 18, Prop. lxvi'. The significance of these changes is that they show us that the original manuscript *De mundi systemate* (*lib. 2us*) was written as a second book (*De motu corporum*) to LL_α rather than LL_β or M, since the numbering of the propositions here and elsewhere in MS Add. 3990 refers always to those of LL_α.[3] Reading the printed version, with altered references (even, as we shall see in a moment, including a reference to Book II), would make it appear that this work was written much later than was actually the case.

Another change of significance occurs in §4 (p. 5), where the manuscript reads:

> Qua autem ratione vires ex motibus et motus ex viribus colligendi sunt, copiosè expositum est in libro superiore.

The last two words have been crossed out and above them are written, in Newton's hand, 'Libris de Motu'. The printed edition contains this correction. In Cajori's version, this sentence reads:

> And how from the motions given we may infer the forces, or from the forces given we may determine the motions, is shown in the first two Books of our *Principles of Philosophy*.

The innocent reader would thus hardly suspect that the tract he was reading was written before Newton had even contemplated either a Book II or a general title other than *De motu corporum*! On the preceding page, §3, Cajori's version further emphasizes this by a reference, 'by Prop. I, Book I, *Princip. Math.*', corresponding to the printed Latin text 'per Prop. I. lib. I. *Princip. Math.*'. In the manuscript Newton had written 'per Prop I lib I'—having in mind *De motu corporum liber primus* or LL_α.

The extent to which Book III of M and the printed editions of the *Principia* derive from this *De mundi systemate* (*lib. 2us*) may be seen in the accompanying table, which lists those paragraphs or sections taken over almost identically in their entirety, and so noted in the manuscript copies belonging to Cotes and Morgan, in which these paragraphs were not even written out but only indicated, together with a reference to the published *Principia*. There are other parts of Book III that derive from *De mundi systemate* (*lib. 2us*), some being phrases and others more extensive, though not full paragraphs or sections, which are noted in the margin of the second English edition (London, 1731). The inclusive words and references to page and line of E_1 are given in the second and fifth columns of the table as they appear in the Cotes and Morgan copies.

There can be no doubt that these paragraphs from MS Add. 3990 are in fact anterior to the corresponding portions of Book III of the *Principia*, since there are corrections, alterations, and additions made in MS Add. 3990 that are incorporated into the text of Book III in M.

[3] On the difference, see Supplement IV.

Examples of large sections of 'De mundi systemate (lib. 2us)' incorporated into M

Folios in MS Add. 3990	Words included	Paragraph nos.: Newton's (Cajori's)	Page nos., 1st Eng. ed.	E_1	E_3
22	Motus medius...tardissimè retrogradum	35 (34)	58–9	428.15–429.12	{423.17–424.2–4}
24–25	Motus autem bini...à Quadraturis ad Syzygias	40 (39)	63–4	429 lin. penult.–430.15	{424.26–425.7}
25–27	Pendent autem effectus...non sint primi à Syzygiis sed tertii	41–4 (40–3)	65–70	430.18–432 lin. ult.	{425.9–427.14}
28	Porrò fieri potest...donec Luna iterum mutet declinationem	46 (45)	72–4	433.4–434.5	{427.18–428.14}
36	Idem colligitur...& inferiorum Planetarum	60 (59)	97	476.5–15	{480.13–481.3}
38	Haec disputavimus...Soli multò propior	63 (62)	102–3	477.22–478.6	{482.6–22}
41–42	Idem denique colligitur...observandi. Si sumantur...in vicinia Solis	65 (64)	110–13	478.12–479.7, 479.16–24	{482.28–483.19, 483.28–36}
42	Si cernerentur...illuminari solent	66 (65)	113–14	479.27–480.6	{484.3–15}
43	vel jubar esse Solis...deriventur	68 (67)	115–19	499.25–501.18	{509.23–511.9}
45–46	Ascensum caudarum...componitur	70 (69)	123–4	504.4 ab ima pag. –505.17	{514.10–29}
46	Caudas autem...terminabitur	71 (70)	124–7	501.19–502.21	{511.10 512.8}

I have referred above to the misleading aspects of Cajori's version. The particular instances that I cited may be traced back to the original English translation which he 'modernized'. The first English translation was published in 1728, the same year as the first Latin edition, but by another publisher, or bookseller, F. Fayram. A second edition was published by Fayram in 1731, reissued in 1737 (according to Gray) 'with a cancel title-page' as a third edition 'Printed for B. Motte and C. Bathurst'. In fact, I own a copy with a title-page also a cancel, described as 'The SECOND EDITION, Corrected and Improved'. It is said to have been 'Printed for J. Robinson'.[4] The text of the second edition was at least twice reprinted before Cajori's edition, in 1803 and in 1819.[5]

The first English edition was based on a Latin manuscript, as Gray surmised.[6] The translator makes this clear by two statements in his preface (which largely deals with the state of astronomy when Newton wrote this work). First, he says that to make this tract 'the more useful to our countrymen, we chuse to give it in a

[4] The title-page reads: *The System of the World, demonstrated in an easy and popular manner. Being a proper Introduction to the most Sublime Philosophy.*

[5] See Gray (1907), *Bibliography*; [Babson Collection] (1950), *Catalogue*, and supplement (1955).

[6] *Bibliography* (1907), pp. 15, 19. Gray, however, was not aware that in the same year (1728) there were published both a Latin and an English version. He misinterpreted the fact that in the second edition of the translation some additions were introduced 'so as to make the matter agree with the printed text'. We shall see below that these alterations consisted of adding references (to propositions) suppressed by the translator and restoring the original text, which the translator sometimes changed when he dropped the references. It is not correct, therefore, to conclude that this type of change 'points to the fact that the translation was made from a manuscript which slightly varied from the one used in the printing'.

careful translation rather than in the original Latin'. Evidently, the translator did not have a Latin printed edition before him, nor did he know one was in prospect. Second, the translator's preface concludes:

> The reader is desired to observe that the references to propositions cited in the original, are omitted, so that many things he will find herein asserted, must be taken upon trust. The reason of this is, that the great work of the Principia in which those propositions were contained, appears to have been put into a very different form after this Tract was written, and many things were added and altered therein. This alteration caused so much confusion in the citations, that it was thought best to leave them all out; one or two excepted which escaped our notice, and have crept in by oversight.

We have seen above that the discrepancy in citation arose from the fact that this was a 'Liber secundus' to LL_{α}, and that the propositions bear the same numbers in MS Add. 3990 as in LL_{α}. But in the printed Latin editions, beginning in 1728, these numbers were altered to conform to the final version (LL_{β}, M, E_1, E_2, E_3). Had the translator had the printed edition of 1728 before him, and not the manuscript or a copy thereof, he would not have written the foregoing paragraph.

Sometimes, when the reference was simply a proposition number within parentheses, the translator could eliminate it without changing Newton's text. At other times, however, the translator had to alter Newton's own words. Thus we find:

> These are astronomical experiments from which it follows by geometrical reasoning, that there are centripetal forces...

But the manuscript contains no mention of 'geometrical reasoning', and reads simply:

> Haec sunt experimenta Astronomica et ex his per Propositiones [*altered by Newton to* per Libri primi Propositiones] tres primas et earum Corollaria consequens est quod dantur vires centripetae...

The printed edition in Latin agrees with the manuscript here, save in matters of punctuation. In the second English edition the reference to Props. I–III and their corollaries was restored, but now a specific reference to the *Principia* was added, so as to read:

> These are astronomical experiments from which it follows, by *Prop.* 1, 2, 3, in the first Book of our *Principles* and their *Corollaries*, that there are centripetal forces...

In a second example, we may see that the Latin printed edition and both of the printed English editions are different from the original manuscript. The two English versions read as follows:

First Edition (1728)	*Second Edition (1731)*
But if the Earth is supposed to move, the Earth and Moon together will be revolved about their common center of gravity. And the Moon will in the same periodic time...	But if the Earth is supposed to move, the Earth and Moon together (by *Cor.* 4. of the *Laws of Motion*, and *Prop.* 57.) will be revolved about their common center of gravity. And the Moon (by *Prop.* 60.) will in the same periodic time...

But Newton's manuscript contains quite different proposition numbers, as follows:

...(per Legum Coroll. 4, et Prop. XXVII)...Et Luna per Prop. XXX eodem tempore periodico...

In the printed Latin version, these references have been altered to 'per Legum Coroll. 4, & Prop. lvii' and 'per Prop. lx', so as to accord with the later numbering of propositions in *M* and the printed editions of the *Principia* rather than the early numbering of these same propositions in LL_α. The second edition thus differs from the first in the restoration of the references to propositions and the change of passages like those just displayed to a more accurate rendering of the Latin text. Since these references contain the numbers assigned by Newton in the *Principia*, rather than those occurring in MS Add. 3990 (the numbers of propositions in LL_α), the revised version still differed in this regard from Newton's manuscript. The preface of the translator was also suppressed, and in its place was put an extract from the beginning of Book III.[7] Two further changes were made. First, in the margin there were put:

References to the *English* Translation of the *Principia* lately published, directing to the several pages in that Translation, where the Things here treated of are proved.

This statement comes from a publisher's announcement, printed on a leaf facing the title-page. A second change, also mentioned in this announcement, was that some asterisks were placed in the margin here and there in order 'to shew how far the Discourse in the *Principia* is the same with what is found in this Treatise'.[8]

The extensive reliance on translations without consultation of the original Latin of Newton's works has produced a widespread attribution to Newton of a phrase and an expression which may not have been written by Newton at all, and may even have been a gratuitous insertion by the translator. I refer to Newton's supposed 'mathematical way',[9] in the following passage; I give the Latin original as found in the printed version (London, 1728), which differs from the manuscript (MS Add. 3990) only in minor details of printing, alongside the Cajori version (pp. 550–1), based on the English translation (London, 1728, 1731).

[7] '*From Sir* ISAAC NEWTON'*s Mathematical Principles*, Lib. III. Introd. In the preceding books I have laid down the principles of philosophy...It remains, that from the same principles I now demonstrate the frame of the SYSTEM of the WORLD. Upon this subject, I had indeed composed the third book in a popular method, that it might be read by many: but afterwards...I chose to reduce the substance of that book into the form of propositions (in the mathematical way) which should be read by those only, who had first made themselves masters of the principles established in the preceding books.'

[8] In the Cajori version, these asterisks were omitted and the page-references were altered to those of his own version of the *Principia*, and were inserted into the text line within parentheses (just like the proposition references) so that they appear to have been part of Newton's text. In §34, and elsewhere, some words and phrases appear in parentheses, in both the eighteenth-century printings and Cajori's. A close examination shows that these were not present in the original.

[9] The existence of such a discrepancy was found independently and called to my attention by Father S. J. Dundon, O.C.D.

Philosophi recentiores aut vortices esse volunt, ut *Keplerus* & *Cartesius*, aut aliud aliquod sive impulsus sive attractionis principium, ut *Borellius*, *Hookius*, et ex nostratibus alii. Ex motus lege primâ certissimum est vim aliquam requiri. Nobis propositum est quantitatem & proprietates ipsius eruere, atque effectus in corporibus movendis investigare mathematicè: proinde ne speciem ejus hypotheticè determinemus, diximus ipsam generali nomine centripetam, quae tendit in centrum aliquod; vel etiam, sumpto nomine de centro, circumsolarem, quae tendit in Solem; circumterrestrem, quae in Terram; circumjovialem, quae in Jovem; & sic in caeteris.

The later philosophers pretend to account for it either by the action of certain vortices, as *Kepler* and *Descartes*; or by some other principle of impulse or attraction, as *Borelli*, *Hooke*, and others of our nation; for, from the laws of motion, it is most certain that these effects must proceed from the action of some force or other.

But our purpose is only to trace out the quantity and properties of this force from the phenomena, and to apply what we discover in some simple cases as principles, by which, in a mathematical way, we may estimate the effects thereof in more involved cases; for it would be endless and impossible to bring every particular to direct and immediate observation.

We said, *in a mathematical way*, to avoid all questions about the nature or quality of this force, which we would not be understood to determine by any hypothesis; and therefore call it by the general name of a centripetal force, as it is a force which is directed towards some centre; and as it regards more particularly a body in that centre, we call it circumsolar, circumterrestrial, circumjovial; and so in respect of other central bodies.

The enormous discrepancy may be seen at once. I give here a very literal (and therefore awkward) translation of this same paragraph, in order to exhibit the liberties apparently taken by the translator:

Modern [more recent] philosophers want either vortices to exist, as Kepler and Descartes, or some other principle whether of impulse or attraction, as Borelli, Hooke, and others of our countrymen. From the first law of motion it is very certain that some force is required. Our purpose is to bring out its quantity and properties and to investigate mathematically its effects in moving bodies; further, in order not to delimit its type hypothetically, we have called by the general name 'centripetal' that [force] which tends toward some centre—or in particular (taking the name from the centre) [we call] circumsolar [that] which tends toward the sun, circumterrestrial [that] which [tends] toward the earth; circumjovial [that] which [tends] toward Jupiter; and so in the others.

In the Cajori edition, in which this tract is printed together with the *Principia*, the title of Book III of the *Principia* has been altered so as to read:

Book Three
SYSTEM OF THE WORLD
(IN MATHEMATICAL TREATMENT)

thus emphasizing a 'mathematical way'. The reader is not told that the phrase in parentheses does not occur in the original Latin text, or even in Motte's translation.

The discrepancy between the Latin and English versions could be the result of one of two factors: possibly the translator[10] took unwarranted liberties, or perhaps he had a manuscript that differs from Newton's in this regard. We are tempted by the latter hypothesis because in the passage just quoted we find the phrase 'from the laws of motion' for 'Ex motus Lege prima'; there is no earthly reason why this reference to 'the first Law of motion' should have been altered. Furthermore, the 'mathematical way' occurs in a portion of the text extensively rewritten by Newton. It just might be the case that in another copy he had made a different alteration.

The Latin paragraph just quoted may be found at the bottom of the first page and top of the second page of the manuscript, both of which are written out in Humphrey Newton's hand. Newton has added 'ut Keplerus et Cartesius' and 'ut Borellius ⌊Hookius⌋ et ex nostratibus alii'. The conclusion of this paragraph (MS Add. 3990, f. 2), as written out by Humphrey, originally read:

Vim aliquam requiri ex motus Lege prima certissimum est. Nobis vis illa ne speciem hypothetice determinemus generali nomine centripeta dicitur.

Then Newton altered and rewrote these two sentences, partly between the lines, so as eventually to produce the version printed in the Latin edition. In his revision of the first of these sentences, he simply changed the word order, beginning the sentence with 'Ex motus Lege prima', and making 'vim aliquam requiri' the conclusion. But the final sentence was considerably expanded from this initial state, in which Newton merely had said at first, 'In order not to delimit its type hypothetically, that force is called by us by the general name centripetal'. Newton crossed out 'dicitur' and wrote in 'diximus ipsam' after 'determinemus', changing 'centripeta' to 'centripetam' and then going on to make further alterations until the sentence achieved its final form.

There are two further examples that are of interest because of the divergence of the English translation from the text of MS Add. 3990 and the printed Latin version. In one of these (page 13 of the first and second English editions: 1728, 1731), there is a passage not found in the printed Latin edition, including a table (based on data sent to Newton by Flamsteed) concerning the 'dimensions' of the orbits of the satellites of Jupiter 'taken by the micrometer and reduced to the mean distance of Jupiter from the Sun, together with the times of their revolutions'. In the manuscript, this table and accompanying text have been cancelled by Newton, which is the reason for their omission in the printed Latin text; but they may not have been cancelled in the manuscript used by the translator.

[10] Cajori argues that the anonymous translator of *De mundi systemate* (*lib.* 2us) must have been Andrew Motte, whose translation of the *Principia* was published in 1729, a year after the translation of 'Newton's system of the world'. Cajori's argument is based on the identical rendering in both works of the word 'comatas' by 'the use of three words, "coma or capillitium"' and on the fact that 'there is a long Latin passage in the *System of the World* which is virtually identical with a Latin passage in Book III of the *Principia*. The translations of these passages into English are so nearly alike, that they most probably are attributable to one and the same translator.' Any such argument must necessarily be inconclusive, however, since Motte would have had plenty of time (perhaps a year) to see how the translator of *De mundi systemate* (*lib.* 2us) would have treated any problem-words or difficult phrases. Cajori was evidently unaware of how many long Latin passages are identical in the two works.

In the other example, the translator presents certain material (practically the whole of page 14 in the first and second English editions) in a form quite different from that of the Latin edition. The English text contains two tables of data which appear in the Latin version as a prose paragraph; there is also a pair of sentences altogether absent from the printed Latin version, reading: '*Cassini* assures us that the same proportion is observed in the circum-saturnal planets. But a longer course of observations is required before we can have a certain and accurate theory of those planets.' Here again, for the third time, we find that a difference between the Latin and English versions occurs where an alteration was made by Newton in his manuscript. The above-mentioned data, and the two sentences just quoted, are part of an afterthought of Newton's, which he entered in his own hand on an otherwise blank verso page of his manuscript; this presentation was, in fact, intended to replace the portion cancelled by Newton, which I have discussed in the preceding paragraph. I believe the evidence thus points again to the existence of a variant manuscript: one in which the addition made by Newton would occur in a form different from that which is found in his own copy, MS Add. 3990. For the English translation differs from Newton's manuscript not only with regard to the tabular form of the data; the two sentences just quoted in English about Cassini, and the need for further observations, prove to be an accurate translation of a conclusion which Newton actually wrote out in the addition to his own manuscript (MS Add. 3990), but which he later cancelled.

The two passages just discussed, and the presentation of Newton's 'mathematical way', are among the most striking differences between the English translation and the Latin version. In the first English edition, however, we have seen that the translator took certain textual liberties in the course of suppressing the misleading references to *LL*.[11] It would thus have been wholly in character for him to have embroidered a little on the simple statement found in MS Add. 3990 concerning Newton's philosophical position. But I doubt whether any translator would have ignored the author's cancellation of one passage and the author's form of a substitute, nor would he have printed the paragraph about Cassini's assurance that (Kepler's) harmonic law holds for Saturn's satellites and the need for further observations, had it been cancelled in the manuscript he was using. But until (or unless) we find a manuscript copy that resembles the English translation with regard to these features, we cannot say for certain who was responsible for the statement concerning the 'mathematical way', Newton or the translator.

There is, however, one further source of information regarding the translator's manuscript. It appears among miscellaneous letters and papers in U.L.C. MS Add. 4005, §4, f. 9, on a single piece of paper, marked 'Copy', and reads as follows:

[11] In the second edition of the translation (1731), an alteration was made of those passages which in the manuscript contain references to LL_{α}, but there was evidently no effort made to check the whole translation against the authorized Latin printed version.

Whereas a treatise of the system of the world said to be translated from a latin treatise on the same subject by the late S[r] Isaac Newton is lately printed for F. Fayram without the name of the translator, Wee the administrators of the late S[r] Isaac Newton do hereby give notice, that wee intend forthwith to publish a treatise of his in Latin on the same subject with several alterations & corrections from his original manuscript.

Tho. Pilkington

28. Sept. 1727—

John Conduitt
Ben. Smith

The first edition in English, published by Fayram, is dated 1728. The date of 1727, given in this document, thus appears to be a copyist's error for 1728, unless possibly the book had been available late in September of 1727, though dated 1728. This announcement or declaration may have the appearance of having been based on hearsay, to the degree that it contains the phrase 'said to be translated from a latin treatise'. But the preface to the first English edition actually does contain a declaration about the presentation 'in a careful translation rather than in the original Latin'. Furthermore, the executors' statement mentions specifically that the 'treatise' has been printed 'without the name of the Translator' and that it 'is lately printed', which would seem to imply that an actual copy had been seen. In any event, this document enables us to assert with conviction that the translator used a manuscript or manuscript copy other than Newton's own draft, then in the keeping of Conduitt. And, equally, this document shows that the Latin edition was published after the English translation had been printed.

SUPPLEMENT VII

A CRITIQUE (BY HALLEY?) OF A PRELIMINARY VERSION OF THE 'PRINCIPIA'

MENTION has been made earlier of evidence, from a critique almost certainly written by Edmond Halley,[1] of the existence of a version of the *Principia* in three books, from which Humphrey Newton copied out the manuscript *M* sent by Newton to Halley (and the Royal Society) for printing. The critique occupies ff. 94–9 of U.L.C. MS Add. 3965 (see Plates 10 and 11), constituting §9 of this manuscript, written on the recto side only,[2] plus an additional f. 93. The latter contains sixteen lines in Newton's hand, arranged into four paragraphs, giving corrections or emendations with reference to page and line, starting with 'p. 1. l. 9. aer duplo densior' {1.8} and ending at 'p. 40 l. 29. retrahitur a' {43.22}. Each entry proves to have been identified correctly as to page and line in E_1. This is hardly worthy of our notice, but no doubt accounts for the ambivalence of the cataloguers of the Portsmouth Collection, who have first described §9 as 'Proposed Corrections to 1st Edn of the Principia', and then have changed this to read, 'probably for Proposed Corrections 1st Edn of the Principia'. A note, later crossed out, contains the further information, 'No reference is here made to the pages, but only to the propositions &c.'. The fact that f. 93 does contain such page references and lacks any number of a proposition no doubt was the cause of the cataloguers' perplexity. But this page in Newton's hand has nothing whatever to do with the remainder of MS Add. 3965, §9.

The other pages contain lists of proposed changes, sometimes with reference to page and line within a fascicle, or a section, or to pages indicated by folios rather than page numbers, and also with reference to Definitions, Laws, Propositions, Lemmas, and Corollaries. For the most part these references do not correspond to any existing manuscript or printed work, although—as we shall see in a moment—some may be identified as comments on parts of manuscripts. This set of lists thus comes to us as a puzzle. I must confess that the enodation of it eluded me again and again, until I finally recognized the need of the assumption that in large

[1] See Chapter IV, §8.

[2] The verso of f. 95 contains some reworkings of Definition I, in Newton's hand, inspired by the critic's note; there is a portion of a proof on f. 96v, also in Newton's hand.

measure the critic was referring to parts (or states) of a pre-*M* manuscript containing Definitions, Laws of Motion, and either all or great parts of the three books comprising the *Principia* as we know it from *M* and E_1. In what follows I shall describe this evidence and give some illustrative examples to show how we may discover the form and extent, and even to some degree reconstruct the text, of this version of Newton's *Principia.* I shall begin with Books II and III because the material is briefer and simpler, and thus enables us to see the nature of the critique and of the manuscripts themselves, before we deal with the complexities of Book I.

All of the references to Book II occur on one side of a single leaf (f. 96). There is no explicit mention of a 'Liber secundus', but the references themselves are unambiguous. The same is true of the references to Book III (f. 98).

The first entry in the list for Book II is 'fol. 2; l. 20', and 'l. 23', corresponding to Prop. II; next Corols. 2 and 4 to Prop. III (referred to only as 'Coroll. 2.' and 'Coroll. 4'), and Prop. IV (referred to as 'Prop. IV. Prob. II.', the designation given in *M* and E_1). There are only two other mentions of a proposition or corollary by number, 'Prop. 8' and 'Prop. 9'. Other entries in the list may be readily identified as relating to Prop. V, Prop. VI, Prop. VII, Lemma II, Prop. VIII, Prop. IX (and Corol. 5), Prop. X (and Example 2), Prop. XVIII, Prop. XIX.

Not all of these suggestions were accepted by Newton, although the last two in the list evidently were:

[fol.] 44. lin. 12. aperui
[fol.] 46. lin. 4. tur, atque hoc ideo q.s.e.ae.e.o.p., e.m.o. exclusus supponitur nisi

as may be seen from {282.3} and {282.27–9}, where the words for 'q.s.e.ae.e.o.p., e.m.o.' are 'quia similis & [et] aequalis est omnium pressio, & [et] motus omnis'; but in E_1 'adeo' is used rather than 'ideo'.[3]

In Prop. X of Book II {252.7 ff., end of first paragraph in E_1} Newton has accepted the suggestion in this list:

[fol.] 21 lin. 12. Nam in lineolis finitae

in which the corrector has at first written 'lineis' and has then altered it to 'lineolis'. But perhaps the most interesting entry for Book II is in Lemma II.

[fol.] 14 lin. 16. post finitas [forsan adjici [*written* adijci] potest: *Finiri enim repugnat* ⌊*aliquatenus*⌋ *perpetuo eorum incremento, aut decremento*]

in which the corrector, like Newton and others of that day, has used square brackets where today we would use inverted commas. Anything about this lemma on moments, a primary document in the fluxionary controversy, has a special significance. From this manuscript remark we learn that 'perhaps there can be added after "finitas" the sentence "Finiri...decremento."' This sentence may be seen to have been added in *M* and E_1 as suggested, but with the ending 'vel decremento' rather than 'aut decremento'; another sentence ('Momenta...esse

[3] On 'adeo' and 'ideo' see the Guide to the Apparatus Criticus of the present edition, §2(*c*).

momenta') was apparently also added by Newton, since it is present between the word 'finitas' and the suggested sentence, 'Finiri...'. But before long Newton had rewritten these two sentences, so as to produce the single sentence found in E_2 and E_3 {243.34–5}.

We have no way of telling how much of this preliminary version of Book II the commentator may have seen. The material included in this single page of remarks covers the first nineteen propositions, corresponding to pages 236–92, or 57 of the 149 pages (pp. 236–383, 400[4]) comprising Book II in E_1; that is, about a third. It is pointless to make conjectures as to whether there may have been another such page or two of comments, corresponding to the remainder of Book II.

On the single manuscript page of comments devoted to Book III (f. 98) the first entry reads:

fol. 1.a. lin. 2. disputari possit. (vel) quibus viz. in rebus philos. disputationes inniti possint.
[lin.] 11. quibus principia prius posita satis intellecta non fuerint, ii

It will be seen that these both refer to the prologue to Book III {386.7, 15–16}, where we find that apparently both suggestions were essentially adopted, since the text of M and E_1 reads:

disputari possit.
quibus Principia posita satis intellecta non fuerint, ii

These portions of lines '2' and '11' occur on text lines 3 and 12 of E_1, indicating that the manuscript of the opening of Book III was much the same in size and presumably, therefore, in content as it is in the printed editions.

Another suggestion accepted by Newton reads:

[fol.] 1.b [lin.] 7 plenâ facie lucentes ultra solem siti sunt, dimidiatâ

This must have been adopted since it is identically the text in M and E_1 {392.24–5}. Another suggestion was

[fol.] 1.b [lin.] 14 et (vel solis c. T. vel) Terrae

which also met with Newton's approval since it occurs identically in E_1 {393.23}, save that 'et' is '&' and 'c. T.' is spelled out as 'circa Terram'. Other entries in the list refer to Prop. IV, Prop. VI, Corols. 2 and 3 to Prop. VIII, Prop. XII, Prop. XIII, and Prop. XVII. Some of these appear to have been adopted by Newton while others were not.

One entry reads:

[fol.] 7.a. Prop. XVII. Axes Planetarū diametris quae ad axes perpēdiculares ducuntur esse minores.

In M and E_1, the corresponding statement is:

Prop. XVIII. Axes Planetarum diametris quae ad eosdem axes normaliter ducuntur minores esse.

[4] In E_1, p. 383 is followed by p. 400.

Thus Newton refused to accept this alteration, or at least all of it. Note, however, that the manuscript note refers explicitly to 'Prop. XVII', showing us that the proposition numbers in Newton's preliminary manuscript version were not quite the same as in M and E_1. There are no numbered propositions in *De mundi systemate* (*liber* 2us), as there are in *De mundi systemate* (*liber* 3^{ius}), to which these manuscript comments thus unambiguously refer.

The final three references in the list are to Prop. XIX and Prop. XXIV. As was the case for Book II, we are unable to tell from this single manuscript page how much more the preliminary version of Book III may have contained. In E_1 the material through Prop. XXIV occupies pages 401–34 of Book III, or just over one-third of the whole (pp. 401–510). But, of the material prior to the discussion of comets (beginning {478.1 ff.} with Lemma IV, in page 474 of E_1), the material on which remarks were made comprises about one-half (thirty-four of seventy-three pages).

Not surprisingly, the comments on Book I and the Definitions and Laws of Motion are closely related to *LL*. The first page (f. 94) is headed 'Fascic. 1^s.'. It contains a full page of suggested alterations, indicated by reference to 'pag. 1', 'pag. 3', 'pag. 7', 'pag. 9', 'pag. 11', 'pag. 13', 'pag. 15' and for a second time 'pag. 15'. For 'pag. 1' there are revisions proposed for Def. '1', for Def. '2 lin. 3', and for Def. '3 lin. 3'. This is clearly a reference to page 1 of *LL* as it appears in MS Dd.9.46, since the suggestions of the critic are written out by Newton in Humphrey's text. To show how the critic's comments actually stimulated some of the revisions of *LL* by Newton, I print side by side the original statement of *LL* and its revision:

Original Def. 1

1. Quantitas materiae est copia seu mensura ejusdem orta ex illius densitate et magnitudine conjunctim. Vas idem plus continet aeris vel pulveris cujusvis qui compressione magis condensatur. Corpus autem duplo densius in duplo spatio quadruplum est. Hanc quantitatem per nomen corporis vel massae designo.

Newton's alterations

1. Quantitas materiae est [copia seu *del.*] mensura ejusdem orta ex illius densitate et magnitudine conjunctim. [⌞Quapropter in eodem vase major⌟ *del.*] [Vas idem plus *del.*] ⌞Vas idem plus [*changed from* majorem]⌟ continet [*changed to* continetur *and then changed back to* continet] aeris vel pulveris ⌞cujuscunque [quantitatum *del.*] quando corpora illa comprimendo⌟ [cujusvis qui compressione *del.*] magis condensa⌞n⌟tur. [& *del.*] ⌞&⌟ corpus [*changed from* Corpus] [autem *del. and* ⌞autem⌟ *del.*] duplo densius in duplo spatio quadruplum est. ⌞Sub nomine verò [*changed from* autem] Corporis vel Massae quantitatem istam ubique intelligo, neglecto ad medium respectu, siquod fuerit, interstitia partium liberè pervadens.⌟ [Hanc quantitatem per nomen corporis vel massae designo, ⌞neglecto medio, siquod sit, quod [perv *del.*] interstitia partium libere pervadit.⌟ *del.*]

To show how these alterations are related to the critic's comments, I give his remarks in full:

1. Quantitas Materiae [cujuscunque] est mensura ejusdem orta ex [illius *del.*] densitate ⌞partium⌟ et magnitudine ⌞totius⌟ conjunctim. [Hinc *del.*] ⌞Quapropter⌟ in eodem vase major est ⌞mihi⌟ quantitas aeris aut pulveris, prout pulvis magis aut minus [compressione *del.*] comprimendo condensatur.

[Hinc etiam *del.*] ⌞Et⌟ corpus duplo densius in duplo spatio quadruplum [est *del.*] erit, sub nomine corporis aut massae quantitatem istam ubique intelligo, neglecto ad medium respectu, si quod fuerit, interstitia partium liberè pervadens.

On the verso of the page of *LL,* Newton has then rewritten this definition as follows:

[Sic Aeris vel pulveris cujuscunque quantitas major in vas idem vehementius comprimendo impellitur. Corpus autem duplo densius in aequali spatio duplum est, inque duplo spatio quadruplum *del.*] ⌞Aer duplo densior in duplo spatio quadruplus est. Idem intellige de ⌞nive et⌟ pulveribus ⌞per compressionem vel liquefactionem condensatis⌟ [comprimendo condensatis *del.*]. Et par est ratio corporum [⌞omnium⌟ *add. and del. and* quae ⌞a viribus⌟ naturae diversimodè condensantur *del., which was originally probably* quae natura diversimodè condensat; *there is also another illegible change here*] ⌞omnium⌟ quae per operationes Naturae diversimode condensantur⌟ neglecto scilicet ad medium respectu, siquod fuerit, interstitia partium liberè pervadens. [⌞Et par est ratio corporum quae natura diversimode condensat⌟ *del.*] Innotescit autem [autem *was originally present but was cancelled, replaced, recancelled, and finally rewritten*] quantitas materiae [*these two words were originally* haec quantitas] per corporis cujuscunque pondus. Nam ponderi proportionalem esse reperi per experimenta pendulorum accuratissimè instituta, uti posthac docebitur. Eandem verò sub nomine *corporis* vel *massae* in sequentibus passim intelligo.

Two further versions of this definition occur on the back of another leaf of the critic's commentary (f. 95ᵛ); yet further revisions are recorded in the Apparatus Criticus {1.8–9}. It is something of a mystery that so much time and effort should have been spent on what seems to us to have been so simple and straightforward a problem of defining 'Quantitas Materiae'.

Newton was evidently less impressed by the critic's note on Def. II.[5] In *LL* this reads as follows:

II. Quantitas motus est mensura ejusdem orta ex velocitate et quantitate materiae conjunctim.

Motus totius est summa motuum in partibus singulis, adeoque in corpore duplo majore aequali cum velocitate duplus est, et dupla cum velocitate quadruplus.

The critic suggested that this be revised so as to read:

adeoque duplus est in corpore duplo majore, sed aequalis velocitatis, et quadruplus in [eodem *del.*] ⌞corpore duplo majore⌟ cum dupla velocitate.

Newton, however, chose to keep his own version. But in Def. III, 'Materiae vis insita', Newton did accept the critic's suggestion and he altered the line written out by Humphrey ('lin. 3') reading 'Estque corpori suo proportionalis neque...' to 'Haec semper proportionalis est suo corpori, neque...' {2.5}.

[5] Newton has crossed out a pair of definitions, 'II. Axis materiae...III. Centrum materiae...', and then renumbered the definition of 'Quantitas motus' from IV to II.

The critic numbered these pages of *LL* not by recto sides (as folios) but as pages, so that the second page of text proper, that is, the second recto page, is designated 'pag. 3' and not page 2. He recommended that Newton introduce changes in Defs. III, V, and VI. Newton accepted the first in part, ignored the second, and began to alter VI as suggested (writing 'ratione causae fortius'), but then crossed out the correction and made a simpler alteration. But the references to definition and line within the Definitions correspond to this page of *LL* exactly, so that there can be no doubt that the critic saw the pages of $LL_{\beta}(1)$ and not the original $\langle LL_{\alpha}^{1-8}\rangle$ which they replaced.

There are no comments on the next recto page (p. 5 in the librarian's enumeration), which would have been the critic's page 5, but there are two emendations suggested for 'pag. 7' (page 7 also in the modern numbering). These are both to the scholium, paragraph 'I' on 'Tempus absolutum verum et Mathematicum' {6.18 ff.}. Both of these were accepted by Newton: 'alioque nomine dicitur Duratio' for 'alio nomine dictum Duratio,' and 'Durationis per motum mensura (seu' for 'Durationis mensura per motum (seu'.

On the next page Newton did not quite adopt the proposed alterations {7.5–7}, and on the following page (11 in the critic's enumeration, 9 in the modern librarian's) Newton adopted almost all of the emendations proposed at line 16 {8.22–4}, ignored the next four (at lines 25, 28, 30, 31), and adopted the following one (line 35), 'motus verus et absolutus definiri nequit per translationem e vicina...' {9.6–7}. A suggestion on the next page, 'participat etiam loci sui motum', was also adopted by Newton. The remaining suggestions were largely accepted, including 'donec revolutiones in aequalibus cum vase temporibus peragendo, quiescat in eodem relativè', altered by Newton on lines 1–2 of 'pag. 15' (which is page 11 in the modern librarian's enumeration) from the original 'quoad usque revolvendo iisdem temporibus cum vase, quiescat in eo relativè'.

I have quoted a number of these instances verbatim to prove that the critic actually saw these very pages. There can be no doubt of it. For not only do the references agree by page and line with *LL*, but the actual proposed new word or expression may be seen to have been written in Newton's own hand.

But the situation is quite different with regard to the Leges Motus. The comments, occupying about a third of a page (f. 95 of MS Add. 3965), are either incorporated into the text or ignored. Thus, to take two examples, the discussion of Lex III as written out by Humphrey follows exactly the critic's revision, 'Quicquid premit vel trahit alterum tantundem ab eo premitur...' {14.5}, as does the portion of the discussion of Corol. 4, 'quod linea horum corporum centra in rectis uniformiter progredientia jungens dividitur ab hoc centro communi in ratione datâ. Similiter &c.' {19.15–17}. Indeed, but for the presence of these remarks on f. 95 of MS Add. 3965, we would not even have known that these lines had undergone any alteration. Hence we may conclude that the critic had not seen the section of 'Axiomata sive Leges Motûs' as it is found in $LL_{\beta}(1)$ (pp. 13–22 in the modern librarian's pagination), but the preceding version ⟨LL_{α}^{1-8}⟩ from which Humphrey had made this copy. Other notes of the critic were directed to Corol. 4 and to the scholium (where Newton had adopted 'textura aliis de causis irregularis' {24.20} and, *inter alia*, 'solet; superata omnia ea resistentia' {27.14} and 'non est hujus instituti' {27.17}).

The remarks dealing with the lemmas (Sec. I, Book I) are harder to interpret. None of them is followed verbatim, but the text as written out by Humphrey is in each case very similar to the critic's suggestion. Most likely, therefore, the critic had seen a prior version, in the alteration of which the spirit if not the letter of the suggestions had been added to Humphrey's copy by Newton. But by the scholium to the lemmas (Sec. I) the text of LL_{α} has been changed, and no longer do we have the new unnumbered pages, but rather the old part of LL_{α}, numbered originally by Newton 9, 10, 11, 12,..., of which every eighth page bears a second number in the new sequence; 9 and 21, 17 and 29,.... It is these pages which the critic has seen, since he suggests a change in the final paragraph of the conclusion, Scholium to Sec. I, a paragraph {38.8–25} that did not originally appear in LL_{α}, and so is inserted on a verso page in Newton's hand, rather than being on a recto page in Humphrey's hand. The critic's emendation {38.12–14} was only in part accepted.

In Sec. II, the critic again saw LL_{α}, since the suggestion for Prop. I, 'et in planis immobilibus consistere &' {38.30–1}, is written into the text by Newton, along with other alterations. Again, the critic's working of the beginning of Prop. IV, Theor. IV, 'Corporum, quae ⌊diversos⌋ circulos aequabili motu describunt, vires centripetas ad centra eorundem circulorum tendere, et esse inter se ut' {43.27–9}, has been accepted by Newton, who could not get it in easily after several tries at alteration and so has written out the whole sentence on the facing verso page.[6] Section III was also seen by the critic, and here too the suggestions adopted by Newton—for example, 'ordinatim applicatam Qv in x' (written 'QU in X' on f. 97 and cancelled) {54.27}—are added in his own hand; in this case the reference to 'pag. 19' corresponds to the numbers assigned to the pages by Newton. That is, in contrast to the earlier parts of the comments, which refer to a portion of *LL* which has no page numbers, here there are actual page numbers for the critic to use, so that he did not have to assign numbers in a system of his own devising.

The final item on f. 97 reads

[p.] 32. prop. 23. donec tandem arcus, vel eò usque ut arcus

On this last page of LL_{α}, the first of these suggestions is written out in Newton's hand.

The critic saw very little of LL_{β}. Indeed, the only reference to LL_{β} in his pages

[6] Curiously, the page '16' (in Newton's enumeration) has above this proposition, 'Theor. 2. Corporibus in circumferentiis circulorum' (the beginning of Theor. 2 in the 'De motu corporum'). But this has been crossed out.

of comments occurs on f. 95, where—tucked away at the bottom of the page discussing the Laws of Motion—there are two suggestions as follows:

Fascic. 4us

p. 1. Prop. 17. lin. 2. et quod vis illius quantitas sit cognita.
p. 2. lin. 3. Orbitarum Ell. vel Orbium Ellipticorum &c.

These refer not to LL_α but rather to LL_β. And, indeed, page 1 (p. 37 in Newton's numbering) of LL_β does contain Prop. XVII, and this first change {63.12–13} is made (almost as written above) in Newton's hand. The verso of this page contains Newton's insertion of a title for 'Artic. IV.' (='Sectio IV'). Newton wrote this as:

Artic. IV.
[continens *del.*]
De Inventione [*replacing* Inventionem] Orbium Ellipticorum, Parabolicorum et Hyperbolicorum [*the three adjectives are changed from the feminine genitive plural* -arum] ex [conditionibus datis *del.*] umbilico dato.

Thus, faced with the two alternatives presented to him by the critic, Newton chose to make the adjectives agree with the masculine genitive plural 'Orbium' and rejected the possibility of altering 'Orbium' to the feminine 'Orbitarum' so as to agree with the adjectives. This occurs at {66.2–3}.

There is one final page of comments (f. 99), referring to a set of pages 35, 41, 43, 45, 47, 49, 51, 55, 57, 61, 63, 65, 67, 69, 71. These do not correspond to LL_α or to LL_β but may be identified as the continuation of LL_α that I have found in ff. 7–14 of MS Add. 3965. I shall not discuss these in any detail, since both the method of indicating the improvements and Newton's way of writing them into Humphrey's copy are now quite evident. But it is fascinating to see that these suggestions on f. 99 confirm the identification of ff. 7–14 as a continuation of LL_α and also give further light on the missing part of LL_α (or $\langle LL_\alpha^{33-40}\rangle$). The first reference (p. 35) is to this missing section and may be readily identified as a part of the introduction to Sec. XI of M and E_1 {160.10–12, 13–15}. This entry is set apart from the next group, which begins with two emendations for 'pag. 41'. These refer to the first page of the set that I have identified as the continuation of LL_α (MS Add. 3965, ff. 7–14), written out by Humphrey and corrected in Newton's hand. And on this page 41, at the very lines (10 and 22) indicated, there occur the phrases which were to have been corrected. They are not corrected on this page, however, but both have been altered by the time of the preparation of M and thus occur corrected in E_1. Presumably, when these and other pages of LL_α were discarded, in completing LL_β (or possibly M), the corrections were introduced in the new copy, but not in the old.

I have mentioned earlier, in describing these eight pages of continuation of LL_α, that p. 41 is followed by a p. 42 containing 'Prop. XXXV. Theor. XVII.' which is Prop. LXVI, Theor. XXVI in E_1. Thus there is omitted the intervening Prop. LXV, Theor. XXV, which appears in M and E_1 (there are no pages of LL_β

that go this far into Book I). But Newton must have had two pages at hand containing this proposition, its proof, and its three corollaries, since they are the subject of the next suggested emendations, but keyed to a 'pag. 43' and 'pag. 45'. But pages containing this material are not now present, nor do the remaining pages correspond to the page numbers and line numbers of the critic. Comparison of the list of proposed alterations to this proposition with *M* and E_1 shows that they were generally adopted. For instance, to cite but one, the form of the beginning of the statement of Prop. LXV, 'Corpora plura quorum vires decrescunt' {168.6}, comes from the critic. The remaining entries contain much that was adopted by Newton. But these corrections must have been made on another manuscript than the eight pages that have been preserved. For instance, not only do the page and line references fail to agree, but the critic makes suggestions which Newton accepted for Corols. 5, 7, 12, 19, and 22 to Prop. XXXVI (=Prop. LXVI of *M* and E_1). In *M* and E_1, this proposition does have twenty-two corollaries, but in the form in which it occurs on ff. 8–10 (Newton's pages 42–4), there is but one corollary, almost the later Corol. 1. Furthermore, the critic refers to this proposition by number as 'Prop. 36', whereas in f. 8 it is called 'Prop. XXXV'. The reason, as we have seen, is that the Prop. LXV of *M* and E_1 is missing from these eight pages. When it was added, each succeeding proposition had its number increased by 1. The final entries are to Props. 39 and 40 (= LXXI and LXXII of *M* and E_1), in Sec. XII of Book I.

SUPPLEMENT VIII

THE TRACT 'DE QUADRATURA' AND THE 'PRINCIPIA'

FROM time to time in Part Three mention has been made of Newton's design to publish *De quadratura* either as a supplement to a second printing or edition of the *Principia* or together with Secs. IV and V of Book I of the *Principia*, on the conic sections, as David Gregory recorded. Only at the very last minute, to judge by Bentley's report to Cotes, did Newton abandon his intention to have a mathematical supplement to the second edition of the *Principia*.[1]

The purpose of such a mathematical supplement becomes clear on a close reading of the text of the *Principia*. For there Newton often assumes certain results of quadrature; that is, the solution of problems of finding areas enclosed by curves, or—to use today's parlance—the methods and results of the integral calculus. For instance, in the course of stating Propositions XLI, LIII, LIV, and LVI of Book I, Newton uses the phrase 'granting the quadratures of curvilinear figures' ('concessis figurarum curvilinearum quadraturis'). Now this phrase not only denotes a mathematical condition necessary for the proof of such a proposition; it raises in the mind of any reader the question of whether Newton means that some day a general method of quadrature may be found which will complete the proof, or whether there already exists such a method. And if the latter, then is it Newton himself who has invented the method or someone else? And if Newton, what is his method? We may thus readily see why Newton should essay a treatise giving the methods of quadrature, or of 'integration', and making plain the mathematical basis of the *Principia* which might otherwise have lain concealed.[2]

In the early 1690s, while recasting the opening dynamical part of the *Principia* (Sec. I), Newton turned again to problems of pure mathematics. In 1691, evidently stimulated largely through his correspondence with David Gregory, Newton wrote

[1] See Chapter IX, §5.

[2] In the Portsmouth Collection (U.L.C. MS Add. 3962, §6, ff. 69–70) there is a reject from the *Principia*—possibly to be dated 1692, or even 1686—which deals with propositions on quadratures of curves (or integration) and was at one time intended to be inserted into the *Principia* in either E_1 or a revision of E_1. This fragment is discussed by Rouse Ball (1893), *Essay*, p. 87, who points out that Newton evidently 'intended to insert at the end of this section [*Principia*, Sec. X, Book I) a classification of algebraical curves whose quadrature could be effected—"tandem ut compleatur solutio superiorum problematum adjicienda est quadratura ffigurarum toties assumpta." Doubtless this is the rule to which he alluded in his letter to Collins of Nov. 8, 1676...' This fragment does not correspond to any published version of *De quadratura*, but to 'an obscure 1676 English text' in MS Add. 3962 on the quadrature of trinomials. It is printed in [Newton] (1967–), *Mathematical papers*, vol. 3, pp. 373–85.

a series of drafts leading up to 'a tract on analytical calculus in sixteen propositions'.[3] Then he changed his mind and decided 'to gather all the mathematical portions of the [*Principia*] in a single compilation, tentatively entitled "Liber Geometriae", divorced from their surrounding physical and mechanical sections'. By the summer of 1693 he had decided instead to produce an abridgement of the work of 1691, 'preserving only the first eleven propositions but adding to the rump an introduction which is the classical statement of his theory of first and last ratios'. This is 'essentially' the *Tractatus de quadratura curvarum*, published by Newton in 1704 as an appendix to the *Opticks*,[4] and reprinted in 1706 with the Latin *Optice*.

I have referred earlier to the edition of *De quadratura*, produced by William Jones in 1711, together with other mathematical writings of Newton's, which, with a preface of Jones's, was reprinted in 1723 with the second Amsterdam reprint of the *Principia* in its second edition.[5] Some time after the second edition had been published, Newton decided to revise and expand the text of *De quadratura*, above all to restore the parts of the 1691 version he had omitted and to add a reply to the criticism that had been raised on the Continent, chiefly by J. Bernoulli. The expanded work was entitled *Analysis per quantitates fluentes et earum momenta*, and (*c.* 1715) was intended to be an appendix to a revised reprint that was planned of the second edition of the *Principia*.

I give here some extracts from the drafts of intended prefaces or introductions, in which an aim of Newton's was to show the propriety of joining together a treatise on the calculus (in the Newtonian form) and the *Principia* (or part thereof). These statements are not only significant here as documentation of Newton's intentions with regard to the later editions or printings of the *Principia* that he planned; they are of relevance because they contain autobiographical statements about the actual methods used in writing the *Principia*.[6] Like all such recollections, these should be taken *cum grano salis*, especially in the light of the controversy over the discovery of fluxions, which introduced—consciously or unconsciously—an unavoidable bias.[7]

[3] [Newton] (1964), *Mathematical works*, intro., p. xv.

[4] Along with the *Enumeratio linearum tertii ordinis*.

[5] See Chapter X, §3. It is a difficult job today to work out the relations of *De quadratura* and the *Principia*, not only on the conceptual level but even as to plans for printing *De quadratura* (in one or another version) with all or part of the *Principia*. D. T. Whiteside informs me that the complexity of the problem is augmented 'by the fact that most of the MSS were ultimately kept by Jones (who was, I believe, to superintend this mathematical appendix) and are now among his papers in Shirburn Castle [the Macclesfield Collection]. For instance, Newton's joint (suppressed) preface to the *Principia* (*c.* October 1712) and a version of *De quadratura* (a revised augment of that published in 1704 [as an appendix to the *Opticks*]) is in Shirburn, together with a Latin revise (headed *Admonitio ad Lectorem*) of the English passage you quote'—document 4, below. On the verso, there is an English draft, reading: 'In the book of Principles explain...To the book of Quadratures prefix a Preface to ye reader confirming what I said in ye Introduction...'

[6] And so should be read in conjunction with other autobiographical statements, as in Supplement I.

[7] These are presented below in the final versions, with only a few changes noted.

1

MS Add. 3968, f. 111

In scribendis Philosophiae Principiis Mathematicis Newtonus Libro hocce de quadraturis plurimum est usus, ideoque eundem Libro Principiorum subjungi voluit. Investigavit utique Propositiones in Libro Principiorum per Analysin, investigatas demonstravit per Synthesin pro lege Veterum qui Propositiones suas non prius in Geometriam admittebant quam demonstratae essent synthetice. Analysis hodierna nihil aliud est quam Arithmetica in speciebus. Haec Arithmetica ad res Geometricas applicari potest, et Propositiones sic inventae sunt Arithmetice inventae. Demonstrari debent Synthetice more veterum et tum demum pro Geometricis haberi.

In writing the Mathematical Principles of Philosophy Newton made very great use of this Book on quadratures, and hence wanted it added to the Book of Principles. In particular he investigated the Propositions in the Book of Principles through Analysis, and after they were investigated he demonstrated them through Synthesis in accordance with the law of the Ancients, who used not to admit their Propositions into Geometry before they had been demonstrated synthetically. Present-day Analysis is nothing other than Arithmetic in species.[8] This Arithmetic can be applied to Geometrical matters, and Propositions thus found are found Arithmetically. They ought to be demonstrated Synthetically in the manner of the ancients and then finally to be regarded as Geometrical.

2

MS Add. 3968, f. 112^v, 112^r

Beneficio hujus methodi didici anno 1664 [vim centr *del.*] vires quibus Planetae primarii retinentur in orbibus circa Solem esse in ratione duplicata distantiarum mediocrium a sole inverse et vim qua Luna retinetur in Orbe circum Terram esse in eadem fere ratione ad gravitatem in superficie Terrae. Deinde anno 1679 ad finem vergente inveni demonstrationem Hypotheseos Kepleri quod Planetae primarii revolvuntur in Ellipsibus Solem in foco inferiore habentibus, & radiis ad Solem ductis areas describunt temporibus proportionales. Tandem anno 1685 [*changed from* 1684] et parte anni 1686 beneficio hujus methodi & subsidio libri de Quadraturis scripsi libros duos primos Principiorum mathematicorum Philosophiae. Et propterea Librum de Quadraturis subjunxi Libro Principiorum.

Anno 1666 incidi in Theoriam colorum, et anno 1671 parabam Tractatum de hac re, aliumque de methodo serierum & fluxionum ut in lucem ederentur. Sed subortae mox disputationes aliquae me a consilio deterruerunt usque ad annum 1704.

By the aid of this method I learned in 1664 that the forces by which the primary Planets are kept in their orbits about the Sun are in the duplicate ratio of their mean distances from the sun inversely and that the force by which the Moon is kept in its Orbit around the Earth is in nearly the same ratio to gravity on the Earth's surface. Then as the year 1679 drew to its close I found the demonstration of Kepler's Hypothesis that the primary Planets revolve in Ellipses having the Sun in the lower focus, and with their radii drawn to the Sun describe areas proportional to the times. At length in 1685 and part of 1686 by the aid of this method and the help of the book on Quadratures I wrote the first two books of the mathematical Principles of Philosophy. And therefore I have subjoined a Book on Quadratures to the Book of Principles.

In 1666 I discovered the Theory of colours, and in 1671 I was preparing a Tract about this thing, and another about the method of series and of fluxions so that they might be published. But some disputations that soon arose put me off from my plan until 1704.

[8] The expression 'in species' signifies 'involving specious [free] variables' or Cartesian algebraic analysis.

3

MS Add. 3968, f. 27ᵛ

This Book I made use of in the year 1679 when I fround [*sic*] the demonstration of Keplers Proposition that the Planets moved in Ellipses, & again in the years 1684, 1685 & 1686 when I wrote the Book of Mathematical Principles of Philosophy [I made much use of it *del.*] & for that reason I have now subjoyned it to that Book.

4

MS Add. 3968, f. 28

The first of the following passages comes from a cancelled paragraph which was apparently replaced by the second paragraph.[9]

Now because by the help of this Method I wrote the Book of Principles I have therfore subjoyned the book of Quadratures to the end of it. And because the methods of series & fluxions are nearly related to one another & were invented in the same year (the year 1665), & were conjoyned by me in the Tracts wch I wrote in those days [*replacing* above 49 years ago], & jointly compose one very general method of Analysis I have here added to the end of this book out of my Letters formerly published, some Propositions for reducing quantities into series...

. . . . .

Now since by the help of this Method I wrote the Book of Principles, I have therefore subjoyned the Book of Quadratures to the end of it. And for the same reason I have subjoyned also the Differential Method. And because several Problems are proposed in the Book of Principles to be solved concessis ffigurarum Quadraturis I have added to the end of the Book of Quadratures some Propositions taken from my letters already published...

5

MS Add. 3968, f. 221ᵛ

Praefatio.

...Et hoc anno [1676] sub autumno ex tractatu praedicto quem scripseram anno 1671, tractatum de quadratura curvilinearum extraxi eodemque plurimum usus sum in componendo libro praecedente de Philosophiae naturalis principiis mathematicis. Et in lib III de Principiis Philosophiae Lem. V, specimen dedi methodi cujusdem [*sic*] differentialis. Quapropter tractatum praedictum [*originally* tractatus duos] de quadratura figurarum subjungere [*written* subjunge] visum est, ut et tractatum de methodo illa differentiali.

Preface.[9]

...And in this year [1676] during the autumn from the aforesaid tract which I had written in the year 1671, I extracted a tract on the quadrature of curvilinear figures and I made very great use of it in composing the preceding book on the mathematical principles of natural Philosophy. And in Lem. V book III on the Principles of Philosophy, I gave a sample of a certain differential method. Wherefore it seemed best to subjoin the aforesaid tract on the quadrature of figures, as also a tract on that differential method.

This 1671 tract, comprising some 160 sheets, still exists and may be found in U.L.C. MS Add. 3960, §14. D. T. Whiteside informs me that the 'quadrature theorems used in the *Principia* can all be found...in the 1671 tract's Problem 8'.

[9] Here, as in document 5 below, the 'differential method' is the method of finite differences.

6

MS Add. 3968, f. 464^{v}

Ad Lectorem.

Interea dum componerem Philosophiae naturalis Principia Mathematica, plura Problemata solvi per figurarum quadraturas quas Liber de Quadratura figurarum mihi suppeditavit. Alia proposui solvenda concessis figurarum quadraturis. Et plurima demonstravi invertendo ordinem inventionis Analyticae. Et propterea Librum de Quadratura Curvarum subjungere visum est quo figuras vel quadravi vel quadrandas proposui, et qui Analysin meam momentorum exhibet qua saepissime usus sum. Et cum in exponenda Cometarum Theoria usus sim methodo mea differentiali, visum est etiam eandem methodum [Libro de Quadraturis *del.*] subjungere.

To the Reader.

During the time that I was composing the Mathematical Principlies of natural Philosophy, I solved many Problems through the quadratures of figures which the Book on the Quadrature of figures supplied for me. I proposed others to be solved, granted the quadratures of figures. And I demonstrated the greatest number by inverting the order of their analytical discovery. And therefore it seems appropriate to subjoin the Book on the Quadrature of Curves by which I either squared figures or proposed them for squaring, and which exhibits my Analysis of moments which I used very often. And since in setting forth the Theory of Comets I used my method of [finite] differences, it seemed best to subjoin this [*lit.* the same] method also.

MS Add. 3968, f. 109

Tractatum de hac Analysi ex chartis antea editis desumptam, Libro Principiorum subjunxi.

A tract on this Analysis, taken from papers published before, I have subjoined to the Book of Principles.

7[10]

MS Add. 3968, f. 102^{v}

In the Xth Proposition of the second Book there was a mistake by drawing the Tangent of the Arch GH from the wrong end of the Arch. But the mistake was rectified in the [last *del.*] ⌊second⌋ Edition. And there may have been some other mistakes occasioned by the shortness of the time in wch the book was written & [partly *del.*] by its being copied by an Emanuensis who understood not what he copied; besides the Press faults. For after I had found [eight or tenn *del.*] ⌊tenn or twelve⌋ of the Propositions relating to the heavens, they were communicated to the R. S. in December [⌊10th⌋ *del.*] 1684, & at their request that the [Book *del.*] ⌊Propositions⌋ might be printed I set upon composing this Book & sent it to the [*changed from* them] ⌊R. S.⌋ in May 1686 [& they ordered it to be printed May 19, *bracketed and del.*] as I find entred in their Journal Books Decem 10 1684 & May 19th 1686. And I was enabled to make the greater dispatch by means of the Book of Quadratures composed some years before [& annexed to this Edition *del.*].

[10] See, *supra*, footnote 4 to Supplement I, §4.

SUPPLEMENT IX

NEWTON'S LEMMAS AND PROPOSITIONS ON THE 'HORARY MOTION OF THE LUNAR APOGEE'

In Chapter IV, §7, I have presented two propositions and two lemmas composed by Newton most likely in 1687 or 1688, presumably after Prop. XLV of Book I had been printed and hence possibly after the whole of the *Principia* had been printed. In §7, the text was given of only the final versions of Prop. [*i*] and Prop. [*ii*], with the beginning of the demonstration of Prop. [*i*], and the final versions of Lemma [*i*] and Lemma [*ii*]. Here I shall show how these propositions and lemmas evolved.

The first draft is in Newton's hand on ff. 108–9, 112.[1] At the top of f. 112^r there occurs a draft statement of Prop. [*i*] in Newton's hand, with many deletions and alterations. A second state, also in Newton's hand, appears on f. 108^r–108^v. The sequence is unmistakable. For instance, in the opening line of text on f. 112, Newton has first written 'Designent A, B syzygias...' and has then changed this to 'Designet ACBD orbem Lunae A & B syzygias'; on f. 108, this alteration has been incorporated into the line so as to read 'Designet ACBD orbem Lunae, A & B syzygias...'. Furthermore, on f. 108 this line continues, 'syzygias, C et D Quadraturas &c ——— atque vis 2 PK...' to show that the copyist (Humphrey Newton) is to insert (in place of the three dashes) a part of another text. In this case, the insert is found on f. 112; a comparison of the two versions shows that they diverge before and after the material indicated by dashes. Hence these manuscript pages exhibit to us a pattern of Newton's habits of work. A rough copy, much worked over, is succeeded by a second version, which is also rewritten and revised before being transcribed by Humphrey; the final version is then altered here and there by Newton himself.

Only the first paragraph of f. 112^r, together with the heading

Prop. Prob.
Invenire motum horarium Apogaei Lunae.

has been rewritten and expanded on f. 108^r–108^v into Prop. [*i*]. The text of the remainder of f. 112^r, a single paragraph occupying about two-thirds of the page, contains a first version of the demonstration of Lemma [*a*], but without a statement

[1] U.L.C. MS Add. 3966.

of the lemma itself marked off or labelled as 'Lemma'. It is interesting to observe that Newton has left a gap ('Et per ea quae in Prop   Lib I ostensa sunt...') for the reference to Prop. XVII, just as it appears in Humphrey's copy (see Chapter IV, §5). The corollary present in Humphrey's copy is not to be found in Newton's draft.

On the verso side of f. 112, Newton has written out in his own hand a much shorter draft of Lemma [*b*] than occurs in Humphrey's final version. As in the case of the draft of Lemma [*a*], there is no title of 'Lemma', nor is the statement of the conditions of the lemma separated from the demonstration. A corollary refers to 'the foregoing Proposition' ('de qua in Propositione priore actum est'), which appears to invoke the long paragraph on f. 112^{r} that was to become Lemma [*a*]; perhaps Newton was contemplating a series of propositions rather than lemmas and propositions.

Proposition [*ii*] also exists in an earlier draft in Newton's hand, on ff. 108^{v}–109^{r}. As in the case of the early and later drafts of Prop. [*i*] and the later draft (in Humphrey's hand) of Prop. [*ii*], Newton's own first version of Prop. [*ii*] is presented without numbers under the heading: 'Prop.   Prob.   '

At first this proposition reads:

Invenire motum medium Apogaei Lunae.

Then the word 'Lunae' has been crossed out, and a new beginning added, so as to make the proposition read:

Prop.   Prob.

Posito quod ⌊Excentricitas⌋ Orbis Lunaris sit infinite parva, Invenire motum medium Apogaei.

It is in this form that we have seen Prop. [*ii*] presented by Humphrey in the final version. These stages may be shown in the accompanying diagram, suggested to me by D. T. Whiteside, who has been largely responsible for establishing the order of all the parts or versions of these texts.

	First draft	Intermediate	Revised draft	
Prop. [*i*]*	112^{r} (par. 1) ⟶	[108^{r}/108^{v} (top)] ⟶	102^{r}–104^{r} (with insert on 107^{v} and figure on 106^{v})	Prop. [*i*]
	112^{r} (par. 2)	⟶	105^{r}–106^{r}	Lemma [*a*]
	112^{v}	⟶	106^{r}–107^{r} (with figure on 106^{v})	Lemma [*b*]
Prop. [*ii*]	108^{v} (bottom)–109^{r}	⟶	104^{r}–110^{r}–111^{r} (with insert on 110^{v} and draft of figure on 108^{v})	Prop. [*ii*]
	In Isaac Newton's hand		In Humphrey Newton's hand, with Isaac Newton's alterations	

* These three undifferentiated paragraphs are all grouped together under the single head ('Prop. Prob.)'.

BIBLIOGRAPHY
INDEX

BIBLIOGRAPHY

AITON, E. J. (1955). 'The contributions of Newton, Bernoulli and Euler to the theory of the tides', *Annals of Science*, vol. 11, pp. 206–23.

(1960) [1962]. 'The celestial mechanics of Leibniz', *Annals of Science*, vol. 16, pp. 65–82.

(1962). 'The celestial mechanics of Leibniz in the light of Newtonian criticism', *Annals of Science*, vol. 18, pp. 31–41.

(1964). 'The celestial mechanics of Leibniz: a new interpretation', *Annals of Science*, vol. 20, pp. 111–23.

(1964*a*). 'The inverse problem of central forces', *Annals of Science*, vol. 20, pp. 81–99.

(1965) [1966]. 'An imaginary error in the celestial mechanics of Leibniz', *Annals of Science*, vol. 21, pp. 169–73.

ANDRADE, E. N. DA C. (1935). 'Newton's early notebook', *Nature*, vol. 135, p. 360.

(1943). 'Newton and the science of his age', *Proceedings of the Royal Society*, vol. 181A, pp. 227–43.

(1950). 'Wilkins lecture, Robert Hooke', *Proceedings of the Royal Society*, vol. 201A, pp. 439–73.

(1953). 'A Newton collection', *Endeavour*, vol. 12, pp. 68–75.

ASTON. *See* [TIXALL LETTERS] (1815).

AXTELL, JAMES L. (1965). 'Locke, Newton, and the elements of natural philosophy', *Paedagogica Europaea*, vol. 1, pp. 235–44.

(1965*a*). 'Locke's review of the *Principia*', *Notes and Records of the Royal Society of London*, vol. 20, pp. 152–61.

(1968). *The educational writings of John Locke: a critical edition with introduction and notes.* Cambridge (England): at the University Press.

[BABSON COLLECTION] (1950). *A descriptive catalogue of the Grace K. Babson collection of the works of Sir Isaac Newton, and the material relating to him in the Babson Institute Library, Babson Park, Mass.* With an introduction by Roger Babson Webber. New York: Herbert Reichner.

A supplement compiled by Henry P. Macomber was published by the Babson Institute in 1955.

BAILY, FRANCIS (1835). *An account of the Revd John Flamsteed, the first Astronomer-Royal; compiled from his own manuscripts, and other authentic documents, never before published. To which is added, his British catalogue of stars, corrected and enlarged.* London: printed by order of the Lords Commissioners of the Admiralty.—(1837). *Supplement to the account of the Revd John Flamsteed, the first Astronomer-Royal.* London: printed for distribution amongst those persons and institutions only, to whom the original work was presented.

Photo-reprint, London: Dawsons of Pall Mall, 1966; abridged.

BALL, W. W. ROUSE (1889). *A history of the study of mathematics at Cambridge.* Cambridge (England): at the University Press.

(1892). *Mathematical recreations & essays.* London: Macmillan and Co.

An 11th edition, revised by H. S. M. Coxeter, was published in 1939, reprinted in 1942.

(1892*a*). 'A Newtonian fragment relating to centripetal forces', *Proceedings of the London Mathematical Society*, vol. 23, pp. 226–31.

(1893). *An essay on Newton's "Principia".* London and New York: Macmillan and Co.

(1893*a*). *A short account of the history of mathematics.* London and New York: Macmillan and Co.

This second edition contains an excellent chapter (XVI, pp. 319–58) on 'The life and works of Newton'.

(1918). *Cambridge papers.* London: Macmillan and Co.

BARROW, ISAAC (1669). *Lectiones XVIII, Cantabrigiae in scholis publicis habitae; in quibus opticorum phaenomenωn genuinae rationes investigantur, ac exponuntur. Annexae sunt lectiones aliquot geometricae.* Londini: typis Gulielmi Godbid, & prostant venales apud Johannem Dunmore, & Octavianum Pulleyn Juniorem.

(1916). *The geometrical lectures.* Translated, with notes and proofs, and a discussion on the advance made therein on the work of his predecessors in the infinitesimal calculus, by J. M. Child. Chicago and London: The Open Court Publishing Company.

A much better (i.e., more literal) translation by Edmund Stone (1735) has also the virtue of completeness, which Child's does not.

[BENTLEY, RICHARD] (1836–8). *The works of Richard Bentley, D.D.* Collected and edited by the Rev. Alexander Dyce. 3 vols. London: Francis Macpherson.

Photo-reprint, New York: AMS Press, 1966.

(1842). *The correspondence of Richard Bentley, D.D., Master of Trinity College, Cambridge.* 2 vols. London: John Murray.

[BERNOULLI, JOHANN] (1955). *Der Briefwechsel von Johann Bernoulli.* Herausgegeben von der Naturforschenden Gesellschaft in Basel. Band I. Basel: Birkhäuser Verlag.

See also [LEIBNIZ AND BERNOULLI] (1745).

BIOT, J.-B., AND F. LEFORT. *See* [COLLINS *et al.*] (1856).

BIRCH, THOMAS (1756–7). *The history of the Royal Society of London, for the Improving of Natural Knowledge, from its first rise. In which the most considerable of those papers communicated to the Society, which have hitherto not been published, are inserted in their proper order, as a supplement to the* Philosophical Transactions. 4 vols. London: printed for A. Millar.

BOAS, MARIE (1952). 'The establishment of the mechanical philosophy', *Osiris*, vol. 10, pp. 412–541.

Section IX deals with Newton.

(1958). 'Newton's chemical papers', pp. 241–8 of [NEWTON] (1958), *Papers & letters.*

See also HALL AND HALL.

BOPP, K. (1916). 'Johann Heinrich Lamberts Monatsbuch, mit den zugehörigen Kommentaren, sowie mit einem Vorwort über den Stand der Lambertforschung', *Abhandlungen der Königlich Bayerischen Akademie der Wissenschaften*, Math.-phys. Klasse, vol. 27, 6. Abhandlung, pp. 1–84.

(1929). 'Die wiederaufgefundene Abhandlung von Fatio de Duillier: De la cause de la pesanteur', *Schriften der Strassburger Wissenschaftlichen Gesellschaft in Heidelberg*, neue Folge, 10. Heft, pp. 19–66.

BOSSCHA, J. (1909). 'La découverte en Australie de l'exemplaire des "Principia" qui a servi à Newton même', *Archives Néerlandaises des Sciences Exactes et Naturelles*, vol. 14, pp. 278–88.

BOYER, CARL B. (1939). *Concepts of the calculus: a critical and historical discussion of the derivative and the integral.* New York: Columbia University Press.

Reprinted by Hafner Publishing Company and again by Dover Publications.

(1949). 'Newton as an originator of polar coördinates', *American Mathematical Monthly*, vol. 56, pp. 73–8.

(1954). 'Analysis: notes on the evolution of a subject and a name', *The Mathematics Teacher*, vol. 47, pp. 450–62.

(1956). *History of analytic geometry.* New York: Scripta Mathematica.

BRASCH, F. E., ed. (1928). *Sir Isaac Newton, 1727–1927. A bicentenary evaluation of his work.* A series of papers prepared under the auspices of the History of Science Society. Baltimore: The Williams & Wilkins Company.

(1952). 'A survey of the number of copies of Newton's *Principia* in the United States, Canada, and Mexico', *Scripta Mathematica*, vol. 18, pp. 53–67.

(1962). *Sir Isaac Newton. An essay on Sir Isaac Newton and Newtonian thought as exemplified in the Stanford collection of books, manuscripts, and prints concerning celestial mechanics, optics, mathematics, and related disciplines as a history of natural philosophy.* Stanford University: Stanford University Press.

BREWSTER, SIR DAVID (1855). *Memoirs of the life, writings, and discoveries of Sir Isaac Newton.* 2 vols. Edinburgh: Thomas Constable and Co.

Photo-reprint from the Edinburgh edition of 1855 with a new introduction by Richard S. Westfall. New York and London: Johnson Reprint Corporation, 1965. (The Sources of Science, no. 14.)

BROUWER, DIRK, AND GERALD M. CLEMENCE (1961). *Methods of celestial mechanics.* New York and London: Academic Press.

[BROWN, ERNEST WILLIAM] (1941). 'Biographical memoir by Frank Schlesinger and Dirk Brouwer presented to the Academy at the autumn meeting, 1939', *Biographical Memoirs*, vol. 21, pp. 243–73. Washington, D.C.: National Academy of Sciences of the United States of America.

BULLEN, K. E. (1962). 'Gift to the university of Newton's *Principia*', *The Gazette* (University of Sydney), May, pp. 36–8.

CAJORI, FLORIAN (1919). *A history of the conceptions of limits and fluxions in Great Britain from Newton to Woodhouse.* Chicago and London: The Open Court Publishing Company.

See also NEWTON (1934).

CHAPMAN, S. (1943). 'Edmond Halley and geomagnetism', *Nature*, vol. 152, pp. 231–7. 'Halley Lecture delivered at Oxford on May 28', including a translation of Halley's ode to Newton.

CHILD, J. M. *See* BARROW (1916).

CLAGETT, MARSHALL, ed. (1959). *Critical problems in the history of science.* Proceedings of the Institute for the History of Science at the University of Wisconsin, September 1–11, 1957. Madison (Wisconsin): The University of Wisconsin Press.

CLARKE, JOHN (1730). *A demonstration of some of the principal sections of Sir Isaac Newton's Principles of natural philosophy. In which his peculiar method of treating that useful subject is explained, and applied to some of the chief phaenomena of the system of the world.* London: printed for James and John Knapton.

CLEMENCE, GERALD M. *See* BROUWER AND CLEMENCE (1961).

COHEN, I. BERNARD (1953). Review of facsimile edition of Newton's *Principia*, *Isis*, vol. 44, pp. 287–8.

(1956). *Franklin and Newton: An inquiry into speculative Newtonian experimental science and Franklin's work in electricity as an example thereof.* Philadelphia: The American Philosophical Society.

Reprint, 1966, Cambridge, Massachusetts: Harvard University Press.

(1960). 'Newton in the light of recent scholarship', *Isis*, vol. 51, pp. 489–514.

(1962). 'The first English version of Newton's "Hypotheses non fingo" ', *Isis*, vol. 53, pp. 379–88.

(1963). 'Pemberton's translation of Newton's *Principia*, with notes on Motte's translation', *Isis*, vol. 54, pp. 319–51.

(1964). '"Quantum in se est": Newton's concept of inertia in relation to Descartes and Lucretius', *Notes and Records of the Royal Society of London*, vol. 19, pp. 131–55.

(1966). 'Hypotheses in Newton's philosophy', *Physis, Rivista Internazionale di Storia della Scienza*, vol. 8, pp. 163–84.

(1967). 'Newton's attribution of the first two laws of motion to Galileo', *Atti del Symposium Internazionale di Storia, Metodologia, Logica e Filosofia della Scienza 'Galileo nella Storia e nella Filosofia della Scienza'*, pp. xxv–xliv.

Collection des Travaux de l'Académie Internationale d'Histoire des Sciences. N. 16. Vinci (Firenze): Gruppo Italiano di Storia della Scienza.

(1967*a*). 'Newton's second law and the concept of force in the *Principia*', *The Texas Quarterly*, vol. 10, pp. 127–57.

The whole issue of *The Texas Quarterly* for Autumn 1967 (vol. 10, no. 3) is devoted to the publication of the papers read in Austin in November 1966 at a Conference on Newtonian Studies. A considerably revised version of the paper on the Second Law has been prepared for the reprint of these papers to be issued by the M.I.T. Press (Cambridge, Massachusetts).

(1967*b*). 'Newton's use of "force", or, Cajori versus Newton: a note on translations of the *Principia*', *Isis*, vol. 58, pp. 226–30.

(1967*c*). 'Dynamics: the key to the "new science" of the seventeenth century', *Acta historiae rerum naturalium necnon technicarum* (Czechoslovak Studies in the History of Science), Prague, Special Issue 3, pp. 79–114.

(1967*d*). 'Galileo, Newton, and the divine order of the solar system', pp. 207–31 of Ernan McMullin (ed.), *Galileo, man of science*. New York, London: Basic Books.

(1969). 'Isaac Newton's *Principia*, the scriptures and the divine providence', pp. 523–48 of Sidney Morgenbesser, Patrick Suppes, Morton White, edd., *Essays in honor of Ernest Nagel: Philosophy, science, and method*. New York: St. Martin's Press.

(1969*a*). 'The French translation of Isaac Newton's Philosophiae naturalis principia mathematica (1756, 1759, 1966)', *Archives Internationales d'Histoire des Sciences*, vol. 72, pp. 37–67.

(1969*b*). 'Newton's *System of the world*: some textual and bibliographical notes', *Physis*, vol. 11, pp. 152–66.

(1969*c*). Introduction, pp. vii–xxii, of *A treatise of the system of the world* by Sir Isaac Newton [a facsimile reprint of the second edition, 1731]. London: Dawsons of Pall Mall.

(1971, in process). *Transformations of scientific ideas: variations on Newtonian themes in the history of science*. Cambridge (England): at the University Press. Scheduled for publication in 1978. The Wiles Lectures, 1966.

(1971*a*, in process). *Newton and Kepler. Aspects of theories of planetary motion in the seventeenth century*. London: Oldbourne Press.

(1971*b*, in process). *Newton's principles of philosophy: inquiries into Newton's scientific work and its general environment*. Cambridge, Massachusetts: Harvard University Press.

AND ROBERT E. SCHOFIELD. *See* [NEWTON] (1958).

AND R. TATON, edd. (1964). *Mélanges Alexandre Koyré*: vol. 1. *L'Aventure de la science*; vol. 2. *L'Aventure de l'esprit*. Paris: Hermann (Histoire de la Pensée, 12 and 13).

See also KOYRÉ AND COHEN.

COLIE, ROSALIE L. (1960). 'John Locke in the republic of letters', pp. 111–29 of J. S. Bromley and E. H. Kossmann, edd., *Britain and the Netherlands*. London: Chatto & Windus.

[COLLINS, JOHN, *et al.*] (1856). *Commercium epistolicum J. Collins et aliorum de analysi promota, etc., ou, Correspondance de J. Collins et d'autres savants célèbres du XVIIe siècle, relative à l'analyse supérieure, réimprimée sur l'édition originale de 1712 avec l'indication des variantes de l'édition de 1722, complétée par une collection de pièces justificatives et de documents, et publiée par J.-B. Biot et F. Lefort*. Paris: Mallet-Bachelier.

COTES, ROGER (1770). *De descensu gravium de motu pendulorum in cycloide et de motu projectilium*. Cambridge (England): T. Fletcher & F. Hodson....

This separate printing is based on the appendix (pp. 73–91) of Robert Smith's posthumous edition of Cotes's *Harmonia mensurarum*... (Cambridge, England, 1722).

See also EDLESTON (1850).

CROMBIE, A. C., ed. (1963). *Scientific change. Historical studies in the intellectual, social and technical conditions for scientific discovery and technical invention, from antiquity to the present*. Symposium on the History of Science; University of Oxford 9–15 July 1961. New York: Basic Books.

DARWIN, CHARLES (1959). *The origin of species. A variorum text*. Edited by Morse Peckham. Philadelphia: University of Pennsylvania Press.

(1964). *On the origin of species. A facsimile of the first edition with an introduction by Ernst Mayr*. Cambridge, Massachusetts: Harvard University Press.

DAVIS, HERBERT (1951). 'Bowyer's paper stock ledger', *The Library. Transactions of the Bibliographical Society*, vol. 6, pp. 73–87.

DE BEER, G. R. *See* MCKIE AND DE BEER (1952).

DE MORGAN, AUGUSTUS (1846). 'On a point connected with the dispute between Keil and Leibnitz about the invention of fluxions', *Philosophical Transactions*, vol. 136, pp. 107–9.

(1848). 'On the additions made to the second edition of the Commercium epistolicum', *Philosophical Magazine*, vol. 32, pp. 446–56.

(1852). 'On the authorship of the account of the Commercium epistolicum, published in the Philosophical Transactions', *Philosophical Magazine*, vol. 3, pp. 440–4.

(1852*a*). 'On the early history of infinitesimals in England', *Philosophical Magazine*, vol. 4, pp. 321–30.

(1854). 'Pemberton and Newton', *Notes and Queries*, vol. 10, p. 181.

(1885). *Newton: his friend: and his niece*. London: Elliot Stock.
Edited by his wife, and by his pupil, Arthur Cowper Ranyard.

(1914). *Essays on the life and work of Newton*. Edited, with notes and appendices, by Philip E. B. Jourdain. Chicago and London: The Open Court Publishing Company.

(1954). *A budget of paradoxes*. New York: Dover Publications.
An unabridged reproduction of the second edition of 1915; the first edition is dated 1872.

DE MORGAN, SOPHIA ELIZABETH (1882). *Memoir of Augustus De Morgan, with selections from his letters*. London: Longmans, Green, and Co.

DESAGULIERS, J. T. (1719). 'An account of some experiments made on the 27th day of April, 1719, to find how much the resistance of the air retards falling bodies. A further account of experiments made for the same purpose, upon the 27th day of July last', *Philosophical Transactions*, vol. 30, pp. 1071–8.

(1763). *A course of experimental philosophy. The third edition corrected*. 2 vols. London: printed for A. Millar.

DE VILLAMIL, R. [1931]. *Newton: the man*. London: Gordon D. Knox.

DIJKSTERHUIS, EDUARD JAN (1961). *The mechanization of the world picture*. Translated by C. Dikshoorn. Oxford: at the Clarendon Press.

See also FORBES AND DIJKSTERHUIS (1963).

DRAKE, STILLMAN (1964). 'Galileo and the law of inertia', *American Journal of Physics*, vol. 32, pp. 601–8.

See also [GALILEO] (1957).

DREYER, J. L. E. (1924). 'Address delivered by the President, Dr. J. L. E. Dreyer, on the desirability of publishing a new edition of Isaac Newton's collected works', *Monthly Notices of the Royal Astronomical Society*, vol. 84, pp. 298–304.

DUGAS, RENÉ (1950). *Histoire de la mécanique*. Neuchâtel: Éditions du Griffon.
Translated by J. R. Maddox as: *A history of mechanics*. Neuchâtel: Éditions du Griffon; New York: Central Book Company, 1955.

(1954). *La mécanique au XVIIe siècle (des antécédents scolastiques à la pensée classique)*. Neuchâtel: Éditions du Griffon.
Translated by Freda Jacquot as: *Mechanics in the seventeenth century (from the scholastic antecedents to classical thought)*. Neuchâtel: Éditions du Griffon; New York: Central Book Company, 1958.

EDLESTON, J. (1850). *Correspondence of Sir Isaac Newton and Professor Cotes, including letters of other eminent men, now first published from the originals in the Library of Trinity College, Cambridge; together with an appendix, containing other unpublished letters and papers by Newton*. London: John W. Parker; Cambridge (England): John Deighton.

ELLIS, BRIAN D. (1962). 'Newton's concept of motive force', *Journal of the History of Ideas*, vol. 23, pp. 273–8.

(1965). 'The origin and nature of Newton's laws of motion', pp. 29–68 of Robert G. Colodny (ed.), *Beyond the edge of certainty. Essays in contemporary science and philosophy*. Englewood Cliffs, New Jersey: Prentice-Hall.

'ESPINASSE, MARGARET (1956). *Robert Hooke*. London: William Heinemann.

FATIO DE DUILLIER, NICOLAS (1949). 'De la cause de la pesanteur. Mémoire présenté à la Royal Society le 26 février 1690. [Reconstitué et publié avec une introduction par Bernard Gagnebin]', *Notes and Records of the Royal Society of London*, vol. 6, pp. 105–60.

FORBES, R. J., AND E. J. DIJKSTERHUIS (1963). *A history of science and technology.* Vol. 1: *Ancient times to the seventeenth century.* Harmondsworth, Middlesex: Penguin Books.

GAGNEBIN, BERNARD. *See* FATIO DE DUILLIER (1949).

[GALILEO] (1957). *Discoveries and opinions of Galileo, including: The starry messenger (1610), Letters on sunspots (1613), Letter to the Grand Duchess Christina (1615), and excerpts from The assayer (1623).* Translated, with an introduction and notes, by Stillman Drake. Garden City, New York: Doubleday & Company (Doubleday Anchor Books).

GERHARDT, C. I. *See* [LEIBNIZ] (1849–63).

GILLISPIE, CHARLES C. (1958). 'Fontenelle and Newton', pp. 427–43 of [NEWTON] (1958), *Papers and letters.*

GRAY, GEORGE J. (1907). *A bibliography of the works of Sir Isaac Newton, together with a list of books illustrating his works.* Second edition, revised and enlarged. Cambridge (England): Bowes and Bowes.

GREENSTREET, W. J., ed. (1927). *Isaac Newton, 1642–1727. A memorial volume edited for the Mathematical Association.* London: G. Bell and Sons.

GREGORY, DAVID (1702). *Astronomiae physicae & geometricae elementa.* Oxoniae: e theatro Sheldoniano.

(1715). *The elements of astronomy, physical and geometrical. Done into English, with additions and corrections. To which is annex'd, Dr. Halley's Synopsis of the astronomy of comets.* In two volumes. London: printed for J. Nicholson...

[——] (1937). *David Gregory, Isaac Newton and their circle, extracts from David Gregory's memoranda 1677–1708.* Edited by W. G. Hiscock. Oxford: printed for the editor.

GREGORY, JAMES CRAUFURD (1832). 'Notice concerning an autograph manuscript by Sir Isaac Newton, containing some notes upon the third book of the *Principia*, and found among the papers of Dr David Gregory, formerly Savilian Professor of Astronomy in the University of Oxford', *Transactions of the Royal Society of Edinburgh*, vol. 12, pp. 64–76.

GRIGORIAN, ACHOTE T. (1964). 'Les études newtoniennes de A. N. Krylov', pp. 198–207 of I. B. Cohen and R. Taton (edd.), *Mélanges Alexandre Koyré*: vol. 1, *L'Aventure de la science.* Paris: Hermann (Histoire de la Pensée, 12).

GROENING, JOHANN (1701). *Historia cycloeidis qua genesis & proprietates lineae cycloeidalis praecipuae, secundum ejus infantiam, adolescentiam & juventutem, ordine chronologico recensentur... Accedunt Christiani Hugenii annotata posthuma in Isaaci Newtonii* Philosophiae naturalis principia mathematica. Hamburg: ap. Gotfr. Liebezeit.

GUERLAC, HENRY (1963). *Newton et Epicure.* Paris: Palais de la Découverte [Histoire des Sciences: D-91].

(1963*a*). 'Francis Hauksbee: expérimentateur au profit de Newton', *Archives Internationales d'Histoire des Sciences*, vol. 16, pp. 113–28.

(1964). 'Sir Isaac and the ingenious Mr. Hauksbee', pp. 228–53 of I. B. Cohen and R. Taton (edd.), *Mélanges Alexandre Koyré*: vol. 1, *L'Aventure de la science.* Paris: Hermann (Histoire de la Pensée, 12).

(1967). 'Newton's optical aether. His draft of a proposed addition to his *Opticks*', *Notes and Records of the Royal Society of London*, vol. 22, pp. 45–57.

HALL, A. R. (1948). 'Sir Isaac Newton's notebook, 1661–1665', *Cambridge Historical Journal*, vol. 9, pp. 239–50.

(1957). 'Newton on the calculation of central forces', *Annals of Science*, vol. 13, pp. 62–71.

(1958). 'Correcting the *Principia*', *Osiris*, vol. 13, pp. 291–326.

(1963). 'The date of "On motion in ellipses" ', *Archives Internationales d'Histoire des Sciences*, vol. 16, pp. 23–8.

AND MARIE BOAS HALL (1959). 'Newton's electric spirit: four oddities', *Isis*, vol. 50, pp. 473–6.

AND MARIE BOAS HALL (1959*a*). 'Newton's "mechanical principles"', *Journal of the History of Ideas*, vol. 20, pp. 167–78.

AND MARIE BOAS HALL (1960). 'Newton's theory of matter', *Isis*, vol. 51, pp. 131–44.

AND MARIE BOAS HALL (1961). 'Clarke and Newton', *Isis*, vol. 52, pp. 583–5.

AND MARIE BOAS HALL, edd. (1962). *Unpublished scientific papers of Isaac Newton. A selection from the Portsmouth collection in the University Library, Cambridge.* Cambridge (England): at the University Press.

HALLEY, EDMOND (1686). 'A discourse concerning gravity, and its properties, wherein the descent of heavy bodies, and the motion of projects is briefly, but fully handled: together with the solution of a problem of great use in gunnery', *Philosophical Transactions*, vol. 16, no. 179, pp. 3–21.

[——] (1932). *Correspondence and papers...Preceded by an unpublished memoir of his life by one of his contemporaries and the 'Éloge' by D'Ortous de Mairan.* Arranged and edited by Eugene Fairfield MacPike. Oxford: at the Clarendon Press.

See also MACPIKE (1939).

HANSON, NORWOOD RUSSELL (1961). 'A mistake in the *Principia*', *Scripta Mathematica*, vol. 26, pp. 83–5.

HERIVEL, J. W. (1959). 'Newton's dynamical method in the tract *De Motu*', *Actes du IXe Congrès International d'Histoire des Sciences*, pp. 488–92.

(1960). 'Halley's first visit to Newton', *Archives Internationales d'Histoire des Sciences*, vol. 13, pp. 63–5.

(1960*a*). 'Newton's discovery of the law of centrifugal force', *Isis*, vol. 51, pp. 546–53.

(1960*b*). 'On the date of composition of the first version of Newton's tract *De motu*', *Archives Internationales d'Histoire des Sciences*, vol. 13, pp. 67–70.

(1960*c*). 'Suggested identification of the missing original of a celebrated communication of Newton's to the Royal Society', *Archives Internationales d'Histoire des Sciences*, vol. 13, pp. 71–8.

(1961). 'Interpretation of an early Newton manuscript', *Isis*, vol. 52, pp. 410–16.

(1961*a*). 'The originals of the two propositions discovered by Newton in December 1679?', *Archives Internationales d'Histoire des Sciences*, vol. 14, pp. 23–33.

(1962). 'Sur les premières recherches de Newton en dynamique', *Revue d'Histoire des Sciences et de leurs Applications*, vol. 15, pp. 105–40.

(1962*a*). 'Newton on rotating bodies', *Isis*, vol. 53, pp. 212–18.

(1962*b*). 'Early Newtonian dynamical MSS', *Archives Internationales d'Histoire des Sciences*, vol. 15, pp. 149–50.

(1964). 'Galileo's influence on Newton in dynamics', pp. 294–302 of I. B. Cohen and R. Taton (edd.), *Mélanges Alexandre Koyré*: vol. 1, *L'Aventure de la science*. Paris: Hermann (Histoire de la Pensée, 12).

(1965). *The background to Newton's Principia. A study of Newton's dynamical researches in the years 1664–84.* Oxford: at the Clarendon Press.

(1965*a*). 'Newton's first solution to the problem of Kepler motion', *British Journal for the History of Science*, vol. 2, pp. 350–4.

HESSE, MARY B. (1961). *Forces and fields. The concept of action at a distance in the history of physics.* London: Thomas Nelson and Sons.

[HILL, GEORGE WILLIAM] (1905). *The collected mathematical works.* 4 vols. Washington: published by the Carnegie Institution.

Introduction by H. M. Poincaré, pp. vii–xviii.

HISCOCK, W. G. (1936). 'The war of the scientists. New light on Newton and Gregory', *Times Literary Supplement*, January 11, p. 34.

See also [GREGORY] (1937).

HOLTON, GERALD (1965). 'The thematic imagination in science', pp. 88–108 of Gerald Holton (ed.), *Science and culture: A study of cohesive and disjunctive forces.* Boston: Houghton Mifflin Company.

This article is reprinted, with some revisions, from Harry Woolf (ed.), *Science as a cultural force.* Baltimore: The Johns Hopkins Press, 1964.

HORSLEY, SAMUEL. *See* [NEWTON] (1779–).

HUXLEY, G. L. (1959). 'Two Newtonian studies, I—Newton's boyhood interests; II—Newton and Greek geometry', *Harvard Library Bulletin*, vol. 13, pp. 348–61.

(1963). 'Roger Cotes and natural philosophy', *Scripta Mathematica*, vol. 26, pp. 231–8.

[HUYGENS, CHRISTIAAN] (1888–). *Oeuvres complètes de Christiaan Huygens*. Publiées par la Société Hollandaise des Sciences. La Haye: Martinus Nijhoff.

Vol. 22: *Supplément à la correspondance, varia, biographie de Chr. Huygens...*, published in 1950.

JEBB, R. C. (1882). *Bentley*. London: Macmillan and Co. (English Men of Letters.)

KARGON, ROBERT (1966). *Atomism in England: From Hariot to Newton*. Oxford: at the Clarendon Press.

[KEPLER, JOHANNES] (1967). *Kepler's* Somnium. *The dream, or posthumous work on lunar astronomy*. Translated with a commentary by Edward Rosen. Madison, Milwaukee, and London: The University of Wisconsin Press.

KOYRÉ, ALEXANDRE (1939). *Études galiléennes*. Paris: Hermann & Cie.

Reprinted, 1966.

(1950). 'La gravitation universelle de Kepler à Newton', *Actes du VIe Congrès International d'Histoire des Sciences* (Amsterdam), pp. 196–211; *Archives Internationales d'Histoire des Sciences*, vol. 4 (1951), pp. 638–53.

(1950*a*). 'The significance of the Newtonian synthesis', *Archives Internationales d'Histoire des Sciences*, vol. 3 [29], pp. 291–311.

(1952). 'An unpublished letter of Robert Hooke to Isaac Newton', *Isis*, vol. 43, pp. 312–37.

(1955). 'Pour une édition critique des œuvres de Newton', *Revue d'Histoire des Sciences*, vol. 8, pp. 19–37.

(1955*a*). 'A documentary history of the problem of fall from Kepler to Newton. De motu gravium naturaliter cadentium in hypothesi terrae motae', *Transactions of the American Philosophical Society*, vol. 45, pp. 329–95.

(1956). 'L'Hypothèse et l'expérience chez Newton', *Bulletin de la Société française de Philosophie*, vol. 50, pp. 59–79.

(1957). *From the closed world to the infinite universe*. Baltimore: The Johns Hopkins Press.

(1960). 'Les queries de l'Optique', *Archives Internationales d'Histoire des Sciences*, vol. 13, pp. 15–29.

(1960*a*). 'Les Regulae philosophandi', *Archives Internationales d'Histoire des Sciences*, vol. 13, pp. 3–14.

(1961). 'Attraction, Newton, and Cotes', *Archives Internationales d'Histoire des Sciences*, vol. 14, pp. 225–36.

(1961*a*). *La révolution astronomique: Copernic, Kepler, Borelli*. Paris: Hermann (Histoire de la Pensée, 3).

(1965). *Newtonian studies*. Cambridge, Massachusetts: Harvard University Press; London: Chapman & Hall.

More than half the volume consists of a previously unpublished study on 'Newton and Descartes', pp. 53–200.

(1966). *Etudes d'histoire de la pensée scientifique*. Avant-propos par René Taton. Paris: Presses Universitaires de France.

(1968). *Metaphysics and measurement: Essays in scientific revolution*. London: Chapman & Hall; Cambridge, Massachusetts: Harvard University Press.

AND I. BERNARD COHEN (1960). 'Newton's "electric & elastic spirit"', *Isis*, vol. 51, p. 337.

AND I. BERNARD COHEN (1961). 'The case of the missing *tanquam*: Leibniz, Newton, & Clarke', *Isis*, vol. 52, pp. 555–66.

AND I. BERNARD COHEN (1962). 'Newton & the Leibniz–Clarke correspondence, with notes on Newton, Conti, & Des Maizeaux', *Archives Internationales d'Histoire des Sciences*, vol. 15, pp. 63–126.

KRILOFF, A. N. (1924). 'On a theorem of Sir Isaac Newton', *Monthly Notices of the Royal Astronomical Society*, vol. 84, pp. 392–5.

LAGRANGE, JOSEPH LOUIS (1797). *Théorie des fonctions analytiques, contenant les principes du calcul différentiel, dégagés de toute considération d'infiniment petits ou d'évanouissans de limites ou de fluxions, et réduits à l'analyse algébrique des quantités finies.* Paris: Impr. de la République [prairial an V].

LARMOR, JOSEPH (1924). 'On editing Newton', *Nature*, vol. 113, p. 744.

LEFORT, F., AND J.-B. BIOT. *See* [COLLINS *et al.*] (1856).

LEIBNIZ, G. W. (1689). 'De lineis opticis, et alia', *Acta Eruditorum*, January 1689, pp. 36–8.
Reprinted in *Leibnizens Mathematische Schriften*, vol. 7, pp. 329–31.

(1689*a*). 'Schediasma de resistentia medii, & motu projectorum gravium in medio resistente', *Acta Eruditorum*, February 1689, pp. 38–47.
Reprinted in *Leibnizens Mathematische Schriften*, vol. 6, pp. 135–44.

(1689*b*). 'Tentamen de motuum coelestium causis', *Acta Eruditorum*, February 1689, pp. 82–96.
Reprinted in *Leibnizens Mathematische Schriften*, vol. 6, pp. 144–61.

[——] (1849–63). *Leibnizens Mathematische Schriften* herausgegeben von C. I. Gerhardt. 2 parts: 7 vols. [Leibnizens gesammelte Werke aus den Handschriften der Königlichen Bibliothek zu Hannover herausgegeben von Georg Heinrich Pertz. Dritte Folge.] Berlin: Verlag von A. Asher & Comp.; Halle: Druck und Verlag von H. W. Schmidt.
Photo-reprint, Hildesheim: Georg Olms Verlagsbuchhandlung, 1962.

(1952). *Theodicy*. Edited with introduction by Austin Farrer. Translated by E. M. Huggard. New Haven: Yale University Press.

[—— AND JOHANN BERNOULLI] (1745). *Virorum celeberr. Got. Gul. Leibnitii et Johan. Bernoulli Commercium philosophicum et mathematicum.* Lausanne and Geneva: sumpt. Marci-Michaelis Bousquet & Socior. 2 vols.

[—— AND SAMUEL CLARKE] (1956). *The Leibniz–Clarke correspondence, together with extracts from Newton's* Principia *and* Opticks. Edited with introduction and notes by H. G. Alexander. Manchester: Manchester University Press.

[—— AND SAMUEL CLARKE] (1957). *Correspondance Leibnitz–Clarke présentée d'après les manuscrits originaux des bibliothèques de Hanovre et de Londres.* Edited by André Robinet. Paris: Presses Universitaires.

LITTLEWOOD, J. E. (1953). 'Newton and the attraction of a sphere', pp. 94–9 of J. E. Littlewood, *A mathematician's miscellany*. London: Methuen & Co.
Reprinted from *The Mathematical Gazette*, vol. 32, 1948.

LOHNE, JOHANNES A. (1960). 'Hooke versus Newton. An analysis of the documents in the case of free fall and planetary motion', *Centaurus*, vol. 7, pp. 6–52.

(1965). 'Isaac Newton: the rise of a scientist 1661–1671', *Notes and Records of the Royal Society of London*, vol. 20, pp. 125–39.

(1967). 'The increasing corruption of Newton's diagrams', *History of Science*, vol. 6, pp. 69–89.

MACOMBER, HENRY P. (1951). 'A comparison of the variations and errors in copies of the first edition of Newton's *Principia*, 1687', *Isis*, vol. 42, pp. 230–2.

(1952). '*Principia* census', *Isis*, vol. 43, p. 126.

(1953). 'A census of the owners of copies of the 1687 first edition of Newton's "Principia"', *The Papers of the Bibliographical Society of America*, vol. 47, third quarter, pp. 269–300.

See also [BABSON COLLECTION] (1950).

MACPIKE, EUGENE FAIRFIELD (1937). *Hevelius, Flamsteed and Halley, three contemporary astronomers and their mutual relations.* London: Taylor & Francis.

(1939). *Dr. Edmond Halley (1656–1742). A bibliographical guide to his life and work arranged chronologically; preceded by a list of sources, including references to the history of the Halley family.* London: Taylor & Francis.

See also [Halley] (1932).

MANDELBAUM, MAURICE (1964). *Philosophy, science and sense perception: historical and critical studies.* Baltimore: The Johns Hopkins Press.

MANUEL, FRANK E. (1959). *The eighteenth century confronts the gods.* Cambridge, Massachusetts: Harvard University Press.

See especially chap. III, 'The Euhemerists and Isaac Newton'.

(1963). *Isaac Newton, historian.* Cambridge, Massachusetts: The Belknap Press of Harvard University Press.

(1968). *A portrait of Isaac Newton.* Cambridge, Massachusetts: The Belknap Press of Harvard University Press.

MARCOLONGO, R. (1919). *Il problema dei tre corpi da Newton (1686) ai nostri giorni.* Milano: Ulrico Hoepli.

MAYR, ERNST. *See* DARWIN (1964).

MCGUIRE, J. E. (1966). 'Body and void and Newton's *De mundi systemate*: some new sources', *Archive for History of Exact Sciences*, vol. 3, pp. 206–48.

(1967). 'Transmutation and immutability: Newton's doctrine of physical qualities', *Ambix*, vol. 14, pp. 69–95.

(1968). 'The origin of Newton's doctrine of essential qualities', *Centaurus*, vol. 12, pp. 233–60.

AND P. M. RATTANSI (1966). 'Newton and the "pipes of Pan"', *Notes and Records of the Royal Society of London*, vol. 21, pp. 108–43.

MCKENZIE, D. F. (1966). *The Cambridge University Press 1696–1712: A bibliographical study.* 2 vols. Cambridge (England): at the University Press.

MCKIE, DOUGLAS, AND G. R. DE BEER (1952). 'Newton's apple', *Notes and Records of the Royal Society of London*, vol. 9, pp. 46–54, 333–5.

MILLER, PERRY (1958). 'Bentley and Newton', pp. 271–8 of [NEWTON] (1958), *Papers and letters.*

MONK, JAMES HENRY (1830). *The life of Richard Bentley, D.D. Master of Trinity College, and Regius Professor of Divinity in the University of Cambridge: with an account of his writings, and anecdotes of many distinguished characters during the period in which he flourished.* London: printed for C. J. G. & F. Rivington.

MORE, L. T. (1934). *Isaac Newton: A biography.* New York, London: Charles Scribner's Sons.

MOTTE, ANDREW (1727). *A treatise of the mechanical powers, wherein the laws of motion, and the properties of those powers are explained and demonstrated in an easy and familiar method.* London: printed for Benjamin Motte.

See also NEWTON (1729) and (1934).

MUNBY, A. N. L. (1952). 'The distribution of the first edition of Newton's *Principia*', *Notes and Records of the Royal Society of London*, vol. 10, pp. 28–39.

(1952*a*). 'The Keynes Collection of the works of Sir Isaac Newton at King's College, Cambridge', *Notes and Records of the Royal Society of London*, vol. 10, pp. 40–50.

NEWTON, ISAAC. (For the editions, reprints, and translations of the *Principia*, see Appendix VIII to the text of this edition.)

(1702). *A new and most accurate theory of the Moon's motion; whereby all her irregularities may be solved, and her place truly calculated to two minutes.* Written by that incomparable mathematician Mr. Isaac Newton, and published in Latin by Mr. David Gregory in his excellent astronomy. London: printed and sold by A. Baldwin.

(1715). '[Recensio libri =] An account of the book entituled *Commercium epistolicum Collinii & aliorum, De analysi promota*, published by order of the Royal-Society, in relation to the dispute between Mr. Leibnits and Dr. Keill, about the right of invention of the new geometry of fluxions, otherwise call'd the differential method', *Philosophical Transactions*, vol. 29, pp. 173–224.

The title of the book given in the *Philosophical Transactions* is not the exact title of the *Commercium epistolicum* [first edition] itself.

A French translation appeared in the *Journal Littéraire*, Nov./Dec. 1715, vol. 6, pp. 13 ff., 345 ff.

(1728). *A treatise of the system of the world.* Translated into English. London: printed for F. Fayram.

(1728*a*). *De mundi systemate liber.* Londini: impensis J. Tonson, J. Osborn, & T. Longman.

(1728*b*). *Optical lectures read in the publick schools of the University of Cambridge, Anno Domini, 1669.* By the late Sir Isaac Newton, then Lucasian Professor of the Mathematicks. Never before printed. Translated into English out of the original Latin. London: printed for Francis Fayram.

(1729). *The mathematical principles of natural philosophy.* Translated into English by Andrew Motte. *To which are added, The laws of the moon's motion, according to gravity.* By John Machin. In two volumes. London: printed for Benjamin Motte.

Facsimile reprint, with introduction by I. Bernard Cohen. London: Dawsons of Pall Mall, 1968.

[——] (1779–). *Opera quae exstant omnia.* Commentariis illustrabat Samuel Horsley. 5 vols. London: John Nichols.

Volume 5 was published in 1785. The whole set of five volumes was reprinted in 1964 by Friedrich Verlag (Günther–Holzboog). Stuttgart–Bad Cannstatt.

(1931). *Opticks or a treatise of the reflections, refractions, inflections & colours of light.* Reprinted from the fourth edition [London, 1730] with a foreword by Prof. Albert Einstein, Nobel Laureate, and an introduction by Prof. E. T. Whittaker, F.R.S. London: G. Bell & Sons.

(1934). *Sir Isaac Newton's Mathematical principles of natural philosophy and his System of the world.* Translated into English by Andrew Motte in 1729. The translations revised, and supplied with an historical and explanatory appendix, by Florian Cajori. Berkeley: University of California Press.

(1952). *Opticks or a treatise of the reflections, refractions, inflections & colours of light.* Based on the fourth edition: London, 1730. With a foreword by Albert Einstein; an introduction by Sir Edmund Whittaker; a preface by I. Bernard Cohen; and an analytical table of contents prepared by Duane H. D. Roller. New York: Dover Publications.

[——] (1958). *Isaac Newton's papers & letters on natural philosophy and related documents.* Edited, with a general introduction, by I. Bernard Cohen assisted by Robert E. Schofield. Cambridge, Massachusetts: Harvard University Press.

[——] (1959–). *The correspondence of Isaac Newton.* Vol. 1, 1661–1675 (1959), vol. 2, 1676–1687 (1960), vol. 3, 1688–1694 (1961), edited by H. W. Turnbull; vol. 4, 1697–1709 (1967), edited by J. F. Scott. Cambridge (England): at the University Press (published for the Royal Society).

[——] (1964–). *The mathematical works of Isaac Newton.* Assembled with an introduction by Dr. Derek T. Whiteside. 2 vols. New York and London: Johnson Reprint Corporation (The Sources of Science).

[——] (1967–). *The mathematical papers of Isaac Newton.* Vol. 1, 1664–1666 (1967), vol. 2, 1667–1670 (1968), vol. 3, 1670–1673 (1969), edited by D. T. Whiteside, with the assistance in publication of M. A. Hoskin. Cambridge: at the University Press.

To be complete in 8 vols.

PATTERSON, LOUISE DIEHL (1949, 1950). 'Hooke's gravitation theory and its influence on Newton. I: Hooke's gravitation theory. II: The insufficiency of the traditional estimate', *Isis*, vol. 40, pp. 327–41; vol. 41, pp. 32–45.

PECKHAM, MORSE. *See* DARWIN (1959).

PEMBERTON, HENRY (1722). 'A letter to Dr. Mead, Coll. Med. Lond. & Soc. Reg. S. concerning an experiment, whereby it has been attempted to shew the falsity of the common opinion, in relation to the force of bodies in motion', *Philosophical Transactions*, vol. 32, pp. 57–68.

(1728). *A view of Sir Isaac Newton's philosophy.* London: S. Palmer.

[——] (1771). *A course of chemistry, divided into twenty-four lectures, formerly given by the late learned Doctor Henry Pemberton.* Now first published from the author's manuscript by James Wilson. London: printed for J. Nourse.

PIGHETTI, CLELIA (1960). 'Cinquant' anni di studi newtoniani (1908–1959)', *Rivista Critica di Storia della Filosofia*, fascicoli II–III, pp. 181–203; 295–318. Firenze: La Nuova Italia Editrice.

(1961). 'Per la storia del newtonianesimo in Italia. I: Newton e la cultura italiana: aspetti e problemi di un importante incontro', *Rivista Critica di Storia della Filosofia*, fascicola IV, pp. 425–34.

(1962). 'A proposito delle ipotesi nella metodologia newtoniana', *Archives Internationales d'Histoire des Sciences*, vol. 15, pp. 291–302.

POINCARÉ, HENRI (1905). Biographical introduction to [George William Hill]: *The collected mathematical works*. Washington: published by the Carnegie Institution.

[PORTSMOUTH COLLECTION] (1888). *A catalogue of the Portsmouth Collection of books and papers written by or belonging to Sir Isaac Newton, the scientific portion of which has been presented by the Earl of Portsmouth to the University of Cambridge*. Prepared by H. R. Luard, G. G. Stokes, J. C. Adams, and G. D. Liveing. Cambridge: at the University Press.

RAVIER, ÉMILE (1937). *Bibliographie des œuvres de Leibniz*. Paris: Librairie Felix Alcan.
Reprografischer Nachdruck der Ausgabe Paris, 1937. Hildesheim: Georg Olms Verlagsbuchhandlung, 1966.

RIGAUD, STEPHEN JORDAN (1844). *A defence of Halley against the charge of religious infidelity*. Oxford: Ashmolean Society.

RIGAUD, STEPHEN PETER (1836). 'Inquiry relative to Dr. Pemberton's translation and illustrations of Newton's *Principia*', *Philosophical Magazine*, vol. 8, pp. 441–2.

(1838). *Historical essay on the first publication of Sir Isaac Newton's Principia*. Oxford: at the University Press.

ed., (1841). *Correspondence of scientific men of the seventeenth century...in the collection of...the Earl of Macclesfield*. In two volumes. Oxford: at the University Press.

ROSENFELD, LEON (1965). 'Newton and the law of gravitation', *Archive for History of Exact Sciences*, vol. 2, pp. 365–86.

[ROYAL SOCIETY] (1947). *Newton tercentenary celebrations*. Cambridge (England): at the University Press.
Contains E. N. da C. Andrade, 'Newton'; Lord Keynes, 'Newton, the man'; J. Hadamard, 'Newton and the infinitesimal calculus'; S. I. Vavilov, 'Newton and the atomic theory'; N. Bohr, 'Newton's principles and modern atomic mechanics'; H. W. Turnbull, 'Newton: the algebraist and geometer'; W. Adams, 'Newton's contributions to observational astronomy'; J. C. Hunsaker, 'Newton and fluid mechanics'.

RUSSELL, J. L. (1964). 'Kepler's laws of planetary motion: 1609–1666', *The British Journal for the History of Science*, vol. 2, pp. 1–24.

SAMPSON, R. A. (1924). 'On editing Newton', *Monthly Notices of the Royal Astronomical Society*, vol. 84, pp. 378–83.

SARTON, GEORGE (1952). *Horus: A guide to the history of science. A first guide for the study of the history of science with introductory essays on science and tradition*. Waltham, Massachusetts: published by the Chronica Botanica Company.

SCHOFIELD, ROBERT E. (1958). 'Halley and the *Principia*', pp. 397–404 of [NEWTON] (1958), *Papers & letters*.

SCOTT, J. F. *See* [Newton] (1959–).

SMITH, [ANDREW] BRUCE (1908). 'An annotated copy of Newton's "Principia"', *Nature*, vol. 77, p. 510.

(1908*a*). 'An annotated copy of Newton's "Principia"', *Nature*, vol. 79, p. 130.

SMITH, DAVID EUGENE (1927). 'Two unpublished documents of Sir Isaac Newton', pp. 16–34 of W. J. Greenstreet (ed.), *Isaac Newton, 1642–1727*. London: G. Bell and Sons.

[SOTHEBY & CO.] (1936). *Catalogue of the Newton papers, sold by order of the Viscount Lymington to whom they have descended from Catherine Conduitt, Viscountess Lymington, great-niece of Sir Isaac Newton*. London: Sotheby & Co.

[——] (1965). *Catalogue of the fine collection of scientific books, the property of Professor E. N. da C. Andrade, F.R.S.* London: Sotheby & Co.

STEWART, AGNES GRAINGER (1901). *The academic Gregories.* Edinburgh and London: Oliphant, Anderson & Ferrier (Famous Scots Series).

STRONG, E. W. (1951). 'Newton's "mathematical way"', *Journal of the History of Ideas*, vol. 12, pp. 90–110.

STUKELEY, WILLIAM (1936). *Memoirs of Sir Isaac Newton's life, 1752: Being some account of his family and chiefly of the junior part of his life.* Edited by A. Hastings White. London: Taylor & Francis.

TATON, R. *See* COHEN AND TATON, edd. (1964).

[TIXALL LETTERS] (1815). *Tixall letters; or the correspondence of the Aston family, and their friends, during the seventeenth century.* With notes and illustrations, by Arthur Clifford Esq. 2 vols. London: printed for Longman, Hurst, Rees, Orme, and Brown....

TRUESDELL, C. (1960). 'A program toward rediscovering the rational mechanics of the Age of Reason', *Archive for History of Exact Sciences*, vol. 1, pp. 3–36.

(1967). 'Reactions of late baroque mechanics to success, conjecture, error, and failure in Newton's *Principia*', *The Texas Quarterly*, vol. 10, no. 3 (Autumn), pp. 238–58.

TURNBULL, HERBERT WESTREN, ed. (1939). *James Gregory: tercentenary memorial volume. Containing his correspondence with John Collins and his hitherto unpublished mathematical manuscripts, together with addresses and essays communicated to the Royal Society of Edinburgh, July 4, 1938.* London: G. Bell and Sons (published for the Royal Society of Edinburgh).

(1945). *The mathematical discoveries of Newton.* London and Glasgow: Blackie & Son.

(1951). 'The discovery of the infinitesimal calculus', *Nature*, vol. 167, pp. 1048–50.

See also [NEWTON] (1959–).

TURNOR, EDMUND (1806). *Collections for the history of the town and soke of Grantham, containing authentic memoirs of Sir Isaac Newton, now first published from the original MSS. in the possession of the Earl of Portsmouth.* London: William Miller.

WEINSTEIN, ALEXANDER (1943). 'Ode on Newton's theory of gravitation by Edmond Halley', *Science*, vol. 97, pp. 69–70.

WESTFALL, RICHARD S. (1958). *Science and religion in seventeenth-century England.* New Haven: Yale University Press.

(1962). 'The foundations of Newton's philosophy of nature', *British Journal for the History of Science*, vol. 1, pp. 171–82.

(1963). 'Newton's reply to Hooke and the theory of colors', *Isis*, vol. 54, pp. 82–96.

(1963*a*). 'Short-writing and the state of Newton's conscience, 1662', *Notes and Records of the Royal Society of London*, vol. 18, pp. 10–16.

(1964). 'Newton and absolute space', *Archives Internationales d'Histoire des Sciences*, vol. 17, pp. 121–32.

(1967). 'Uneasily fitful reflections on fits of easy transmission', *The Texas Quarterly*, vol. 10, no. 3 (Autumn), pp. 86–102.

(1967*a*). 'Hooke and the law of universal gravitation', *The British Journal for the History of Science*, vol. 3, pp. 245–61.

WHISTON, WILLIAM (1716). *Sir Isaac Newton's mathematick philosophy more easily demonstrated: with Dr. Halley's account of comets illustrated. Being forty lectures read in the Publick Schools at Cambridge. By William Whiston, M.A. Mr. Lucas's Professor of the Mathematicks in that University. For the use of the young students there.* In this English edition the whole is corrected and improved by the author. London: printed for J. Senex and W. Taylor.

(1728). *Astronomical lectures, read in the Publick Schools at Cambridge. Whereunto is added a collection of astronomical tables; being those of Mr. Flamsteed, corrected; Dr. Halley; Monsieur Cassini; and Mr. Street. For the use of young students in the university.* The second edition corrected. London: printed for J. Senex, W. and J. Innys, J. Osborne and T. Longman.

WHITE, R. J. (1965). *Dr Bentley. A Study in academic scarlet.* London: Eyre & Spottiswoode.

WHITESIDE, DEREK THOMAS (1961). 'Patterns of mathematical thought in the later seventeenth century', *Archive for History of Exact Sciences*, vol. 1, pp. 179–388.

(1961*a*). 'Newton's discovery of the general binomial theorem', *Mathematical Gazette*, vol. 45, pp. 175–80.

(1962). 'The expanding world of Newtonian research', *History of Science*, vol. 1, pp. 16–29.

(1964). 'Isaac Newton: birth of a mathematician', *Notes and Records of the Royal Society of London*, vol. 19, pp. 53–62.

(1964*a*). 'Newton's early thoughts on planetary motion: a fresh look', *British Journal for the History of Science*, vol. 2, pp. 117–37.

(1966). 'Newton's marvellous year: 1666 and all that', *Notes and Records of the Royal Society of London*, vol. 21, pp. 32–41.

(1966*a*). 'Newtonian dynamics', *History of Science*, vol. 5, pp. 104–17.
An essay-review of HERIVEL (1965), *Background*.

See also [NEWTON] (1964–), [NEWTON] (1967–).

WIGHTMAN, W. P. D. (1953). 'Gregory's "Notæ in Isaaci Newtoni Principia Philosophiæ"', *Nature*, vol. 172, p. 690.

(1955). 'Aberdeen University and the Royal Society', *Notes and Records of the Royal Society of London*, vol. 11, pp. 145–58.

(1957). 'David Gregory's commentary on Newton's "Principia"', *Nature*, vol. 179, pp. 393–4.

WILSON, JAMES (1771). Biographical preface to Henry Pemberton: *A course of chemistry, divided into twenty-four lectures. Now first published from the author's manuscript.* London: printed for J. Nourse.

ZEITLINGER, H. (1927). 'A Newton bibliography', pp. 148–70 of W. J. Greenstreet (ed.), *Isaac Newton, 1642–1727, A memorial volume*, edited for the Mathematical Association. London: G. Bell and Sons, 1927.

INDEX